INNOVATIONS IN MINING BACKFILL TECHNOLOGY

PROCEEDINGS OF THE 4TH INTERNATIONAL SYMPOSIUM ON MINING WITH
BACKFILL / MONTREAL / 2-5 OCTOBER 1989

# Innovations in Mining Backfill Technology

*Edited by*
**F.P.HASSANI & M.J.SCOBLE**
*McGill University, Canada*

**T.R.YU**
*Kidd Creek Mines Ltd, Canada*

A.A.BALKEMA / ROTTERDAM / BROOKFIELD / 1989

## ORGANIZING COMMITTEE

| | |
|---|---|
| Symposium chairman | Dr Ferri Hassani, McGill University, Canada |
| Honorary chairman | Dr Ed Thomas, University of New South Wales, Australia |
| Co-chairman | Jacques Nantel, Noranda Technology Centre, Canada |
| Technical committee | Dr Thian Yu, Kidd Creek Mines Ltd, Canada |
| Secretary | Dr Malcolm Scoble, McGill University, Canada |
| Finance | Dr Hani Mitri, McGill University, Canada |
| Technical mine tours | Doug Ames, Ontario Ministry of Labour, Canada |
| Guest program | Fabian Miller, Noranda Technology Centre, Canada |
| Publicity and proceedings | Perla Gantz, CIM, Canada |
| Information, registration and accomodation | John Gaydos, CIM, Canada |

*The texts of the various papers in this volume were set individually by typists under the supervision of each of the authors concerned.*

Authorization to photocopy items for internal or personal use, or the internal or personal use of specific clients, is granted by A.A.Balkema, Rotterdam, provided that the base fee of US$1.00 per copy, plus US$0.10 per page is paid directly to Copyright Clearance Center, 27 Congress Street, Salem, MA 01970. For those organizations that have been granted a photocopy license by CCC, a separate system of payment has been arranged. The fee code for users of the Transactional Reporting Service is: 90 6191 985 1/89 US$1.00 + US$0.10.

Published by

A.A.Balkema, P.O.Box 1675, 3000 BR Rotterdam, Netherlands

A.A.Balkema Publishers, Old Post Road, Brookfield, VT 05036, USA

ISBN 90 6191 985 1

© 1989 A.A.Balkema, Rotterdam

Printed in the Netherlands

*Innovations in Mining Backfill Technology, Hassani et al. (eds), © 1989 Balkema, Rotterdam. ISBN 90 6191 985 1*

# Contents

# 3 Modelling and design

# 4 Rockburst control

# 5 General topics

## 6 Soft rock mining

## 7 System design

# 8 General topics

# 1 Review

*Innovations in Mining Backfill Technology, Hassani et al. (eds), © 1989 Balkema, Rotterdam. ISBN 90 6191 985 1*

# Backfill research in Canadian Mines

J.E.Udd
*Mining Research Laboratories, Energy, Mines and Resources, Ottawa, Ontario, Canada*

ABSTRACT: The Canadian mining industry has been innovative in its approach to the use of backfill in underground mines. Outstanding advances have been made in the technologies of using cemented mill tailings and smelter slags. Present research is focussed on these, as well as on: densified and paste fills; cemented waste rock; frozen fills; fly ash substitutes; and post-consolidation.

In this paper, an overview is made of the history of the development of backfill technology in Canada, and of the work presently in progress. The needs of the industry for future research are also identified.

## 1 INTRODUCTION

Waste rock from mining operations has been used to provide unit support and to fill openings for hundreds, if not thousands, of years. In the ancient greek silver mines at Laureion, pillars were constructed of stone to support the back(1). Agricola, in his classic work "De Re Metallica" refers to waste rock being thrown into the openings among the timber(2) - evidently not for support, but rather, for waste disposal purposes.

Pillars constructed of rock (sometimes quarried), or concrete, or timber, have been used continuously in mining up to present times. In a historical context, however, the use of waste rock or other materials in order to fill openings for support purposes is a relatively recent development - necessitated by large scale mining methods, and facilitated by modern milling practices.

The development of the flotation process late in the 19th century and early in the 20th had monumental implications for both mining and milling practices. The rapid application of flotation technology on a global scale was probably one of the driving forces in the transformation of the minerals industry from selective to mass production methods. The milling processes which were implemented in the present century produced, as waste, a material which is probably the cheapest and easiest to handle for purposes of filling mine openings.

Prior to the availability of mill tailings, waste rock or alluvial sands or gravels were commonly used for fills. Crushed smelter slag was sometimes used, depending on its availability. At the beginning of the century, the common mining methods involving fill were rill stoping, cut-and-fill, and shrinkage (with broken ore as temporary fill)(3).

The same technology, with several variations, was reported in the 1940's(4).

During the 1930's and 1940's, equipment was developed which led to very rapid improvements in mining practices. Notable amongst these were the introduction of scrapers in the 1930's and the development of tungsten-carbide bits after the Second World War(5). By the 1950's, trackless mining was being introduced.

Concurrently, in the long-established mines, mining had proceeded to greater depths. Both the local ground conditions and the sizes of the stopes being mined necessitated the use of fill for support. During the 1950's "sand" fill (hydraulic tailings) came into much more widespread use in Canadian mines, particularly in the massive deposits of the Sudbury Basin in which a conversion was being made from square set mining to cut-and-fill.

Initially, "sand" fill was used to replace waste rock in filling the timber sets(6). Later, after the development of consolidated fills, it became possible to eliminate the labour-intensive and extremely costly square-set method almost entirely from Canadian practice. The word "almost" is used advisedly, for, as will be mentioned later, there is still one important operation at which square-setting is in use in 1988.

During the 1930's some notable work with backfill was conducted at the Horne Mine of Noranda Mines Ltd.(7). With the local geological materials being unsuitable for the backfill that was needed for pillar mining operations, experiments were conducted on reverberatory furnace slag to which had been added pyrrhotite tailings. The result was a very strong fill which has never been duplicated. This author recommends a reading of the paper listed in the references, by Patton(7), as a "classic" in the development of innovative fill technology.

A paper in the same volume(8), by the staff of Cominco's Sullivan Mine, at Kimberley, B.C., is also recommended reading because of the diversity of filling practices which are described. At the time of writing, in 1957, surface gravel, development waste, waste from caving, and concentrator float and sulphide tailings were all being used to provide the large tonnages necessary to fill open stopes prior to pillar recovery.

In the mines of the Sudbury Basin, in which many of the technical innovations in backfilling practice originated, the original practice was to fill shrinkage and cut-and-fill stopes with waste rock. In 1935, a mixture of sand and gravel, from overburden, were used for the first time at the Falconbridge mine(9). In the late 1940's and early 1950's the use of classified mill tailings was introduced.

In the cut-and-fill stopes of the mines of the Falconbridge organization, the practice was to use elm plank floors, for scraping purposes, on top of the sand fill. In the late 1950's, however, concrete working floors were laid when the fill was suitable for aggregate(9). A very key development, reported in 1959(10) was: "Recent experiments indicate that tailing fill and cement can be mixed on surface and delivered to a stope through the tailing fill system to produce an adequate mining floor when laid to a thickness of 4 inches".

At Inco, the use of Portland cement to stabilize sand fill was begun in 1960(11). After a period of extensive tests the technique was introduced into practice in the Frood mine in 1962. Subsequently, the use of cemented fill became standard practice in all of the mines of the Sudbury basin and throughout the Canadian industry.

The use of cemented fill has had a very great impact on Canadian mining practices. With its introduction in the Sudbury mines in the early 1960's, the immediate benefits realized were through improvements to cut-and-fill mining practices and pillar recovery operations. Square-set mining disappeared as cut-and-fill practices were applied to pillar mining. Also in Sudbury, the newly-developed backfill technology was closely associated with new mining methods: undercut-and-fill, at Inco(11), and post pillars, at Falconbridge(12). Subsequently, it has become possible to apply bulk mining methods, principally blasthole, in instances where more selective methods could have been used previously(13). The strengths of cemented fills, and their free standing properties, has enabled mine planners to increase level intervals and stope dimensions. Cemented backfill has been essential to the reductions in mining costs which have been achieved in the 1980's.

Table 1. Some important dates in connection with backfill in Canada

1903   -   first flotation plant in Canada, using the Elmore oil process, Le Roi Number 2 mine, Rossland, B.C. (Ref: History of Milling in Canada", 6th Commonwealth Mining & Metallurgy Congress Vol. "The Milling of Canadian Ores", 1957, p. 25).

1933   -   experiments with reverberatory furnace slag and pyrrhotite tailings at the Horne Mine, Noranda, Quebec(7).

1935   -   introduction of sand and gravel backfilling at Falconbridge Mine, Falconbridge, Ontario(9).

1948   -   replacement of waste rock by "sand" (tailings) fill at Frood Mine, Inco, Sudbury, Ontario(6).

1959   -   experimental use of cemented tailings for cut-and-fill stope floors at Falconbridge(10).

Table 1. (Cont'd)

1960  -  tests on the addition of
          Portland cement to stabilize
          hydraulic fill at Inco(11).


2 VARIATIONS ON PRESENT BACKFILL PRACTICE
  AND KEY TECHNOLOGICAL ISSUES

During the twenty-five years, or so, that
cemented backfills have been used in
Canadian mines, there has been much
experimentation to determine the optimum
mixes for particular applications.
Cement is expensive, and the cost
implications from achieving even slight
reductions in the quantities added are
very large.  In 1983, in Inco's mines in
the Ontario Division alone, it was
estimated that over two million tonnes of
cemented backfill were placed underground
in a normal operating year(14).  The
quantities being placed at present by the
entire Canadian industry must be several
times that amount.

No attempt is made in this paper to
review all of the backfilling practices,
or the variations thereon, which are in
use in Canada.  There are perhaps as many
of these as there are individual mines.
Numerous references can be found in the
literature, particularly the "CIM
Bulletin", very quickly.

Nonetheless, the following examples are
to illustrate some of the present trends
in backfill research in Canadian mines
and some of the key technical issues.
The list is not necessarily complete and
reflects this writer's opinion only:

2.1 Optimizing the use of cement

The design of backfill as a source of
support is now regarded as an important
part of the process of mine design.  In
view of the costs of cement and
additives, however, it is essential that
there should be tight controls over the
quantities used.

To date, these have been established as
the result of trial and error.  As but
one, and only one, example, at the
Strathcona Mine, of Falconbridge, the
following ratios of cement to tailings
were in use in 1983(13):
  - 1:8 for containment plugs for
    blasthole stopes, for 0,3m floors in
    cut-and-fill stopes, and for 1,5m
    layers over timber sill mats above
    new mining horizons;
  - 1:32 to 1:40 for bulk pours in
    cut-and-fill stopes;

  - 1:16 to 1:40 for bulk pours in
    blasthole stopes.

For pours in stopes, the mixes selected
depended on mining geometries and design
calculations.  No doubt similar approaches
are followed by other mine operators.

2.2 Alternatives to cemented mill tailings

Because of the cost of cement there are
powerful economic incentives to find
acceptable substitutes.  In Canada,
especially at the Kidd Creek Mine, at
Timmins, Ontario, much work has been done
on the use of cemented rockfill.  Since
1982, it has been reported(15), ground
blast furnace slag has been substituted
for between 30% of 60% of the Portland
cement previously used.

In theory, given an adequate and
available supply of suitable waste rock, a
cemented rock fill should not only be less
expensive than cemented tailings but also
substantially stronger and of a higher
modulus.  There are potential problems to
be overcome in placement, however, in
order to prevent zoning, segregation, and
degradation of particles.

In Australia, considerable work has been
done to assess the properties of pozzolans
as potential substitutes for cemented
backfills.  Pozzolans are materials which
can provide silica, which, in the presence
of water and calcium hydroxide, react to
form cemented hydrated calcium silicates.
Potentially, smelter slags and fly ash are
pozzolanic.  According to work reported by
Thomas, et al(16), some Canadian slags
have demonstrated pozzolanic potential.
Apart from the work at Kidd Creek, this
writer is unaware of any large scale
attempts in Canada, at present, to use
smelter slag as a backfill.  As mentioned
earlier, the work at Noranda in the 1930's
is a landmark(7) in this area.

2.3 Emplacement systems and "total" fill

Each of the several materials used for
backfilling present unique problems from a
materials handling point of view.  Coarse,
or large-sized fills, including waste
rock, gravel, slag, and alluvial sand are
usually placed either by waste passes
which lead directly to the stope to be
filled, or by mechanical means including
hauling or conveying, or by a mixed-mode
system.  Such systems are expensive.

Mill tailings, on the other hand, are
almost always developed hydraulically
through systems comprised of boreholes and
pipelines.  The enormous advantage of
using hydraulic fill is that the delivery

5

systems are much less costly and are much more flexible. This flexibility is only useful in medium to large mines, however, in which bulk pours can be arranged and accommodated. For small operations the only alternative may be to place fill by small-scale very-expensive methods, possibly even still involving physical handling. Two of the research thrusts mentioned in this paper are directed to special applications: one in which very small quantities are needed; and the other a situation when only dry fill can be used because of the solubility of the waste material (salt).

Apart from the problem of integrating mine and mill production schedules so that adequate quantities of fill will be available when these are needed, there are five principal technical problems in using mill tailings for backfill. First, in order to achieve adequate percolation rates so that the water may be discharged and the fill stabilized as quickly as possible, the finest fractions of the tailings, or slimes, must be removed by classification. Second, mine pumping systems must be capable of handling the large quantities of water involved. Third, the percolation of water through the fill inevitably results in some leaching of the cement which may be mixed to bond it. Fourth, it is very difficult to reduce zoning and achieve a consistency of the poured mixes. Finally, hydraulically-placed mill-tailings can involve a lot of mess and resulting clean-up.

At present, research is underway which, hopefully, will result in improvements to practices in many of the problem areas. The technology of emplacing densified fill, if successfully developed to the scale at which it can be applied to large-stope bulk mining operations, will have a very large and rapid payback for the industry.

If, further, a technique can be developed by which all of a mill's tailings output, including slimes, can be safely disposed of underground, a major step will have been made towards solving one of the industry's important environmental problems.

2.4 Stabilization of today's and yesterday's backfill

In our present practices, elaborate precautions must be taken to ensure that pours of liquid fills are adequately contained until drainage of water precludes any possible danger of liquefaction. Escaping backfill, or overburden, can cause tremendous damage and possible loss of life.

To safeguard against any possibilities of this occurring, elaborate bulkheads and fill fences are usually constructed across the lower accesses into stopes to be filled. Further, it is not uncommon for the initial pours of fill behind such barricades, to be enriched by additional cement in order to achieve much greater strengths. These structures are known as fill plugs.

The use of densified fills may make the use of such elaborate and costly structures unnecessary. Before one can eliminate these with complete confidence, however, it will be necessary to establish that there is no liquefaction potential.

Because the techniques of hydraulic backfill have been developed only in the last forty years, or so, it is still possible to find cases when pillars in older parts of long-established mines are adjacent to stopes filled with unconsolidated sands, gravels, or waste rock. The challenge to the operator is to recover these pillars with complete safety and taking the greatest possible advantage of the most recent mining technology.

One of the present thrusts in the industry, then, is to determine the ways in which loose fill can be consolidated.

2.5 Use of backfill in the north

In Canada's far north, the climate is such that an entirely different approach is needed for successful backfilling. Because of the cold, which is often intense, the handling of large quantities of water would cause enormous problems. Hence the handling of hydraulic fill is considered to pose almost insurmountable problems (from an economic point of view).

Water can be handled in small quantities however, and under closely controlled and monitored circumstances. This has led to the development of a frozen backfill at Cominco's Polaris operation. Waste rock, quarried on surface, is sprayed with water as it is placed in completed stopes(17). Once frozen, it has been found that the adjacent panels can be mined quite successfully.

Mining in the north in the future will present many technical and operational challenges. Backfilling will be high on the list of priorities for research.

3 BACKFILL RESEARCH IN CANADA IN 1989

Principally as the result of a series of

Mineral Development Agreements (MDAs) signed between the Canadian Federal Government and most of the provincial and territorial governments (of which there are 12), there has been a substantial increase from the mid 1980's onwards in the quantity and quality of mining research. Most of the projects which were approved through the MDA's were conceived by the industry to address its needs and were destined for delivery by the industry - often acting in partnership with governments, consultants, and academe. The "bottom line" is that a lot of research has moved from the laboratory into the mine, and from bench scale to field-demonstration scale.

Backfill research has been one of the important areas addressed in agreements between the federal government and three provincial governments; namely, Manitoba, Ontario and New Brunswick. In the mid 1980's, backfill research was also done in Saskatchewan through a specific federal program known as "START" (Short Term Assistance for Research and Technology).

All of the projects included in these programs are described briefly in this paper. Collectively, the various levels of government are probably committing about $1 million annually to backfill research. The industry, through its own in-kind contribution to MDA projects, is probably at least matching that amount and possibly expending twice as much.

Nor is the only research being done through the MDAs. Many individual mining companies are undoubtedly enhancing their practices and techniques through their own resources. Unfortunately, there is no simple way to ascertain the totality of the industrial effort. Recently(18), the Mining Research Directorate of the Ontario Mining Association estimated that about $100 million is now being expended annually on mining research in Canada. If that is the case, it is probably a fair guess to estimate that about 10% of that amount, or $10 million/annum, is being committed to backfill research.

The dimension of research does not end there, however, since research is also in progress in some Canadian universities. Both McGill and Queen's Universities, at Montreal, Quebec and Kingston, Ontario, respectively, have been associated with some of the MDA backfill projects. The Universities of Saskatchewan and Waterloo, at Saskatoon, Saskatchewan and Waterloo, Ontario, respectively, have also been involved with the potash backfill research.

There may be others also in the field, but the writer has not been made aware of these. Naturally, any omissions are regretted and, if there are some, the writer would appreciate being advised of this.

The following are brief descriptions of the projects presently in progress through the various MDAs.

3.1 Evaluation of methods for delayed backfill consolidation - Manitoba MDA

While the use of cemented and consolidated fills is now common, this has been a feature of Canadian mining practice for only about 30 years. Previously, unconsolidated alluvial sands, gravels, waste rocks, slags, and many combinations of these were used. In many areas, depending upon the availability of mill tailings and other considerations, such is still the case.

Thus, it is not uncommon presently for a mining engineer to have to contend with the problem of mining pillars between stopes filled with unconsolidated fill. The alternatives which are available to stabilize such loose fills include pressure grouting and the percolation of cemented mixtures by gravity.

In this project, which was the subject of a contract in the amount of $100,000 to the Hudson Bay Mining and Smelting Company Ltd., a review was to be made of the cost effectiveness of the methods which are available for delayed consolidation of backfills. This was to have been followed by field trials using preferred methods and the establishment of predictability criteria. There would also have been the development of theoretical models and the correlation of field results with these. The major objective is to develop practical techniques for the in-situ consolidations of fills.

At an advanced stage of the work, however, the company was unable to proceed further with the field experimentation because of the lack of a suitable experimental stope in the Flin Flon area.

The following two reports were produced on those portions of the project which were completed:
1. "Stabilization of Soils and Backfill - A Review of Mining and Civil Applications".
2. "Laboratory Investigations of HBMS Smelter Slag, Progress Report No. 1".

3.2 In-Situ determination of dewatered tailing fill properties projects - Ontario MDA

The most common backfilling method is to

emplace mill tailings in the form of a slurry. On curing, the water which is used to transport the solid particles to the stopes must be pumped from the mine as the fill consolidates and ages. Cement, added to increase the strength of the fill, is leached away as the water percolates downwards.

The successful use of higher-density paste-type fills would offer a number of improvements to mine operators. First, with much less water being used, there would be significant reductions in pumping costs and cement losses. Second, because of the increased retention of cement and higher density, the fill would attain higher strengths more rapidly. Third, this, in turn, would simplify the methods which are used to design the structures used to contain the fills (i.e., bulkheads and fill fences). Fourth, there would be an improvement in the handling of slimes and in costs of clean-up underground.

Two projects, involving alternative technologies are now nearing completion in the province of Ontario:

1. At Inco, in the Sudbury basin, work has been in progress for some time at the Levack Mine to design a system which will permit the delivery of high-density fills directly through pipelines. A CANMET contract, in the amount of $112,000, is in place to accelerate this work.

At Inco a surface mixing plant and a gravitational system is used to deliver the paste fill to a test stope. The initial results, at depths at less than 1,200 feet, have been very successful. Monitoring instrumentation has been installed and the trial stope filled. Mining is in progress.

Based on the early successes, and even before completion of the project, consideration has been given to extending the technology to openings at greater depths.

The Inco Project is now nearing completion. A draft final report has been received and is being reviewed by CANMET staff.

Some of the significant findings from the report are as follows:

i) The strengths of paste fill samples were found to be about double those of samples of hydraulic backfill with similar cement contents.

ii) The uniaxial compressive strengths of paste-fill samples recovered in-situ were about 80% higher than those of laboratory-prepared samples.

2. At Dome Mines Ltd. an alternative approach to the delivery of paste fill is being investigated. Through a CANMET contract, in the amount of $152,580, the company is evaluating the potential use of a device known as the "tailspinner". Operating much like a centrifuge, the tailspinner receives liquid backfills at normal pulp densities (about 60% solids by weight). On delivery, the water is spun from the fill and removed. An extruded paste is emplaced.

At Dome, in-situ monitoring of fill behaviour is now in progress. Laboratory studies of the behaviour of paste fill are complete and a report has been received from McGill University, the sub-contractor on the project. A debriefing seminar on this aspect of the work will be held when the final report becomes available for general distribution.

A draft final report for the Dome project is now being prepared. A series of accidents at the mine during 1988, however, prevented access to the experimental stope for several months. These delayed the completion of this project, as well as another entitled "Liquefaction Potential of Dense Backfill". The final report from Dome was expected by the end of 1988.

Falconbridge Limited, another sub-contractor on this project, has also submitted a draft final report on a survey of World Paste Fill Practices. The Falconbridge research group is currently assessing the feasibility of using paste fill for some of their mining operations.

3.3 In-Situ monitoring and computer modellilng of a cemented sill mat and confines during tertiary pillar recovery - Ontario MDA

In cut-and-fill mining, the intervening pillars between previously-mined stopes are recovered during secondary extraction. Sill pillars, between the mining blocks and the levels, are recovered during a final, or tertiary, stage. The entire process of extracting all of the ore between levels may involve several years.

During this process, however, mining practices and economic conditions are constantly changing. The results can be great departures from original plans and large variations between the properties of fills in contiguous openings.

In order to provide increased confidence in both design methodology and extractive practices, a project, involving both in-situ monitoring of ground conditions during the extraction of sill pillars and computer modelling for predictive and back-analytical purposes, was initiated

8

with Falconbridge Ltd. The contract, under the Canada/Ontario MDA is valued at $154,720.

At the time of writing, the project has been completed and a draft final report is being reviewed.

Some of the significant findings of the research are:

1. The results of computer simulations indicated a close correlation between the behaviour of in-situ fill and that predicted by the computer model.

2. The computer model has now been used to evaluate sill mats, other than at the trial area, at Falconbridge's Strathcona Mine.

A debriefing session will be held as soon as the final report becomes available for release.

## 3.4 Use of cemented fills for controlling violent failure in pillars - Ontario MDA

In the room-and-pillar mines in the near-horizontal tabular uranium deposits of Elliot Lake backfill was not considered to be necessary since the vast mined out areas remained quite stable. Commencing about 1985, however, there was much increased rockbursting and failures of the rib pillars in the area near the boundary pillar between the Denison and the Rio Algom mining operations. The area affected was more-or-less in the centre of the previously mined part of the ore-bearing conglomerate reef.

Experience has shown that the area affected by rockbursting in a room and pillar mining operation can grow rapidly and become extensive. The only practical remedial action may be to pour backfill around the pillars. This seems to be a method of increasing the post-yield strengths of pillars and, consequently, of limiting the growth of the failure zones.

To study the use of tailings backfill as a means of stabilizing an area which is in the process of fracturing, a project, with a contract value of $610,000, was initiated with Denison Mines Ltd. In this research, the stabilities of pillars are being monitored as a selected area is backfilled. Monitoring involves both stress measurement and microseismic techniques. A microseismic system, belonging to CANMET, has been installed in the designated area of the mine for the purpose of the study.

The research is now well-advanced, with the previously-mined stopes of the test panel area having been filled with about 120,000 tons of deslimed tailings consolidated with iron ore blast furnace slag. The area immediately up-dip from the backfill area has been seismically active, with local rockbursting pillar spalling and heaving of the floor. Because of this, the microseismic system was redeployed to provide better coverage of the active area. Denison Mines also decided to expand the area that would be filled as the backfilled panel is less seismically active than the surrounding area.

## 3.5 In-situ properties of backfill alternatives in Ontario Mines - Ontario MDA

In spite of the fact that a wide range of materials has been used as backfills in mines (i.e., alluvial sand, waste rock, mill tailings, slags, and mixtures of these), very little is known concerning the relative merits or demerits of these. There is a need to determine the properties of various backfill alternatives and to establish general engineering specifications.

To accomplish this, a contract, in the amount of $470,000 was signed with Falconbridge Ltd. Much of the work is being carried out at the Kidd Creek operations, at Timmins, Ontario.

In the research, which commenced in 1987, various types of backfill are being emplaced in openings which have been surrounded by monitoring instrumentation. The results of this large-scale comparative study should permit a quantification of the support characteristics of fills. Further, the relationships between laboratory and field properties, once established, will permit the establishment of specifications.

At present, both laboratory testing of various binder alternatives (including slags and flyash) and physical modelling trials are proceeding according to schedules. The installation of instrumentation and field trials are also underway. Laboratory trials on copper slag and fly-ash binders, as well as field trials involving layered fill, Reiss Lime, Slag and Monolithic Packing Materials have been completed. Field Instrumentation trials are continuing at Kidd Creek Mines. Additional trials are planned.

An evaluation of anhydrides as binder alternatives is also being carried out by McGill University under sub-contract. A report on this was expected by January 31, 1989.

It is now anticipated that the entire project will be completed early in 1990.

The results should be reliable
specifications for various filling
materials. Additionally, less expensive
alternatives to present methods and
approaches may result.

## 3.6 3-D numerical models for simulation of bulk mining at depth - Ontario MDA

During the past decade, especially, there
have been rapid advances in the
analytical tools which are available to
rock mechanics specialists. Numerical
modelling techniques have taken the place
of experimental stress analysis and are
now used for engineering design purposes.

The computing requirements for the
larger models, however, can be far beyond
the capabilities of most organizations.
For this reason "mine wide" models are
very rare and are mostly the property of
large international-scale consulting
organizations.

There is a need, both to advance the
technology which is available and to
investigate ways in which it can be
transferred to smaller scale computers.
By doing so, it would become available to
the smaller organizations which do not
presently possess the specialized skills
necessary.

In order to develop a sophisticated
three-dimensional model, applicable on a
very broad scale, and suitable for
simulating a wide variety of mining
conditions including non-elastic and
post-failure behaviour of a rock mass, a
contract in the amount of $1,000,000 was
signed with Inco Ltd. The company will
not only develop the highly sophisticated
model but will also calibrate it and
refine it by making frequent reference to
actual in-situ conditions and
measurements.

The project is progressing according to
schedule and within budget.

A review of bulk mining at depth was
completed in June, 1988, and it is
anticipated that a report will be
available shortly. Two interim progress
reports for the first year of the project
(1986/87), were available for general
distribution late in 1988. The reports
treat the numerical modelling and the
instrumentation aspects of the work.

Two interim progress reports, covering
work completed during the second year,
(1987/1988) are expected shortly. One
will cover numerical modelling while the
second concerns instrumentation.

Simultaneously, the development of a
two-dimensional plasticity model, capable
of simulating the failure zones around

excavations and localized shearing, is
nearly completed. The development of a
three-dimensional plasticity model will
commence in 1989 and will require about 18
months to complete. A technology transfer
seminar/workshop on the two-dimensional
plasticity model will take place in the
spring or summer of 1989. The final
details will be established after
consultations with the contractor.

## 3.7 Liquefaction potential of dense backfill - Ontario MDA

One of the greatest concerns of any mine
operator using mill tailings as a
backfilling material relates to its
liquefaction potential. Fine-grained
materials, with a high moisture content,
can liquefy under dynamic loading
conditions. In a worst case scenario, a
seismic disturbance could cause the fill
in a recently-filled stope to liquefy,
break the bulkheads due to the resulting
sudden increasing pressure, and to flood
out into the openings below. The results,
as at Belmoral (but with overburden rather
than fill) could be catastrophic.

In order to define the engineering
parameters involved, and to study such
behaviour of fill, and particularly
densified fill, a contract in the amount
of $125,250, was signed with Dome Mines
Ltd. The objectives were to study the
liquefaction potential of dense backfill,
and to develop procedures for determining
the safe limits for various types of fill
materials.

The project is now nearly complete. A
draft final report is being prepared by
Dome and was expected to be completed by
the end of 1988. A report on the
Laboratory and field tests has been
received from McGill University, the
sub-contractor on the project. This
report is currently being reviewed by
CANMET staff.

The results of the research should
assist the industry in establishing safe
limits for evaluating the liquefaction
potential of dense backfills.

## 3.8 Use of backfill in New Brunswick potash mines - New Brunswick MDA

Potash mining in Canada is essentially a
"one-pass" type of operation. Rooms are
mined in a series of passes using highly
mechanized boring machines. The rate of
advance is very rapid and total extraction of mineral probably averages about
40%. The intervening pillars between
rooms are not mined, nor is backfill

used. The present economics of potash mining are said to preclude the use of fill. The extraction ratio is low by design in order to provide long-term stability both of the rooms and of the overlying strata.

In the long-run, however, the low extraction ratio will result in a loss of reserves.

A second problem is that the potash is interbedded with salt. Because of contamination with other minerals this salt is not usable for any purpose. After separation during milling, therefore, it is transported to storage piles on the surface. In the future the ultimate disposal of the waste salt will pose a number of environmental concerns.

The project in New Brunswick was designed to address both of these concerns. Under a $214,740 contract with the federal government the Denison Potacan Potash Company is evaluating the stabilizing effect of waste salt as a backfill in mined openings.

In the first phase of the work, completed in 1987, a study was made of the engineering properties of waste salt backfill, and of the effects of additives on strength. The costs and benefits of alternative stowing procedures have been assessed. Finally, using numerical methods of stress analysis, determinations were made of the effects of backfilling upon convergence and the creep of mine openings. The results have shown that at least ten years are required before backfill provides roof support.

A debriefing session for the first phase of the project was held on September 4, 1987.

At the time of writing, a second phase of the work, at a cost of $199,130 was in progress. This aspect of the project was scheduled for completion by the end of March, 1989. To date: the instrumentation around a trial stope has been installed; computer modelling to identify suitable mining geometries and permissible ground reactions for mining long secondary stopes in pillars; and laboratory determinations of the properties of highly consolidated fills, are in progress.

A third phase, at a contracted value of $150,500, was also in progress and scheduled for completion in June, 1989. In this phase, CANMET is contributing $63,370 to the work while the company is contributing the remainder. A microseismic monitoring system, to identify the reactions of the hanging wall to mining, will be installed after an evaluation of presently-available technology has been completed. A comparison will also be made of the outputs resulting from computer simulations of a standard mining sequence using both the GEOROC and VISCOT codes.

3.9 Backfill projects - Saskatchewan MDA

The federally-funded mining research which has taken place in recent years in the Saskatchewan potash mines can be divided into two phases. In the first of these, which preceded the Mineral Development Agreements by about two years (i.e., 1983 to 1985), approximately $443,000 was committed to mining research through the START (Short-Term Assistance for Research and Technology) program.

Two of these projects related directly to possible uses of backfill in Saskatchewan potash mines. The first project entitled "Determination of Engineering Properties of Waste Salt for Backfilling Underground Potash Mines", was the subject of a $25,250 contract with the firm RE/SPEC Ltd. The second, called "A Field Test Program to Evaluate the Use of Waste Salt Backfill in Saskatchewan Potash Mines" was the subject of a $120,000 contract with Central Canada Potash. Both projects were concluded in the mid 1980's and the information derived from them has provided valuable background data for the work presently in progress in New Brunswick. The details may be found in the Proceedings of a CANMET seminar which was held in Saskatoon(18).

3.10 Pneumatic backfill system for small deposits - a project originally planned for the Yukon Territory

For almost half a century a substantial proportion of Canada's silver production was derived from deposits worked by United Keno Hill Mines, Limited, near Elsa, in the Yukon Territory. The deposits, of which there are many in the area, are generally small, high-grade and surrounded by weak wall rocks. Mining was small-scale and labour-intensive. The mines at Elsa were probably the last in the country in which square set mining was still employed.

Because of the small stopes, low production, and high costs, however, the future viability of mining in the district was threatened. With silver prices having remained static for several years, it follows that given ever-increasing costs, profitability could only be improved through increasing productivities

11

and reducing unit costs. The only effective way to do this, in the long term,
was to replace square set mining with the
less expensive cut-and-fill technique.

At Elsa, however, exceptional impedi-
ments to making such a conversion were
caused by the small sizes of the
individual stopes and mines. Not only
would conventional pipeline fill delivery
systems be too expensive but also the
exceptionally small quantities required
from time to time at a large number of
geographically dispersed locations would
present unsurmoutable operating
problems. The capital and operating
expenditures needed to cope with the
water accompanying hydraulic fill would
also be severe. The same is probably
true for many small mining operations.

With these constraints as background,
it was proposed to make field trials of a
new pneumatic small-scale and portable
fill delivery system. In concept, fill
would be delivered to the machine in a
relatively "dry" form. Cement would be
added at the machine and water injected
as a spray at the nozzle as the fill is
blown into the stope. The proposed
research would have brought the concept
through to a fully operating technique
under sub-arctic conditions.

The project was proposed jointly by the
company and CANMET for possible inclusion
in the second round of an MDA between the
federal and Yukon territorial
governments.
Unfortunately, continuing financial
losses forced the closure of the
operation very early this year (1989).
The project, because of a general
applicability to small-scale mining
operations, remains as a high-priority
item for future work.

4 SUMMARY

In this paper, the author has attempted
to capture the flavour and the excitement
of the backfill research which is
currently in progress in Canada. Because
of the funding which has been available
through various jurisdictions of
government, Canada is becoming an
important player internationally in
backfill research. All of the projects
now underway are designed to discover
better ways in which backfill can be
designed and used and then emplaced for
support purposes. The view of backfilling in Canada has matured from one of·
it being a problem of waste disposal and
providing a mining floor, to that of

accepting that properly engineered backfill is an essential part of mine design
and long-term mine stability.

As the approach has matured, the
research has taken a more innovative tack.
The field-scale comparative studies of
fill alternatives and the mine-scale
field-validated predictive models, for
example, are projects of which any country
would be proud. Truly, these, and the
others described, are on the leading edge
of backfill technology.

5 REFERENCES

Manchester, H.N., "An Illustrated History
of Mining and Metallurgy", McGraw Hill,
New York, 1922, p. 14.
Agricola, Georgius, "De Re Metallica",
translation by Herbert Clark Hoover and
Lou Henry Hoover, Dover Publications
Inc., New York, 1950, p. 126 (Book V).
Hoover, Herbert C, "Principles of Mining",
McGraw Hill, New York, 1909, pp 107-118.
Peele, Robert, "Mining Engineer's
Handbook", John Wiley & Sons, New York,
Third Edition, 1941, Vol. I, pp 10-237 -
10-274 (Filled Stopes); 10-274 - 10-297
(Shrinkage Stopes).
Rice, H.R., "Introduction to Mining
Methods in Canadian Mines", Mining in
Canada, 6th Commonwealth Mining and
Metallurgical Congress, Northern Miner
Press, 1957, pp 43-46.
Brock, A.F. and Taylor, W.J., "Square-Set
Stoping at the Frood Mine of the
International Nickel Company of Canada,
Limited", Mining in Canada, 6th
Commonwealth Mining and Metallurgical
Congress, Northern Miner Press, 1957,
pp 181-186.
Patton, F.E., "Backfilling at Noranda",
Mining in Canada, 6th Commonwealth
Mining and Metallurgical Congress,
Northern Miner Press, 1957,
pp 229-236.
Staff, "Stope Filling at the Sullivan
Mine of the Consolidated Mining and
Smelting Company of Canada, Limited",
6th Commonwealth Mining and Metallur-
gical Congress, Northern Miner Press,
1951, pp 237-242.
Anon, "The Falconbridge Story", Canadian
Mining Journal, Vol. 81, No. 6, June,
1959, pp 132-133.
Ibid., p. 145.
McCreedy, J. and Hall, R.J., "Cemented
Sand Fill at Inco", CIM Bulletin, July,
1966, pp 888-892.
Cleland, Roy S. and Singh, K.H.,
"Development of 'Post' Pillar Mining at
Falconbridge Nickel Mines Limited", CIM

Bulletin, April, 1973, pp 57-64.

Bharti, S., Udd, J.E., and Cornett, D.J., "Ground Support at Strathcona Mine", Underground Support Systems, CIM Special Volume 35, Montréal, 1987, pp 13-26.

Landriault, D., "Preparation and Placement of High Density Backfill", Underground Support Systems, CIM Special Volume 35, Montréal, 1987, pp 99-103.

Yu, T.R., "Ground Support with Consolidated Rockfill", Underground Support Systems, CIM Special Volume 35, Montréal, 1987, pp 85-91.

Thomas, E.G., Nantel, J.H., and Notley, K.R., "Fill Technology in Underground Metalliferous Mines", International Academic Services Ltd., Kingston, Ontario, 1979, ISBN 0-920912-00-1, pp 203-225.

Jubb, J.T., Udd, J.E., and Doyle, D.M., "Technological Challenges in the 90's", CANMET Division Report MRL 88-19(OP), May, 1988, 15 pages.

Whiteway, Patrick, "The Pressure Is On", The Northern Miner Magazine, December 1988, pp 30-33

*Innovations in Mining Backfill Technology, Hassani et al. (eds), © 1989 Balkema, Rotterdam. ISBN 90 6191 985 1*

# Fill research at Mount Isa Mines Limited

A.G.Grice
*Mount Isa Mines Limited, Mount Isa, Australia*

ABSTRACT: Fill research activities at Mount Isa Mines cover a wide range of topics and include the design, production and placement of multiple fill types into the underground operation. This paper summarises research since 1983 and discusses the introduction of new fill types, production methods and investigations into bulkhead stability. Discussion on alternate cementing products is also included.

## 1. INTRODUCTION

Mount Isa Mines Limited currently produces 5.5 million tonnes of Copper ore and 4.5 million tonnes of Silver-Lead-Zinc ore (hereafter referred to as Lead ore) per annum.

Copper ore is produced from the 1100 and 1900 Orebodies to the south of the central shaft complex, entirely by open stoping. The deeper 3000/3500 Copper Orebodies are located directly below the central shaft complex and initial access development has commenced. To the north, Lead ore is extracted by cut-and-fill, benching and open stoping techniques. Detailed descriptions of operations in Copper and Lead areas at Isa Mine are available in Alexander (1981) and Goddard (1981) respectively. At Hilton Mine, 20 km north of Isa Mine, production of Lead ore has commenced in cut-and-fill stopes with trial open stoping due to commence in 1989. Further details about Hilton operations are available in Black (1986).

A variety of fill types and placement methods are used at Mount Isa to suit the different mining methods and performance requirements of the fill. Where the fill is required only to fill the extracted void and to provide regional support, uncemented fills are used. However, the majority of fill is subsequently re-exposed and therefore must be cemented.

Filling is an integral component of the mining operations and the research activities are targeted at the optimisation of ore extraction by improvements in fill design and techniques. Research covers the major areas of fill type and system design, production, placement and performance.

## 2. FILL TYPES

Details of fill types and production methods are described in McKinstry (1989) and Neindorf (1983) and only a brief description of the fill types will be presented here.

Hydraulic fill is produced from de-slimed and de-watered mill tailings at the Wet Fill Station. The hydraulic fill is specified to contain not more than 10% of less than $10\mu$ sized fines and to have a slurry density of 69% solids by weight (specific gravity of the solids = 2.9).

Cemented hydraulic fill (CHF) is produced at the same plant by adding 3% by weight of Portland Cement and a slurry containing 6% by weight of Copper Reverberatory Furnace Slag (CRFS) to the hydraulic fill. Thomas (1973) discusses the original design concepts for cemented fill and the strength specification is given below.

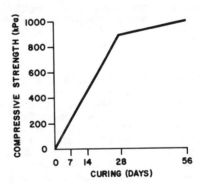

Figure 1. Fill strength specification

Rejects from the Heavy Medium Separation Plant can be added at around 25% by weight to both the cemented and uncemented fill slurries. The reject material is a coarse gravel sized component referred to as aggregate and the fill types are cemented aggregate fill (CAF) and hydraulic aggregate fill respectively.

All the above four fill types are reticulated by gravity to the underground workings via vertical boreholes and connecting pipelines. A quarry, crusher and conveyor system transfers larger dry siltstone rock to the Copper areas where it is added to cemented hydraulic fill at the top of the stope to produce cemented rockfill. The design of this system was reported by Mathews (1973).

Dry aggregate is placed via the rockfill conveyor system to the Copper areas. Dry waste materials can be trucked and tipped from surface via fill passes to shallow Lead stopes.

Sizings curves for the fill types are given in Figure 2.

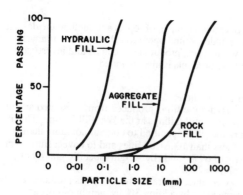

Figure 2. Fill particle sizings

## 3. DESIGN

Design is concentrated into three areas; assessment and refinement of new and existing fill types, investigation of alternate components used in fill production, and design of plants and delivery systems for new mining areas.

## 3.1. CEMENTED ROCKFILL DESIGN

Cowling (1983) and Mathews (1973) discuss earlier aspects of cemented rockfill design, the filling system for most stopes in the Copper areas. More recent developments in modelling are covered in Cowling (1989). Application of the new modelling techniques has increased the volumetric filling rate with reduced cement consumption and drainage times, resulting in significant reductions to stoping cycle times.

It was proposed by Swan (1985) that the strength of cemented fills could be predicted from component gradings and cement addition rates by determination of the binder number. This was investigated and confirmed for Isa fill types.

## 3.2. AGGREGATE FILL INTRODUCTION

McKinstry (1989) describes the production and placement of aggregate and hydraulic fills containing aggregate as a low cost bulking agent. Because of the relatively small size of the shale content of the aggregate, the strength of the fill is the same as cemented hydraulic fill.

Pumping trials carried out in a surface test pipeline showed that up to 30% of aggregate could be added to the hydraulic fill slurries and successfully delivered to stopes in the Lead areas. The cemented aggregate fill showed improved pumping characteristics due to the lubricating effects of the finer cement components. It has been observed more recently that pipe wear by uncemented hydraulic aggregate fill is excessive.

Cemented aggregate fill segregates when placed into Lead stopes. The aggregate is deposited around the footwall in a well cemented zone and little aggregate reaches the extremities of the stope.

Aggregate can be delivered through small diameter drill holes which permits the better distribution of aggregate within stopes. Two new filling methods have been trialled using this capability and one is now in common use in the Copper areas.

The differential filling technique involves the selective placement of a cemented hydraulic fill beach against a relatively small future exposure and the bulk of the remaining void is filled with aggregate. Figure 3 illustrates the differential filling concept which is being employed in late stage pillar stopes.

Figure 3. Differential filling

Aggregate was trialled as a replacement for rockfill in a method called In-situ CAF. Dry aggregate and cemented hydraulic fill at 1:1 ratio was introduced at the the top of the stope in a similar manner to the cemented rockfill method. Mining a trial drift back through the centre of the fill mass showed that large loose zones of uncemented aggregate had been washed out by the hydraulic fill against the stope limits, reducing the security of the subsequent exposures. In addition, small pits were constructed on surface which confirmed that penetration of cemented hydraulic fill into dry aggregate was minimal. For these reasons, the method was not investigated further. Figure 4 illustrates the technique.

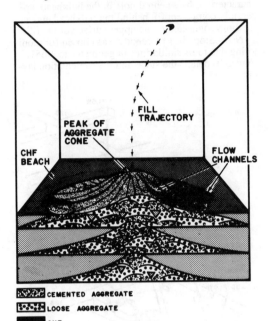

CEMENTED AGGREGATE
LOOSE AGGREGATE
CHF

Figure 4. In situ CAF

Dry aggregate has an angle of repose of around 37° and the shape of the cone at the top of a filled stope does not tightly fill or provide any crown support. Moderate application of hydraulic fill or water to this cone causes local slumping, reducing the rill angle to around 5° and permits tight filling to the back of the stope.

## 3.3. ALTERNATE CEMENTS

Portland Cement is the highest cost component of cemented fill, accounting for upto 85% of the costs. Investigations into further partial or total replacement have continued to evaluate other options.

The pozzolanic characteristics of finely ground CRFS are activated by the release of free lime (CaO) as a by product of the hydraulic Portland Cement reaction. This pozzolanic reaction commences around a week after fill placement and continues through the first three months. To increase the cost/performance ratio, addition of lime, commercial binders and the total replacement of Portland Cement have been investigated.

The potential benefit from lime addition is that only very small addition rates are required because the lime is a catalyst, activating free lime present on the pozzolanic surface.

Investigations into total replacement of Portland Cement at laboratory and bulk scale have indicated major potential benefits. Unfortunately, the fill production and delivery systems currently do not permit these fill types to be made at Mount Isa. The main problem is that the lean lime mixes (less than 2% by weight) are very susceptible to segregation effects during placement, resulting in zones of poor quality fill.

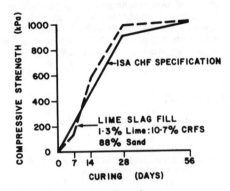

Figure 5. Lime slag fill performance

## 3.4. NEW FILL SYSTEMS

New fill systems are required for the Hilton Mine and the 3000 / 3500 Copper Orebodies and are discussed below.

### 3.4.1. HILTON MINE

A fill re-pulping plant has been commissioned at Hilton Mine to supply hydraulic fill for the initial cut-and-fill Lead mining operations. Excess fill slurry from the Isa Mine Wet Fill station is placed into dams and allowed to drain via an aggregate base. After a few days the material can be handled with a front end loader and is stockpiled close to the Hilton ore handling area. Trucks bringing in Hilton ore are then back loaded with fill. A single plant operator drives a front end loader and places the fill on a ramp to be sluiced by high pressure water monitors into the densifying tank via a tramp screen. Hydraulic fill at up to 70% solids by weight content is produced.

A decision has been taken to construct a concentrator on site to handle the increase in production from open stoping. Design work has commenced to provide cemented fill. The fill from the concentrator will require ongoing additional material to be supplied from Isa Mine.

Hydraulic fill placement in the cut-and-fill stopes has been successfully introduced. Small open stopes are planned with relatively small plan areas. The drainage characteristics of the fill will be critical to the success of the system. The smaller stopes will re-

17

quire a faster production, fill, drain and cure cycle time. This will be achieved by the placement of higher density fills to reduce the drainage water and an increase in strength permitting earlier exposure.

### 3.4.2. 3000/3500 OREBODY

Preliminary stoping designs indicate 30 metre square high rise open stopes. A high performance fill material is required to achieve the high production targets and faster cycle times will be essential. Fill design work has commenced with a broad goal of producing denser high strength fill with dry and wet fill components to achieve the necessary placement rates.

## 4. PLACEMENT

### 4.1. FILL BULKHEAD SECURITY

Hydraulic fill is retained in the stopes by 0.45 metre thick concrete brick bulkheads. Figure 6 illustrates the standard bulkhead design.

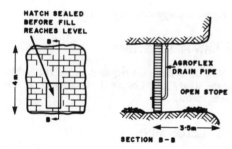

Figure 6. Typical fill bulkhead

These bulkheads permit drainage of the fill while retaining the material as it consolidates and cures. Over a period of eighteen months three of these fill bulkheads failed in separate incidents. In order to identify and solve the engineering problems all aspects of stope drainage and fill bulkhead design were investigated. Only a broad overview of the projects is presented here.

There were two major areas of research. The first concerned the behaviour of fill during placement and drainage by monitoring and modelling. The second investigated the behaviour and modelling of bulkheads under loading conditions. The University of Queensland, Department of Civil Engineering and the Commonwealth Scientific and Industrial Research Organisation (CSIRO), Division of Geomechanics were involved in aspects of both of these projects.

### 4.2. STOPE DRAINAGE TRIALS

The failure of N651 Stope bulkhead occurred whilst filling with cemented aggregate fill. The fill surface was approximately 65 metres above the bulkhead, which was in turn, 90 metres above the base of the stope. The stope had a narrow plan area of 17.5 m by 25 m and was therefore subjected to a rapid rise of level during filling. The failure was characterised by a central hole in the bulkhead, apparently competent fill behind the bulkhead and a sinkhole connected to the upper surface of the fill.

N673 Stope in 14 Orebody was chosen for monitoring due to its similarities in geometry to N651. Figure 10 details the stope and instrumentation layout.

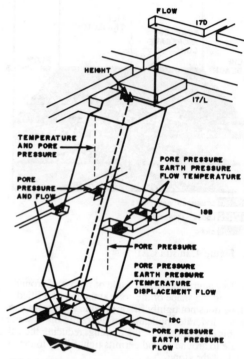

Figure 7 Instrumentation of N673 Stope

Some aspects of the project are covered by Patterson (1987). The extraction level bulkheads were instrumented to record pore water and earth pressure loadings. Magnetic flow meters were installed at each bulkhead and on the inlet fill slurry pipeline to monitor water flows. Daily manual readings were used to monitor the height of fill. All instruments were linked to a data acquisition system and results were recorded at 15 minute intervals throughout the 82 days of the project on a personal computer.

During the project, 49200 tonnes of cemented aggregate fill were placed in 31 pours over 38 days and filled a total volume of 33300 m³. Fill was placed in average run lengths of 5.25 hours with 18.75 hours of resting ("drainage") at an average density of 68.5% solids by weight and ranged between 63.8% and 69.8%. The slurry contained 19500 tonnes of water and an additional 1300 tonnes for flushing the lines.

Of the 20800 tonnes of placed water, only 6550 tonnes or 31% drained out. At the end of filling, Day 38, 78% of the total drainage was complete. A total of 76% of the water reported to the lowest horizon. The percolation of the drainage water varied throughout the filling at a rate proportional to the height of the fill surface, in accordance with Darcy's Law. The average permeability was 0.014 metres per hour. At the end of measurable drainage, the moisture content of the fill was 35% with a saturation of around 90%.

At the completion of filling, an inspection of the upper fill surface revealed the existence of a sinkhole, approximately 150 mm in diameter, and of indeterminate volume. Figure 8 is a photograph of the sinkhole.

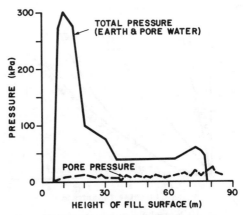

Figure 9  Pressure results in N673 Stope

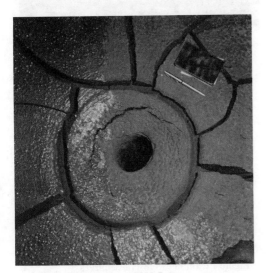

Figure 8. Sinkhole in N673 Stope

It was also observed that the bulkhead on the intermediate sublevel vertically below this point had leaked a quantity of fill. The discovery of the sinkhole was unexpected since the perceived failure mechanism had been linked to the rate of filling in the stope and N673 had been filled at a conservative rate for these reasons. This is discussed in detail with the modelling work.

Pressure loadings measured at the bulkheads are reported in figure 9. These values are taken from the southern bulkhead on 19C and record the vertical components of pressure acting on the wet surface of the bulkhead.

## 4.3. STOPE DRAINAGE MODELLING

In conjunction with the field trials a project to model the drainage of fill was initiated using the ISAACS 2 dimensional seepage and pressure model The model is described in Isaacs (1983). It was validated and calibrated using the field project data. ISAACS presents numerical drainage and pressure data and graphical output of pressure contours. The model highlighted a number of significant characteristics of fill placement. Cowling (1988) discusses details of the results obtained from this application.

During fill placement, drainage of water occurs continuously through the fill and the highly permeable bricks. The model assumes that drains are either open or closed and hence pore pressure loadings on a draining bulkhead are zero. In practice, the bulkhead has a high permeability and the field trials measured peak values of 100 kPa acting on the wet surface of the bulkhead. Pore pressure distribution is a function of the permeability and the local hydraulic head. The highest pressures are recorded at the the base of the stope furthest from the drain. Since the drainage system is dynamic, the friction heads generated by seepage flow reduce the peak values to well below geostatic pressures. However, when a bulkhead is blocked, the drainage path to the nearest draining bulkhead is effectively increased and the blocked bulkhead becomes subjected to the peak pore pressure values in the stope, modelled and measured at between 250 to 300 kPa.

## 4.4. BULKHEAD TRIALS

A series of tests were initiated to establish the performance of full sized concrete brick bulkheads. A literature search indicated a wide variation in design criteria for bulkhead loadings and design. Figure 12 illustrates a range of guidelines from three publications, Peele (1942), Garrett (1961) and AS 1475 (1977) together with an earlier modelling

19

exercise carried out at Mount Isa.

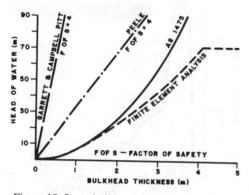

Figure 10. Some bulkhead design curves.

Three full sized bulkheads were built and tested underground and a modelling project with CSIRO Division of Geomechanics was initiated.

### 4.4.1. TEST BULKHEADS

A standard bulkhead, as per Figure 8, was built in a short dead end drive underground. The bulkhead was deliberately sealed on the internal surface to permit pressure build up. One valve was connected into a 150 mm pressure water line, fed from a point higher up in the mine. A pressure transducer was included in the water line and a frame holding a series of linear displacement transducers was installed against the external face of the bulkhead. All instruments were connected to a data acquisition system. Each test was videotaped and controlled from a short distance away using a monitor.

Three bulkheads were tested, the first terminating at 460 kPa when leakage rate equalled maximum water delivery rate. The second test reached 220 kPa before cracking was observed and water flow equalised. For the third test the water supply was changed to greatly increase the flow rate to counteract leakage and the bulkhead was successfully destroyed at a pressure of 750 kPa. A photograph of this bulkhead following the test is shown in figure 11.

Figure 12 details the performance response of the three bulkheads and indicates that inelastic behaviour commences at low pressures of around 100 kPa as the onset of cracking from the formation of the voussoir arch commences. Analysis of the videotape of the failure of the third test showed that initial failure occurred as shearing around the perimeter, followed by tensile cracking through the bricks and mortar. The peak value of 750 kPa represents a hydraulic head of around 75 metres.

Figure 11. Final test bulkhead

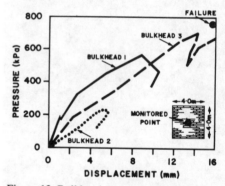

Figure 12. Bulkhead response curves

Most significantly, the tests showed that pressure build up was only possible if the bulkheads were sealed because of the high permeability of the bulkheads. A sealed bulkhead is subjected to much higher pore water pressure loadings than one which is permitted to drain freely.

### 4.4.2. BULKHEAD MODELLING

The results of the modelling aspects of the project are described in Cowling (1989) and their implications to the failure mechanism are covered in the following section.

### 4.4.3. PROPOSED BULKHEAD FAILURE MECHANISM

By considering the results of the field test work and the modelling, a number of conclusions could be

20

made regarding bulkhead security. A free draining bulkhead is subjected to minimal pore pressure loadings due to the very low hydraulic gradient through the bricks. If this bulkhead has been sealed, however, pore pressures will be higher in response to the longer drainage path to adjacent draining bulkheads and the lack of friction head component from seepage. The earth pressure loadings are dependent upon the geometry and shear strength of the fill material. The maximum earth pressure is encountered during the first placement of fill slurry, but following the rapid consolidation of fill, this value steadily drops and is independent of subsequent filling.

Even in the worst conditions of blocked drainage and poor bulkhead siting, the total stress on the bulkhead appears insufficient to cause failure. The observation of sinkhole features as described above point to an alternate mechanism of failure.

Observation of fill leakage from the bulkhead below the sinkhole in the first test stope suggests an erosion mechanism for the development of sinkholes. Leakage of sand from the bulkhead causing the removal of fill will result in the growth of an erosion pipe into the fill against the direction of seepage flow. If this pipe continues to grow upwards and connects with free water on the top surface of the fill, then suddenly geostatic pressure loading will occur on the bulkhead, causing rupture and loss of fill from the stope by erosion of the sinkhole. Figure 13 demonstrates this mechanism.

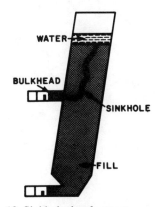

Figure 13. Sinkhole development

The two main contributing factors of this failure mechanism can be managed effectively to ensure the security of bulkheads. Firstly, inspections of all bulkheads in stopes are carried out twice per shift to detect leakages of fill and a procedure to stop filling and permit fixing of the leak has been introduced. Secondly, it has been shown Cowling (1989) that for Isa fill types, production and delivery of fills at around 72% density would eliminate the pond of free water on top of the fill, thereby eliminating the major pressure hazard.

## 4.5. DENSER FILL PROJECT

Work is underway to investigate the changes required to place denser fill underground. A series of projects have been initiated to look at aspects of production and delivery of these fills and includes automatic process monitoring systems, new and enhanced instrumentation and modifications to the fill plants.

Laboratory testing has indicated that these reductions in water content significantly increase the fill strength.

## 5. CONCLUSIONS

Fill research at Mount Isa Mines has resulted in major benefits to the operations. New materials and advanced modelling techniques have been used to increase the rates of fill placement and reduce unit costs. Work on cement replacement continues to show potential for further savings. Questions regarding bulkhead security have been addressed and management procedures implemented. The production and delivery of denser fills by gravity reticulation are seen as essential to the economic extraction of Hilton Mine and the 3000 / 3500 Orebodies; these will also offer advantages to current operations.

## 6. ACKNOWLEDGEMENTS

Mount Isa Mines Limited is thanked for permission to publish this paper. The work reported here is the result of the contributions of many members of staff and their assistance is gratefully acknowledged.

## 7. REFERENCES

Alexander, E. & Fabjanczyk, M.W.. 1981. Excavation Design Using Open Stopes for Pillar Recovery in the 1100 Orebody at Mount Isa, SME-AIME International Conference on Caving and Sublevel Stoping, Denver.

Black, B.N. and Mutton, B.K. 1986. The development of Hilton Mine, 1947 - 1985. 13th CMMI Congress, Singapore.

Cowling, R. 1983. Experience with cemented fill stability at Mount Isa Mines. Proceedings of the International Symposium on Mining with Backfill, Luleå.

Cowling, R. 1988. Numerical modelling of fill drainage within a stope, Proceedings of the ninth conference on Numerical methods in Geomechanics, Innsbrück.

Cowling, R. 1989. Computer and Modelling simulations in Fill Design, Fourth International Symposium on Mining with Backfill, Montréal.

Garrett, W.S. & Campbell Pitt, L.T. 1961. Design and Construction of Underground Bulkheads and Water Barriers, 7th CMMI Congress, Vol 3, pp 1285-1301, Johannesburg, SAIMM.

Goddard, I. 1981 .The Development of Open Stoping in Lead Orebodies at Mount Isa Mines Limited, SME-AIME International Conference on Caving and Sublevel Stoping, Denver.

Isaacs, L.T. & Carter, J.P. 1983. Theoretical study of pore water pressure developed in hydraulic fill in mine stopes, Trans. Instn. Min. Metall. (Sect. A: Min. Industry), 92.

Mathews, K. & Kaeshagen, F. 1973. The Development and Design of a Cemented Rock Filling System at the Mount Isa Mines, Australia, Jubilee Symposium on Mine Filling, pp 13-23, Australian Institute of Mining and Metallurgy.

Mckinstry, J.D. 1989. Backfilling operations at Mount Isa Mines Limited, Proceedings of the fourth International Symposium on Mining with BackFill, Montréal.

Neindorf, L.B. 1983. Fill Operating Practices at Mount Isa Mines, Proceedings, Symposium 'Mining with Backfill', University of Luleå.

Patterson, J. 1987. Secrets of the Open Stope, MIMAG, Vol 1, MIM Holdings, Brisbane.

Peele, J. 1941. Mining Engineers Handbook, John Wiley & Sons, Inc. Canada.

Standards Authority of Australia, 1977. AS 1475 Concrete Blockwork in Buildings. Brisbane.

Swan, G. 1985. A new approach to cemented backfill design, CIM Bulletin, December.

Thomas, E.G. 1973. A Review of Cementing Agents for Hydraulic Fill, Jubilee Symposium on Mine Filling, pp 65-75, Australian Inst. of Mining & Metallurgy, Mount Isa.

*Innovations in Mining Backfill Technology, Hassani et al. (eds), © 1989 Balkema, Rotterdam. ISBN 90 6191 985 1*

# Developments in Sweden of the rock mechanics of cut and fill mining

N.Krauland
*Boliden Mineral AB, Boliden, Sweden*

ABSTRACT: Cut and fill mining is used in 13 of the 16 mines of Boliden Mineral AB. Systematic failure observations and in situ measurements in the Näsliden Mine provided insight into the type and magnitude of the rock mass reactions, the function of the fill and the mechanisms governing stope stability. Aiming at developing numerical models as a planning tool in mining, the necessary modelling technique was developed and the agreement of model reactions with mine behaviour examined. The application of the acquired understanding of cut and fill mining was applied to practical mining problems. These comprise the prediction of future mining conditions, evaluation of the various types of cut and fill mining and of the effectiveness of the various support measures. Ongoing and future developments are outlined.

## 1.    GENERAL

### 1.1    Boliden Mineral AB

Boliden Mineral AB operates 16 mines in Northern and Central Sweden. In 13 of these 16 mines cut and fill mining (C+F) is used. The tonnage produced in the C+F mines is only 22 % of Boliden´s total tonnage; the revenue from the C+F mines amounts, however, to 76 % of Boliden´s total revenue.

### 1.2    Cut and fill mining

The deposits which are mined by C+F consist of small irregular orebodies with a wide variation in dip and ore value with favourable to very unfavourable rock conditions.

C+F has been chosen as it permits high recovery and small dilution; it allows also to cope with difficult rock conditions and complicated orebody geometries.

C+F was developed by mechanization into a highly efficient mining process in recent years. Mining operations are concentrated to a small number of stopes. This made the mining process very sensitive to disturbances, however. Boliden Mineral AB decided therefore to investigate systematically the rock mechanics of C+F. These studies were conducted at the Näsliden Mine and are referred to as the Näsliden project. The aim of the Näsliden project was to develop the base of rock mechanics knowledge required for mine planning, operation and continued development of C+F. In the following the main results of these investigations, their applications to practical mining problems as well as plans for further development are described.

### 1.3    Rock mechanics

Figure 1 describes schematically the situation around a stope from a rock mechanics point of view: the loading of the orebody above and below the backfilled stope increases as mining progresses. When the stope approaches an above lying backfilled excavation, the remaining ore becomes a horizontal ore pillar (referred to as a sill pillar) which is subjected to very high stresses. When the stresses approach the strength of the rock mass failure will occur either in the sill pillar or in the surrounding sidewall.

The following questions are essential for the stability of the stopes and have to be clarified therefore
* stresses and deformations around the stope
* the function of the fill
* the effect of the support measures

## 1.4  Geomechanical situation

The orebody in Näsliden consists of sulphide ore of medium to high strength. The sidewall rocks consist of metamorphized sedimentary and volcanic rock types of good strength. Near the orebody these rocks are alterated resulting in chloritization and sericitization. These alteration zones are weakness zones which distinctly influence the stability of the stopes. This situation is typical for most sulphide orebodies in Sweden.

## 2.  INVESTIGATION OF THE ROCK MECHANICS OF C+F MINING – THE NÄSLIDEN PROJECT

## 2.1  Investigations

The systematic investigation of C+F mining began with measurements and observations underground simultaneously with the start of mining in Näsliden in 1970. The aim of the field investigations was to map the stresses and deformations around C+F stopes and to clarify the mechanisms that govern stope stability.

In 1977 development of numerical models, mainly finite element models, started. This development was carried out by the University of Luleå, The Royal Institute of Technology in Stockholm and Boliden Mineral AB as a common research project of the Swedish Rock

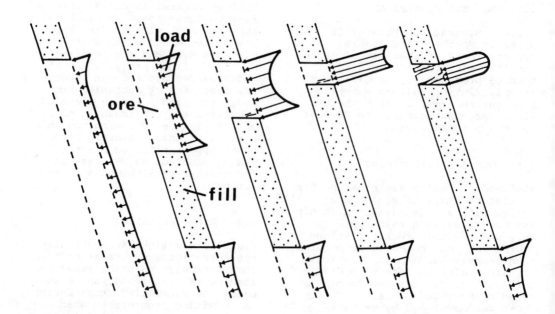

Figure 1. Increase of stresses in the roof of the stope with progress of mining

Mechanics Research Foundation. The aim of this project was to determine the degree of agreement of FE-models with mine behaviour and to establish the suitability of such models as a planning tool in mining. This development work has been reported extensively at the Conference on the application of rock mechanics to C+F mining in Luleå 1980 (Stephansson, Jones, eds., 1981).

## 2.2    Results

### 2.2.1 Rock mechanics of C+F mining

Failure mechanisms around a stope: The roof above the stope is subjected to high horizontal stresses. At the elevation of the backfilled excavation the hanging wall and footwall are destressed. In general the measured deformations comply well with elastic behaviour of the rock mass. The stress distribution,

the deformations and failure modes depend strongly on the interaction between hanging wall, roof and footwall. In figure 2 this mechanism is shown schematically. The high horizontal stresses result in nearly horizontal compressive strains in the roof. The soft alteration zones are deformed more than the ore. The large strains in dip direction within the alteration zone induce tensile strains in dip direction also in the adjacent orebody above the stope. This results in several alternative failure mechanisms, which depend on the ratio of the deformation and strength properties of the units involved (Krauland 1975, 1984).

a) Punching of the roof into hanging wall or footwall:
If the strength of the hanging wall or footwall is lower than that of the orebody then the roof will punch into the sidewall. Thereby the maximum horizontal stresses that can occur in the roof are

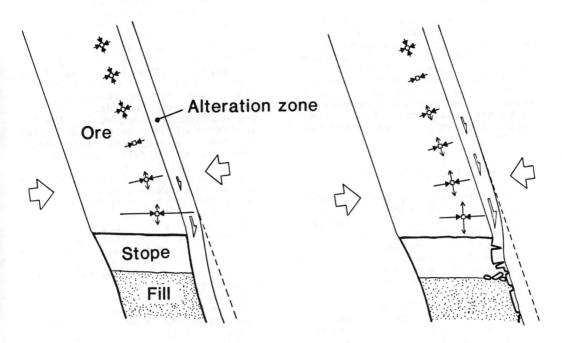

Figure 2. Influence of alteration zone in hanging wall on strain distribution in stope roof: a. inhibited deformation of alteration zone in dip direction; b. increased deformation of alteration zone in dip direction (facilitated e.g. by hanging wall failure)

limited. If large deformations in dip direction occur within the alteration zone (e. g. as a consequence of a fallout from the hanging wall or footwall), this may result in progressive roof failure, propagating far above the original stope roof (figure 2b). If, however, the deformations of the alteration zones in dip direction can be limited, (e. g. by supporting the sidewalls by rock bolting, fill etc) then favourable conditions for roof stability are created (figure 2a).

b) Failure of the roof:
If the punching strength of the sidewalls is larger than the compressive strength of the roof, then failure will occur in the roof, often by slabbing; this may occur suddenly by bursting. If there are very weak materials in the alteration zones increased tensile strains in dip direction will be induced in the orebody above the roof; this reduces the load bearing capacity of the roof.

Fill: The stresses and deformations of the fill are well described by means of an analytical model of a bin with converging walls, developed for this purpose (Knutsson 1981). In the case of Näsliden one third of the measured horizontal stresses are due to the convergence of hanging wall and footwall; two thirds are due to gravity stresses of the fill.

2.2.2 Modelling technique

Geology: The geology was strongly simplified in the model. Geological and tectonic elements that are important for the mechanical behaviour of the mine have to be considered. In the case of Näsliden these were the weak alteration zones, but also the occurrence of chlorite in part of the footwall (northern half of the orebody). In the chlorite zone the roof punched into the footwall; this prevented overloading of the roof. Where the footwall consisted of strong felsites failure of the roof occurred instead.

Rock mass properties: The influence of joints on the deformation properties of the rock mass can be considered well by the approach by R. Goodman (Goodman 1976, Borg 1981). The determination of realistic strength properties for joints is still difficult and requires experience and good engineering judgement (Barton 1981).

Model geometry: In Näsliden two-dimensional linear elastic models were used with satisfactory result. Three dimensional model calculations conducted for checking the 2D-approximation error showed very good agreement in early mining stages; in late mining stages deviations can amount to about 30 % (Borg 1983).

Model type: In some models the weak alteration zones between orebody and sidewall rocks were simulated by two columns of joint elements on each side of the orebody. The simulation of the stress distribution and of the displacements was improved thereby, and a meaningful interpretation of the observed failure phenomena was facilitated (Groth, Jonasson 1981, Krauland et al. 1981). Without knowledge of the governing failure mechanism it would have been impossible to develop a model that describes these processes adequately. The inclusion of joint elements implies however a considerably increased time and cost effort; also increased quantitative uncertainty of the model reaction is to be expected due to increased uncertainty in determining joint properties.
Examination of the model reactions showed that joint elements influence the immediate surrounding only. The behaviour of the mine as a whole is well described by elastic models. The choice of the model type depends therefore on the problem to be examined. Investigations of failure mechanisms or research tasks require often detailed modelling. In mine planning on the other hand, time and cost are often decisive factors. For many mine planning tasks elastic models are sufficient and therefore to be preferred.

# 3. APPLICATIONS OF ROCK MECHANICS

It is not before our rock mechanics knowledge is used to solve practical problems that benefit arises to mining. Three areas of applications can be distinguished:

* prediction of future mining conditions as a basis for mine planning
* evaluation of the various types of cut and fill mining from a rock mechanics point of view and criteria for their selection with regard to rock conditions
* evaluation of support measures.

## 3.1 Prediction of future mining conditions

The prediction of future mining conditions is an important help in mine planning in both technical and economic respect. It is especially important to know when an essential adjustment of the mining method and/or of the support measures to the mining conditions or even termination of mining (as the worst case) is to be expected.

T. Borg (Borg 1983, Borg et al. 1983) developed therefore a prediction method which meets the needs of practical mining. The principle of the method is based on transferring past experience to the future by means of models. The mining situation is simulated by FE-models. By using a failure criterion it is indicated when the reactions of the model (stresses or strains) reach critical values, which correspond to certain failure stages. These are defined on the basis of experience.

In the case of Näsliden two failure criteria were used, namely the Coulomb criterion and the extension strain failure criterion suggested by Stacey (1981). The failure criteria are calibrated by simulating such mining situations in which characteristic and important failure phenomena have occurred. The calibration of these failure criteria includes therefore not only rock conditions, but also the effects of the mining method and the support. The range of validity of the prediction is therefore defined by the available range of experience.

Rock stress phenomena were observed in the stope that was mined out first in Näsliden. They could be divided into three stages:

1. Brittle, often violent, failure of the roof due to high horizontal compressive stresses. Moderate rock bolting sufficient.
2. Failure of the face, often in the shape of large wedges. Frequent scaling and intensive rock bolting necessary.
3. Intensive failure phenomena in the roof, also at large distance from the face. Stable roof cannot be achieved by scaling and conventional rock bolting. Termination of mining, change to different mining method or to different support method.

Predictions of future mining conditions have been carried out so far for the Långdal Mine, which is located under a river, and for the Renström Mine, in order to facilitate long term planning. Such predictions were also carried out for the Zinkgruvan Mine of the Company Vieille Montagne (Borg et al. 1984).

Recent developments concerned the prediction of sill pillar strength. Borg and Agmalm 1988 analysed the strength of sill pillars in Näsliden using large scale rock mass strength values derived from earlier mining stages in Näsliden, applying the pillar strength formula by Obert and Duvall 1967. This approach was pursued further by comparing the pillar design approach suggested by Hoek and Brown 1980 with the Obert and Duvall pillar formula (Nyström 1988). Both approaches showed good agreement with observed average pillar strength. Andersson 1987 determined rock mass strength from laboratory tests and field studies at Zinkgruvan and applied the Hoek and Brown failure criterion to the prediction of sill pillar strength with satisfactory result.

## 3.2 Types of cut and fill mining

Cut and fill mining is a mining method which is very flexible with regard to rock conditions and orebody geometry. The favourable support action of the backfill forms the basis for the various types of C+F mining according to figure 3 (Krauland et al. 1984). The sidewalls are the abutments for the roof. Good stability of the sidewalls is therefore a pre-requisite for the stability of the stope. Experience has shown that considerable improvement of stope stability can be achieved in C+F mining by decreasing stope height. From a rock mechanics point of view this means that rock bolts as a means of support are replaced by backfill in the lower parts of the sidewalls. Thereby the stope area that has to be supported by rock bolts is decreased. In drift and fill mining the exposed rock surface in the stope is reduced further. In undercut and fill also the rock in the roof is replaced by stabilized fill.

In drift and fill one of the sidewalls consists of fill. To avoid the costly construction of fill fences the fill is stabilized by the addition of 5 % cement. In UC&F the roof consists of stabilized fill. It has to be selfsupporting

up to a span of at least 8 m. When the width of the orebody exceeds 8 m, drifts with a width of 4 m are used. Figure 4 shows UC+F in the Garpenberg Mine.

It is appropriate to compare the support action of backfill with that of rock bolts. The average rock bolt density is 2-4 m$^2$ per rock bolt. The horizontal fill pressure near the fill surface is approximately 60 kPa. This corresponds to the support action of one rock bolt per 3 m$^2$. Important differences exist, however, in the support action between rock bolts and backfill, to the advantage of backfill:

1. Fill pressure is evenly distributed in contrast to the point loads of rock bolts.

2. The support action of rock bolts decreases as soon as failure occurs in the anchors. In contrast, the support action of backfill increases when the displacements of the sidewalls into the stope increase.

3. There simply is no space available, into which the rock mass could move after failure.

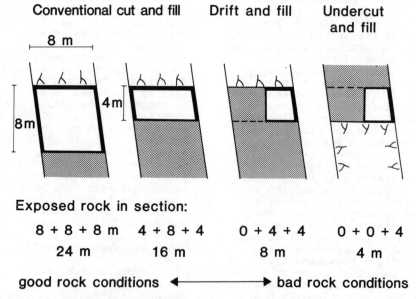

Figure 3. Decrease of stope circumference supported by rock bolting under increasingly difficult rock conditions: a. C&F, large stope height; b. small stope height; c. drift and fill (large ore thickness); d. underhand cut and fill (UC&F)

## 3.3 Improvement of stope stability

In addition to the various types of C+F mining there exist a number of support measures in C+F to improve stope stability. These are summarized in Table 1. In general it is easy to calculate the costs for these measures. Determination of their efficiency has been a matter of personal judgement to a large degree. Calculations, experiments and practical experiences resulted in a more objective appraisal of the potential and limitations of the various support measures, also given in Table 1. It is apparent that considerable further development is necessary in this area. Several of these measures are being investigated at present. Some of these measures have been tested with promising results such as cable bolting (Ahlenius 1987) and destressing (Krauland, Söder 1988).

## 4. SUMMARY AND FURTHER DEVELOPMENT

In summary we can state that systematic observations of failure phenomena and measurements underground resulted in insight into the type and magnitude of rock mass reactions in C+F mining as well as an understanding of the mechanisms that govern stope stability.

The numerical models facilitated on improved interpretation of the in situ measurements; they also allowed to quantify the influence of the various input parameters; also, numerical models proved to be useful tools for the application of rock mechanics to mine planning.

Further development is orientated towards the following areas important for practical mining:

* Systematic evaluation of available experience from C+F mining. By backcalculation of interesting mining cases in Swedish mines a data base of rock mass strength data is to be created.

* Further development of C+F mining; at present a major study of underhand cut and fill mining is being carried out at the Garpenberg Mine. The aim is to achieve a better understanding of this mining method and to work out a specification of fill properties for different applications of this method.

* Further development of the

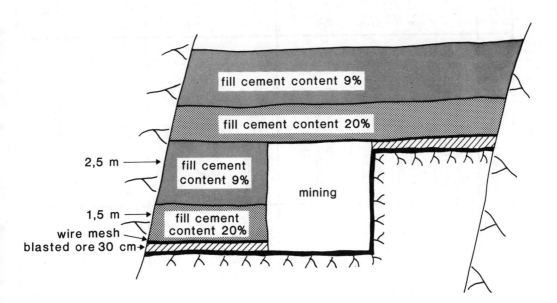

Figure 4. Underhand cut and fill in the Garpenberg Mine

Table 1: Measures to improve stope stability

| Principle | Stabilizing measure | Effectiveness | Future development |
|---|---|---|---|
| Leaving ore as pillar support | Decreasing level spacing (final height of backfilled excavation) | Effective at moderat level spacing, large ore losses | Model calculations combined with economic evaluations |
| | Pillars on strike (similar sill pillars) | Efficient at moderate pillar spacing moderate range of influence of pillars shown by calculations (Borg 1983) large ore losses | Dimensioning of pillars to be based on rock mass data determined by back calculation |
| | Pillars on dip | | |
| | Vertical yielding pillars (post pillar mining) | Very efficient under suitable conditions (orebody geometry (Cleland, Singh 1973) | |
| Change of material properties | Increase of fill stiffness | Only small influence on roof stresses if convergence is small (Pariseau 1981) | |
| | Destress blasting | Very efficient in rock with brittle behaviour and under high stresses (Krauland et al 1979, Karwoski et al 1979) | Dimensioning blasting process, rock mechanics models and field investigations |
| Influencing failure | Rock bolting | Well proven by experience | Considerable improvement to be expected by improved understanding of support mechanisms and improved support material. Mainly field investigations. |
| | Cable bolting of roof | Positive but limited experience (Fabianczyk 1982, Ahlenius 1987) | As above. Mainly field investigations. |

method for the prediction of future mining conditions in order to define better its scope and limitations.

* Investigation and continued development of the support measures given in Table 1. The most important areas for further development are
 - destress blasting, mainly effect and dimensioning of blasting
 - improved understanding of the support effect of rock bolting

The development achieved so far resulted in considerable improved confidence in rock mechanics as a useful tool in mining. We hope that the planned developments and applications of rock mechanics will increase this confidence further.

REFERENCES

Ahlenius, A. 1988: An evaluation of cable bolting in the Renström Mine (in Swedish), Proc. Rock Mechanics Meeting, Stockholm: Swedish Rock Engineering Research Foundation, 197-210

Andersson, L. 1987: Pillar strength and pillar stability at Zinkgruvan. An introductory field study and analysis (in Swedish), M.Sc. Thesis, Luleå University of Technology 1987:077 E 97 p.

Barton, N. 1981: Estimation of in-situ joint properties, Näsliden Mine. In Stephansson, O., Jones, M.J., (Eds.) 1981: loc. cit., 186-192.

Borg, T. 1981: The Näsliden Project - FEM modelling strategies. In Stephansson, O., Jones, M.J., (Eds.) 1981: loc. cit., 196-203.

Borg, T. 1983: The prediction of rock failure in mines with application to the Näsliden mine in Northern Sweden. Doctoral thesis, 1983:26 D, Luleå University, Sweden, 131 p.

Borg, T., Agmalm, G. 1988: Experiences from a prediction of the mining conditions at the Näsliden Mine. Proc. Rock Mechanics Meeting, Stockholm: Swedish Rock Engineering Research Foundation, 73-86

Borg, T., Krauland, N. 1983: The application of the finite element model of the Näsliden Mine to the prediction of future mining conditions. In : Granholm, S. (ed.): Internat. Symp. Mining with Backfill, Luleå, June 7 - 9, 1983, Rotterdam: Balkema, 309-318.

Borg,T., Röshoff, K., Stephansson, O. 1984: Stability prediction of the Zinkgruvan Mine, Central Sweden. In: Brown, E.T., Hudson, J. A. (eds.): ISRM Symposium, Design and Performance of Underground Excavations, Cambridge September 3-6, 1984, London: British Geotechnical Society, 113-121.

Cleland, R. S., Singh, K. H. 1973: Development of "post" pillar mining at Falconbridge Nickel Mines Limited. CIM Bull., 66, April 1973, 57-64.

Fabjanczyk, M. 1982: Review of ground support practice in Australian underground metalliferous mines. Proc. Aus. I.M.M. Conference, Melbourne,Vic., August 1982, 337-349.

Goodman, R.E. 1976: Methods of geological engineering in discontinuous rocks. Saint Paul, Minnesota: West Publishing Co.,472 p.

Groth, T., Jonasson, P. 1981: Application of the BEFEM code to the Näsliden Mine models. In Stephansson, O., Jones, M.J., (Eds.) 1981: loc. cit., 226-232.

Hoek, E. Brown, E.T. 1980: Underground excavations in rock. London: The Institution of Mining and Metallurgy. 527 p.

Karwoski, W. J., McLaughlin, W. C., Blake, W. 1979: Rock preconditioning to prevent rock bursts - report on a field demonstration. Rep. Invest. US Bur. Mines 8381,47 p.

Knutsson,S. 1981: Stresses in the hydraulic backfill from analytical calculations and in-situ measurements. In Stephansson, O., Jones, M.J., (Eds.) 1981: loc. cit., 261-268.

Krauland, N. 1975: Deformations around a cut and fill stope experiences derived from in situ observations (in Swedish). Proc. Rock Mechanics Meeting in Stockholm, February 1975. Swedish Rock Mechanics Research Foundation, Stockholm pp. 202-216.

Krauland, N. 1984: Rock mechanics investigations of the cut

and fill mine Näsliden by means of field observations and FE-models. In: Wittke,W. (ed.): 6. Nationales Felsmechanik Symposium, Aachen, April 3-4, Geotechnik, Sonderheft 19, 139 -147.

Krauland, N., Nilsson, G., Jonasson, P. 1981: Comparison of rock mechanics observations and measurements with FEM-calculations. In Stephansson, O., Jones, M.J., (Eds.) 1981: loc. cit., 250-260.

Krauland, N., Nilsson, G., Magnusson, I. 1986: Backfilling in the Boliden mines - why and how? Erzmetall, 39, 547-551.

Krauland, N., Strandberg, R., Bjarnholt, G. 1979: Destress blasting in the development drift of the Guttusjö Mine ( in Swedish). Swedish Detonic Research Foundation, Report DS 1979:14, 5 p.

Krauland, N., Söder, P.E. 1988: Rock stabilization by destress blasting (in Swedish). Proc. Rock Mechanics Meeting, Stockholm: Swedish Rock Engineering Research Foundation, 137-156

Nyström, A. 1988: The final stages of mining at the Näsliden Mine , M.Sc. Thesis (in Swedish), Luleå University of Technology 1988:125 E  120 p.

Obert, L., Duvall W.I. 1967: Rock mechanics and the design of structures in rock. New York: John Wiley & Sons

Pariseau, W. G. 1981: Finite element method applied to cut and fill mining. In Stephansson, O., Jones, M.J., (Eds.) 1981: loc. cit.,284-292.

Stacey, T.R. 1981: A simple extension strain failure criterion for fracture of brittle rock. Int. J. Rock Mech. Min. Sci. & Geomech. Abstr., 18, No. 6,469-474.

Stephansson, O., Jones, M.J., (Eds.) 1981: Proceedings of the Conference on the Application of Rock Mechanics to Cut and Fill Mining, Luleå, June 1 - 3, 1980, London: Institution of Mining and Metallurgy, 1981, 376 p.

*Innovations in Mining Backfill Technology, Hassani et al. (eds), © 1989 Balkema, Rotterdam. ISBN 90 6191 985 1*

# The experimental and practical results of applying backfill

J.Palarski
*Technical University, Gliwice, Poland*

ABSTRACT: The paper describes achievements with backfill in Polish mining where presently about 15% of coal production come from backfill longwall faces. The process of preparing and transporting different types of fill mixtures is described, and problems connected with gravitational transport are discussed. Such ways of reducing pressure and flow velocities as the use of an additional hopper, installation of horizontal resistance pipes and change of pipe diameter are presented. Technologies of filling workings and in particular the role and structure of fill stoppings are described. The use of finely-grained fill material in thick seam winning for creating an artificial roof is discussed. The carried out field and laboratory tests have shown that fly-ash and finely-graind waste material can be used for compacting workings. The influence of backfill, especially its strain and compactness on deformations of the surface and roof subsidence is analysed.

## 1. INTRODUCTION

The use of hydraulic backfill in Silesia was mentioned for the first time at the turn of the 20th century. Pneumatic stowing was introduced in the 1920s as one of the first solutions in the world. The development of hydraulic and pneumatic stowing took place mainly in coal mining when from 1962 to 1967 about 40% of coal production were obtained from backfill longwall faces. At the moment a considerable interest in backfill technology can be observed due to the natural environment conservation and to the fact that most seams exploited with caving have been worked out.

Recently about 15% of coal has been mined from hydraulic backfill longwall faces and about 2% from pneumatic backfill longwall faces. Apart from that Polish mining dumps tallings and fly-ashes in old workings using technologies similar to those used in pneumatic and hydraulic backfilling. Table 1 presents quantitive characteristic of waste dumping and backfill technologies.

## 2. CONDITIONS FOR THE USE OF BACKFILLING

Main reasons for the use of back-filling are as follows:

- conservation of the rock mass and the surface,
- need for mining thick seams which for full exploitation should be divided into layers/the layers being mined one by one from the roof to the floor or vice versa/,
- waste dumping in old workings,
- improvement of safety in cases of such mining hazards as high pressure of the rock mass, temperature, and gas outflow.

## 3. PREPARATION OF FILL MIXTURE

Sand and waste rock are main fill materials as it is seen in Table 1. Those materials are supplied by sand mines, treatment plants, or dumps and are stored in slope, shaft or slit tanks /Figure 1/. Fill materials are mechanically removed or washed out from a tank by water with a pressure reaching 10 bars. Next the material flows through special screens into channels and then into a hopper. Rocks of larger size are crushed and through successive screens they also get to the hopper. If necessary water is supplied to the hopper. When a mixture of different materials is used for backfilling then either they are mixed just before they get into a tank or they are fed into a hopper from two chambers through two respective

Table 1. Development of fill technologies in Polish coal-mining.

| Type of fill | Average production | Number of faces | Number of mines | Length of pipelines | Type of fill material | Amount of used material |
|---|---|---|---|---|---|---|
| | mill.t/a | – | – | km | – | mill.m$^3$/a |
| Backfill | 32,3 | 270 | 39 | 1100 | sand<br>waste<br>fly-ash<br>slurry<br>tailing | 25,1<br>3,4<br>1,0<br><br>1,1 |
| Pneumatic stowing | 3,4 | 53 | 13 | 7,8 | waste | 2,5 |
| Caving | 155,2 | 559 | 64 | – | – | – |

channels. The mixture preparation in new mines is fully automated.

## 4. GRAVITATIONAL TRANSPORT OF BACKFILL MATERIALS

Gravitational hydrotransport of materials is used in Polish coal mining. Only in exceptional situations, when a horizontal pipeline is very long, gravitational hydrotransport is helped by pumps. Gravitational parameters of hydrotransport, and in particular its efficiency, flow velocity, pressure at different points of a pipeline, can be influenced by:
- space arrangement of the pipeline,
- pipe diameter of both vertical and horizontal pipes,
- properties of the mixture, in particular its concentration and choice of its components.

The use of backfill is particularly difficult in deep mines. High pressure and large flow velocities, which lead to quick wear of pipes and to the risk of pipe bursts, are observed in such mines. In case of the occurrence of too high pressures and flow velocities in pipes the following ways of reduction are used /Figure 2/:
- use in a shaft of pipes of smaller diameter than that used in horizontal workings,
- installing a system of horizontal resistence pipes in a vertical pipeline at a depth determined for a given pipeline,
- use of a second hopper at a certain depth of a shaft /the hopper divides a pipe into two parts: one with the free fall and the second part with the full flow/
- use of reducers, which slow the flow, in a vertical pipeline.

The process of backfilling starts with flushing of a pipeline. It takes 5 to 15 minutes depending on its depth and length.

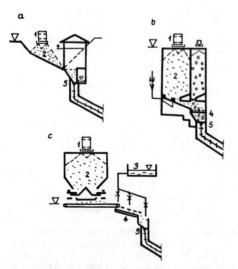

Fig.1 Tanks for backfill materials a/ slope tank, b/shaft tank, c/ slit tank, 1. tank car, 2. tank, 3. water tank, 4. screen, 5. hopper with pipeline

34

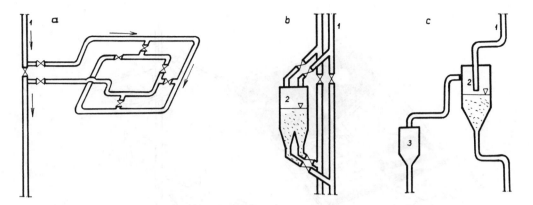

Fig.2 Devices for reducing pressure and velocity.
a/ resistance lay-by, b/ secondary hopper, c/secondary hopper with safety outflow.
1. vertical pipe, 2. secondary hopper, 3. safety hopper

During the final stage of flushing, fill material is fed into the pipeline and mixture concentration is increased up to a required value. Generally under normal work conditions the mixture efficiency ranges from 500 to 1200 $m^3$/h, and with concentration rate of the mixture of 1 $m^3$ of the material to 1 $m^3$ of water it is possible to fill 250-600 $m^3$/h. This is a very high efficiency but also a great amount of water, which has to be drained off from the workings. In Polish coal-mines in a shaft there are at least two pipelines and usually four, one of them being a reserve. The pipeline can branch in a bottom shaft and in horizontal workings.

It is worth noting, however, that during hydrotransport the mixture flows in only one pipeline from the surface to a working. Steel pipes, pipes with a basalt or rubber lining are used in the main sections of the pipeline. Whereas in winning faces steel,

rubber, plastic, or fibreglass pipes are installed.

## 5. TECHNOLOGY OF FILLING WORKINGS

In a longwall, a pipeline branches out every 8 - 12 m, and through those branches the mixture is directed to a space which is to be filled. Different types of fill stoppings are used in longwalls depending on a type of support, degree of mechanization, and seam thickness. Fill stoppings, both those paralled to the longwall face and those parallel to workings consist of a steel or wooden construction, ropes or chains and juta fabric /Figure 3/.
Fill stopping parallel to the face are left in the backfill, except for new self-advancing fill stoppings used together with a mechanized support. Maximum distance

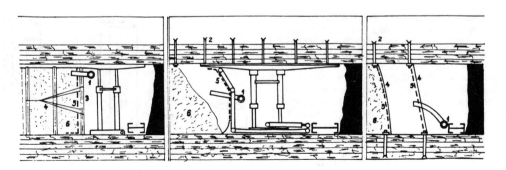

Fig.3 Types of fill stoppings
1. pipeline, 2. bolt, 3. wooden prop, 4. rope, 5. juta fabric, 6. fill material

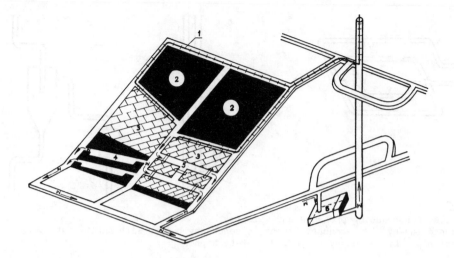

Fig.4 Diagram of backfill exploitation
1. pipeline, 2. longwall, 3. backfill space, 4. settling tank in coal,
5. settling tank in backfill, 6. collecting tank

between the face and backfill supporting the roof cannot exceed 8 m. When a seam is horizontal, is is difficult to fill the space compactly up to the roof and after moving the stopping, in case of self-advancing stoppings the backfill material slumps outwards and thus it is impossible to meet the condition of an 8 m distance. Water from the mixture flows through the jute fabric and next along open troughs laid at floor level parallel to the pipeline; then it passes through settling and collecting tanks before being pumped to the surface /Figure 4/.

## 6. STRAIN OF FILL AND ROOF DEFORMATION

The final value of roof subsidence depends on the strain of fill and technology of filling. Figure 5 shows backfill material strain at 2,5 MPa pressure and final roof subsidence measured at a depth of 970 - 1000 m for different types of material - The underground measurements were carried out in backfill longwalls of a similar seam thickness - 2,4 - 2,6 m, angle of seam 10 - 12°, and backfill with 3,2 m. It has been concluded from the measurements that the final roof subsidence on backfill is 35 - 45% bigger than the material strain. Such results prove imperfection of backfill technology. It can be seen from Figure 5 that the smallest roof subsidence can be achieved with backfill technology using a

mixture of fly-ashes and tailings. This is so because in case of this types of backfilling, the fill is compact and water is almost completely absorbed by surrounding rocks and chemically bound by fly-ashes. Thus, there is no washing out of fine grains by water. Backfilling with coarse-grained material /d 50 mm/ however should be treated as dumping waste and not as compact filling.

## 7. THE USE OF TAILINGS AND FLY-ASH FOR FILLING

Apart from typical technologies of backfill with fly-ashes, slurry or tailings there

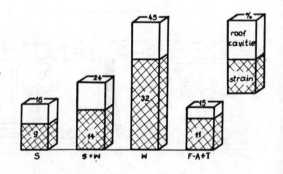

Fig.5 Roof subsidence and backfill strain
S-sand, W-waste, F-A fly-ash, T-tailing

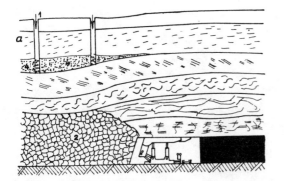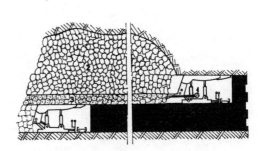

Fig.6 Ways of filling with finely-grained waste
a - injection of mixture into a void, b - making artificial roof
1. pipeline, 2. caving, 3. artificial roof, 4. void in rock mass

are some new technologies of depositing
those waste materials underground
/Figure 6/.

During exploitation with caving, strata
deformation occurs at certain height above
the rubble. Those strata are of different
strength and thus the rate of their
deformation is different. As a result of
that some voids are formed. It is possible
to fill these voids with waste through
boreholes. While using this technology of
filling voids inside the rock mass, the
deformation on the surface is 20 - 30%
smaller than in case of exploitation with
caving. In case of winning thick seams
there is a possibility of compacting the
caving in the roof layer with a mixture of
water, fly-ashes, and slurry. Under the
influence of chemical processes and the
pressure of the rock mass, the rocks
undergo binding and after some time become
an artificial roof for the exploitation in
lower layers. Underground observations
prove that in many cases it is enough to
fill cavings to the height of 0,50 - 0,70 m
to obtain a layer of a compact artifical
roof. A similar solution is used while
exploiting thick seams of small strength
when they occur under buildings. Such
conditions make it necessary to win a seam
in layers from the roof to the floor. The
upper layer is filled with stones either
hydraulicly or pneumaticly and next this
fill is compacted with finely-grained
waste, fly-ashes, and addition of up to 3%
of cement. Presently compacting with finely
grained material is extensively used. It
improves ventilation, eliminates the risk
of coal self-ignition, and deposits waste
materials.

*Innovations in Mining Backfill Technology, Hassani et al. (eds), © 1989 Balkema, Rotterdam. ISBN 90 6191 985 1*

# Backfilling on gold mines of the Gold Fields group

N.Kamp
*East Driefontein, Driefontein Consolidated Limited, RSA*

SYNOPSIS: The paper describes the methods employed in placing backfill in deep level gold mines with narrow, gently dipping, tabular orebodies.

The whole process is covered including aspects of preparation, distribution systems and methods of placement.

The effect of backfilling on current mining practice is examined and some implications for the future are predicted.

## 1. INTRODUCTION

The gold mines of the Gold Fields group are presently mining at depths in excess of 3 000m below surface. Future mining is planned to exceed depths of 4 500m. The major problems encountered with mining at depth are rock stress and heat. Backfill was introduced to help combat these problems.

The group has extensive experience with bulk mining backfill operations gained on its base metal mines. However this knowledge is not directly applicable to mine wide filling systems for narrow tabular orebodies.

Backfilling of massive excavations is done for different reasons to those on a deep level gold mine and therefore the properties required of the backfill are different.

The nature of the orebody, the depth below surface and the mining methods employed, all give rise to a host of different conditions that exist on a deep level gold mine. It was for those reasons that backfill trials were first introduced on two of our gold mines. These pilot systems helped develop the full scale mine-wide systems that are currently in operation on the gold mines of the group.

## 2. BACKFILL MATERIALS

The backfills being placed in gold mines are produced from two main sources. The largest source is from metallurgical plant tailings and to a lesser degree from crushed and milled development waste rock.

The different types of backfill are characterised according to their particle size distribution and optimum placement properties. The properties of the placed fill depend mainly on the water/solids ratio. Better placement characteristics are achieved with course-grained, well graded backfills (IH CLARK).

## 3. PREPARATION

On Gold Fields mines backfill is prepared from metallurgical tailings in two ways:
Dewatered total plant tailings.
Deslimed tailings.

### 3.1 Dewatered Slimes

The method was introduced on West Driefontein in 1979 as a result of the work done on Western Holdings gold mine. To date some 250 000 tons of material has been placed using this method.

### 3.2 Deslimed Tailings

This method is favoured by the majority of the group gold mines for the following reasons:-
1. The material is readily available.
2. It can be easily distributed.
3. It has adequate in-situ load bearing capabilities.

4. The skilled labour required to operate the system is minimal.
5. The capital cost is relatively low.

## 4. DEWATERED SLIMES BACKFILL (West Driefontein)

The essential features of this system are:-
1. Repulping the plant tailings filter cake to the required density.
2. Gravity feed to underground storage dams.
3. Dewatering by means of a Centrifuge.
4. Placement of dewatered slime by a positive displacement pump.
5. Method of placement and confinement of the fill in the stope.

centrifuge operates between 1200 and 1 500 revolutions per minute depending on the relative density of the feed. (Figure 2)

28% (solids by mass) is spun off and discharged through ports at the top of the machine. The waste product containing particles of less than 44 microns is pumped to a waste disposal dam. With most of the water and fine fraction now removed the underflow material emerges as backfill consisting of 78% solids by mass and 58% +44 micron at a relative density of between 1,9 and 1,98. This material then drops onto worm screws which convey it via a launder to the placement pump.

The placement pump is a double acting hydraulic driven concrete placer type pump. It is rated at 8m³/hour at a pressure of

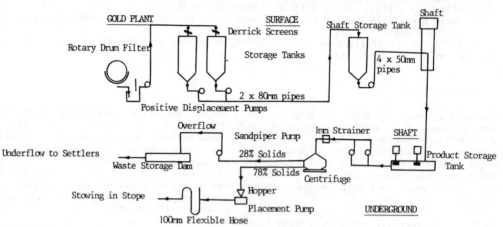

FIGURE 1

DEWATERED SLIME SYSTEM - SCHEMATIC LAYOUT

The general arrangement for processing the fill both on surface and underground is illustrated in figure 1.

Slime is supplied from the rotary drum filters and is repulped to the required density of between 1,65 and 1,70. It is then pumped to storage tanks passing through screens to remove debris and particles which might affect the operation of the centrifuge. Ferrous sulphate is added to neutralize any residual cyanide in the slime.

From the storage tank the slime gravitates down 25mm open ended high pressure ranges to the various underground storage tanks. These are prefabricated mild steel tanks which can be moved forward and thereby reduce the time of forward moves of the dewatering plant.

The stowing station consists of a strainer to remove particles greater than 1mm, the centrifuge and the placement pump. The

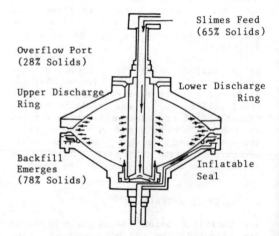

FIGURE 2 : CROSS SECTION OF CENTRIFUGE

100 bar and has a maximum operating distance of 300m.

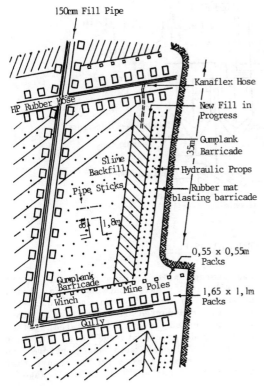

150mm Fill Pipe

Kanaflex Hose

New Fill in Progress

Gumplank Barricade

Hydraulic Props

Rubber mat blasting barricade

HP Rubber Hose

Slime Backfill

Pipe Sticks

35m

1,8m

0,55 x 0,55m Packs

1,65 x 1,1m Packs

Gumplank Barricade

Mine Poles

Winch

Gully

FIGURE 3

STOWING OF DEWATERED SLIME - STOPE LAYOUT

4.1 Placement (Figure 3)

The fill is contained in plank paddocks which extend down-dip to the top edge of the gully packs.

Internal support consists of 150mm pipe sticks at 1,8m centres, with 3 rows of hydraulic props down the face. This enables the panel to be filled within 8 m of the face.

The stoping width averages just over 1m and each station is able to supply fill to three or four 35m panels.

4.2 Advantages of the system

1. No special desliming process is required on surface.
2. Special dewatering paddocks are not required and containment is relatively simple and cheap.
3. Minimal in-stope water drainage with no fines loss.
4. The risk of backfill runaways is low.
5. Marginally superior initial strength.

4.3 Disadvantages

1. Limited pumping distance of the placement pump.
2. Initial capital cost is high.
3. Skilled labour is required to operate the plant and accounts for approximately 60% of the cost.

This system is still used on a limited scale on West Driefontein to backfill isolated remnant stopes, the main reason being to usefully employ the capital invested in the dewatering equipment.

5. DESLIMED TAILINGS BACKFILL

The first full-scale preparation plant was commissioned on East Driefontein in July 1986. The plant is capable of producing 2 700 T.P.D. of backfill by multistage cycloning of the filter residue. (Figure 4.) The plant receives the run of mill tailings in a stock tank, from where it is pumped to a three-stage hydrocycloning processor.

Cycloning provides a high density underflow 1,65 - 1,7 with all coarse and high density material reporting to backfill. Batches of backfill are made up in one of two stock tanks, where ferrous sulphate is added to neutralize residual cyanide. Quality control tests are carried out before the backfill is pumped into the distribution plant.

Typical standards required are:-
Free Cyanide          <0,02 g/l
Pulp Density          1,65 - 1,75 kg/m$^3$
Particle size         <15% - 45 micron
The overflow from the cyclones is thickened before being pumped to the slimes dams.

5.1 Slimes dam stability

As the extent of backfilling increased problems were anticipated due to the rise in the proportion of overflow material. The effect of the cyclone overflow on the stability of the slimes dam walls was the subject of an investigation carried out by Consulting Engineers on one of the dams at East Driefontein.

They concluded that there should be no great difficulties associated with constructing slimes dams with the cyclone overflow product using existing technology and methods. The one aspect that may have

41

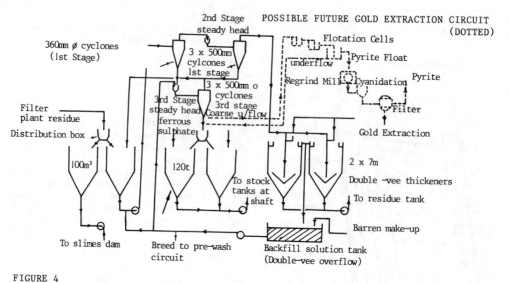

FIGURE 4

FLOWSHEET (SCHEMATIC) FOR EAST DRIEFONTEIN PREPARATION PLANT

to be considered is the rate of rise of the dam, and that this did not really present a problem, as this rate may still be more than that actually adopted for an equivalent dam of total tailings.

## 5.2 Gold recovery from backfill

The cycloned residue used for backfilling contains upgraded gold values in the course underflow product. An investigation was carried out to determine the feasibility of recovering gold from the backfill before it was sent underground and lost forever.

Various methods were investigated (Van Niekerk and Uys) and it was found that a treatment route based on roasting and leaching of sulphide flotation concentrate, was one method that could be economically viable. On one mine it was established that at a conservative gold price of R25/g and a low backfill head grade of 0,3 g/t, the capital outlay would be repaid within two years.

## 5.3 Surface distribution plant

The backfill product is pumped from the preparation plant to storage silos at the shaft. The silos are equipped with compressed air diffuser rings for agitation, and facilities to dump sub standard material are provided, should the need arise.

The backfill material is gravity fed from the storage silos, via surge cones in the distribution plant, and down the shaft. Depending on which system is in use, the fill is transported directly to the stope via open-ended ranges, (dedicated system) or into underground storage dams. The pipe lines are flushed with water before and after the passage of slime to ensure that they remain clear and free from blockage.

## 6. BACKFILL RETICULATION SYSTEMS IN USE ON GROUP GOLD MINES

Backfill is conveyed in open-ended pipe lines over vast distances in the gold mines of the group. Vertical distances of 2 500m and horizontal distances of 3 500m are typical. The pipe sizes vary from 38mm N.B. to 50mm N.B. schedule 80 steel in the shaft and between 50 N.B. and 114mm N.B. schedule 40 steel on the levels. At full capacity over 100 000 t.p.m. of backfill can be placed.

## 6.1 Dedicated pipe lines from surface

This system allows each mining section to have its own open-ended dedicated column. Each column is fed directly from the distribution plant on surface, down the surface shaft, across to the sub-vertical shaft, down the sub-vertical shaft out the level to the stopes. (Figure 5).

Variations of this basic concept have been introduced and are as follows:

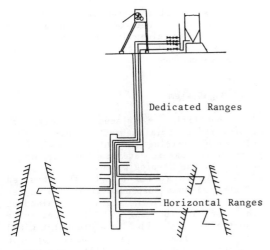

Dedicated Ranges

Horizontal Ranges

FIGURE 5

DEDICATED RANGES FROM SURFACE

## 6.3 Pipes in boreholes

Dedicated pipe lines are placed in bore holes drilled parallel to the sub-vertical shaft. This has the advantage of keeping the sub-vertical shaft clear of backfill ranges and thus reduce shaft down time when ruptured columns have to be repaired. This variation is suitable for low pressure systems where HDPE piping can be used, but does entail the additional cost of the boreholes.

## 6.4 Storage dams at the top of each longwall. (Figure 7)

Backfill storage dams, similar to those described above, are situated above each longwall. Each dam is fed by a dedicated column from surface from where the backfill is distributed to the longwall in columns placed in bore holes between levels.

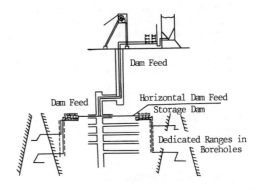

Dam Feed

Dam Feed

Horizontal Dam Feed

Storage Dam

Dedicated Ranges in Boreholes

FIGURE 7

STORAGE DAM AT TOP OF EACH LONGWALL

## 6.2 Storage dams at the top of the sub-vertical shaft (Figure 6)

In this system two columns feed the backfill from the surface storage plant, down the surface shaft and across to one or two storage dams. Each dam is agitated by mechanical agitators. Dedicated columns from these dams then convey the backfill down the sub-vertical shaft, out on the level and into the stope.

## 6.5 Communication

In the smaller systems, telephone communication between the stope and the operator at the distribution centre, is used. This type of communication is reliable and works as well when only a few pipe lines are being used. However for larger systems, indicator lights used with telephones, as a back-up were found to work well. The operator in the stope signals "start" or "stop" to the distribution centre. Indicator lights alert the operator at the distribution centre who then acknowledges the signal and gives the light indication back to the operator in the stope. The telephone provides back-up for emergencies.

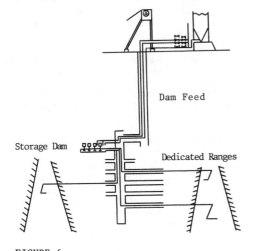

Dam Feed

Storage Dam

Dedicated Ranges

FIGURE 6

STORAGE DAM AT TOP OF SUB VERTICAL SHAFT

### 6.6  Summary

A typical deep level gold mine would require some 350km of piping for a fully dedicated reticulation system to be employed. The mines of the group generally favour the variations of the concept for the following reasons:-
  1.  Excessive pipe wear in the multiple shaft columns which results in shaft down time and production delays.
  2.  Long periods spent on flushing operations.
  3.  Reduced capacity of each pipe line when multiple fillings of small quantities to numerous working places, take place.
  4.  Less piping required for the other systems, resulting in fewer burst columns and less dust contaminating the intake air.
  Note:  The saving in the cost of piping is offset by the additional cost of the dams and boreholes.
The following advantages are present in the overall concept:-
  1.  Most of the mechanical components are sited on surface, making for ease of supervision and maintenance.
  2.  The systems underground use relatively low pressure piping made of standard materials which are readily available.
  3.  Large vertical and horizontal distances can be covered.
  4.  The system is simple to operate, requiring the minimum amount of skilled labour.

## 7.  BACKFILL PLACEMENT

### 7.1  Paddock System

The backfill is placed in pre-erected paddocks which are typically 35m long with an average stoping width of 1,4m.
The paddocks are constructed in various ways, this being the most arduous part of the whole operation. Geotextile fabric is laid inside a framework of wire mesh or gate stulls, supported by packs, mine poles and or props.
Filling can take place from the start of the night shift to the end of the day shift with a break during the blast period.
Flocculant is added at the discharge end of the pipe line to assist with the rapid de-watering of the backfill. The run-off water is led off to join the rest of the mine's dirty water system.
During the fill-time of approximately 16 hours, some 100 cubic metres of a fill at a relative density of 1,65 can be placed.
Drilling and blasting operations start on the day filling is complete. Blasting continues until the span from the face to the backfill is between 6 and 8 m. The panel is then cleaned and swept so that paddock construction can begin and the fill cycle re-commence.

### 7.2  Bag System

The backfill is placed in bags of geotextile material behind one or two rows of rapid yielding hydraulic props. The size of the bag is determined by the mining cycle and the stoping width.
The bag is rolled out to the required length and then suspended in such a way as to ensure that all the slack is retained at the top of the bag. Sufficient side constraint is provided to avoid bulging and so ensure that the bag expands upwards against the hanging when filled.
Backfill slurry is introduced to the bag via a perforated filler tube installed along the length of the bag and flocculant is added to assist with the dewatering of the backfill. Some problems have been experienced with bags of inadequate strength and draining characteristics, something that is aggravated by the presence of a high percentage (10%) of ultra fine fraction.
Research work is being carried out to remove the ultra fine fraction (-10 micron) from the backfill material, which has a marked, deleterious effect on the percolation and settling properties of the backfill.
The ultra fine fraction can be removed by two stage cycloning using desliming cyclones, producing a deslimed product grading below 2% -10 micron at a maximum recovery exceeding 40%. The percolation and settling properties of extensively deslimed tailings are superior to those of conventional backfill.
This product has a larger mean particle diameter and lower viscosity compared to conventional backfill. However transport of the fill and increased pipe erosion could cause problems. Pilot plant tests continue.

### 7.3  Mining System (Figure 8)

The mining system incorporates the use of hydraulic props, gate stulls and profile props. A three blast/fill cycle is employed which enables the panel to be blasted three times before the fill cycle takes place and at no stage is the fill more than 5m from the face. With this system higher face advances are possible because the fill acts as an effective barricade which facilitates cleaning, and drilling and backfilling can be carried out simultaneously.

The bag system fits in well with the cyclic nature of mining, enabling higher face advances to be achieved. A further major advantage of the system is that the fill can be kept close to the face allowing maximum benefits to be derived.

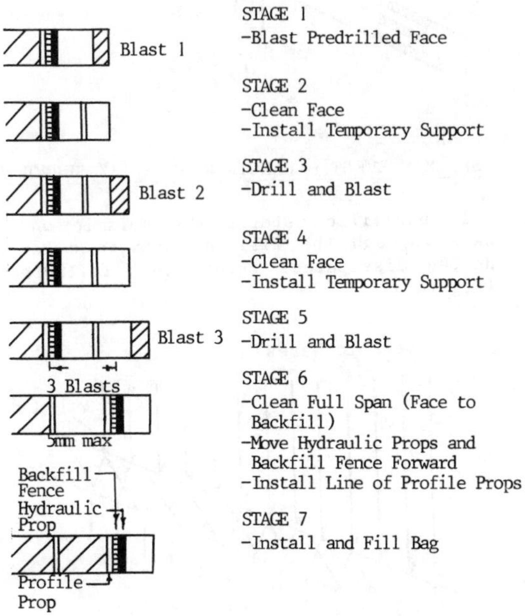

Blast 1
Blast 2
Blast 3
3 Blasts
5mm max
Backfill Fence
Hydraulic Prop
Profile Prop
STAGE 1
-Blast Predrilled Face
STAGE 2
-Clean Face
-Install Temporary Support
STAGE 3
-Drill and Blast
STAGE 4
-Clean Face
-Install Temporary Support
STAGE 5
-Drill and Blast
STAGE 6
-Clean Full Span (Face to Backfill)
-Move Hydraulic Props and Backfill Fence Forward
-Install Line of Profile Props
STAGE 7
-Install and Fill Bag


FIGURE 8

BACKFILLING CYCLE

## 8. POTENTIAL BENEFITS OF BACKFILL

### 8.1 Ground Control

Strata control difficulties increase in direct proportion to the increased depth of mining operations. The increase in the incidence and magnitude of seismic events is of major concern.

Rockbursts are associated with energy changes and consequently the severity of the rockburst problem can be reduced by designing mining layouts and systems for minimum volumetric convergence.

This can be achieved in three ways:-

1. Reducing the stoping width.
2. Leaving stabilizing pillars.
3. Backfilling the mined out area.

or a combination of all three.

The reduction in stoping width is not always possible, as a minimum width is required to conduct mining operations. The presence of multiple reef bands further complicates matters.

Regularly spaced stabilization pillars have proved to be successful on several deep level gold mines, but this method does result in at least 15% of the reef being permanently locked up. In multi reef situations the percentage of ore left in-situ could be unacceptably high.

### 8.2 Backfill and energy release rates

Backfilling limits the closure in stopes and therefore reduces the energy release rate, which is a function of elastic closure and field stress. Mechanical closure metres placed inside the fill in a stope at East Driefontein have indicated a reduction in closure of approximately 45% when compared with a station outside the fill. (Figure 9)

240
200
160
120
80
40
0
Closure (mm)
Closure and ride station (outside the fill)
Mechanical Closure meters (inside the fill)
25
50
75
100
125
150
Time (days)


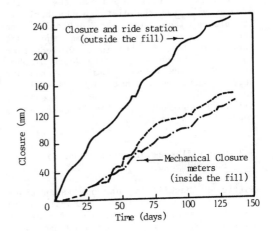

FIGURE 9

Energy release rates in excess of $40MJ/m^2$ lead to increasing dangers from rockbursts.

At depths greater than 2 000m, some form of regional support is required to achieve ERR values of less than $40MJ/m^2$. If 80% of the area is filled an E.R.R. of less than $40MJ/m^2$ can be achieved in 1m stoping widths at a depth of up to 3 000m. (Figure 10)

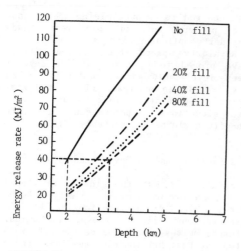

FIGURE 10

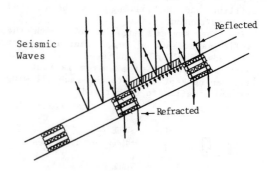

The above results depend on the quality of the fill and the distance the backfill is placed from the face. Placing the backfill effectively 5m from the face instead of 10m, results in a 10% reduction in the ERR value at a mining depth of 3km. At depths below 3 300m, stabilizing pillars become necessary, but by backfilling between the pillars the number and size of pillars required, can be significantly reduced.

8.3 The influence of Backfill on Rock-burst damage.

Sufficient backfilling has now taken place on group mines to enable rockburst damage in filled panels to be compared with damage caused by rockbursts in unfilled panels.

Observations have shown that the damaging effects of rockbursts in backfilled panels are greatly reduced. Damage is generally limited in extent and severity and can usually be attributed to the following:-
1. Geological disturbances.
2. Long leads/lags between panels.
3. Insufficient support in the face area.
4. Incorrect backfilling practice.

Severe and extensive damage has been observed in adjacent unfilled panels. In some cases the damage is so severe that panels have had to be abandoned.

In unfilled panels seismic waves are reflected at the footwall or hangingwall interface. The reflection of these waves is in part responsible for the violent ejection of blocks of ground into the excavation. (Figure 11)

In backfilled stopes the seismic waves pass through the fill and are to a large degree dissipated, rather than reflected. (Figure 12)

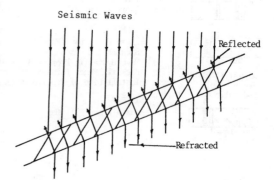

FIGURE 12

SEISMIC EVENT - HANGINGWALL - 80% BACKFILLED

8.4 Ventilation

The advantages of backfill with regard to environmental conditions, are well known and well documented. Theoretical exercises using computer programmes, have been carried out in order to quantify the benefits of backfill under various underground conditions. These results have compared favourably with those obtained from actual undergound observations.

8.5 Observed Benefits

Two working places were studied on East Driefontein. Stopes with similarities in area mined, number of panels and gullies etc. were chosen. The one stope selected had been 30% backfilled and the other not at all.

## 8.6 Results

|                    | Backfilled Stope | Non Backfilled Stope |
|--------------------|------------------|----------------------|
| Air Quantity       | 16,5m³/S         | 18,8m³/S             |
| Intake Temperature | 26,3/26,8°C      | 29,5/29,9°C          |
| Sigma Heat Pickup  | 2,64 kj/kg       | 6,35 kj/kg           |
| Face Velocities    | 0,88 m/s         | 0,31 m/s             |
| Dust               | 127 p/ml         | 158 p/mℓ             |

In order to make a more realistic comparison between the two stopes, the non-backfilled stope was assumed to have the same intake temperature as the backfilled stope. The measured heat pickup was added to the Sigma heat content at the intake to give a resultant (projected) face temperature from which kata and specific cooling power values could be determined. The results were as follows:-

|                        | Backfilled Stope | Non Backfilled Stope |
|------------------------|------------------|----------------------|
| Face Temperature       | 27,0/27,1°C      | 27,9/28,0°C          |
| Kata                   | 15,7             | 10,8                 |
| Specific Cooling Power | 328 w/m²         | 262 w/m²             |

As a further exercise these stopes were compared using theoretical results from a computer programme and a good correlation was obtained. (Kata 14,8 vs 15,7).

The measured results show that the presence of backfill in a stope can lead to substantial improvements in environmental conditions on the working faces. In the above working place the backfill present was only 30% and the quantity through the stope was 12% less, yet the improvement in environmental conditions was significant as measured by a wet kata improvement of 45%. (Figure 13)

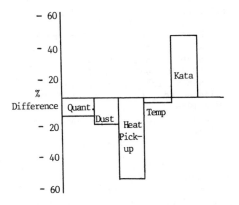

BACKFILLED STOPE (30%) VS NON-BACKFILLED STOPE

## 8.7 Fire Control

The incidence of fires is greatly reduced. Much less timber is used and the presence of backfill inhibits the spread of local fires. Back areas, where 65% of all major fires start, will be sealed off.

## 8.8 Gold Recovery

Overall gold recovery is improved due to:-
1. The elimination or reduction in the size of stability pillars.
2. Higher extraction of reef particularly in a multi level situation.
3. Fewer areas having to be abandoned due to major rock falls.
4. Less fines loss in back areas and timber packs. Current sweepings will further reduce the loss of fines in foot-wall cracks. The quality of sweepings will be better when high pressure water jetting is used in conjunction with backfill.
5. Additional gold which can be recovered from the deslimed tailings.
6. Less waste dilution due to improved strata control.

## 8.9 Safety

Safety is improved due to:-
1. The reduced damage caused by rockbursts.
2. Better hangingwall control and fewer falls of ground.
3. Improved environmental conditions.
4. Less access to old areas.
5. Reduced fire risk.

## 8.10 Productivity

Improved productivity is brought about by:-
1. Fewer production delays due to falls of ground.
2. Improved environmental conditions.
3. A more efficient support system requiring less handling and transport of materials.
4. A more efficient cleaning cycle when used in conjunction with high pressure water jetting.
5. Improved morale and safety.

## 9. CAPITAL EXPENDITURE

The capital expenditure required to introduce a minewide backfill system into a typical deep level gold mine is approximately R15,5 million in 1988 money terms.

The potential for capital savings however are significant, the major features being:-
1. Smaller shafts and fan installations and a reduction in the number and size of airways.
2. Smaller refrigeration and pumping installations.
3. Fewer material cars and smaller timber yards and storage bays.
The potential savings in capital for a typical deep gold mine is approximately R54 million.

## 9.1 Working Expenditure

Theoretically this should be less as a result of:-
1. Reduced cost of temporary and permanent support.
2. Improved productivity and safety.
3. Less power required for ventilation, refrigeration, hoisting and pumping operations.

In practice it has not been achieved due to:
1. The tendency to still use conventional pack support, with backfill, coupled with the relatively high cost of backfill bags.
2. One of the major areas for reducing labour is in the stopes and this would follow if there was increase in face advance in areas using backfill.To date there has not been a significant increase in face advance in stopes on backfill. It is felt however, that given the improved environmental conditions and other benefits associated with the use of backfill, improved productivity will be achieved as resistance to change is overcome and the advantages of backfill are fully realised.
3. Most of the power savings will be achieved in the long term modified layouts. In the shorter term however, additional water has to be handled $\pm 15\%$, with pumping costs being further increased by the increased wear on the clear water pumps.

## 9.2 Revenue

The single greatest source of increased revenue is the additional gold that becomes available for mining, which would otherwise be locked up in stabilizing pillars. On East Driefontein this additional minable gold is valued at approximately R5 billion.

## 9.3 Integration with other technologies.

A stoping system based on backfill lends itself to be integrated with other technologies such as hydropower and trackless mining. The advantages of a mechanised support system are combined with improved cleaning and sweeping operations provided by high pressure water-jetting. Hydropower can also be used to drive hydraulic jackhammers and set hydraulic props. Future developments will see hydropower driving other pieces of machinery such as winches, fans and pumps. When added to these technologies, trackless mining will further improve the overall efficiency of stoping operations.

## 10. CONCLUSION

The future of the South African gold mining industry will depend on the ability to mine at ever increasing depths.
Backfill as a means of support will play an increasingly important roll in helping to overcome some of the major problems associated with mining at these depths.
The system is easily intergrated with other new technologies, providing further advantages and greater rewards. Add to this increased safety and improved morale, and backfilling will go a long way in helping to solve some of the problems facing these deep mines of the future.

## REFERENCES

Bruce, M.F.G. & Klokow J.W. S.A.I.M.M. Publication "Backfill in S.A. Mines".
Clark, I.H. S.A.I.M.M. Publication - "Backfill in South African Mines". Table 1 (Page 28).
Close, A.J. & Klokow J.W. The development of the West Driefontein tailings backfill project. Association of mine managers, circular 1/85.
de Jongh, C.L. The potential of backfill as a stop support in deep gold mines. - Gold 100, S.A.I.M.M. 1986.
Kirsten H.A.D. and Stacey T.R., "Hangingwall behaviour in tabular stopes subjected to seismic events". S.A.I.M.M. - May 1986.
Klokow, J.W. Gold Fields technical seminar (1987).
Loubsher, D.S. S.A.I.M.M. Publication "Backfill in S.A. Mines".
Moore, B. "New Backfill technology for deep level mines". (1985)
Peverett, N. Internal Research.
Pothas, W.J.C. S.A.I.M.M. Publication "Backfill in S.A. Mines".
Robbertze, G.J. The development of hydrau-

lic backfilling at East Driefontein gold
mine. - A.M.M. Circular 1/84.
Stewart, J.M. & Clark I.H. & Morris A.N.
"Assessment of fill quality as a basis
for selecting and developing optional
backfill systems for S.A. gold mines".
Gold 100, S.A.I.M.M. (1987)
van Niekerk, C.J. & C.J. Uys. "Backfill
in S.A. mines". S.A.I.M.M. publication.

2 Laboratory testing

*Innovations in Mining Backfill Technology, Hassani et al. (eds), © 1989 Balkema, Rotterdam. ISBN 90 6191 985 1*

# Analysis and modelling of sill pillars

R.J.Mitchell & J.J.Roettger
*Queen's University, Kingston, Ontario, Canada*

ABSTRACT: Sill Pillars are structural elements upon which the safety and economics of overhead cut-and-fill mining in moderate width steeply dipping ore zones depend. Traditional sill design appears to have developed from experience and timber sill mats are generally used to ensure sill stability. Centrifuge model studies are combined with equilibrium analysis in this paper to provide insight into sill pillar behaviour. This research indicates that stable sill pillars can be designed with confidence and that the general use of timber mats may be an unnecessary expense.

## INTRODUCTION

Sill pillars are commonly used to support uncemented backfill in steeply dipping ore zones of limited width. Such sills are normally cast from cemented sand backfill materials, generally without reinforcements but often underlain by a timber mat. The loading conditions on a cemented sand artificial sill and the potential failure modes are examined in this paper using analytic results combined with centrifuge model studies. The use of sill reinforcements and a wedge shaped sill geometry in place of timber mats is discussed.

### Loading conditions and failure modes

Figure 1 shows a sill pillar at the critical operational stage, that is just after the underlying ore lift is removed. It will be supporting a non-uniform vertical stress, $\sigma_v$, of unknown magnitude and lateral closure stresses, $\sigma_c$, which build up as rock deformations develop and which are difficult to estimate accurately. Sills are normally of such length that the problem can be idealized as a two-dimensional plane strain stability problem. These external stresses are resisted by bending and shear stresses in the sill.

Possible modes of failure are sill slippage, crushing or caving, sill shear,

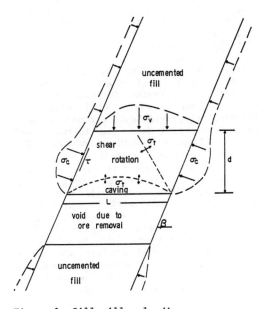

Figure 1 Sill pillar loadings

rotational shear or flexural failure. Sill slippage would be a likely failure mode only if the wall rocks are relatively smooth such that movement can occur in the contact. Slippage is estimated to occur when

$$\gamma H = \delta E \, \ell\mu/L^2 \, (\sin^2\text{ß} - 0.15 \sin^2\text{ß}) \qquad (1)$$

where $\gamma$ is the unit weight of the
  uncemented fill
  H is the height of uncemented fill
  $\delta$ is the estimated wall closure
  E is the stiffness modulus of
    cemented sill
  L is the sill width (HW to FW
    distance)
  $\mu$ is the coefficient of friction
    between the sill and the rock
    walls.
  $\ell$ is the total contact length
    between sill and wall rocks
  ß is the HW/FW dip

Sill crushing would occur if the closure stress exceeds the plane strain unconfined compressive strength (PSS) of the cemented sill material. If the expected rock movement is estimated, it is possible to engineer the cemented sill so that the lateral prestress is a considered design condition, giving $\sigma_c = E \, \delta/L$. The order of magnitude of the sill stiffness modulus, E, may be controlled by the cemented content used and it is suggested that $\sigma_c$ should be less that 50% of the PSS of the cemented fill. Closure stress will, of course, increase the flexural stability of the sill.

Irregularities in the wall rocks due to blasting of the ore are likely to be sufficient to provide interlocking with the sill and prevent sill slippage. Alternatively, overblasting could be used to key the sill into the footwall and prevent slippage in this contact. The general case, then, would be a rough wall condition and subsequent analyses consider failure of sills under rough wall conditions. With rough wall conditions the vertical stress will be reduced by arching of the overlying uncemented fill and it is necessary to estimate this stress. Indications from theoretical considerations, model studies and field measurements are that arching between the HW/FW rocks will reduce the vertical stresses to something in the order of

$$\sigma_v = \frac{\gamma L}{2K \, \tan\phi} \qquad (2)$$

where $\gamma$ is the unit weight of the fill and K is a constant often taken to be unity. It is prudent to assume that this stress acts uniformly on the sill although it is likely that the footwall supports a considerable portion of the adjacent stress.

A wide thin sill would, quite obviously, be susceptible to flexural (bending) failure due to the relatively low tensile strength of cemented tailings. Using standard flexural formula for a fixed end uniformly loaded beam, failure is predicted when

$$\left(\frac{L}{d}\right)^2 > 2 \; (\sigma_t + \sigma_C)/w \qquad (3)$$

where $\sigma_t$ is the tensile strength of the cemented sill, w is the uniform loading which should include the self-weight of the sill (ie., $w = \sigma_v + d\gamma$) and d is the sill depth.

A thick narrow sill might be more apt to cave or undergo side wall shear failure. If it is assumed that caving would extend to a stable arch of height L/2 (semi-circular arch), then all unreinforced sills should be designed with d > L/2 and caving would develop when

$$L\gamma > 8 \, \sigma_t/\pi = 2.5 \, \sigma_t \qquad (4)$$

From equilibrium, block sliding of the sill due to side shear failure occurs when

$$(\sigma_v + d\gamma) > 2 \; (\tau_f/\sin^2\text{ß}) \, (d/L) \qquad (5)$$

where $\tau_f$ is the shear strength in the fill-wall rock contact zone and ß is the HW/FW dip angle.

Rotational failure is most likely to develop when shearing resistance in the hanging wall contact is low due to poor quality HW rocks and/or low HW dip angles which allow separation in this contact. For lower dip angles, a simple approximate prediction may be arrived at by assuming a tensile failure at -ß° and complete separation in the HW contact. Making some allowance for the potential for an increased value of $\sigma_v$ on the hanging wall side, rotational failure would develop when

$$(\sigma_v + \gamma d) > \frac{\sigma_t}{2L} \frac{d^2}{(L - d \cot\text{ß}) \, \sin^2\text{ß}} \qquad (6)$$

The above analyses apply to unreinforced cemented tailings sill pillars. Reinforcements (steel wire, geogrid systems) can be used to improve the tensile or flexural performance of the sill. Reinforcing elements could also be fastened to the walls, particularly the hanging wall to prevent wall shear or rotational failures. A structural mat is used as a separate support but should be designed to be compatible with the cemented fill.

The Dickenson Mines Ltd. fill mat is typical of a timber structural component and a standard design is available for stopes up to 5m in width. Round wood stulls on 2m centres act as independent beams with free end conditions and support flat lagging covered by screens on which the sill is poured. Bending compatibility between the mat and the sill must be considered if these two structural components are to be complementary in supporting the overburden uncemented fill. Then

$$\frac{(w-q)L^4}{384 \ (EI)SILL} = \frac{5qL^4S}{384 \ (EI)STULL} \qquad (7)$$

where $w = (\sigma_v + d\gamma)$ is the total unit loading
$q$ = the unit loading supported by the stulls
$S$ = stull spacings.

Then

$$q = w\left(\frac{1}{1 + 5 \ \dfrac{S(EI)SILL}{(EI)STULL}}\right) \qquad (8)$$

To accommodate the possibility of longitudinal shear, the standard shear formula gives, at failure

$$\sigma_v \ LS > 1.3 \ (tA)s \qquad (9)$$

where $(tA)s$ is the shearing resistance (sectional area times the allowable shear strength) of the stull. For flexural failure

$$w \ \frac{SL^2}{t^3} > 0.785\sigma_s \qquad (10)$$

where $\sigma_s$ is the tensile (flexural) strength of the stull material and t is the diameter of the model stull.

Centrifuge modelling of sill pillars and sill mats

Acceleration of a model in the geotechnical centrifuge causes prototype gravitational stresses to be simulated by the inertial stresses created by centrifugal force. Since $\gamma = \rho a$ and the centrifuge scale factor is $\lambda = a/g$ (where $a$ is the centrifugal acceleration induced), the induced unit weight is $\gamma = \rho\lambda g$. The model then experiences a self-weight stress increase proportional to the scale factor and, thus, the induced stresses are identical to a prototype of linear scale $\lambda$ times the model scale. Actual volumes, hence areas, remain at the model scale, however, and forces (stress x area) are reduced, in the model, according

to the square of the linear scale. Thus, for example, in flexure, the loading scale is unity, the length scale is $\lambda$, the moment scale is $\lambda^2$, and the fibre stress scale is unity (as required for similitude). Thus, similitude can be achieved for all reinforced centrifuge models or sill mats by scaling structural elements on an area basis. This allows both the spacing and size of elements to be varied in order to produce similitude with a prototype. In order to have similitude of the vertical stress distribution it is considered necessary to have the uncemented overburden fill extend to a height of at least 2L above the sill.

The materials listed on Table 1 were used in a variety of combinations to form models representing plain, reinforced and mat supported horizontal sills of 4m to 5m width at a model scale of about 50. The model sills were cast nominally 30mm to 60mm in depth, representing sills of 1.5m to 3m in thickness. Other scale equivalents including wall roughness are noted on Table 2. Models were cast at the rate of two per day and cured for 28 days in a moist room. On the 28th day, each model was mounted in a centrifuge strongbox and conditioned for several minutes at a centrifuge speed of about 60 rpm (producing accelerations of $\lambda = 10$ g). The centrifuge speed was then increased at a rate of about 2 rpm per minute until a sill failure condition was achieved. Closure stresses were not applied to these models in order to represent the weakest case for transverse shear or flexural failures. Model tests were carried out in two series: the first series used 20:1 T:C sills (5% cement by mass of dry tailings) to study the effects of HW/FW dip, sill depth and embedded

Table 1 Properties of Model Materials

| MATERIAL | SIZE | STRENGTHS MEASURED, kPa | | | TEST METHOD |
|---|---|---|---|---|---|
| | | Compressive | Shear | Tensile | |
| 7:1 T:C | cast | 2050 | 1000 | 400 | unconfined compression, |
| 20:1 T:C | control | 250 | 120 | 50 | shearbox and direct |
| 30:1 T:C | samples | 200 | 90 | 40 | tension |
| wood dowel | 2.1mm diam. | N/A | 12.5 MPa | 125 MPa | direct tension (dry) |
| wood dowel | 2.1mm diam. | N/A | N/A | 190 MPa | 3 point bending (dry) |
| steel rebar | 2.3mm diam. | N/A | 150 MPa | 390 MPa | direct tension |
| wood stulls | 4.7mm diam. | N/A | 8.7 MPa | 78 MPa | direct tests (wet condition) |
| wood lagging | 2.5mm square | N/A | 4.3 MPa | 38 MPa | direct tests (wet condition) |
| steel wire | 0.4mm diam. | N/A | N/A | 550 MPa | direct tension |

55

Table 2 Scaled Equivalents

| ITEM | MODEL | PROTOTYPE (TYPICAL) |
|---|---|---|
| linear scale | 1/50 | 4 to 5m sill width |
| wall roughness | milled 10mm rounds at 5mm | ± 0.13m breakage on 0.5m (moderately rough wall) |
| wood dowel reinforcement | 2.1mm diameter at 20mm c-c free end condition | 0.10m round timbers on 1m centres to 0.14m round timbers on 2m centres |
| steel rebar reinforcement | 2.3mm diameter at 20mm c-c free end condition | 0.12m diameter steel on 1m centres to 40mm diameter steel on 0.1m centres |
| steel wire reinforcement | 0.4mm diameter at 20mm c-c covered by light plastic mesh anchored to walls | 0.1m x 0.1m mesh of common steel reinforcing wire anchored to rock walls |
| reinforcement cover | placed 5mm to 10mm from base of sill | 0.25 to 0.5m cover (from sill base) |
| timber support mat | 4.7mm dowels on 40mm c-c (free ends) with 2.5mm lagging at 20mm c-c covered by geotextile | Dickinson mat: 0.3m round stulls on 2m centres with 0.1m lagging on 0.8m centres screen (fabrene cover) |

Table 3 Sill Depth and Dip Angle in Reinforced Models

| MODEL No. (∂°) | REINFORCEMENT TYPE AND COVER | $\frac{d}{L}$ | SCALE FACTOR λ AT FAILURE | PROTOTYPE EQUIVALENT WIDTH, M | FAILURE MODE |
|---|---|---|---|---|---|
| 1 (60) | wood dowel 5mm cover | 0.3 | 55 | 5.5 | sill rotation dowel shear |
| 2 (60) | wood dowel 5mm cover | 0.6 | 80 92 | 9.2 | caving sill shear |
| 3 (60) | steel rebar 5mm cover | 0.3 | 77 108 | 10.8 | caved to rebar sill rotation and rebar bending |
| 4 (60) | steel rebar 5mm cover | 0.6 | 77 132 | 13.2 | caving to rebar sill rotation and rebar bending |
| 5 (90) | wood dowel 5mm cover | 0.3 | 52 | 5.2 | flexural failure |
| 6 (90) | wood dowel 5mm cover | 0.6 | 84 | 8.4 | dowel shear and caving stable arch |
| 7 (90) | steel rebar 5mm cover | 0.3 | 68 137 | 13.7 | caved to rebar sill shear and rebar bending |
| 8 (90) | steel rebar 5mm cover | 0.6 | 110 | 11.0 | caved to rebar sill shear and rebar bending |

Note: All models of strike length L = 100 mm

reinforcements; the second series used a typical 70° HW/FW dip and investigated the use of anchored reinforcements and timber mats. All model sills were overlain by about 250mm of uncemented fill.

A third series of model tests was carried out to investigate the effects of wall closure and a wedge shaped sill geometry on sill stability. A smooth wall condition, created with the finished (treated) side of the formply in contact with the sill and overlying fill, was used in this series of model tests.

Table 3 contains data on the centrifuge sill model series designed to study HW/FW dip, sill depth and reinforcements. The failure of model 1 is shown on Figure 2 where, as in all of these models, the cemented sill is spray painted white for ease of in-flight visual observations. This failure developed suddenly, without prior sill cracking, by dowel shear and sill rotation. Even with the wall roughness used, the HW contact provides little resistance to slippage and rotation about the FW contact appears to be a primary failure mode. Model 2 experienced sill caving just prior to sill collapse.

Figure 2 Failure of Wood Reinforced Sill

Models 3 and 4, which contained steel rebar reinforcements suffered caving of material below the steel reinforcements at a scale factor of about 80, followed by rebar bending as the sill collapsed. The failure of model 4 is shown on Figure 3 and it can be noted that the rebar is still supporting the overlying materials although plastic deformation has occurred.

56

Figure 3   Failure of Steel Reinforced Sill

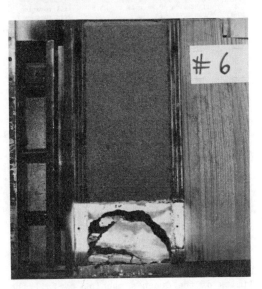

Figure 4   Flexure and Caving Failures

Model 5 exhibited a classic flexural failure with rupture of the wooden dowels but model 6, having a deeper sill, suffered caving with shear of the dowels while the cemented sill maintained a stable arch at a scale factor of $\lambda = 77$. Photographs of these model sills are shown on Figure 4. Both of the 90° HW/FW sill models with steel rebar exhibited caving to the rebar as noted on Table 3 and considerable flexural cracking in the sills as deformation developed. Models with vertical walls generally exhibit flexural, rather than rotational, failure as a result of the better shearing resistance in the HW contact but this improvement appears to be offset by larger surcharge stress, $\sigma_v$, such that HW/FW dip does not appear to be a major factor in sill pillar design. Wood (timber) reinforcing placed in the sill appears to marginally increase the sill stability and steel reinforcements provide substantial support. The wooden members failed in shear while the steel members eventually succumbed to flexural failures. If such reinforcements were to be used in practice, hollow or flanged steel sections would be recommended. Doubling the sill depth from 1.5m to 3m prototype depth increased the stability by a factor of over 40%.

The model tests listed on Table 4 were designed to compare high strength and low strength sills with and without mat supports. High cement content unreinforced sills are commonly used in practice and can be cost-effective where there is good wall rock anghorage and moderate closure strains. High closure strains could result in lateral crushing of these relatively stiff sills. A dip of 70° and a model sill depth of 30mm were selected for the tests. This sill depth produces reasonable compatibility in bending between the sill and timber mat, giving q - 0.4w in equation (8) for a typical ratio of $El_{STULL} - 8El_{SILL}$.

57

Table 4 Cement Content and Mats in Centrifuge Models

| MODEL No. (T:C) | REINFORCEMENT TYPE AND COVER | $\frac{d}{L}$ | SCALE FACTOR $\lambda$ AT FAILURE | PROTOTYPE EQUIVALENT WIDTH, M | FAILURE MODE |
|---|---|---|---|---|---|
| C1 (20:1) | timber support mat | 0.38 | 160 | 12.8 | sill rotation stull bending |
| C2 (20:1) | none | 0.43 | 35 50 80 | 6.5 | caving cracking shear failure |
| C3 (7:1) | none | 0.40 | 125 | 10.0 | sill rotation |
| C4 (30:1) | none | 0.38 | 53 | 4.2 | caving, sill rotation |
| C5 (30:1) | timber support mat | 0.38 | 68 115 | 9.2 | sill cracking flexural failure |
| C6 (30:1) | light wire 10mm cover | 0.38 | 68 | 5.5 | sill rotation wire rupture |
| C7 (7:1) | timber support mat | 0.38 | 240 | 19.2 | No failure |
| C8 (7:1) | light wire 10mm cover | 0.38 | 88 215 | 17.2 | sill cracking wire rupture |

Note: All models at ß = 70°, L = 80 mm

The value of the timber mat is demonstrated by the results from models C1 and C2. Models C3 and C4 show that the 7:1 T:C sill is over twice as stable as the 30:1 T:C sill despite the fact that the strengths of these two mixes are different by an order of magnitude - the rotational shear mode of failure involves both shear and tensile cracking and is a complex mode. Model C5 shown on Figure 5 demonstrated that it is possible to fail a timber mat below a low cement content sill but it was not within the normal working limits of the Queen's machine to fail the combination of a timber mat and 7:1 T:C sill (model C7). Models C6 and C8 were wire reinforced to represent prototype anchored wire mesh. Failure, in both cases, developed by sill cracking followed by sill rotation and rupture of the reinforcing wires. This type of reinforcement, however, was more effective with the 7:1 T:C sill than with the 30:1 T:C sill, giving improvement factors (over the plain sills) of 1.7 and 1.3, respectively, for the two cement contents.

Perhaps the main conclusion from these two model test series should be that the tradition of using timber mats beneath sill pillars is a good, albeit expensive,

Figure 5 Failure of Timber Mat

solution to a complex design problem. The data do indicate, however, that engineered designs combining rationally based choices of T:C content with appropriate reinforcements can produce safe alternatives to the timber mat. Unreinforced and unsupported cemented tailings sill pillars having a d/L ratio as low as 0.4 can be used for spans up to 5 meters provided that closure strains are sufficiently small to eliminate the potential for sill crushing.

Table 5 Wedge Shaped Sills with Smooth Walls

| MODEL No. (T:C) | CLOSURE STRAIN % | $\frac{d}{L}$ (AVG) | SCALE FACTOR $\lambda$ AT FAILURE | PROTOTYPE EQUIVALENT WIDTH, M | FAILURE MODE |
|---|---|---|---|---|---|
| S1 (12:1) | 0.3 | 0.50 | 30 | 2.5 | sill slippage |
| S2 (12:1) | 0.5 | 0.63 | 131 | 10.5 | sill slippage |
| S3 (12:1) | 1.0 | 0.75 | 200+ | 16+ | no failure |
| S4 (20:1) | 0.7 | 0.50 | 108 | 8.6 | rotational shear |
| S5 (20:1) | 1.0 | 0.63 | 132 | 10.5 | rotation slippage |
| S6 (20:1) | 1.1 | 0.75 | 197 | 15.8 | caving, shear |

Note: All models at ß = 65° and L = 80 mm

Table 5 lists the six wedge shaped sill models tested with smooth wall contacts and various closure strains. The photograph on Figure 6 shows the arrangements for applying and measuring closure strains and a sill pillar that has failed by slippage. Note that the base of the sill is perpendicular to the wall rocks while the top of the sill is horizontal. The sill FW depth is then equal to the sill hanging wall depth plus L cos ß. This provides a greater contact length on the footwall (usually better quality rock) and helps prevent sill rotation. The same effect could be achieved by pouring a deeper horizontal sill (d = L, for example) but caving of the subbase would still be a potential failure mode.

From the results on Table 5 it can be noted that closure strains in excess of 0.5% were required to prevent sill slippage with the smooth walls. Using an average E value of 150 MPa for the 12:1 T:C material, the data for model S2 would indicate a coefficient of friction, from equation 1, of about 0.40. This would appear reasonable (i.e. tan $\phi$ = 22°). With rough walls, the risk of a failure due to slippage would be negligable providing some rock closure does occur.

The data from models S4 (Table 5) and C2 (Table 4) may be used to infer that the wedge sill design is more stable than the horizontal sill design. It is clear from Table 4 that increases in the sill depth to length ratio has a significant effect in increasing sill stability. This ratio should be maintained, in practice, at a value not less than 0.5.

Figure 7 shows the eventual failure of sill S6. The closure strain of 1.1% caused some noticeable transverse cracking in the sill but this did not appear to reduce performance under the imposed vertical stress.

Comparisons of analytic and model results

With the rough wall condition, fill arching develops in the overlying uncemented fill and the surcharge stress is calculated, from equation (2) with $\gamma$ = 20kN/m$^3$, L = 5m, K = 1, and $\phi$ = 33°, to be 77 kPa. The value of d$\gamma$ is 30 kPa for a 1.5m deep sill and 60 kPa for a 3m deep sill. An average value of $(\sigma_v + d\gamma)$ = 110 kPa can be used for approximate prediction of the unit sill loading at a scale factor

Figure 6   Wedge Sill with Smooth Walls

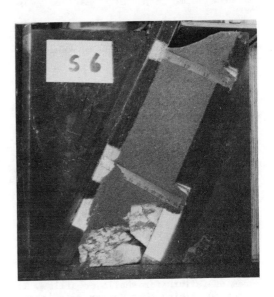

Figure 7   Failure in Sill S6

of 50. Thus, to predict the failure scale factor, w = $(\sigma_v + d\gamma)$ = 2.2$\lambda$ is used to calculate values on Table 6. Consideration of the flexural and shearing resistance in the free end wood dowel reinforcements would add factors of $\lambda$ = 108 and $\lambda$ = 20, respectively, as indicated in brackets on Table 6, to all cases where the failure mode required rupture of these reinforcements. Sills containing the free end steel rebar reinforcements failed by caving of the cover and bending of the

59

Table 6  Prediction for Unreinforced
         Models

| FAILURE MODE | DERIVED FORMULA | PREDICTED VALUE OF λ AT FAILURE | | | |
|---|---|---|---|---|---|
| | | (d/L) = 0.3 | | (d/L) = 0.6 | |
| | | 7:1 T:C | 30:1 T:C | 20:1 T:C | 30:1 T:C |
| Sill flexural | $\lambda = (\sigma_t/1.1)\,(d/L)^2$ eqn. (3) with $\sigma_c = o$ | 33 (140) | 3 (110) | 16 (124) | 13 (120) |
| Sill caving | $\lambda = 1.4\,\sigma_t$ eqn. (4) | 560 | 56 (76) | 70 (90) | 56 (76) |
| Sill shear ß = 90° | $\lambda = 0.91\,\sigma_t(d/L)$ eqn. (5) | 273 | 25 (45) | 66 (86) | 50 (70) |
| Sill shear ß = 60° | $\lambda = 3.64\,\sigma_t(d/L)$ eqn. (5) | N/A | 100 (120) | 260 (280) | 200 (220) |
| Sill rotation ß = 60° (not applicable to ß = 90°) | $\lambda = \dfrac{1}{\dfrac{1.1}{\sigma_t}\left(\dfrac{L}{d}\right)^2 - \dfrac{0.58}{\sigma_t}\left(\dfrac{L}{d}\right)}$ | 39 (59) | 4 (24) | 24 (44) | 19 (39) |
| timber mat flexural failure | $\lambda = 2(\sigma_x\,\text{MPa})$ eqn. (10) | 156 | 156 | 156 | 156 |

Note: Numbers in brackets indicate expected improvements due to wood
      reinforcements.

rebar. These bars can theoretically perform in flexure to an ultimate stress level given by $\lambda = 340$ but large deformations are indicated at levels of about 50% of this ultimate. Detailed predictions of failure for the anchored steel reinforced sill pillars are beyond the scope of the present paper.

From the data in Table 6 it is concluded that sill flexure and sill rotation are critical failure modes for typical unreinforced sill pillars. Deep unreinforced sills of reasonably high cement content or sill reinforcements are required to avoid these types of failures. Sill shear and caving are also potential modes of failure for shallower sills. It is not surprising that timber mats became common in sill pillar installations.

The question of how these predictions compare with centrifuge model results is addressed in Table 7. The best predictions are associated with timber mat flexural support (models C1 and C7) and the poorest predictions with plain sill flexural support (models C2 and C4). The better than predicted flexural performance of plain sills is likely due to the assumption of a uniform vertical stress when the real loading condition is likely much less severe. The use of timber mats is further justified by the results of the plain sill models. The very poor prediction of sill flexural performance is not directly applicable to the sloped HW/FW conditions in these models.

Wood reinforced sill rotation predictions are conservative by a factor of about 2 (models 1 and 2) due possibly to the assumption of negligible HW contact shearing resistance. Wood reinforced sill shear predictions (models 5 and 6) are fairly accurate and the steel rebar reinforced sill ultimate load predictions are quite reasonable (models 3,4 7 and 8).

Models C6 and C8 were designed to evaluate the likely performance of light mesh reinforced prototype sills and the tentative prognosis is very good. Light mesh reinforcements could be placed in the sill at a lesser cost than constructing a continuous timber mat and would seem to be quite effective. Further model testing of this economical alternative to timber mats is certainly warranted.

Conclusions

Data from 22 centrifuge model tests on sill pillars with various types of reinforcements and/or support systems have been compared to analytical performance predictions with the following conclusions:

1. Sill pillar slippage is a potential mode of failure if the rock walls are relatively smooth. In such cases a minimum closure stress of about 50% of the failure stress is needed to prevent slippage.

2. Analytical predictions of the flexural performance of thin horizontal plain cemented tailings sill pillars are not very accurate. This is of little practical significance however, since such sills are not effective and might be dangerous. Other analytical predictions compare favourably with centrifuge model test results.

3. Sill rotation appears to be a prevalent failure mode for HW/FW dips of 70° or less and can only be prevented in horizontal sills by the use of reinforcements or other support systems.

4. Wedge shaped plain sill pillars having a base perpendicular to the rock wall interface and a horizontal top surface provide greater resistance to sill rotation and remain stable for spans of over 8 meters with a depth to width ratio of 0.5

Table 7 Predictions and Model Results

| MODEL NO. (δ°) | MODEL DESCRIPTION | d/L | PREDICTED FAILURE | OBSERVED FAILURE | COMMENTS |
|---|---|---|---|---|---|
| 1 (60) | wood dowel in 20:1 T:C | 0.3 | sill rotation λ = 25 | sill rotation λ = 55 | model stronger by factor of 2 |
| 2 (60) | wood dowel in 20:1 T:C | 0.6 | sill rotation λ = 44 | caving, sill shear λ = 92 | model stronger by factor of 2 |
| 3 (60) | steel rebar in 20:1 T:C | 0.3 | caving, λ = 70 rebar bending at λ = 170 | caving, λ = 77 rebar bending at λ = 108 | rebar bending progressive |
| 4 (60) | steel rebar in 20:1 T:C | 0.6 | caving, λ = 70 rebar bending at λ = 170 | caving, λ = 77 sill rotation at λ = 132 | sill supported by bending in rebar |
| 5 (90) | wood dowel in 20:1 T:C | 0.3 | sill shear λ = 53 | sill flexure λ = 52 | dowels sheared |
| 6 (90) | wood dowel in 20:1 T:C | 0.6 | sill shear λ = 86 | sill caving, dowel shear at λ = 84 | sill caving predicted at λ = 90 |
| 7 (90) | steel rebar in 20:1 T:C | 0.3 | rebar bending λ = 170 | rebar bending λ = 137 | caving observed at λ = 68 |
| 8 (90) | steel rebar in 20:1 T:C | 0.6 | rebar bending λ = 170 | rebar bending λ = 188 | caving observed at λ = 110 |
| C1 (70) | timber mat 20:1 T:C | 0.38 | mat flexure λ = 156 | mat flexure λ = 160 | very good prediction |
| C2 (70) | plain 20:1 T:C | 0.43 | sill flexure λ = 10 | sill shear λ = 80 | some caving at λ = 50 |
| C3 (70) | plain 7:1 T:C | 0.40 | sill flexure λ = 82 | sill rotation λ = 125 | sill rotation predicted, λ = 135 |
| C4 (70) | plain 30:1 T:C | 0.38 | sill flexure or rotation, λ < 10 | caving, sill rotation, λ = 53 | poor prediction |
| C5 (70) | timber mat 30:1 T:C | 0.38 | mat flexure λ = 156 | mat flexure λ = 115 | sill cracking at λ = 68 |
| C6 (70) | anchored wire in 30:1 T:C | 0.38 | no formal prediction | wire rupture at λ = 68 | reinforcement carrying load |
| C7 (70) | timber mat in 7:1 T:C | 0.38 | sill and mat flexure at λ = 238 | no failure at λ = 240 | stable sill to 25m width |
| C8 (70) | anchored wire in 7:1 T:C | 0.38 | no formal prediction | wire rupture at λ = 215 | some cracking at λ = 88 |

5. The traditional use of timber sill mats is supported by both analytical and model research. To ensure flexural compatibility, however, timber mats should be designed using site specific information.

6. Reinforced sill pillars, particularly sills with light mesh reinforcements anchored to the walls, appear to be viable and economical design alternatives to the use of timber sill mats.

Model sill pillar failures were produced at scale factors ranging from λ = 50 to λ = 215, representing prototype sills of 4m to 22m in width and having HW/FW dips from 60° to 90°. This covers a wide range of practical conditions in order that the analytical and model test results presented in this paper can be of direct use in mine design. Sill pillar design is a complex engineering problem that has not previously been addressed in the mining geotechnical literature. The safety and economy of overhand cut-and-fill operations in moderate width stopes depends on creating stable sill pillars between mining levels.

Acknowledgements

Financial support from the Natural Sciences and Engineering Research Council (NSERC) for centrifuge modelling of mine backfill structures is greatly appreciated. The interest of the Canadian mining community in promoting research to improve their understanding of sill pillar behaviour and design is most encouraging. The authors also thank Mr. J.D. Smith of John D. Smith Engineering Associates, Kingston for many informative discussions on sill pillar behaviour and for suggesting the wedge design shape.

*Innovations in Mining Backfill Technology, Hassani et al. (eds), © 1989 Balkema, Rotterdam. ISBN 90 6191 985 1*

# Use of a flue-gas desulphurization by-product for stabilizing hydraulic mine backfill

H.T.Chan, H.M.Johnston, L.Konecny, R.D.Hooton & R.Dayal
*Civil Research Department, Ontario Hydro, Toronto, Ontario, Canada*

ABSTRACT: A study was undertaken to investigate the feasibility of using a flue gas desulphurization (FGD) by-product from a coal-fired generating station as a binder for hydraulic mine backfill. This by-product was obtained from the limestone (lime) injection into the furnace (LIF) desulphurization process. The strength of a number of specimens stabilized with the LIF material was determined and compared with specimens stabilized with cement. It was found that the strength development of the LIF specimens was slower than that of the cement specimens. At 68% and 65% pulp density, some LIF-stabilized specimens, with percolation rates greater than 10 cm/h, yielded 56-day compressive strength equal to or greater than that of 6% cement-stabilized specimens.

## 1. INTRODUCTION

Stabilized hydraulic mine backfill has been used in Canada for about 30 years. Portland cement is usually used as a stabilizing agent, but increasing cement costs have initiated research into ways to reduce cement content, or to replace it with alternative cementing agents. A number of alternative agents, such as natural pozzolans, blast furnace slags, and fly ash, have been studied for many years with different degrees of success, (Thomas, et al, 1979; Nantel and Lecuyer, 1983; Chan, 1984; Yu and Counter, 1988) and the search for new stabilizing materials for mine backfill is continuing.

This paper describes the results of a laboratory research program undertaken to investigate the feasibility of using a new material, a flue-gas desulphurization (FGD) by-product from a coal-fired electric power generating station, as a binder material for the stabilization of hydraulic mine backfill.

This FGD by-product was obtained from a pilot-scale FGD process evaluation experiment conducted at Ontario Hydro's Lakeview generating station (GS) in the summer of 1987. The limestone (lime) injection into the furnace (LIF) technology was studied at the Lakeview GS as part of Ontario Hydro's effort to reduce acid gas emission (Mozes, et al,

1988). The LIF is a yet commercially unproven process in which the sorbent (crushed limestone or hydrated lime) is injected directly into the furnace of a coal-fired boiler. The LIF reaction by-products, consisting of fly ash, sulphation by-products and unreacted sorbent, is subsequently removed from the flue-gas stream using fabric filters or electrostatic precipitators.

## 2. LIF MATERIAL CHARACTERIZATION

Three LIF samples were obtained, LIF-1 and LIF-2 from limestone injection and LIF-3 from a hydrated lime injection. These samples of LIF material were characterized with respect to their physical and chemical properties. Particle-size distribution, as determined by mechanical analysis, is shown in Figure 1 for sample LIF-2. All three samples exhibited similar grain-size distributions with approximately 95% of the material being smaller than 70 $\mu$m and approximately 82% of the grains less than 9 $\mu$m in diameter. The per cents retained on the 45 $\mu$m sieve (ASTM C311) were 26.4, 22.4 and 26.4 respectively. The Blaine fineness (ASTM C204) were 445, 426 and 494 m$^2$.kg$^{-1}$ respectively. Losses on ignition at 750$^b$C (ASTM C311) were 9.11, 6.73 and 4.16% respectively. The specific gravity of the LIF material was again similar for all three samples and was

63

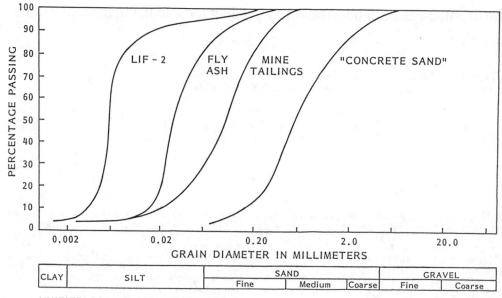

UNIFIED SOIL CLASSIFICATION

FIG. 1   GRAIN-SIZE DISTRIBUTION OF SAND, TAILINGS, FLY ASH AND LIF-2

measured to be 2700, 2670 and 2630 kg.m$^{-3}$ respectively, slightly higher than the typical Lakeview fly ash value of 2450 kg.m$^{-3}$. From X-ray diffraction, all three LIF samples were found to contain quicklime (CaO), anhydrite (CaSO$_4$), calcite (CaCO$_3$), quartz (SiO$_2$), magnetite (Fe$_3$O$_4$), mullite (Al$_6$Si$_2$O$_{13}$) and alumina-silicate glass. The quartz, mullite, magnetite and glass are normal constituents of Lakeview fly ash.

The heats of reaction of the LIF materials were measured using a styrofoam-jacketed nickel calorimeter and stirrer (60 rpm rotational speed). In each test, 40 g of the solid LIF material was allowed to react with various amounts of water and the exothermic reaction temperatures recorded over time. The calculated heats of reaction ($\Delta$H) were -50 cal.g$^{-1}$, -54 cal.g$^{-1}$, and -63 cal.g$^{-1}$ for samples LIF-1, LIF-2, and LIF-3, respectively. Under similar test procedures, a sample consisting of 100% free lime produces a $\Delta$H value of approximately -233 cal.g$^{-1}$ (Williams, 1988). Using this value as a reference, the three LIF samples were estimated to contain 21%, 23%, and 27% free CaO. These values correspond well with the CaO contents determined by X-Ray Diffraction (XRD) analysis as shown in Table 1.

The free CaO content of the LIF material is reflected by the measured pozzolanic activity index. After seven days, the measured strength of the LIF-cement mortar was comparable to that of plain cement mortar (99.7% of control) while at 28 days, the strength of the LIF/cement was measured to be 108.5% of

Table 1.   Chemical Composition of LIF Samples*

| Element | NBS** Fly Ash (1633) | Lakeview Fly Ash | LIF-1 | LIF-2 | LIF-3 |
|---|---|---|---|---|---|
| Calcium | 42 800 | 25 400 | 279 000 | 240 000 | 277 000 |
| Magnesium | 18 325 | 5 200 | 20 260 | 6 600 | <3 000 |
| Sodium | 3 250 | 1 300 | 2 700 | 2 700 | 2 600 |
| Potassium | 17 600 | 12 800 | 7 000 | 10 000 | 8 000 |
| Iron | | 98 400 | 62 000 | 89 000 | 54 000 |
| Aluminum | 116 000 | 118 000 | 48 700 | 63 000 | 60 000 |
| Silicon | C | 212 300 | C | C | C |
| Strontium | 1 600 | 1 000 | 750 | 840 | 830 |
| Arsenic | C | 76 | 49 | 58 | 36 |
| Barium | 3 175 | 700 | 370 | 530 | 450 |
| Chromium | C | 170 | 83 | 112 | 87 |
| Manganese | 491 | 260 | 140 | 142 | 160 |
| Selenium | C | 5 | 7 | 6 | 11 |
| Titanium | 6 200 | C | 2 500 | 3 300 | 3 200 |
| Uranium | 14 | 7 | 7 | 8 | 8 |
| Vanadium | 217 | 170 | 127 | 130 | 110 |
| CaO (% free) (free) | C | 0.34 | 19 | 19 | 27 |
| Carbonate (as CO$_3$⁼) | C | C | 81 000 | 57 300 | 39 500 |
| Sulphate | C | 79 800 | 78 900 | 61 500 | 69 300 |
| unburnt-C | C | C | 37 000 | 32 400 | 19 600 |

\* all concentrations in µg/g
\*\*average of 2 replicates
C not determined

the control cement mix. However, the LIF/cement mix required approximately 6% more water than did the control specimen.

A complete elemental analysis of the LIF material was conducted and the results listed in Table 1. Calcium, $SO_4^{-2}$, and $CO_3^{-2}$ were determined on acid digested samples, unburnt carbon was measured by high temperature thermal combustion, and the CaO content obtained by XRD. The concentrations of all other elements were obtained by instrumental thermal neutron activation analysis. Included in Table 1, for comparison purposes, are the elemental analyses conducted on a representative sample of Lakeview fly ash and the NBS (National Bureau of Standards) fly ash (#1633). The composition of the LIF material is similar to that of fly ash, as would be anticipated, with the exception of the high calcium content. The increase in the Ca content by an order of magnitude (and to a lesser extent Mg) is due to the use of limestone/lime as the sorbent material in the scrubbing process. Again this is reflected in the quantity of free lime associated with the LIF material (19-20%) compared to less than 1% found in the sample of Lakeview fly ash.

The chemical characteristics of the LIF by-product were also evaluated with respect to its leachable fraction using the current Ontario Ministry of the Environment (MOE) Leachate Extraction Procedure (monofill). In this procedure a 20:1 deionized water to solid (LIF material) mixture was tumbled for 24 h at a rotational speed of 10 rpm. The extracts were filtered through 0.45 $\mu$m Milipore filters prior to chemical analysis. The results obtained for the three samples of LIF by-products are shown in Table 2. The sixth column in Table 2 lists the current suggested concentration limits for leachate generated by materials considered for use as an unrestricted backfill material in Ontario. These limits were derived from the Ontario Drinking Water Objectives: one times the maximum permissible health-related objectives and 5 times the maximum objectives based on the aesthetic quality of the water (Golomb, 1986). For comparative purposes, results are also shown for a sample of Lakeview fly ash. From Table 2 it can be seen that the LIF material falls within the MOE backfill guidelines with the exception of barium (2.0-2.6 $mg.L^{-1}$) and chromium (0.1 $mg.L^{-1}$) which are slightly higher than the suggested maximums of 1.0 $mg.L^{-1}$ and 0.05 $mg.L^{-1}$ respectively.

Table 2 also lists the measured pH values for the leachate extracts. The alkaline nature (pH = approximately 12) of the leachate could also be considered a beneficial characteristic of a binder material in some circumstances as it would aid in the neutralization of acidic mine tailings.

## 3. EVALUATION OF LIF AS BINDER

Stabilized backfill specimens were prepared using either "concrete sand" or Falconbridge mill tailings obtained from Sudbury, Ontario as a filler material. The tailings material is greyish black in colour and has a similar grain-size distribution as a silty sand (Figure 1). In the preliminary testing program, all three samples of LIF were evaluated as the binding materials. For strength comparison purposes, specimens were also formed using 3% type-10 Portland cement as the binding agent. The pulp density of these specimens was 70% (ie 70% solids and 30% Toronto tap water).

In making a specimen, the mix was poured into a split mold, 5.1 cm (2") in diameter and 10.2 cm (4") in height. A glass plate was placed at the bottom end of the mold and any free water from the mix was allowed to drain from the bottom. Usually, only a few millilitres of free water was lost, because, at 70% pulp density, most of the mixing water was used in the slaking of CaO to $Ca(OH)_2$. The prepared specimens were then allowed to cure for 7, 14 and 28 days (100% humidity and 23°C) before testing.

The strength of the test specimens were measured by the standard ASTM uniaxial (unconfined) compression test

Table 2. Leachate Quality

| Constituent | Lakeview Fly Ash | LIF-1 | LIF-2 | LIF-3 | Suggested Ontario Guidelines for Backfill Use |
|---|---|---|---|---|---|
| Arsenic (As) | 0.03-0.04 | <0.001 | <0.001 | <0.001 | 0.05 |
| Barium (Ba) | 0.5-0.9 | 2.6 | 2.6 | 2.0 | 1.0 |
| Boron (B) | 4.9-9.6 | 5.1 | 5.1 | 2.5 | 5.0 |
| Chloride (Cl⁻) | 4-16 | 57 | 49.3 | 46.4 | 1250 |
| Chromium (Cr) | 0.14-0.22 | 0.1 | 0.1 | 0.1 | 0.05 |
| Copper (Cu) | <0.001 | <0.1 | <0.1 | <0.1 | 5.0 |
| Iron (total Fe) | <0.05 | <0.1 | <0.1 | <0.1 | 1.5 |
| Manganese (Mn) | <0.05 | <0.1 | <0.1 | <0.1 | 0.25 |
| Nitrate ($NO_3^-$ as N) | 0.01-0.34 | C | C | C | 10.0 |
| Selenium (Se) | 0.03-0.16 | 0.012 | 0.014 | 0.009 | 0.01 |
| Silver (Ag) | <0.001 | <0.001 | <0.001 | <0.001 | 0.05 |
| Sulphate ($SO_4^{-}$) | 515-986 | 1400 | 1250 | 1120 | 2500 |
| Zinc (Zn) | <0.02 | <0.1 | <0.1 | <0.1 | 25 |
| Cyanide (CN⁻) | 140.02 | <0.02 | <0.02 | <0.02 | 0.2 |
| PCB | 0.0001 | <0.001 | <0.001 | <0.001 | 0.003 |
| pH | 11- 12 | 12.5 | 12.5 | 12.5 | - |

all concentrations in mg/L
C not done
*Golomb, 1986

(ASTM D-2166) and all tests were conducted in triplicate.

The results of the unconfined compression tests (average values) are summarized in Figure 2. The proportions of the mix components are on dry-weight basis and are identified (eg LIF(L): Tailings (T): Water (W), 1.25 : 1 : 0.96).

The results presented in Figure 2 indicate that:

(1) samples LIF-1 and LIF-2 (limestone sorbent) were stronger binders than LIF-3 (lime sorbent).

(2) for the same mixing proportions, LIF-1 specimens were usually stronger than the LIF-2 specimens, with the exception of the L : T : W (1.25 : 1 : 0.96) mix.

(3) most specimens stabilized by LIF-1 and LIF-2 yielded higher strength than those stabilized with 3% Portland cement. However, it should be noted that a much larger percentage of LIF (the binder) was used in the mix for stabilizing.

Based on the preliminary results, LIF-3 was not as effective a stabilizer as the 3% cement binder. It was therefore decided that further study using this

material as a binder was unjustified. However, LIF-1 and LIF-2 both showed potential as binding agents. Owing to budget and time considerations, LIF-2 was arbitrarily selected for the second-phase testing program to gain further appreciation of the behaviour of LIF materials for use as a stabilizing agent.

In order to minimize the number of specimens for strength measurements, results of percolation tests were first performed on a number of candidate mixes. The percolation test was done according to a similar method described by Thomas et al (1979). In this procedure, the mix is poured into a plastic tube ($\emptyset$ = 2.5 cm) to a height of 30 cm. A water column of equal height is then placed above the sample giving a hydraulic gradient of 1. The rate of change in the water level is then monitored over time. In preparing the mixes, both 68% and 65% pulp density were used. These pulp densities are commonly used by the mining industry. For comparison purposes, the percolation rates of the 70% pulp density mixes used in the preliminary testing program were measured.

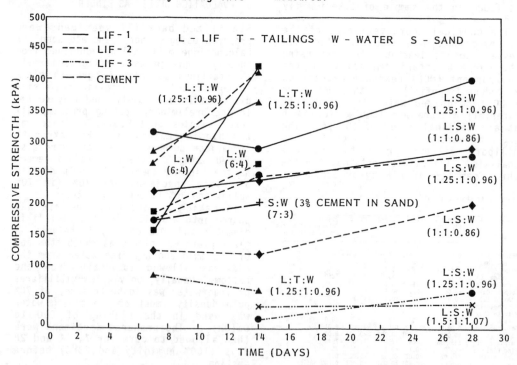

FIG. 2    PRELIMINARY STRENGTH DATA OF STABILIZED SPECIMENS AT 70% PULP DENSITY (EXCEPTING TWO SPECIMENS OF LIF AND WATER AT 60% PULP DENSITY)

Table 3. Percolation Rate Measurements on Candidate Mixes

| LIF-2 | Ash | Cement | Tailings | Sand | Water | Pulp Density (%) | Percolation Rate (cm/hr) |
|---|---|---|---|---|---|---|---|
| 1.25 | | | 1.0 | | 0.96 | 70 | 2.3 |
| 1.25 | | | 1.0 | | 0.96 | 70 | 2.5 |
| 1.25 | | | 1.0 | | 0.96 | 70 | 2.5 |
| | | 0.05 | 1.0 | | 0.45 | 70 | 12.7 |
| 1.0 | | | 1.0 | | 0.94 | 68 | 3.8 |
| 0.6 | | | 1.0 | | 0.75 | 68 | 4.5 |
| 0.5 | | | 1.0 | | 0.71 | 68 | 7.6 |
| 0.4 | | | 1.0 | | 0.66 | 68 | 8.9* |
| 0.35 | | | 1.0 | | 0.64 | 68 | 8.9 |
| 0.3 | | | 1.0 | | 0.61 | 68 | 10.2 |
| 0.5 | 0.5 | | 1.0 | | 0.94 | 68 | 5.7 |
| 0.3 | 0.3 | | 1.0 | | 0.75 | 68 | 6.0 |
| 0.25 | 0.25 | | 1.0 | | 0.71 | 68 | 6.1 |
| 0.2 | 0.2 | | 1.0 | | 0.66 | 68 | 7.6* |
| 0.15 | 0.15 | | 1.0 | | 0.61 | 68 | 9.1 |
| 0.8 | | | | 1.0 | 0.85 | 68 | 6.4 |
| 0.7 | | | | 1.0 | 0.80 | 68 | 7.0 |
| 0.6 | | | | 1.0 | 0.75 | 68 | 10.2 |
| 0.5 | | | | 1.0 | 0.71 | 68 | 11.2 |
| 0.4 | | | | 1.0 | 0.66 | 68 | 15.2* |
| 0.3 | 0.3 | | | 1.0 | 0.75 | 68 | 9.1 |
| 0.25 | 0.25 | | | 1.0 | 0.71 | 68 | 9.9 |
| 0.2 | 0.2 | | | 1.0 | 0.66 | 68 | 10.4* |
| 0.15 | 0.15 | | | 1.0 | 0.61 | 68 | 10.4 |
| 0.4 | | 0.02 | 1.0 | | 0.66 | 68 | 8.9* |
| 0.4 | | 0.01 | 1.0 | | 0.66 | 68 | 8.9* |
| | | 0.033 | 1.0 | | 0.49 | 68 | 22.9 |
| | | 0.066 | 1.0 | | 0.50 | 68 | 21.6* |
| 1.0 | | | 1.0 | | 1.08 | 65 | 6.3 |
| 0.7 | | | 1.0 | | 0.92 | 65 | 6.3 |
| 0.6 | | | 1.0 | | 0.80 | 65 | 6.5 |
| 0.5 | | | 1.0 | | 0.81 | 65 | 8.5 |
| 0.4 | | | 1.0 | | 0.75 | 65 | 9.0* |
| 0.38 | 0.38 | | 1.0 | | 0.94 | 65 | 5.7 |
| 0.35 | 0.35 | | 1.0 | | 0.92 | 65 | 6.4 |
| 0.30 | 0.30 | | 1.0 | | 0.86 | 65 | 7.6 |
| 0.25 | 0.25 | | 1.0 | | 0.81 | 65 | 10.0 |
| 0.2 | 0.2 | | 1.0 | | 0.75 | 65 | 10.8* |
| 0.7 | | | | 1.0 | 0.92 | 65 | 8.6 |
| 0.5 | | | | 1.0 | 0.81 | 65 | 10.8 |
| 0.4 | | | | 1.0 | 0.75 | 65 | 12.4* |
| 0.35 | 0.35 | | | 1.0 | 0.92 | 65 | 9.2 |
| 0.30 | 0.30 | | | 1.0 | 0.86 | 65 | 9.6 |
| 0.25 | 0.25 | | | 1.0 | 0.81 | 65 | 10.0 |
| 0.2 | 0.2 | | | 1.0 | 0.75 | 65 | 12.0* |

*mixes selected for strength measurements

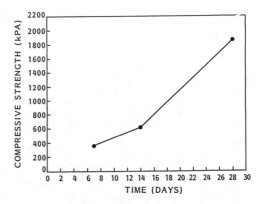

FIG. 3 COMPRESSIVE STRENGTH RESULTS OF SPECIMENS MADE OF LIF, FLYS ASH AND WATER [LIF-1: FLY ASH: WATER (1:1:1), PULP DENSITY 66%]

Table 3 is a summary of the percolation rate results. As anticipated by the correlation between the amount of LIF in a mix and the increase in compressive strength of the specimen, the percolation rate decreases with an increase in the amount of LIF-2 in the mix. The selection of the mixes for additional strength tests are based on a compromise between the measured percolation rates of the mixes and the anticipated compressive strength of the specimens. The desirable percolation rate of a selected mix should be greater than 10 cm per hour (4"/h), which is recommended for a stabilized mine backfill (Thomas et al, 1979) and the compressive strength of the specimen should be similar to that of a specimen stabilized with 3 to 6% cement, a quantity often used in the mining industry.

A bituminous fly ash from Ontario Hydro's Lakeview GS was used as a component in some mixes. The fly ash is similar to a silt material in grain-size distribution (Figure 1). As indicated in Figure 3, fly ash and a LIF would render compressive strength much higher than these shown in Figure 2.

For this part of the testing program, cylindrical specimens, 7.6 cm in diameter and 15.3 cm in height were used. In preparing the specimens, the free mixing water was not allowed to drain away from the bottom of the cylindrical mold. This procedure differed from that used in the preliminary testing program. Since this procedure is commonly used by the mining industry in Ontario (Hopkins, 1988) it was adopted in the second phase of the testing program.

Studies by Mitchell and Wong (1982) have shown that the strength of stabilized mine backfill specimens depend on the procedure of sample preparation, curing, testing etc. In the additional laboratory tests, the same procedures were used in preparing and curing the backfill specimens. Therefore, the comparison of the strength data of various specimens is valid.

The strength data obtained from the second phase testing program are summarized in Figures 4 and 5. A number of conclusions may be obtained from these results.

(1) Specimens stabilized with cement displayed higher early strength than other specimens. However, the 56-day strength values were higher for the specimens with binders other than cement. The slow strength development is typical of specimens stabilized with materials containing CaO.

(2) Mixes with tailings gained higher strength than did identical mixes with sand. The differences were more pronounced in mixes containing LIF-2 and fly ash. On the contrary, mixes with

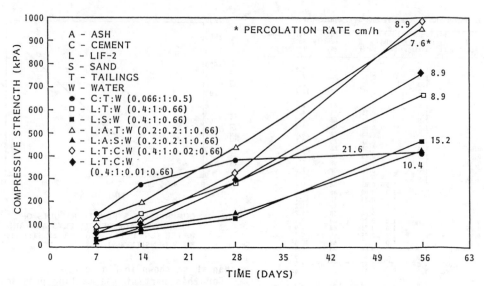

FIG. 4   COMPRESSIVE STRENGTH RESULTS OF STABILIZED SPECIMENS
(PULP DENSITY = 68%)

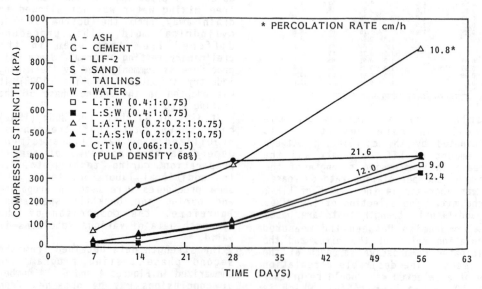

FIG. 5   COMPRESSIVE STRENGTH RESULTS OF STABILIZED SPECIMENS
(PULP DENSITY = 65%)

tailings had smaller percolation rates than sand mixes.

(3) The addition of 1% to 2% cement to the LIF and tailing materials (L : T : C : W mixes) did not seem to improve the early strength development of the specimens. However, the cement did increase the 56-day strength.

(4) As expected, mixes with higher pulp density (ie lower water content)

displayed higher strength than those with the same proportions of solid components but lower pulp density (higher water content).

(5) Based on the results summarized in Figures 2, 4 and 5, higher strength for the specimens could probably have been obtained if LIF-1 had been used as the binder.

## 4. SUMMARY

A laboratory research program was undertaken to investigate the feasibility of using a flue gas desulphurization (FGD) by-product from a coal-fired electric power generating station as a binder for hydraulic mine backfill. Specifically, this by-product was obtained from the limestone (or lime) injection in the furnace (LIF) process.

A number of specimens were prepared and tested for their strength at 7, 14, 28 and 56 days. The results were compared with those specimens stabilized with about 6% type-10 Portland cement. The comparison of data indicates that, generally, the specimens stabilized with the LIF material, with or without fly ash, show higher 56-day strength than the cement-stabilized specimens.

Limited laboratory data obtained in this study demonstrated that it would be possible to design a hydraulic mine backfill mix stabilized with a LIF material to satisfy the recommended rate of >10 cm/hr. The same mix would also render a 28-day or 56-day strength greater or equal to that of specimens stabilized with about 6% cement. The slow strength development is characteristic of binding agents containing CaO. Whether or not the slow strength development of the LIF- binder is acceptable would depend on the actual strength/time requirements of the stabilized mine backfill at specific underground sites. Hence, it is suggested that the use of the LIF as a binder for mine backfill would need to be studied on a site-specific basis with regard to mixing, placing and strength development.

The leaching characteristics of the LIF was determined to evaluate its environmental acceptability as a binder. Results of the leaching tests (Table 2) indicate that the LIF leachate quality for the majority of elements tested is within the suggested Ontario guidelines for backfills. Therefore, it is unlikely that the use of the LIF material as a binder for hydraulic mine backfill would be of significant environmental concern.

## ACKNOWLEDGMENTS

Laboratory tests for this study were peroormed by the technical support staff of the Civil Research and Chemical Research Departments. Special thanks are extended to Mr. Jim Jones, who prepared and tested most of the backfill specimens.

Mr. T.J. Carmichael, Mr. G.S. Kellay and Dr. A.T. Jakubick of the Civil Research Department provided support during the course of this investigation.

The tailings were supplied by Mr. Phil Hopkins and Mr. M. Beaudry of Falconbridge Mines in Sudbury. project. Their assistance is gratefully acknowledged.

## REFERENCES

Chan, H.T. 1984. A study of partialcement replacement by lignite fly ash in cement-stabilized mine backfills. Ontario Hydro Research Division Report No. 84-141-K.

Golomb, A. 1986. MOE leachate extraction procedure. Ontario Hydro Research Division Report No 86-196-K.

Hopkins, P. 1988. Private communications with Mr. Hopkins of Falconbridge Mines in Sudbury, Ontario.

Mitchell, R.J., and Wong, B.C. 1982. Behaviour of cemented tailings sand. Canadian Geotechnical Journal, 19(3), pp. 289-295.

Mozes, M.S., Mangal, R. and Thampi, R. 1988. Sorbent injection for $SO_2$ control: A. sulphur capture by various sorbents, B:Humidification. Ontario Hydro Research Division Report No. 88-63-K.

Nantel, J., and Lecuyer, N. 1983. Assessment of slag backfill properties for the Noranda Chadbourne project. CIM Bulletin, 76(849), pp. 57-60.

Thomas, E.G., Nantel, J.H., and Notley, K.R. 1979. Fill technology in underground metalliferous mines. International Academic Services Limited, Kingston, Ontario, 293 p.

Williams,P. 1988. Personal communications. Ontario Hydro Research Division.

Yu, T.R., and Counter, D.B. 1988. Use of fly ash in backfill at Kidd Creek Mines. CIM Bulletin, 81(909), pp. 44-50.

*Innovations in Mining Backfill Technology, Hassani et al. (eds), © 1989 Balkema, Rotterdam. ISBN 90 6191 985 1*

# An experimental study to investigate the effect of backfill for the ground stability

U. Yamaguchi
*University of Tokyo, Tokyo, Japan*

J. Yamatomi
*Akita University, Akita, Japan*

ABSTRACT: At the 3rd International Symposium on Mining with Backfill which was held in Luleå, Sweden, 1983, authers presented a paper titled "A consideration on the effect of backfill for the ground stability". In which authers concluded that the effect of back-fill is to limit the deformation of rock, to restrict the progression of failure to the farther extent, and consequently, to generate a slow and moderate "dilatant" failure in the rock mass around the opening. An experimental study to support the above conclusion has been lately performed. This paper is the report of the experimental study and the discussion.

The experiment is the uniaxial compression test using a small cylindrical test piece of rock which is surrounded with backfill sand in a heavy steel vessel. Test piece of rock is compressed and the stress-strain curve is observed during the compression. Obtained stress-strain curve is not different up to the peak stress of the failure, compared with the usual uniaxial compression test without any backfill sand surrounding the test piece. However, after the maximum load of compression, namely the failure, easier negative slope of the stress-strain curve and also higher residual stress are observed in comparison with them obtained from the usual uniaxial compression test.

These results of the experimental study suggest that by backfill in certain density and the dilatancy of rock, confined stress state is made around a rock pillar and/or the rock mass around the opening filled with backfill material. And then, an increase of the re-sidual strength of rock and a generation of slow and moderate "dilatant" failure on the pillar and/or in the rock mass around the opening are expected.

The above suggestion was also numerically confirmed.

## 1 INTRODUCTION

At the 3rd International Symposium on Mining with Backfill which was held in Luleå, Sweden, 1983, authers presented a paper titled "A consideration on the effect of backfill for the ground stability".
In which authers concluded:

1) Supporting effect of backfill has gradually come into existence after the sur-rounding rock deformed sufficiently and squeezed into opening. At least, at the moment when the fill material is filled into the opening, no stress should be naturally found in the fill material except a light stress due to the gravity.

2) Even if the filled material is com-pressed, it is rather act to restrict the movement of rock and limit the deformation of surrounding rock mass than to support the rock stress directly,

3) Limiting and restricting the movement and deformation of rock in the farther ex-

tent is to lead the rock to be a softer failure.
And then, during the slow and moderate failure, rock will increase the volume by cracking or plastic deformation, stress re-lease zone namely "Trompeter" zone will be built in the surrounding rock mass around the opening, stresses around the opening shall be restricted in the newly developed arched rock mass and stabilized.

In recent, an experimental study to sup-port the above conclusion has been conducted. This is the report of the experiment and the discussion.

## 2 BACKFILL EXPERIMENT EMPROYED WITH A STEEL CYLINDER

Steel cylinders showing in Fig.1 were em-ployed for the backfill experiment. Cylin-ders were 200 mm in the outside diameter, 150 mm high and had a cylinderical hole in

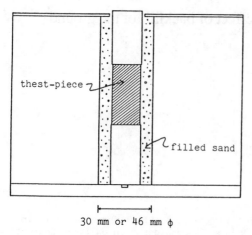

thest-piece

filled sand

30 mm or 46 mm φ

Fig.1 Steel cylinder employed.

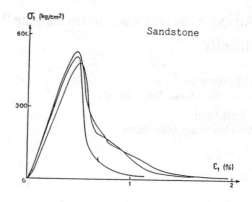

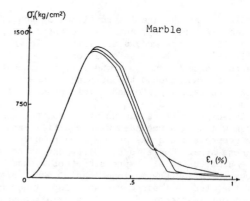

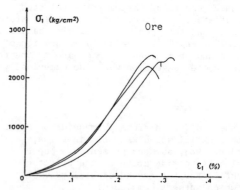

Fig.2 Stress-strain curves at a constant strain rate.

the center. 30 mm and 46 mm diameter hole cylinders were prepared. Cylindrical test-piece of rock was settled in the center hole, sand as the backfill material was filled in the cylindrical space between the cylinder wall and the test-piece, and the test-piece was uniaxially compressed by a compression machine.

Rocks tested were Tako sandstone, Akiyoshi marble and Yanahara pyrite ore. Lumpes of these rocks were drilled for making test-pieces. And then, cylindrical test-pieces were shaped into 25 mm in diameter and 50 and 100 mm long. They were dried in air of 45 degree centigrade for 48 hours and stored in a desicator until an experiment began.

Mechanical properties and the stress-strain curves of these rocks tested are shown in Table 1 and Fig.2 to 3. Strain rate for the uniaxial compression employed for the above was constant and $2 \times 10^{-4}$ mm/s. Even this strain rate, the failure mode of Yanahara pyrite ore was in violent and the failure control was not possible after the strength failure.

Differential stress-strain diagrams of these rocks obtained by confined compression tests are also shown in Fig.4. On the Yanahara pyrite ore, it is also impossible to

Table 1. Mechanical properties of rocks

| Item | Unit | Tako sandstone | Akiyoshi marble | Yanahara ore |
|---|---|---|---|---|
| Compressive strength, Sc | MPa | 54.4 | 132 | 264 |
| Young's modulus, E | ×10⁵ MPa | 1.26 | 5.39 | 11.7 |
| Poisson's ratio, ν | − | 0.128 | 0.312 | 0.198 |
| Aparent specific gravity, ρ | g/cm³ | 2.25 | 2.70 | 4.90 |

72

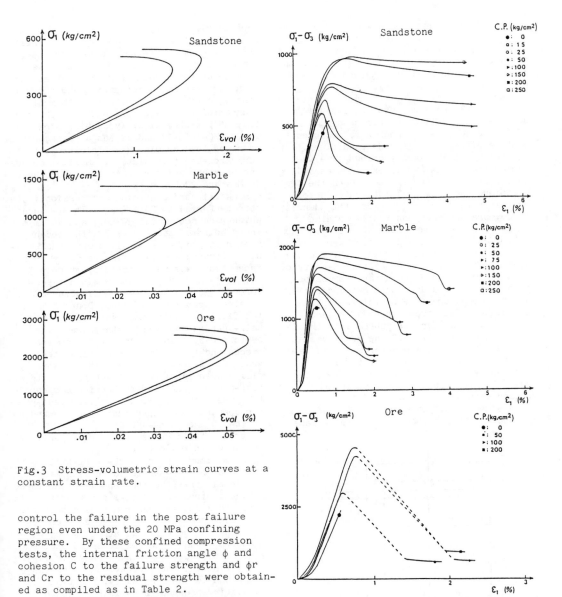

Fig.3 Stress-volumetric strain curves at a constant strain rate.

control the failure in the post failure region even under the 20 MPa confining pressure. By these confined compression tests, the internal friction angle φ and cohesion C to the failure strength and φr and Cr to the residual strength were obtained as compiled as in Table 2.

For the backfill material, Toyoura standard silica sand of which 75 percent of

Fig.4 Results of confined compression tests.

Table 2. Mechanical constants of rocks obtained by confined compression tests.

| | Item | Unit | Tako sandstone | Akiyoshi marble | Yanahara ore |
|---|---|---|---|---|---|
| Peak | Internal friction angle, φ | ° | 27 | 36 | 60 |
| Peak | Cohesion, C | MPa | 16. 8 | 33. 0 | 33. 1 |
| Residual | Internal friction angle, φr | ° | 36 | 43 | 63 |
| Residual | Cohesion, Cr | MPa | 3. 3 | 6. 6 | 23. 0 |

73

total weight is under size to 0.25 mm screen was mostly employed. To the Yanahara pyrite ore, however, undersize product screened by 2 mm screen of real backfill material sampled from the Yanahara underground mine was employed.

When a test-piece was set in the center of the steel vessel and above mentioned fill sand was poured into the surrounding space of test-piece, two methods were employed for getting different confining conditions. The first is that 40 times tappings by a wooden hammer around the steel vessel on each 15 g sand poured into the space between the test-piece and the inner wall of center hole. By this method about 1.6 g/cm³ fill density of Toyoura sand was constantly obtained. However, by this same method, only 1.49 to 1.57 g/cm³ fractuated densities were reached to the Yanahara fill material. By the way, 1.66 g/cm³ was the apparent fill density of backfill measured on columnar samples obtained from the underground cut and fill stope of Yanahara mine by thin-wall tube technique. The second method for sand packing is that sand is poured into the space under vibration of the steel vessel by an electro-vibrating hand-pick machine and about 1.5 g/cm³ fill density is also constantly obtained to the 30 mm φ center hole. To the Yanahara ore sample, the second method was not applied.

On some tests, measurements of confining pressure acting on the side of test-piece were carried out by using pressure gauges cemented on the side of test-piece. These pressure gauges were hand-made. Because, the similar type of commercial gauge was not usable due to the damage at the connecting point between the leading wire and the sensing part. Fig.5 shows the schematic diagram of the pressure gauge. On the back-side of pressure disk, Kyowa wire strain gauge KFC-1D16-11 was pasted and the drawing point of leading wire was Alaldite-coated for water-proof insulation. Maximum measuring pressure and strain of this gauge were 10 MPa and 2000 μ strain.

Backfill sand confining compression tests were performed to 25 φ × 50 mm test-pieces which were set in the 30 mm φ and 46 mm φ inner holes of steel vessel with wooden hammer tapped densely filled sand and hand-pick vibrated softly filled sand. Each of these tests was represented by symbols 30 φD, 30 φS and 46 φD. 46 φS was not performed. 30 and 46 φ show the diameter of hole of steel vessel each. Also, D and S mean the filling conditions, dense and soft. Half-buried compression tests were also performed by using 25 mm φ × 100 mm test-pieces and the 30 mm φ inner hole steel vessels. A test-piece was set and densely filled with sand to a half of the length of test-piece. Full length burried compression test was performed to 100 mm long test-piece for comparison. These tests were marked as 30 φD-100 and 30 φD-100/2.

Stress-strain curves of these experiments are shown in Fig.6 to 8.

The minimum stress in the post failure region of stress-strain curve is named "residual stress" of the specimen and these residual stresses of rocks are tabulated in every rock and the experimental condition. Confining pressures of filled sand corresponding to the residual stress which was measured by small gauges cemented on the side of test-piece are also shown in Table 3.

Omitting the total behaviour of pressure gauge due to the complexity, the typical and simplified pressure behaviour is shown in Fig.8, superposed with the stress-strain curve. No data of pressure was obtained to the Yanahara pyrite ore. Some comments on it shall be mentioned later.

From these experiments, it is conclusively resulted that residual stress and confining pressure by filled sand are risen with the fill density of sand. Inner diamer of the center hole of vessel is also effective that smaller diameter is larger stress and higher pressure. In the larger diameter hole, absolute value of the condensation of sand is larger than the smaller hole and therefore, the reaction to the deformation after the failure is to be smaller.

On the length of test-piece, the longer

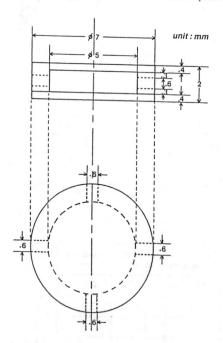

Fig.5 Diagram of pressure cell.

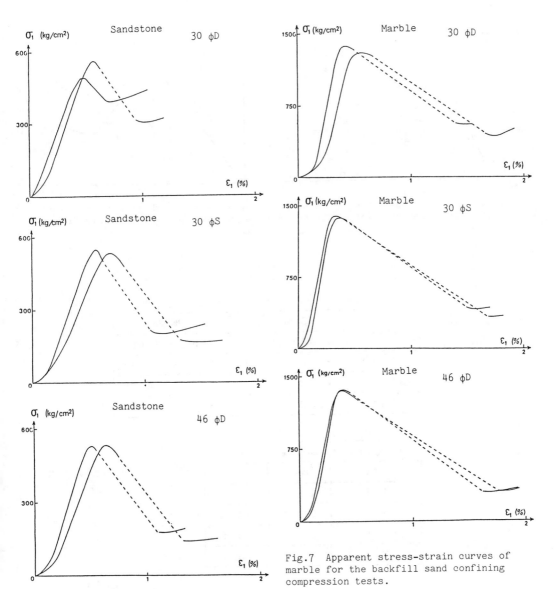

Fig.6 Apparent stress-strain curves of sandstone for the backfill sand confining compression tests.

Fig.7 Apparent stress-strain curves of marble for the backfill sand confining compression tests.

shows smaller residual strength and no residual stress is observed on the half-burried test-piece. However, it was very much interested that several test-pieces of half-filled experiments were broken by shear failure on the not filled part of test-piece and no apparent crack was observed on the burried part of it.

Higher confining pressure was observed on Akiyoshi marble rather than it on Tako sandstone. The reason is supposed to be the larger dilatancy after the failure of the marble. No residual stress and no confining pressure by filled material were observed in all the tests of Yanahara pyrite ore. Dilatancy of Yanahara pyrite ore is much smaller than the other rocks and also not enough fill density was reached are the reason considered. Even on the Yanahara pyrite ore, larger deformation much after the failure will be effective to the residual stress and pressure rises. Backfill to these kinds of rock should be carefully considered.

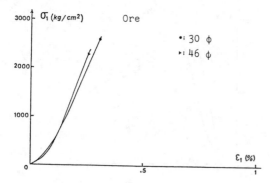

Fig.8 Apparent stress-strain curves of ore for the backfill sand confining compression tests.

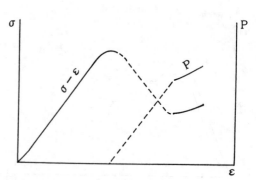

Fig.9 Schematic diagram of pressure measured on test-piece and the stress-strain curve.

Table 3. Residual strength and the corresponding pressure refered to the confining compression test.

a) Residual strength and the corresponding pressure.

| Experiment | Sandstone | | Marble | |
|---|---|---|---|---|
| | $\sigma'$ (MPa) | P' (MPa) | $\sigma'$ (MPa) | P' (MPa) |
| 30 $\phi$ D | 3 1 . 5 | 4 . 2 | 5 2 . 8 | 7 . 6 |
| 30 $\phi$ S | 1 4 . 6 | 1 . 7 | 3 4 . 3 | 4 . 1 |
| 46 $\phi$ D | 1 6 . 9 | 1 . 2 | 3 5 . 0 | 2 . 9 |

b) Confining compression test.

| Confining pressure | Sandstone | Marble |
|---|---|---|
| MPa | Mpa | Mpa |
| 1 . 5 | 1 9 . 2 | — |
| 2 . 5 | 2 5 . 1 | 3 9 . 4 |
| 5 . 0 | 3 5 . 4 | 6 0 . 2 |
| 7 . 5 | — | 7 5 . 6 |

Conclusions of the experiment descrived above are:

1) The strength failure of rock is not effected with and without fill surrounded.

2) The residual strength of rock increases with backfill that the higher strength the higher fill density and smaller compaction of fill material is. And larger dilatant rock shows more increased residual strength.

3) Fill material confines test-piece after the test-piece was failed and deformed, and the confinement reaches to some MPa. were obtained.

3 NUMERICAL SIMULATION OF THE EFFECT OF BACKFILL

In the preceding section, it is found that the supporting effect of backfill appears after the strength failure of the miniature pillar (cylindrical test-piece) and causes the increase of its residual strength. Due to the inelastic dilatancy, the failed miniature pillar forced out the surrounding fill which applied gradually increasing confining pressure against the radial expansion of the pillar. In addition, it is found the residual strength of the backfilled miniature pillar increases with the increases of the filling density.

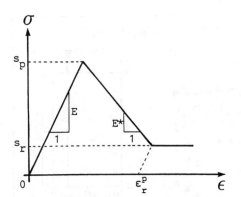 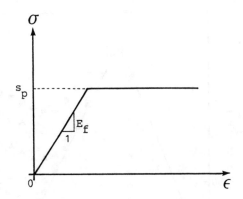

Fig.10 Simplified complete stress-strain diagrams of Akiyoshi marble and the backfill sand which are used for numerical simulation.

Table 4. Material constants of rock and sand used for FEM simulations.

| Material constants | Akiyoshi marble | Densely packed sand | Lightly packed sand |
|---|---|---|---|
| E (MPa) | 40110 | varied | varied |
| $\nu$ (-) | 0.31 | 0.3 | 0.3 |
| Sp (MPa) | 131. | 1.5 | 1.0 |
| $\phi$ ( ° ) | 36.0 | 42.6 | 35.1 |
| E* (MPa) | -3216 | 0 | 0 |
| Sr (MPa) | 4.7 | 1.5 | 1.0 |

In order to substantiate the experimentally obtained results and clarify the supporting effect of backfill in more detail, the authers have carried out a couple of numerical simulations. The numerical simulation tool employed is elasto-plastic FEM taking strain-softening of failed rock into consideration.

Although several kinds of rock were used for the aforementioned laboratory experiments of the miniature pillars, the rock type used for the numerical simulations was Akiyoshi marble alone and its complete stress-strain diagram was simplified by a tri-linear curve as shown in Fig.10 for the convenience of calculation; whereas the sand was assumed as a perfectly elasto-plastic material whose stress-strain curve was a bi-linear curve as presented in Fig.10. The mechanical properties of the idealized Akiyoshi marble and two kinds of the sand, i.e., the densely filled and lightly filled sand

were tabulated in Table 4.

The stress-strain relation used for the numerical simulations is briefly summarized as follows.

1) Rock and sand behave elastically prior to strength failure or yielding. The elastic behaviors of rock and sand are characterized by Young's modulus: E and Poisson's ratio: $\nu$.

2) Strength failure or yielding occurs when the applied stresses satisfy the Mohr-Coulomb's criterion, which involves another two material constants, i.e. the angle of internal friction: $\phi$ and uniaxial compressive yield strength: Sp.

3) Once the yield criterion is satisfied, the yielded rock loses its loading capacity in proportion to the amount of the inelastic deformation and reaches the state of residual strength. This can be visualized by the stress-strain diagram of rock presented in Fig.10 and expressed by the following equations.

$\sigma = Sp + H'\epsilon^P$ in the strain-softening state : $\epsilon^P \leq \epsilon^{Pr}$

$\sigma = Sr$ in the residual strength state : $\epsilon^P \geq \epsilon^{Pr}$

where $\sigma$ and $\epsilon^P$ are the equivalent stress and equivalent plastic strain, respectively, and Sr is the uniaxial compressive residual strength of rock. In addition, $\epsilon^{Pr}$ and H' are defined by

$\epsilon^{Pr} = (Sr - Sp)/H'$

$H' = EE^*/(E - E^*)$

where E* is the coefficient of the negative slope indicated in Fig.10.

On the other hand, the behavior of the yielded sand can be represented by

$\sigma = Sp (= Sr)$ with $E^* = H' = 0$.

4) Based on the isotropic hardening/softening hypothesis and the associated flow rule, the incremental stress-strain relation in the state of post-yielding have been

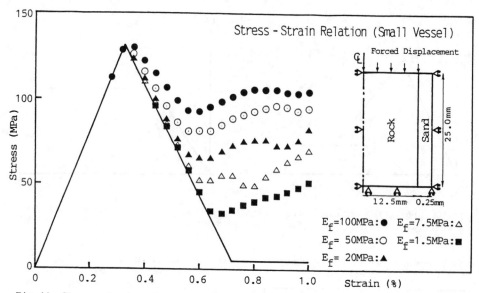

Fig.11  Stress-strain relations of Akiyoshi marble in the small diameter hole of steel vessel.

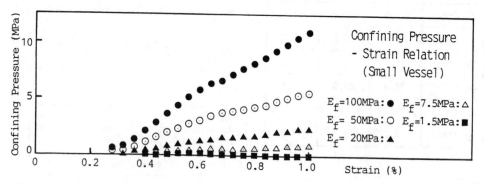

Fig.12  Confining pressure-strain relations of the same situation to Fig.11.

derived from the mathematical theory of plasticity.

5) The material constants describing the elasto-plastic characteristics of rock and sand are found to be Sp, $\phi$, $E^*$ (or H'), and Sr in addition to E and $\nu$. These constants except for the elastic constants of the sand were determined by laboratory experiments and listed in Table 4.

Fig.11 to 14 shows the results of the FEM numerical simulations. Several computations were carried out and it was found that the strength characteristics and/or post-yielding behaviors of the backfill sand had little effect on the results of the calculations, mostly because the very limited yielding was occurred in the sand. Hence, the elastic stiffness of the sand, was changed in the range of 1.5 MPa to 100 MPa, which could be expected to increase with the increase of the packing density.

Forced displacements were applied step by step on the upper end of the miniature pillar and induced stresses and confining pressures were calculated. The solid lines shown in Figs.11 and 13 are the complete stress-strain curves of the idealized Akiyoshi marble and the symbols indicate the stress-strain relationships of the backfilled miniature pillar obtained by the FEM simulations. There can be found the significant deviations of the stress-strain relations after the strength failure. Namely, the negative slope of the backfilled miniature pillar becomes more gentle and its residual strength increases remarkably with

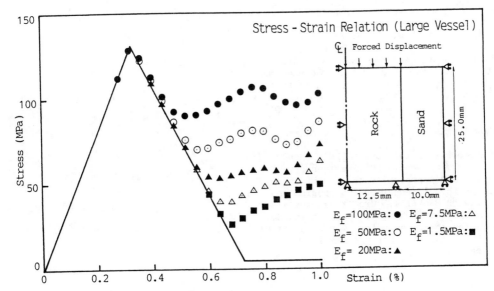

Fig.13  Stress-strain relations of Akiyoshi marble in the large diameter hole of steel vessel.

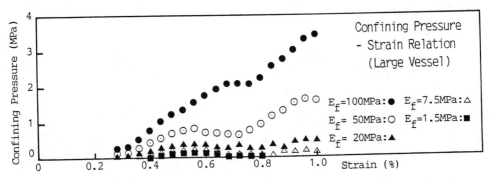

Fig.14  Confining pressure-strain relations of the same situation to Fig.13.

the increase of the elastic stiffness of the sand. The Young's modulus of the sand was estimated to be 1.5 to 7.5 MPa based on the rough coincidence of the calculated and experimentally obtained residual strength of the miniature pillar.

Symbols in Figs.12 and 14 show the increasing of the lateral confining pressure applied on the side of the miniature pillar; the confining pressure of the sand is negligible small prior to the strength failure but thereafter it begins to appear and gives rise to the increase of the residual strength. However, the computed confining pressure of the sand with the Young's modulus of 1.5 to 7.5 MPa was found to be significantly less than that of the experimental results. This could be caused by the

idealization of the mechanical models of the rock and sand, especially, the lack of the compaction effect of the sand which can be expected to increase the stiffness of the sand as the lateral expansion of the miniature pillar forcing out the surrounding backfill.

4 FINAL CONCLUSION

An experimental study to investigate the effect of backfill for the ground stability has been conducted. An analytical discussion on it has been also performed. And then, authors' conclusion on the effect of backfill which was presented and discussed in the last Symposium on Mining with

Backfill is verified. The conclusion is that the effect of backfill is to limit the deformation of rock, to restrict the progression of failure to the farther extent, and consequently to generate a slow and moderate "dilatant" failure in the rock mass around the opening and the ground is finally stabilized.

## ACKNOWLEDGMENT

The authors wish to express their thanks to Prof. F. P. Hassani, McGill University, for offering the opportunity to present this paper at this International Symposium on Mining with Backfill. Gratitudes are also due Messrs. K. Yanaga, T. Sugimoto and S. Sumida for their laboratory works and also Mrs. K. Sugiura for her preparation works of this paper.

## REFERENCES

Yamaguchi, U. & J.Yamatomi, 1984. A consideration on the effect of backfill for the ground stability. Proc.of the Internat. Symp.on Mining with Backfill. Luleå. 7-9 June 1984:443-451.
Yamatomi, J. & Y.Kotake, 1986. Pillar control and the effects of back-filling support at Kosaka mine. Int.J.Rock Mech. Min.Sci.& Geomech.Abstr. Vol.23.No.1:41-53.

*Innovations in Mining Backfill Technology, Hassani et al. (eds), © 1989 Balkema, Rotterdam. ISBN 90 6191 985 1*

# Fill-induced post-peak pillar stability

G.Swan
*Falconbridge Limited, Sudbury Operations, Onaping, Ontario, Canada*

M.Board
*Itasca Consulting Group, Minneapolis, Minn., USA*

ABSTRACT: Some laboratory experiments are reported which demonstrate the ability of relatively soft, cemented fills to promote stable post-failure in extremely brittle pillars. In an attempt to induce unstable failure, a soft compression testing machine was used. The available evidence points to an explanation requiring vertical splitting and a buckling failure mode in the pillars. A set of numerical experiments were performed using FLAC to examine the possibility of alternative explanations. In these experiments the pillar, fill stiffness and strength properties were varied as well as the hardening rule for the backfill. The results in this case show that where unconfined pillars exhibit less brittle behaviour, a 30:1 cemented fill can increase post-peak strength to a level in excess of 50% peak strength. This is simply achieved through passive resistance of the fill-to-pillar expansion as the pillar fails in shear mode.

## 1 INTRODUCTION

One of the difficulties in assessing the performance of fill as a passive support system derives from practical considerations: on a cut-by-cut basis fill rarely occupies the entire stope volume. Furthermore delays occur in placing the fill such that non-elastic wall convergence is never fully utilized. Clearly these factors will diminish the supportive effect that fill can have on a loaded pillar or an undermined hangingwall. The question is, to what extent is an idealized analysis, which neglects such factors, valid?

An example of this type of analysis is due to Blight and Clarke (1983). After classifying their fill using the terms "soft" and "stiff", it is shown that in either case confining stress in excess of 1MPa can be derived from fill surrounding a pillar undergoing compressive loading. For a rock pillar of quartzite it is then shown experimentally that considerable post-peak strength (in the case of a "stiff" fill, even peak strength) is derived from such a confining stress. This is explained on the basis of fill pressure mobilizing additional shearing resistance through the rock's internal friction angle.

These results are quite surprising when it is considered that:
1) complete failure curves for the same rock, with oil as the confining medium, show that confining stress well in excess of 1 MPa is required to increase strength significantly, Figure 1.
2) the rock pillar and fill surround are loaded simultaneously from an unloaded condition. In reality the stope is mined, leaving the pillar loaded in excess of virgin stress levels, and fill is added subsequently as noted above.

In order to clarify the situation, some additional work is necessary - both experimental and, considering the complexity of the problem, with a numerical model.

This paper describes results from a series of pillar confinement tests using cemented mine tailings. The tests were conducted on laboratory-size specimens of an extremely brittle quartz conglomerate core. The influence of cement content and fill tightness is examined. A 2-D constitutive model of the fill-pillar combination is then described which is solved numerically using FLAC (a 2-D large strain finite difference model available through Itasca Consulting Group, Inc.) to obtain a simulation of some aspects of the experiment.

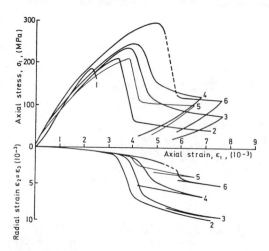

**Figure 1** Complete stress-strain behaviour of Witwatersrand Quartzite at different oil confining pressures, after Hojem et. al., 1975.

Curve 1. Uniaxial Compression
Curve 2. $\sigma_2 = \sigma_3 = 3.45$ MPa
Curve 3. $\sigma_2 = \sigma_3 = 6.90$ MPa
Curve 4. $\sigma_2 = \sigma_3 = 13.80$ MPa
Curve 5. $\sigma_2 = \sigma_3 = 27.60$ MPa
Curve 6. $\sigma_2 = \sigma_3 = 27.60$ MPa

## 2 EXPERIMENTS

The experimental design for achieving a loading test on a fill-confined rock core followed closely that described in Blight and Clarke (1983). An important distinction, discussed in detail below, may have been our use of a comparatively soft loading machine with stiffness 0.38 GN/m. This was deliberately chosen in order to promote violent failure of the rock specimen - a feature of the mine pillars from which the core was taken. In brief, the experimental procedure was as follows:

1) rock core, diameter 44mm and diameter to height ratio of 0.5 was prepared.

2) the core was positioned concentrically in a thick steel ring with inside diameter D (see Table 1) and a cemented tailings mixture at a moisture content of 20% was compacted around the core. The degree of compaction was such to achieve a void ratio of 0.60 (a result based upon in-situ measurements);

3) the sample was sealed in plastic and cured for 7 days at room temperature;

4) the sample was removed from its seal and loaded in a 500kN press at a strain rate of $2 \times 10^{-4}$/min. The platen/sample configuration is shown schematically in Figure 2;

5) in some samples the post-failure appearance of the confined core was observed by sectioning both ring and backfill after the test.

A summary of the complete testing programme is given in Table 1. Note that the influence of the initial gap, proximity of the steel ring, cement content and fill layering were all investigated in the tests. However the test scope was limited in the following respects:

1) no tests were repeated;

2) load measured was aggregate load rather than true pillar load. On the basis of small strain in the fill (< 20 x $10^{-3}$), these loads are essentially identical.

Table 1 - Summary of Testing Programme

| Test Code | Ring Dia. D (mm) | Initial Gap (mm) | Cement Content (%) | Remarks |
|---|---|---|---|---|
| 11 | 203 | 88.0 | 0 | unconfined test |
| 21 | 203 | 4.4 | 9.1(=10:1) | 5% gap |
| 22 | 203 | 0.0 | 9.1 | tight |
| 31 | 203 | 4.4 | 0.0 | uncemented, 5% gap |
| 32 | 203 | 0.0 | 0.0 | uncemented, tight |
| 41 | 203 | 1.8 | 6.3(=15:1) | 2% gap |
| 42 | 203 | 6.3 | 6.3 | tight |
| 51 | 203 | 1.8 | 6.3/9.1/6.3 | layered, 2% gap |
| 52 | 203 | 0.0 | 6.3/9.1/6.3 | layered, tight |
| 61 | 146 | 1.8 | 6.3 | 2% gap |
| 62 | 146 | 0.0 | 6.3 | tight |
| 71 | 203 | 0.0 | 9.1 | backfill only |

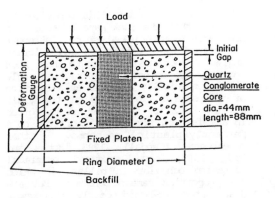

Figure 2   Sectional diagram showing test configuration.

not have been instrumental in controlling failure where control is evident. The controlling mechanism does however require that:

1) the fill be cemented,

2) a minimal gap exist between fill and platen.

Referring to Figure 3, it can be seen that the second of these requirements serves to ensure non-violent behaviour during the immediate post-peak unloading. Appreciable residual strengths - upwards of 50% failure stress - can still be attained when this requirement is relaxed. The cement content appears to determine the level of residual strength attainable for any given gap.

## 3 TEST RESULTS

The raw load-deformation results from each of the 12 tests were corrected for interface and platen deformation effects. For the purposes of the present study, tests coded 41, 42 and 51 were rejected because pillar failure occurred on pre-existent weakness planes. The remaining data were classified into three types:

1) unstable, no post-failure strength

2) unstable but some post-failure strength

3) stable with significant post-failure strength,

and plotted in the form of stress-strain curves, Figure 3.

The sectioned specimens produced a surprising observation: little or no barrelling had occurred. In fact the cores were split into vertical columns, some of which were broken normal to their axes and crushed. No failure surfaces were observed in the backfill at or close to the rock interface.

## 4 DISCUSSION

A careful examination of the stress-strain curves, Figure 3, confirms the unexpected result that violent failure, typical of an unconfined test and depicted by a vanishing post-peak load, can be controlled without resorting to either a stiff machine or a "stiff" fill. The result from Test 71, curve 8 in Figure 3, clearly shows that vertical stresses in the fill were negligible so that significant lateral stresses could

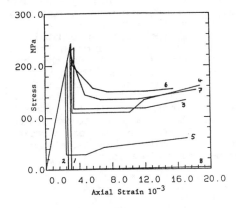

Figure 3   Complete stress-strain curves recovered from pillar-backfill confinement tests.

Curve 1. Test 11 } unstable, no
      2. Test 32 } post-failure
Curve 3. Test 21 } unstable,
      4. Test 52 } some
      5. Test 61 } post-failure
Curve 6. Test 22 } stable
      7. Test 62 } post-failure
Curve 8. 10:1 cemented fill only

A possible mechanism which seems capable of explaining these observations is one involving buckling of the multiple slender columns comprising a split-failed core. In this case only moderate lateral stresses are required to maintain a stable condition. The generation of a vertical column "splitting" failure mode, as distinct from the more usual shear failure, may be due to reduced end confinement effects under transient (dynamic) loading conditions. Such conditions may in turn be unique to brittle, stiff rock/soft loading machine combinations. At this time however further work is required to clarify the details of such a mechanism.

## 5 TWO-DIMENSIONAL STRAIN-SOFTENING ANALYSIS OF RIB PILLARS USING THE FLAC CODE

Studies of rib pillar stability were conducted using the two-dimensional explicit finite difference code FLAC. Details concerning FLAC can be found in Cundall and Board (1988). The code has three features which are required for examination of the reinforcing effects of backfill. First, the code is formulated for large-strain problems - i.e., the geometry of the body is updated and necessary rotational corrections to the stresses are determined as the body undergoes plastic stain. The ability to examine large strains is essential in analyzing the confining effects of backfill, since the fill has porosity which must be closed prior to development of significant load-bearing capacity. Second, the code allows modelling of strain-softening behaviour.

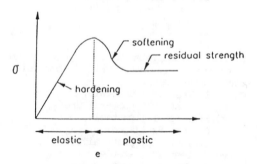

FIGURE 4 - Generalized Form of Stress-Strain Curve, Rock Pillar.

Figure 4 illustrates the meaning of softening for mine pillar compression. The ability of pillars to support load after peak strength is well known. The slope of the post-peak portion of the curve and the residual strength can vary from an elastic-brittle response to a purely plastic response.

The FLAC code models this behaviour by assuming that the rock mass behaves as a Mohr-Coulomb material whose peak strength is defined by its cohesion and friction angle. Once peak strength has been reached, the cohesion and friction angle may be empirically adjusted as a function of the strain, in excess of the yield strain, to allow the material to soften in any fashion from elastic-brittle to a perfectly-plastic response. Other researchers (Cundall, 1988, Hobbs and Ord, 1988) have produced numerical softening behavior in pillars in which the properties remain constant, but are distributed in-homogeneously throughout the pillar. In the present model, the material is assumed to be homogeneous.

Finally, the code allows simulation of slip planes within the body upon which the elements composing the grid may slide or separate. Slip along the planes is governed by a Mohr-Coulomb criteria in which the cohesion and friction angles are those of the slip plane.

## 6 FLAC SIMULATIONS

FLAC was used to perform a series of numerical experiments on hard rock pillars to determine the effects of the following parameters on the stress-strain response:

1) pillar strength and deformation properties,

2) slip planes on hangingwall and footwall contacts,

3) backfill.

Two sets of simulations were conducted: the hangingwall and footwall contacts were either bonded or allowed to freely slip. For each of these base cases, three primary runs were made: no backfill, tight backfill, and backfill with a gap left to represent porosity within the fill. In order to examine the effect of pillar loading stiffness, the pillar deformation modulus was either 15 or 55 GPa. In fact two different sets of assumptions were made regarding pillar properties: a "weak" property set characterized by low modulus and cohesion, and a "strong" set. The country rock was considered to be elastic.

Two models were used to represent the backfill. In each, the fill was treated as a Mohr-Coulomb material with properties of a 30:1 cemented sandfill (friction=35°, cohesion=0.1 MPa, bulk modulus=110 MPa, shear modulus=37 MPa). Porosity in the fill was modelled by placing a 15% gap at the top of the fill pour to represent the available volumetric strain required until significant load-bearing capacity was reached. In the second model, the fill was placed tight to the back and was treated as a non-porous material with 30:1 fill properties. In both cases, the fill still acted as passive resistance to the expansion of the pillar.

Figure 5 shows the starting mesh used for the FLAC simulations.

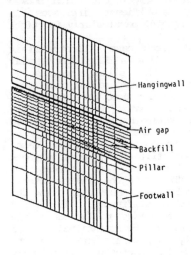

FIGURE 5 - Plot of Initial Finite Difference Grid Used For Pillar Problems.

A grid of 400 elements was used to define the pillar and hangingwall and footwall rock. The pillar was considered to be composed of strain-softening material whose behaviour is controlled by the variation in cohesion as a function of strain beyond the yield point. Figure 6 illustrates the variation chosen for these parameters. The hangingwall and footwall was composed of elastic material. The pillar stress-strain curve was obtained by applying a constant vertical velocity to the top and bottom surfaces, causing the compression of the pillar.

The softening of the pillar was produced by linearly adjusting the cohesion as a function of plastic strain as shown in Fig. 6. Parametric variation of the strain, over which softening occurs, was felt to be beyond the scope of the present work. The choice of residual strength from 10% to 25% of peak strength was based

on the experiments reported in Section 3.

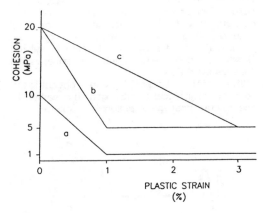

FIGURE 6 - Variation in Cohesion With Plastic Strain to Produce Softening.

## 7 RESULTS

### 7.1 Case 1: Simulations with Bonded Contacts.

Figures 7 and 8 present the results of the simulations in terms of numerically produced stress-strain curves for the pillars. These were produced by averaging the axial stress across the pillar as well as the axial displacement, which was then converted to strain. Each figure shows the results of three simulations: a single pillar without backfill; the pillar backfilled, leaving a 15% gap to represent porosity in the fill; and the pillar backfilled with a tight, non-porous fill. In each case, the fill was considered to be a 30:1 cemented fill with properties defined previously.

As seen in Fig. 7, the unconfined pillar loads to a peak strength of roughly 80-85 MPa at a strain of approximately 0.0025 (or a closure of approximately 20mm for a 7m pillar). The load then drops to its residual strength of 5 MPa with continued strain. The effect of backfill is quite evident from these plots. First, there is a small but distinguishable increase in peak strength of about 5 MPa. The most dramatic effect occurs, however, in the post-peak behavior. The backfill with a 15% gap results in an increase in the residual strength to roughly 50% of the peak strength. The residual strength reaches a constant value, at roughly 2% strain (140 mm closure), which was the extent of this simulation. The failure of the pillar itself is a classical form of hourglassing as demonstrated by the principal stresses in Figure 9 which shows a confined, central core with destressed walls.

The tight backfill can potentially develop significant strength increases in the pillar post-peak strength range.

As shown in Fig. 7, continued dilation of the pillar actually results in hardening behaviour at increasing strain. It is felt, however, that the assumption used here is unrealistic - i.e., the fill cannot have zero porosity. The simulation does indicate, however, the potential, most beneficial effect of fill and emphasizes the advantages of fill placement with minimal porosity.

These calculations closely resemble laboratory tests reported above, which showed little effect of fill on peak strength of a sample but which showed potentially significant increases in residual strength, particularly for tight fills.

The "weak" rock case (illustrated in Fig. 8) shows that a decreased initial stiffness of the pillar results in significantly different pillar behaviour. The lesser cohesion used here results in a lower peak strength (about 40-45 MPa). This peak strength can be adjusted independent of the modulus simply by increasing the cohesion. In this case, the effect of fill on the peak strength is quite dramatic - an increase of roughly 30-35% due to the greater lateral pillar displacement resulting from the reduced pillar stiffness. The rapid dilation of the "weak" pillar causes the peak strength to be maintained over a wide (nearly 10%) strain range, for the fill with a 15% porosity, and actually results in pillar hardening behaviour with tight fill.

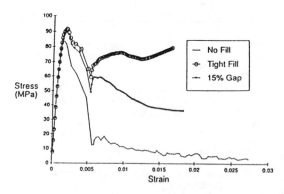

FIGURE 7 - Stress-Strain Curves for a Bonded Hangingwall and Footwall (E = 55 GPa, a = no backfill, b = fill with 15% gap volume, c = tight fill/no porosity).

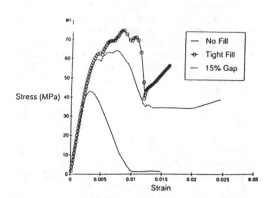

FIG. 8 - Stress-Strain Curves for a Bonded Hangingwall and Footwall (E = 15 GPa, a = no backfill, b = fill with 15% gap volume, c = tight fill/no porosity).

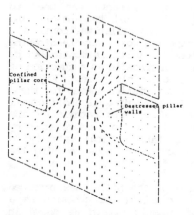

FIGURE 9 - Principal Stress Distribution Illustrating Confinement of Pillar Core and Hourglassing.

## 7.2 Case 2: Simulations with Slipping Contacts

A series of simulations were conducted to examine the effects of frictional contacts at the hangingwall and footwall. The contacts were assigned a friction angle of 20° and cohesion of 100 kPa to reflect the strength of a clay gouge coating. The results of these simulations for the "strong" rock assumption are shown in Figure 10. The peak strength of the pillars is approximately the same as in the bonded runs, but the post-peak response appears to be controlled by the slip of the contacts.

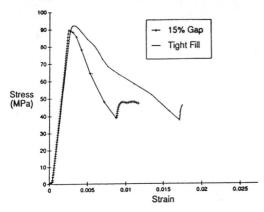

FIG. 10 - Stress-Strain Plot for "Strong" Rock and Slipping Contacts, For (a) Tight, Non-Porous Fill and 15% Volume Porosity.

In both "strong" and "weak" rock assumptions, the post-peak response shows a nearly linear decrease in load to the residual strengths of roughly 1/3 its peak value. The apparent effect is not as significant here due to the overriding influence of slip on the contact planes. However, allowing slip does have a substantial influence on failure mode.

## 8 CONCLUSIONS

The following conclusions can be drawn from the previous studies.

1. Under experimental conditions, which otherwise produce violent uncontrolled pillar failure, "soft" backfill confinement can act as a passive support to achieve a degree of control. A full explanation remains to be found.

2. Numerical experiments using FLAC show assumed inital pillar stiffness has a large influence on the effect of backfill on pillar peak strength. The pillar strength is proportionately increased due to passive resistance of the fill-to-pillar expansion. For a "strong" rock assumption, the fill has little effect on peak strength. For a "weak" rock, peak strength may be dramatically increased.

3. In both physical and numerical experiments, backfill has a significant influence on pillar post-peak behaviour. The FLAC results show that for a "strong" pillar, porous fill promotes a residual strength of roughly 50% of the peak. For a "weak" pillar, the fill results in a response closely approximating elasto-plastic behavior.

4. Slip at the loading contacts modifies the failure mode in FLAC experiments. Further work is required to fully simulate the physical experiments.

## 9. ACKNOWLEDGEMENT

All the experimental work described in this paper was completed while the senior author was employed by CANMET, Energy, Mines and Resources, Canada. We acknowledge the facilities and technical assistance from staff of the Elliot Lake Laboratory.

REFERENCES

Blight, G.E. and Clarke, I.E. (1983) Design and Properties of stiff fill for lateral support Proc. Int. Symp. Mining & Backfill/Luleå/June 7-9.

Cundall, P. A. and Board, M.P.,(1988) A microcomputer program for modelling large-strain plasticity problems. In: 6th Int. Conf. Numerical Methods in Geomechanics, Innsbruck, Austria.

Cundall, P.A., (1988) Numerical experiments on localization in granular assemblies. Presented at workshop on Limit Analysis of Bifurcation Theory in Geomechanics, Karlsruhe, Feb. 1988; published in: Ingenieur Archiv.

Hobbs, B.E. and Ord, A., (1988) Validation of an explicit code for shear band formation in frictional-dilatant materials. Ibid.

Hojem, J.P.M., Cook, N.G.W., and Heins, C. (1975) A stiff, two meganewton testing machine for measuring the work-softening behaviour of brittle materials. S. African Mech. Eng., Sept. 25.

*Innovations in Mining Backfill Technology, Hassani et al. (eds), © 1989 Balkema, Rotterdam. ISBN 90 6191 985 1*

# The influence of material composition and sample geometry on the strength of cemented backfill

A.W.Lamos & I.H.Clark

*Chamber of Mines Research Organization, Johannesburg, RSA*

ABSTRACT: The effect of material composition and backfill geometry on the strengths of cemented backfills, used in South African gold mines, has been determined. The results of the investigation on materials highlight the importance of the backfill size distribution, the cement type and -content and the water/solids ratio. These findings were incorporated into an empirical model, which allows the prediction of fill material strengths. The initial strengths of samples of differing width/height ratio are independent of the sample geometry. Cemented backfills are significantly stiffer at high stresses, than are their uncemented counterparts.

## 1 INTRODUCTION

The very high rock pressures encountered in deep level mines require support systems, sufficiently strong to allow the safe and economic mining of the ore bodies. Hydraulically placed backfill has become accepted as an essential element of these systems and is utilized for both local and regional support purposes.

The addition of cement to tailings backfill produces materials of relatively high initial stiffness, which are well suited to the following mine support applications :

* the provision of local hangingwall support
* the support of middlings in multi-reef mining
* the support of hangingwall in room and pillar mining to permit the extraction of secondary pillars, and
* the provision of working surfaces and roadways for machinery.

The cemented backfill properties, which affect underground support applications, include not only backfill composition and the resultant compressive strength, but also the geometry of the backfill body, which usually determines the degree of confinement, including self-confinement through friction and/or external containment (Thomas, 1969; Thomas et.al., 1979; Vickers, 1983; Clark, 1986).

The objective of this publication is to present and discuss the research findings on :

* the influence of material composition on the compressive strength of cemented backfills
* the effect of different width/height ratios on the stress-strain behaviour of cemented backfill bodies.

The presented research findings may find applications in the design of mine support systems, utilizing cemented backfill. The effect of the material compositions on the strengths of cemented backfills were, therefore, incorporated into an empirical model, which allows the estimation of the compressive strengths of cemented backfills of given composition.

## 2 CHARACTERISTICS OF CEMENTED BACKFILL

The addition of cement to quartzitic tailings backfill results in a material which provides relatively high strength at low strain values. From Figure 1 it is apparent that this characteristic is markedly different from the stress-strain behaviour of a cohesionless backfill: they remain inactive at strain values at which cemented backfills already carry load. The curves shown in Figure 1 are typical results of numerous laboratory tests and apply, both, to unconfined and confined specimens. To illustrate the difference between the two backfill types in one graph, the strain is plotted on a logarithmic scale.

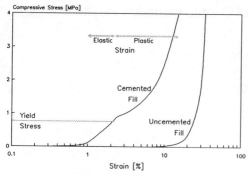

Figure 1. The stress-strain characteristics of cemented and cohesionless backfills.

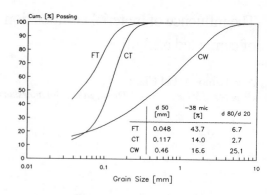

Figure 2. Size distributions and distribution parameters of the three backfill types.

The yield stress of cemented backfills is a significant material property: it divides the stress-strain curve into an elastic and a plastic domain. Up to this level of stress, the strain is of an elastic nature and recoverable; above the yield stress, the material deforms plastically, with a relative decrease of the elastic strain component to a residual minimum.

## 3 LABORATORY TESTING OF CEMENTED BACKFILLS

### 3.1 Backfill composition

Four compositional variables were identified as important to cemented backfill strength:

* backfill type, i.e. particle size distribution
* cement content
* water/solids ratio, i.e. water content
* cement type

The aggregate and cement materials, used in this investigation, are briefly described below.

1. Backfill types. Three different backfill types were used to determine the influence of the tailings size distribution on the strength of cemented backfills:

* full plant tailings
* cyclone-classified tailings
* comminuted waste

All these materials are presently being used as backfills in the South African gold mining industry. As a result of their different size distributions, these materials have different placement porosities (Clark, 1986). The size distributions of these backfill types are shown in Figure 2.

2. Binders. Five different binders were investigated, three of which contained pozzolans (cement extenders). The first three cements listed below are used in cemented backfills in gold mines.

* Ordinary Portland Cement (OPC).
* OPC/Pulverised Fuel Ash (PFA), 1:2, a commonly used binder.
* Portland Blastfurnace Cement (PBFC), an interground mixture (1:1) of OPC
    and Slagment (ground blastfurnace slag).
* OPC/PFA, 1:1.
* Lime*

---

* Lime exhibits some cementitious properties (Fulton, 1977), but when tested with backfill, it was found ineffective as a binder and further work was discontinued.

---

### 3.2 Sample geometry

A suite of cemented backfill samples were tested to determine the influence of geometry on the strength of cemented backfill bodies. The samples covered a range of width/height ratios from 0,5 to 14; a confined compression test simulated an effectively infinitely wide backfill sample.

### 3.3 Laboratory Procedures

A number of laboratory techniques were used to determine the effect of material composition on the strength of cemented backfills. The task involved the preparation of specific backfills, their curing and the subsequent measurement of their compressive strengths.

1. Test samples.  In accordance with standard
practice for comparing different materials (Thomas,
1969; Vickers, 1983), cylindrical test samples of a
width/height ratio of 0,5 were used for those
unconfined compression tests.  The samples were
84 mm high and had a diameter of 42 mm.

The geometry-related samples had square cross-sec-
tions, were 48,8 mm high and varied in their width.

2. Sample curing.  All backfill samples were cured
in sealed stainless-steel moulds at 30°C.  The curing
periods were selected following standard practice at :
3, 7, 14, 28 and 90 days.  All samples were fully
saturated when tested.

3. Compression testing.  All samples, except one,
were subjected to unconfined, uniaxial compression
to determine their compressive strengths.  The "infinite" width/height ratio sample was tested in confined compression in a steel cylinder.  The tests
were done in a 2 MN hydraulic press, using a 4,5 t
high precision load cell.  The displacement rate was
kept constant at 100 mm/h.

From the stress-strain curves, the following values
were extracted : the maximum compressive stress,
the yield stress, the strains at these stresses and the
tangent modulus of elasticity.

4 RESULTS AND DISCUSSION

4.1 Backfill composition

A material, consisting of full tailings and 10 % OPC
(in dry mass), with a water/solids ratio of 0,39 was
selected as a relative standard against which all
compositions were compared.

The maximum compressive strength (MCS), as
measured in uniaxial, unconfined compression, was
selected as the criterion for comparing the materials.
It is regarded as the single most useful measure for
comparing the relative strengths of cemented backfills (Thomas et.al., 1979; Thomas, 1969; Fulton,
1977); this parameter, at the chosen width/height
ratio of the samples (i.e. 0,5), produces a clearly
defined peak value.

1. Effect of cement content.  Figure 3 shows the
maximum compressive strengths of the compositions
comprising full tailings with increasing amounts of
OPC; the water/solids ratio of the samples were kept
constant at 0,39.  The cement content of a sample
was defined as the total binder content of the solids,
in mass per cent.  For comparison with the 20 %
full tailings composition, a sample of comminuted
waste, also containing 20 % OPC, was tested; its
water/solids ratio was equivalent to the full tailings
at a w/s ratio of 0,39.

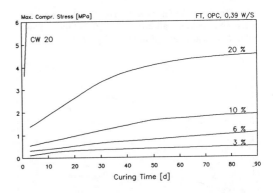

Figure 3.  Maximum compressive strengths of compositions with a range of cement contents.

An equivalent amount of cement (here: 20 %) in a
backfill type of much lower porosity, such as
comminuted waste as compared to full tailings,
produces a material of significantly higher strength
(CW 20).  In mining applications, additions of less
than 3 % OPC to cemented backfills were found to
be ineffective and amounts above 10 % show dimin-
ishing returns (Thomas et.al., 1979).

2. Effect of water/solids ratio.  Figure 4 shows the
maximum compressive strengths of materials, prepared to a range of water/solids ratios.  The backfill
type is full tailings, except for sample CT 28, which
is a classified tailings material, included for comparison.  All samples contained 10 % OPC.

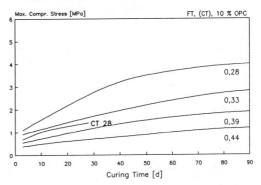

Figure 4.  Maximum compressive strengths of
samples prepared to a range of water/solids ratios.

The water/solids ratios in this group of sample
compositions are equivalent to a range of relative
slurry densities between RD 1,80 and RD 2,00 (w/s
0,44 and 0,28 , respectively).

This graph clearly demonstrates the advantage of
operating at a high slurry density (i.e. a low water
content of the backfill), as the strength increases
more than threefold over the test range.

91

3. Effect of backfill size distribution. The three tested compositions all contained 10 % OPC. The relative strengths are plotted in Figure 5.

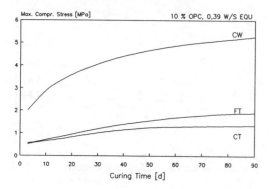

Figure 5. Maximum compressive strengths of compositions comprising different backfill types.

Two factors, viz. placement porosity and water demand account for the wide range of strengths observed in this group. In the comminuted waste sample, the available amount of cement needs to fill considerably less pore space than in a full tailings sample; the cement crystals are, furthermore, of a stronger type, owing to the reduced water/cement ratio.

4. Effect of cement type. In this group, four different binders are compared, using full tailings and a total binder content of 10 %. The results are shown in Figure 6.

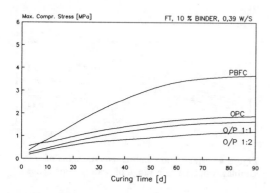

Figure 6. Maximum compressive strengths of compositions using different binders.

The admixture of PFA to Portland cement in the commonly used ratio: OPC/PFA 1:2, reduces the backfills maximum compressive strength to about two thirds of that reached with OPC. The relatively high strength of Portland Blastfurnace Cement (PBFC) is probably due to the fine grind of the (50 %) OPC fraction.

Figure 7 shows the compressive strengths of the three backfill types, using a OPC/PFA mix of 1:2. The same relationships between the materials apply, although at lower strength levels.

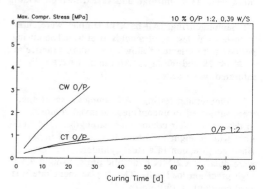

Figure 7. Maximum compressive strengths of different backfill types, using an OPC/PFA 1:2 binder blend.

The yield strengths of the materials amount to about 75 % of their maximum compressive strengths; this result has some important implications in the design of cemented backfills, as it defines the limit of elastic loading.

The strains at the maximum compressive strengths lie mostly between 1 and 2 per cent and tend to decrease with higher material strength. Of the 165 measurements, the lowest and highest recorded values were 0,55 and 3,83 per cent, respectively.

Elevated curing temperatures accelerate the gain in strength of cemented backfills, provided that full saturation is maintained and the temperature does not exceed a critical level (Fulton, 1977). For commonly used backfills, 50°C is an approximate safe limit (Thomas, 1969). If the environmental conditions are such, that this temperature is exceeded, the backfill may not reach its full strength and the composition will have to be reconsidered.

4.2 Backfill geometry

As the compressive strength of the samples were unimportant to the outcome of the experiment, it was decided to select a material, which would provide a compressive strength of about 1,2 MPa after 14 days curing.

A selection of the acquired stress/strain curves are shown in Figure 8.

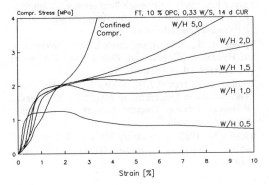

Figure 8. Stress/strain responses of cemented backfill samples of increasing width/height ratio.

This graph shows that all cemented backfill samples, tested in this series, essentially exhibit identical stress-strain behaviour in the elastic region, up to their respective yield stresses. The slight decrease in stiffness with increasing w/h ratio can be attributed to the larger sample sizes and the technical difficulty of uniformly loading wide samples.

As the material is progressively crushed at higher strains, the stress/strain response changes according to the sample's width/height ratio. There is an increase in the ultimate stiffness with increasing self-confinement, up to a maximum, at an effectively infinite w/h ratio. For this material, the stress/strain curves become monotonic at a w/h ratio of about 1,5.

## 5 COMPOSITIONAL STRENGTH MODEL

The correlations between the strengths of cemented backfills and their material composition can assist in the design of cemented backfill systems. For this purpose, the effects of the compositional variables on the material strength have been analysed and incorporated into an empirical model. The unconfined compressive strength (in MPa) of a desired composition may be estimated according to the following relationship:

$$UCS = e^{P1 + \left(P2\left(\frac{OPC}{W} + P3\frac{PFA}{W} + P4\frac{PBFC}{W}\right)\right) \times \left(1 + P5\frac{CT}{NCS} + P6\frac{CW}{NCS}\right) + P7\frac{NCS}{W}}$$

where P1 = -2,15  P2 = 5,65  P3 = 0,67
      P4 = 1,60  P5 = -0,07  P6 = -0,34
      P7 = 0,21

OPC = Ordinary Portland Cement
PFA = Pulverised Fuel Ash
PBFC = Portland Blastfurnace Cement
W = Water
CT = Classified Tailings
CW = Comminuted Waste
NCS = Non-Cement Solids

( all material masses in g )

Limiting Ranges: Water/Cement :      2 - 10
                 Total Solids/Water:  2 - 5
(It is understood that the above equation does not yield the dimension [MPa], results have to be multiplied with a unity factor of that dimension.)

Within the given limits, this model allows the estimation of the 28-day compressive strengths of cemented backfills, cured at 30°C and 100 % humidity.

In practice, the equation can be used in various ways, for instance:

* rock mechanics considerations demand a cemented backfill of a particular strength to meet a given requirement. With the available backfill type, the required cement amount is calculated, using the chosen cement type and water content.
* the effect of varying the amount of water and cement, added to a particular backfill type can be calculated.

The validity of the model has been tested by comparing the model strength predictions against observed values. This showed that the model predictions are (statistically) within ± 25 % of the observed values.

Model predictions need to be corrected when backfills are cured under different environmental conditions.

It must be remembered that such a design equation is not intended as a definitive strength predictor, but rather as a general guide to the strengths which may be expected from materials of certain compositions. Laboratory testing, albeit much narrower in scope, should not be omitted.

## 6 CONCLUSIONS

### 6.1 Backfill composition

Sixteen different cemented backfill materials were tested to determine the influence of their composition on the unconfined compressive strengths. It is concluded that:

93

* the material composition strongly influences the strength of cemented backfill,
* materials with low porosities produce stronger backfills with a given amount and type of cement,
* water content is a critical determining factor in the strengths of cemented backfills; a reduced water content significantly increases backfill strength,
* an increased cement content increases backfill strength,
* Portland Blastfurnace Cement was found to be the most effective of the binders tested,
* cemented backfills can be subjected to a static load during curing with no detrimental effect, provided that the material yield strength is not exceeded at any stage.

From the findings, the requirements for high strength in cemented backfill can be listed as:

* Low backfill porosity
* Minimum water
* Adequate active cement
* Adequate curing time
* Slightly elevated curing temperature

## 6.2 Backfill geometry

With regard to the geometry of backfill bodies, it can be stated that:

* the compressive strength of backfills in the initial, elastic response region, is independent of the backfill body geometry,
* the compressive strength of cemented backfills at high stresses increases to a specific maximum which is dependent on the width/height ratio; it is significantly higher than that of equivalent uncemented materials.
* the compressive strength of placed backfill is determined by the minimum lateral dimension of the body. The effective width/height ratio, therefore, relates to the minimum lateral dimension.

It can be assumed, that all cemented backfills exhibit this fundamental behaviour at their respective magnitudes. Within the limits of practical backfill applications, there is no interfering mechanism.

## ACKNOWLEDGEMENT

The work described in this paper forms part of the research programme of the Chamber of Mines of South Africa Research Organization.

## REFERENCES

Clark I.H., 1986. The Properties of Hydraulically Placed Backfill. Johannesburg, South Africa: Symposium on Backfilling in Gold Mines, SAIMM/AMMSA.

Fulton F.S., 1977. Concrete Technology, A South African Handbook. Johannesburg, South Africa: Portland Cement Institute.

Thomas E.G., 1969. Characteristics and Behaviour of Hydraulic Fill Material. PhD Thesis. University of Queensland, Australia.

Thomas E.G., Nantel J.H., Notley K.R., 1979. Fill Technology in Underground Metalliferous Mines. Kingston, Ontario, Canada: International Academic Services Limited.

Vickers B., 1983. Laboratory Work in Soil Mechanics. London, Great Britain: Granada Publishing Limited.

*Innovations in Mining Backfill Technology, Hassani et al. (eds), © 1989 Balkema, Rotterdam. ISBN 90 6191 985 1*

# New materials technologies applied in mining with backfill

R.F.Viles & R.T.H.Davis
*Technik International Ltd, Birmingham, UK*

M.S.Boily
*Foseco Canada Inc., Guelph, Canada*

ABSTRACT: The paper describes recent advances in the chemistry of backfill binders.

Cement stabilisation of hydraulic backfill material has come of age. The most commonly used cement material has been ordinary Portland cement (OPC). However, given the very large volumes of backfill placed, considerable savings have been achieved by the incorporation of cheaper pozzolanic materials. These include on-site produced slags, ground granulated blast furnace slag, together with formulated activators.

Systems are described and the effects upon backfill properties discussed. The control parameters and laboratory methods used in the development of these binders are outlined.

Pozzolanic materials are generally characterised by very slow early strength development, followed by the attainment of good ultimate strength. In the backfill compositions described, it is shown that it is possible to reduce the binder content in addition to using one of lower cost, whilst maintaining backfill properties.

Hydraulically placed backfill is necessarily placed at high water contents -up to 40% by weight. As much as 30% of the total initial fill volume is lost by dewatering. Classification of tailings is usually employed to facilitate this loss. Materials technology challenges these requirements for cemented fill and the future of backfill binder systems is discussed.

## 1 INTRODUCTION

In recent years, materials technologists, working alongside the mining engineer, have adapted technologies learned in a variety of industries to solve specific problems in key production areas of mining. Substantial technical, safety and cost benefits have resulted. Examples have been the high yield grouts, used for monolithic packing in coal mines, and the organic and inorganic fixing materials used in rockbolting.

The design characteristics of a backfill material will depend upon geological conditions, mining layout and the value of the ore being extracted. In situ strength required may vary from sufficient cohesion to remain in place, through to high compressive strengths to resist high stresses.

The backfill system should be environmentally acceptable and, above all, cost effective.

This paper examines two aspects of materials technology, currently being applied to backfill, and discusses the test methods used in their evaluation.

## 2 FORMULATION OF LOW COST BINDER SYSTEMS

The considerations to be applied to the design of the backfill include:

* Level and rate of attainment of strength

* Availability/cost of materials

* Composition and nature of fill

* Composition of mine water

It is not the purpose here to describe how the mining engineer determines physical backfill requirements (Cowling et al 1983; Sijing Cai 1983; Cabrera et al 1984) but to look at the results which emerge given those targets.

The use of pozzolans as part replacement for ordinary Portland cement (OPC or NPC) is well documented by those concerned with civil engineering and backfill practice (Thomas 1978; Cowling et al 1983; Nieminen and Seppanen 1983; Cabrera et al 1984; Spearing and Wilson 1988; Clarke 1988).

Pozzolanic materials such as ground granulated blast furnace slag (GGBS) and fly ash (PFA) have sometimes been regarded as 'inferior', cheap reactive diluents for OPC. Whilst PFA/OPC blends, in particular, may give rise to lower early strengths than OPC, superior ultimate performance is possible. GGBS based systems will outperform 100% OPC in many respects. Used in conjunction with carefully designed activator systems OPC can be eliminated.

Some of the principles of pozzolan activation can be illustrated.

2.1 Alkali Control

Pozzolans usually require alkali for activation. OPC contains some free calcium oxide, and further amounts are released on hydration. Thus OPC is itself an activator (Plowman 1984).

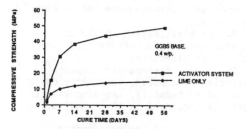

Figure 1. LIME ACTIVATION COMPARED WITH ACTIVATOR SYSTEM.

Figure 1 shows the use of an optimum concentration of alkali (lime) used alone in conjunction with a moderate activity GGBS. This is compared with a two component activator system.

It should be well noted that when additional activators are included, the optimum alkali level may change considerably.

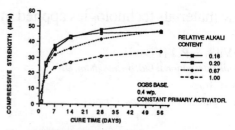

Figure 2. EFFECT OF ALKALI CONTROL.

Figure 2 demonstrates the effect of relatively small changes in alkali concentration in a 3-component binder system based upon GGBS. Note that there is an optimum level for both early and later strength development.

2.2 Primary activator level

For a constant alkali concentration, and a constant background ionic activity in solution, the effect of altering a primary activator can be examined.

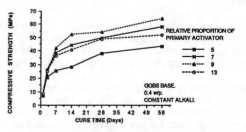

Figure 3. EFFECT OF PRIMARY ACTIVATOR.

Figure 3 shows that a saturation level was found in a 4 component mix, again based upon GGBS. Note the apparent discontinuities in some strength development curves. Sometimes GGBS can (genuinely) appear to hydrate in a stepwise manner, with lag phases and periods of activity.

2.3 Pozzolan variance

Pozzolans from some sources are active in their own right whereas others, such as non-ferrous slags, may require very high activator additions. The response to a given activator may vary from source to source. The dominant features are:

* Surface area

* Oxide analysis

96

In the case of slags the glass content and age of the material are also significant (Thomas et al 1979).

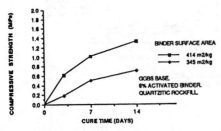

Figure 4. EFFECT OF SURFACE AREA.

Figure 4 shows the direct effect of surface area, where GGBS was used in a quartzitic rockfill.

The setting of quality standards is important. One supplier provided two batches of GGBS which performed very differently (see fig. 5). The reasons for such variance were compositional, having been sourced from different furnaces, and possessing different surface areas.

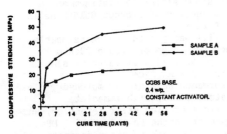

Figure 5. EFFECT OF POZZOLAN VARIANCE.

## 2.4 Activator variance

The precise nature of the activator is also important. Figure 6 demonstrates how different sources of essentially the same chemical can influence the outcome.

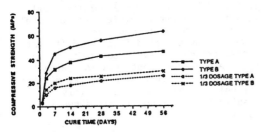

Figure 6. INFLUENCE OF PRECISE SOURCING OF PRIMARY ACTIVATOR.

## 2.5 Variation in fill material

The major aspects are:

* Particle size distribution and shape

* Chemical impurities

* Interaction with the activator/binder

Most of the features desirable in concrete aggregates (Dolar-Mantuani, 1983) should appertain to a suitable backfill material. Physically this means a tough crystalline angular or subangular material, with a size distribution which permits good compaction (Thomas et al 1979:25 - 48).

Chemically there is greater tolerance than could be allowed in concrete, provided that the backfill is reasonably consistent. Intrinsic acidity/alkalinity can be taken care of by adjustment of the binder/accelerator. The presence of cation exchange minerals (eg clays) can be very significant as was the case in figure 7. Here a binder system devised for a quartz sandfill behaved differently with a crushed argillaceous rockfill. Whilst later strengths did recover, a modification to the binder formulation was necessary.

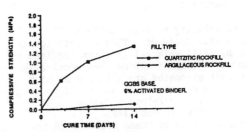

Figure 7. EFFECT OF FILL MATERIAL.

The mine water or fill material can be contaminated with accelerators or retarders. However under most circumstances it is sufficient to examine a specific fill material and binder in conjunction with tap water.

## 3 RAPID SETTING, DRAINAGE FREE, HYDRAULIC FILL

The pulp density of hydraulic fill is usually maximised so as to limit the adverse effects of excess water and its resultant drainage. These include:

* Demand on dewatering and pumping facilities

97

The provision, operation and maintenance of such systems is not merely arduous and time consuming, but very expensive.

* Cycle time

Allowance of time for drainage and subsequent setting can significantly prolong the stoping cycle. This can be a significant production cost determinant.

* Volume yield/placement capacity

Dewatering reduces the effective capacity of the backfill plant and will further increase cycle time by way of increasing placement time.

* Segregation/liquefaction potential/strength of fill

The integrity of the fill may be questioned if drainage is partially inhibited.

* Escape of cement and backfill fines

Drainage inevitably puts fines and cement particles where they are least welcome, giving rise to problems such as cemented sumps and loss of gold sweepings.

Permeability of fill is the key to successful drainage. Thus preparation of the backfill material is often required to remove the fine fraction. Paradoxically the best strengths are obtained at high binder levels where the binder is as fine as possible, therefore reducing permeability (Thomas 1978).

One solution to these problems is to eliminate the need for drainage by locking up the water within the backfill mass.

Rapid setting, rapid hardening grout systems have been developed for monolithic packing in coal mines (Beale & Viles 1984, Bexon 1986). At 72% water content by weight (92% by volume), these grouts can achieve compressive strengths of 1MPa in 2 hours and 3 MPa within 24 hours. Such compositions have been successfully tested as backfill binders, but found to be costly.

A system has been developed (Smart 1988) for use with hydraulic tailings or sand fill, which retains the elements of a binder/activator as described in section 2. However, in addition, an accelerator

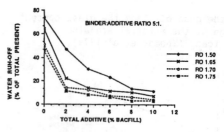

Figure 8. RUN-OFF, RELATIVE DENSITY & ADDITIVE DEMAND.

(TEKFILL) is added underground which causes rapid setting of the backfill, arresting water run-off. As subsequent hardening takes place, water thus held becomes bound up within the cement matrix.

Preliminary laboratory investigations demonstrated the importance of maintaining relative density of the backfill material as high as possible for maximum effect and economy (Figure 8).

First trials of this system were carried out in narrow tabular ore bodies: a test stope at Impala Platinum and a working section of Libanon Gold Mine in South Africa.

The fill was placed inside permeable geotechnical bags of around 20cu.m capacity each, at a distance of some 2m from the working face.

Performance underground matched laboratory tests, with immediate indications of high water retention, and ultimate strengths in excess of 1 MPa. Trials in the working stope demonstrated that the placed backfill easily withstood blasting after 48 hours, as well as subsequent blasts as the face advanced. No sign of deterioration was observed in samples obtained 3 weeks after placement.

The next step was to transfer this technology to Canada and test it in a different mining environment:a VRM stope at INCO's Creighton Mine. Since every raw material was different to that used in S.Africa, a thorough laboratory test programme was first carried out to 'tune' the system to local conditions.

When using cemented fill two different binder ratios are employed at Creighton. First a higher strength mix is poured to form a plug followed by a regular mix for the bulk of the stope. Laboratory results for the plug mix are shown in Table 1.

TABLE 1. TECHNIK LABORATORY TEST RESULTS.
COMPARISON WITH A STANDARD CONTROL "PLUG MIX".

| | | CONTROL | TEKFILL |
|---|---|---|---|
| PULP DENSITY (70% SOLIDS) | | 1.94 | 1.94 |
| TAILINGS / BINDER | | 10:1 | 7.5:1 |
| PLACED DENSITY | | 2.17 | 2.03 |
| WATER RETENTION | 2 Hours | 77.2 | 90.8 |
| (% OF ORIGINAL) | 6 Hours | 63.4 | 90.4 |
| | 24 Hours | 50.5 | 90.0 |
| PLACED COMPOSITION (kg/m³) | | | |
| BINDER + TAILINGS | | 1784 | 1465 |
| BINDER | | 162 | 172 |
| WATER | | 386 | 565 |
| WATER LOSS kg/m³ PLACED | | 382 | 56 |
| U.C.S. MPa (psi) | 0.5 Days | - | 0.12 (17) |
| | 1 Day | - | 0.19 (28) |
| | 3 Days | 0.14 (20) | 0.29 (42) |
| | 7 Days | 0.35 (51) | 0.66 (96) |
| | 14 Days | 0.62 (90) | - |
| | 21 Days | - | 1.06 (154) |
| | 28 Days | 0.89 (129) | - |

Whilst the activator level was adjusted to cope with the increased binder dilution rate, it was decided to hold the accelerator (TEKFILL) level constant at 1.3% for the bulk stope pour. The backfill to binder ratios were recalculated to allow for the increased water retention, so as to maintain the binder content per unit volume at a similar level to that of normal cemented fill.

TABLE 2. COMPARISON OF LABORATORY & FIELD DATA. TEKFILL.

| | PLUG MIX | | STOPE FILL | |
|---|---|---|---|---|
| | LAB | FIELD | LAB | FIELD |
| TAILINGS / BINDER | 7.5:1 | | 25:1 | |
| PLACED DENSITY | 2.03 | 1.96 (5) | 2.03 | 2.00 (7) |
| PLACED WATER CONTENT (% BY WEIGHT) | 28.1 | 36.3 | 28.1 | 34.1 |
| PLACED WATER CONTENT (kg/m³) | 570 | 713 | 570 | 684 |
| BINDER CONTENT (kg/m³) | 172 | 146 | 56 | 51 |
| U.C.S. MPa (psi) 1 Day | 0.19 (28) | 0.11 (16) | 0.05 (7) | |
| 3 Days | 0.29 (42) | 0.26 (38) | 0.09 (13) | 0.13 (19) |
| 7 Days | 0.66 (96) | 0.59 (85) | 0.16 (24) | 0.29 (42) |
| 28 Days | >1.06 (154) | 1.03 (144) | 0.34 (50) | 0.52 (76) |

Performance in the mine exceeded targets (Table 2). At a slightly lower pulp density than anticipated, the rapid set of the plug fill and the overall performance of the stope fill, were better in practice than in the laboratory.

The carefully controlled pour was carried out in stages so as to monitor progress. The first stage of the plug was poured behind a partial barricade 1.5m high. 358 tons were poured in 1 3/4 hours with little sign of run-off water. Sufficient strength had developed to allow

personnel to stand on the surface almost immediately after placement. Once the barricade was completed the remainder of the plug pour of 1,131 tons proceeded. There was little run-off observed through the drainage system. The main stope fill was similarly completed stage by stage.

Samples taken from the stope showed that:

* Objective 7 day uniaxial compressive strengths were exceeded in about 5 days

* Run off was restricted to less than 5% of total water

* Volume loss was reduced by over 30% of normally placed hydraulic backfill

In practical terms benefits include:

* Time savings (fill time/turn around)

* Increased backfill plant capacity

* Water pumping and maintenance costs reduced

* Reduced loss of fines

* Reduced loss of binder

* Simpler, faster barricade or containment possible

* Reduced need to plug leaks

* Improved safety and environment

* Elimination of flocculants

The possibility of using unclassified tailings remains to be examined.

4   TEST METHODS REVIEW

4.1.   Initial pozzolan/cement evaluation

The potential of a cement system can be screened by carrying out unconfined compression tests on 40mm or 100mm cubes prepared with a 0.4 water/powder ratio. A particle size analysis/surface area measurement must be made and chemical analysis by X-ray fluorescence (XRF) and x-ray diffraction (XRD) is usually helpful in completing the assessment. XRF will yield the oxide composition, whereas XRD will give a reasonable estimate of glass content and reveal crystalline species present.

## 4.2. Laboratory determination of backfill performance

There are no shortcuts. The fill has to be prepared in a manner which as closely as possible follows mine practice. Fresh tailings/binders should be mixed in the proportions intended, for a period that typically reflects pipeline travel time. For strength testing a 2:1 aspect ratio cylinder is recommended.

It is important that cylinders of sufficient diameter to include a representative sample are used. 150mm is preferred for most rockfill/pastefill materials, whereas 75mm is adequate for tailings fill.

To permit drainage the cylinder is fitted with a perforated base plate and stood on a bed of sand. Volume shrinkage is compensated by the attachment of a collar to the top of the cylinder.

Curing is usually at $20^0$C, 100% RH. This can be achieved by storage in a conditioned room or by enclosing samples in plastic bags in a room maintained at $20^0$C. It is acknowledged that practically temperatures may differ from $20^0$C, and this should be taken account of in any critical analysis. Typically, the ultimate strength of OPC systems will be reduced in elevated temperature conditions. Early strengths are generally increased by a temperature rise and in some pozzolanic systems a considerable improvement at all ages may be seen.

TEKFILL drainage data was obtained after filling cylinders 18" high by 6" diameter, checking the weight loss with time.

Triaxial compression data and assessments of liquefaction potential for TEKFILL are currently being determined in accordance with methods indicated elsewhere (Aref & Hassani 1988). Early work shows that TEKFILL behaves as a solid at about 12 to 24 hours.

## 5 THE FUTURE FOR CEMENT BACKFILL DEVELOPMENT

The limiting factor often controlling use of cemented backfill is its cost. This must be balanced carefully against the losses resulting from not using such a system. The emphasis must be placed upon the formulation of a cost effective system on a mine to mine basis. Reduction in mine development and maintenance costs need to be estimated in addition to benefits from improvements in direct productivity.

Low or zero drainage hydraulic fill is only in its infancy and there remains a good potential for further advances. The application of such systems in conjunction with rapidly constructed foam barricades (Clarke & Haarenen 1989) is an exciting prospect.

### ACKNOWLEDGEMENTS

The authors gratefully acknowledge the kind assistance provided by the staff and management of the following mining companies: JCI, Anglo American, GFSA, Gencor and Inco.

The findings presented are based upon field trials to date and the procedures described have not necessarily been adopted by these companies.

## REFERENCES

Aref, K & Hassani, F.P 1988. Strength characteristics of paste backfill. In Proc. Conf. on Applied Rock Engineering. Newcastle-upon-Tyne. pp 4 - 8.

Beale, J & Viles, R.F 1984. Cement compositions. UK patent. GB 2 123 808.

Bexon, R. 1986. Monolithic packing. Min. Engr. May 1986. pp 512 - 519.

Cabrera, J.G et al 1984. The relevance of PFA properties on the strength of concrete. Proc. 2nd Intl. Conf. on Ash Technology & Marketing. London. p 303.

Clarke, D.M & Haarenen, A.E 1989. Introduction and uses of cementitious foam in the underground operations of INCO Ltd. Proc. 9th CIM Underground Operators Conference. Sudbury. paper no.23.

Clarke, I.H 1988. Evaluation of cemented tailings backfill in South African Mines. SAIMM Spec. Publ. no. SP2. pp 77 - 89.

Cowling, R., Auld, G.J & Meck, J.L 1983. Experience with cemented fill stability at Mount Isa Mines. Proc. Intl. Symp on Mining with Backfill. Lulea. pp. 329 - 340.

Dolar-Mantuani, L. 1983. Handbook of concrete aggregates. New Jersey. Noyes publications.

Nieminen, P & Sepparen, P. 1983. The use of blast furnace slag and other products as binding agents in consolidated backfilling at Outokumpu Oy's Mines. Proc. Intl. Symp. on Mining with Backfill. Lulea. pp 49 - 58.

Orchard, D.F 1976. Concrete Technology volume 3. Properties and testing of aggregates. London. Applied Science Publishers.

Plowman, C. 1984.  The Chemistry of PFA
    in Concrete; a review of current
    knowledge.  Proc 2nd Intl. Conf. on
    Ash Technology & Marketing.  p. 437.
Sijing Cai. 1983.  A simple and
    convenient method for design of
    strength of hydraulic fill.  Proc.
    Intl. Symp. on Mining with Backfill.
    Lulea.  pp 405 -412.
Smart, R.M 1988.  Patents applied for.
Spearing, A.J.S & Wilson, W.B.  The
    design of a backfill system wth
    particular emphasis on Randfontein
    Estates Gold Mining Company (W) Ltd.
    SAIMM.  Spec. Publ. no SP2.  pp 153 -
    165.
Thomas, E.G 1978.  Fill permeability and
    its significance in mine fill
    practice.  Proc.  Symp.  Mining with
    Backfill.  Sudbury.  pp 139 - 145.
Thomas, E.G, Nantel, J.H & Notley, K.R.
    1979.  Fill Technology in underground
    metalliferrous mines.  Kingston.
    International Academic Services Ltd.

# 3 Modelling and design

*Innovations in Mining Backfill Technology, Hassani et al. (eds), © 1989 Balkema, Rotterdam. ISBN 90 6191 985 1*

# Application of high-strength backfill at the Cannon Mine

C.E.Brechtel & M.P.Hardy
*J.F.T.Agapito & Associates, Grand Junction, Colo., USA*

J.Baz-Dresch & J.S.Knowlson
*Asamera Minerals (US), Inc., Wenatchee, Wash., USA*

ABSTRACT: High-strength backfilling in the B-NORTH Zone at the Cannon Mine is being used in sublevel bench-and-fill mining of gold ores. The high strength backfill is required to minimize surface subsidence and to provide safe operating conditions during pillar mining between backfill pillars that are carrying portions of overburden weight. Pillar recovery between the backfill pillars has been very successful. Undercutting of the B-NORTH Zone between July, 1987 and November, 1988 has resulted in measured backfill stresses of 2.28 MPa, or 68 percent of full overburden weight. Low water content and high proportions of coarse aggregate allow in-place strengths of 5.66 MPa to be achieved, with cement contents between 5 and 6 percent.

## 1 INTRODUCTION

High-strength backfill is being employed in a sublevel bench-and-fill mining method to achieve high recovery of gold ores at the Cannon Mine. The mine, located in central Washington, USA, produces up to 1400 mtpd of ore, and is a joint venture of Asamera Minerals (U.S.), Inc. (the operator) and Breakwater Resources, Inc. of Vancouver.

The high-strength backfill is required because of constraints imposed by the location and geology of the ore zone. These constraints are listed below.

• The high grade of the ore made leaving pillars for ground support undesirable.
• The location of the ore zone at the outskirts of the city of Wenatchee, WA and a lack of control of surface rights required that the mining produce very little subsidence and that backfilling assure the long-term stability of the surface topography.
• The B-NORTH ore zone is shallow, wide, and thick. The large width-to-depth ratios (up to 2.4) would require that the backfill carry a portion of the overburden weight over the long-term.
• Very weak rock strata in the overburden and a significant thickness of alluvium at the surface made it very probable that failure in the hanging wall materials above a large, unfilled void would have the potential of propagating to the surface.
• The relatively complex shape of the deposit required multilift mining to maximize recovery and limit the number of remnant ore blocks.

The design and emplaced backfill strengths are similar to the highest reported in published literature; however, the cement content and the in-place cost of the backfill are quite low. The use of a low water content backfill with clean, well-graded aggregate maximizes the strength produced per unit of cement.

This paper discusses the design of backfill materials and the structural aspects of the backfilling scheme. Design of the backfill was based upon analysis of backfill loading under the worst case of full overburden support. Further analysis was performed to project performance of the backfill during pillar mining and to compare these projections to field data collected during the backfilling operations. Finally, deformation and stress change data collected during pillar mining is presented to compare the design basis to actual response of the backfill to full pillar recovery.

## 2  BACKGROUND

### 2.1  Geology

The Cannon Mine is currently extracting ore from several mineable zones within silicified arkosic sediments that have been faulted and folded. The B-NORTH Zone was the first mined and is the subject of this article. Silicification produced strong ore, surrounded by weak unsilicified sediments. Major structural disturbance has apparently broken the silicification into the individual zones, separating them by thick, weak shear zones of up to 12 to 30 m thick.

The shape of the B-NORTH Zone is illustrated by the isometric diagram of mineable blocks in Figure 1. The silicified sediments trend north-south and plunge south. Surface topography is relatively flat with the overburden depth increasing from 61 m in the north to 152 m in the south. Shear zones/faults form the east and west boundaries of the ore and sheared claystone beds occur in the immediate hanging wall. Ott, et al. (1986) present a detailed geologic description of the B-NORTH Zone.

### 2.2  Mining Method

The mining method, sublevel bench-and-fill, has been described by Kelly (1986) and is similar to systems employed at the Kerreti Mine in Finland and the Tara Mine in Ireland, reported by Koskela, et. al (1983) and Oram (1985), respectively. The ore is divided into large sublevel blocks, as illustrated in the isometric diagram of Figure 1. Stopes are typically 7.3 m wide and 19.8 m high, however, some stopes in the northern area were mined 29 m high. Figure 2 illustrates the recovery of benches between the drilling and mucking developments at nominally 15 m high levels. Three vertical lifts are common with a maximum of four lifts. Development drifts were driven the full width of the stope to facilitate the bench blasting. Rock mechanics aspects of the primary mining design were described by Brechtel, et. al (1987).

The system employed a vertical mining sequence where primary stopes were mined locally up to the hanging wall. Cemented backfill was hauled to the stopes by trucks with telescoping beds. The fill was dumped into the stope from the drilling level access and the low slump backfill allowed the trucks to drive over the freshly placed fill as it advanced towards the stope end. The cemented fill floor became the mucking level for the next overhand lift. A detailed description of the backfill system was presented by Baz-Dresch (1987).

The backfill was tightly packed at the hanging wall to minimize the vertical displacement prior to development of overburden support. Cemented fill was placed in the last 3 m high drill level using LHD's. Tight packing was achieved using a small, 1 m$^2$ ramming plate attached to an LHD bucket by 2 m of structural steel. This technique has been very successful. Visual examination of many stopes suggested that an interface zone between the hanging wall roof and backfill was achieved where backfill touched the roof in some areas, and was generally within 5 to 7 cm of the roof in others.

Problems in backfill placement have occurred due to segregation of coarse aggregate materials as the fill flowed down the slope which generated zones of loose cobbles at the toe of the pile and along the stope ribs. These areas have caused some minor problems during pillar recovery, and had the net effect of reducing the pillar load bearing area. Maximum aggregate size was reduced to six cm to help reduce the segregation.

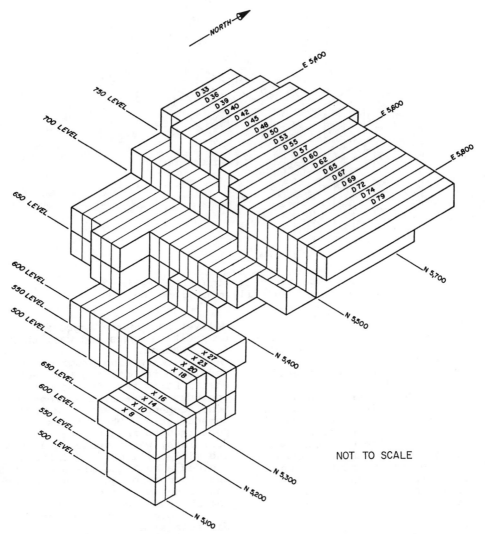

NORTH

E 5,400
E 5,600
E 5,800

750 LEVEL
700 LEVEL
650 LEVEL
600 LEVEL
550 LEVEL
500 LEVEL

650 LEVEL
600 LEVEL
550 LEVEL
500 LEVEL

D 33
D 36
D 39
D 40
D 42
D 45
D 48
D 50
D 53
D 55
D 57
D 60
D 62
D 65
D 67
D 69
D 72
D 74
D 79

X 27
X 23
X 20
X 18
X 16
X 14
X 10
X 8

N 5,700
N 5,500
N 5,400
N 5,300
N 5,200
N 5,100

NOT TO SCALE

Figure 1. Isometric drawing of mineable blocks in the B-NORTH Zone
(coordinates and level elevations are in feet, although not to scale).

## 3  BACKFILL DESIGN

### 3.1  Strength Specification

The specified laboratory strength
of the cemented backfill of 8.3 MPa
was based upon the following design
requirements.
• The backfill would provide full
  overburden support to assure
  long-term prevention of surface
  subsidence.
• Backfill pillars must support the
  full overburden weight, the
  weight of undercut blocks of ore
  suspended between cemented fill

pillars, and the weight of uncemented backfill in adjacent secondary stopes during pillar
recovery.
• The in-place backfill pillar
  strength should be 5.8 MPa.  The
  difference in laboratory versus
  in-place strength reflects a
  factor of safety to account for
  the impact of segregation and
  size effects in the field.
The integration of various criteria and constraints in the operation resulted in the life of the
mine being strongly divided into
phases where, first, primary mining

107

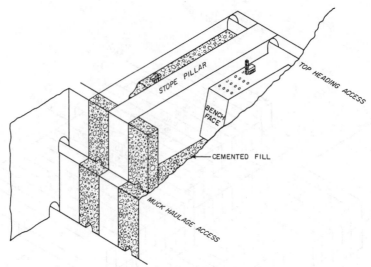

Figure 2. Isometric drawing illustrating the
sublevel bench-and-fill mining method.

would dominate then, later, pillar
or secondary mining would dominate.
Production rate requirements and
structural constraints made it
impossible to develop a retreating
schedule of pillar recovery.
Because of this, some secondary
stopes would be mined next to cemented backfill pillars that would
be supporting portions of overburden weight. The uncertainty of
the pillar mining schedule and the
potential of mining next to open
backfill highwalls carrying overburden loads were prime factors in
the specification of the high
strength.

## 3.2 Backfill Mixture Design

Design of the backfill mixture was
influenced by the availability of
bulk materials, the requirement for
relatively high strength, and
requirements to minimize cost. To
meet the strength requirement at
minimum cost, the bulk materials
had to be coarse and fine aggregates typical of concrete mixtures
so that the proportion of cement
could be minimized. A clean sand
resource was available on-site,
however, the costs to produce suitable coarse aggregates from on-site
materials were high. An off-site
commercial aggregate source could

deliver the coarse aggregate for
competitive costs and was, therefore, chosen.
A mixture testing program was
developed to optimize the backfill
cost. Parameters investigated
included coarse/fine aggregate content, cement content, and water:
cement ratio. Other tests were
performed to investigate the use of
pozzolans, water reducing agents
and mill tailings. Optimization
yielded a backfill mixture with 55
percent coarse aggregate, 40 percent fine aggregate, 5 to 6 percent
cement, and a water:cement ratio of
1:1. Table 1 lists the aggregate
particle size.
Minimization of materials cost was
most sensitive to the cement content which could be reduced by
increasing the amount of coarse
aggregate in the mix. The strength
was very sensitive to both cement
and coarse aggregate content. Figure 3 shows the results of the mixture tests and illustrates the
sensitivity of strength to coarse
aggregate. The mixture proportions
were selected on the basis of these
tests.

## 3.3 Backfill Quality Control

Quality control testing was performed to assure that backfill
delivered to the stopes met the

Table 1. Backfill aggregate description.

| Grain Size (mm) | Percent Passing | |
| | Coarse Aggregate | Fine Aggregate |
| --- | --- | --- |
| 75 | 100 | 100 |
| 50 | 90 | 100 |
| 35 | 70 | 100 |
| 25 | 40 | 100 |
| 19 | 20 | 100 |
| 13 | 10 | 100 |
| 9.5 | 4 | 90 |
| 4.8 | 1 | 87 |
| 2 | 0 | 74 |
| 1 | 0 | 48 |
| 0.5 | 0 | 20 |
| 0.1 | 0 | 1 |
| 0.74 | 0 | 1 |

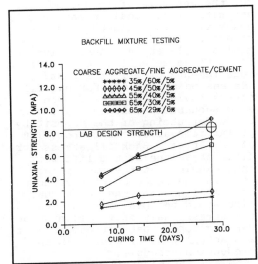

Figure 3. Effect of coarse aggregate content on backfill strength.

design specification. The fine aggregate was "pit run" material to minimize handling cost. The commercial coarse aggregate was screened to remove all +70 mm and all -5 mm material. Periodic size analysis was performed to verify that the materials conformed to the general requirements. Compressive strength samples were also prepared from backfill in the delivery trucks on route to the stopes.

These strength tests were performed to verify that the mixture would meet the lab strength specification on a production basis.
Compression test samples were placed in a hot water bath for 48 hours after preparation to accelerate the rate of curing. A linear least-squares correlation, developed between the 28-day strength and the accelerated cure strength, indicated that the 28-day strength would be approximately 1.25 times the accelerated cure with a regression coefficient of 0.91. Test data developed for backfill going into 28 lifts of primary stopes, indicated that the mean laboratory strength for individual stopes ranged between 2.48 and 15.01 MPa, and averaged 8.45 MPa. Variation of the stope means is illustrated by the frequency histograms in Figure 4. When scaled for the accelerated cure, only 21 percent of the stopes had mean test strengths below the laboratory design strength of 8.3 MPa and only 10 percent of the stopes had test strengths with less than 85 percent of the design strength.

3.4  In-Stope Measurements

Core samples were taken from several stopes in an attempt to characterize the in-place strength of the cemented backfill. The samples, nominally 15 cm in diameter, were collected from two different stopes where backfill had been in place for more than 30 days. Strength of the samples varied from 23.44 MPa to 2.05 MPa. The mean strength, excluding two extremely high values, was 5.66 MPa with a standard deviation of 2.00 MPa. Figure 5 shows frequency histograms illustrating the variation in strength for the in-stope sampling.
Segregation caused zones of structural weakness within the backfill, and therefore, the mass strength of the backfill pillars was unknown. The coring revealed planar zones of inhomogeneous backfill corresponding to interruptions of the fill cycle where the slope face was allowed to harden. Segregation along these surfaces produced weak zones with little cementing, and coring through these cobble zones

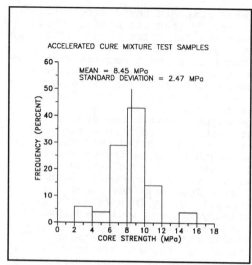

Figure 4. Variation of stope mean test strengths from accelerated cure samples.

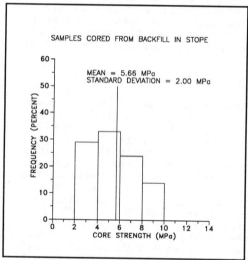

Figure 5. Variation of core sample strength taken from fully cured backfill in two stopes.

was difficult. The cores tested represent the strongest zones of the backfill.

## 4  DESIGN ANALYSIS

The backfill quality control sampling and in-stope sampling had indicated that a consistently high-

strength mixture had been manufactured by the backfill plant and that in-place strengths were consistent with the desired range of strength. However, since 50 percent of the test samples cored from the stopes did not meet the design criteria of 5.8 MPa and the mass strength was not known, more detailed analyses were undertaken to assess whether the strength being achieved in-stope would be adequate. The analysis was conducted using a three-dimensional, boundary element computer model.

The boundary element computer code was used to model the buildup of stress in cemented backfill pillars. This particular code, EXPAREA, is based upon a formulation by Starfield and Crouch (1972), which considers only the stress in the vertical direction and calculates transfer of overburden loads in response to changing extraction. EXPAREA has been employed primarily for the analysis of seam-type deposits (e.g., coal, trona, etc.); however, it was used to simulate stress change due to the primary mining cycle at the Cannon Mine and found to agree well with overcoring measurements of pillar stress.

Extension of its use for backfill stress analysis was based upon a 24 step sequence developed to simulate the undercutting of the deposit by pillar removal at the bottom lift of any stope. Backfill stress for lifts above the bottom lift were then examined at the appropriate step in the sequence corresponding to the point when the individual lift would be open. In this way, the stress analysis for each stope was developed that corresponded to its actual time of mining in the production schedule.

The simulation is illustrated in Figure 6, which shows the mesh corresponding to the mine at the end of primary mining. The mine layout is simulated by 3.7 m square elements identified by an integer code representing the material type. Types 0, 2, 3, 4, 5, 6, and 7 correspond to rock pillars with an elastic modulus of 6138 MPa. Blank spaces correspond to the mined primary stopes. The mined elements are changed to backfill in Step 2 of the simulation. Any vertical

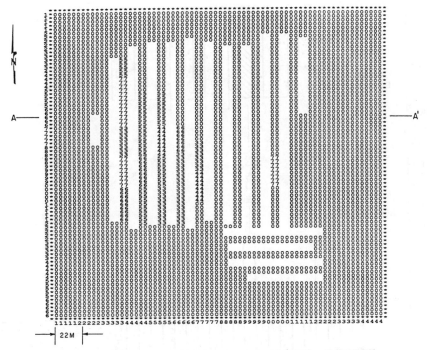

Figure 6. Boundary element mesh simulating B-NORTH mining at the end of primary production.

displacement that occurred in sub-sequent steps of the simulation was elastically coupled to the backfill.

A series of materials properties assumptions were made that would produce a worst-case simulation of backfill loading in the 24 step simulation. Analyses were per-formed at full pillar extraction to establish an overburden modulus that would produce near-full over-burden loads in the pillars. An elastic modulus of 345 MPa produced 92 percent of full overburden weight on the backfill at full pil-lar extraction. Rock pillars were assigned a value of 6138 MPa based upon adjustments of lab data for rock quality and pillar dimensions. Backfill was assumed to be linear elastic with a modulus of 1862 MPa, based upon the results of the mix-ture design tests.

The objective of the analysis was to estimate stresses in the back-fill pillars during the point in the production sequence where each secondary stope would be open and stability of the backfill highwall would be most important. Results

of the stress analysis are shown for step 15 in Figure 7 to illus-trate transfer of stress to the backfill and remaining rock pillars. Locations of individual stopes can be derived from Figure 1. Figure 7 also illustrates the effect of two steps in the mine planning taken to increase the structural integrity of the mine during pillar removal.

• Mining of pillar D57 was delayed until late in the sequence to leave a stiff abutment in the center of the B-NORTH Zone until the late stages of pillar mining. This was designed to minimize overburden deformation as long as possible, thereby reducing back-fill stresses in many stopes.

• Cemented backfill was placed in the secondary stope D62 to create a triple backfill pillar in the middle of the B-NORTH Zone. The triple pillar would separate the ore zone into two panels to iso-late any potential problems. Each panel would have separate access so that operations could be conducted independently, if necessary.

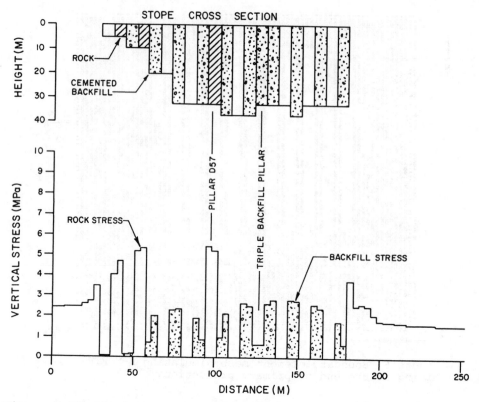

Figure 7. Vertical stress levels and stope cross section for
Step 15 of the 24-step mining simulation.

Analysis of backfill pillar stability was performed by calculating
the probability of failure assuming
the strength was given by the tests
on cored backfill materials. For
each secondary stope mined, the
average pillar vertical stress was
calculated at the point in the 24
step simulation when each lift of
each stope would be open. The
stress calculated by the model only
considered the overburden loading.
Loads due to undercut rock, self
weight, and adjacent uncemented
backfill were calculated based upon
the mining geometry and added to
the stress calculated by the model.
Both stress and strength were
assumed to be normally distributed
with standard deviations of 10 and
36 percent of the value, respectively. The standard deviation for
the stresses was assumed, but the
value used for the strength corresponds to that measured on samples
taken from the cemented fill

pillars. Using this approach, the
probability of failure can be calculated for the stress level determined in the model and the assumed
strength of 5.66 MPa. Figure 8
shows the probability of failure
versus time from the beginning of
pillar mining in July, 1987. The
figure indicates that probability
of failure begins relatively low
and increases with time as the
amount of undercutting produces
higher stress levels. Considering
the level of uncertainty in mass
strength of the backfill, the six
stopes with probabilities of failure greater than 20 percent were
considered to have a risk
sufficient to require close observation during mining.

This type of analysis provided
valuable feedback to the production
staff. First, it identified the
stopes and time within the production schedule where the potential
for problems would be greatest.

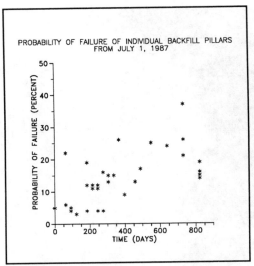

PROBABILITY OF FAILURE OF INDIVIDUAL BACKFILL PILLARS
FROM JULY 1, 1987

Figure 8. Probability of failure in individual stopes versus scheduled time of mining.

Second, the risk due to stress levels could be compared to the gold resource tied up in each stope. Risk due to the projected backfill loading could be managed by changing the production schedule for particularly rich stopes.

## 5 BACKFILL PERFORMANCE

### 5.1 Pillar Mining From the B-NORTH

During the period of July, 1987 through November, 1988, virtually the entire B-NORTH Zone was undercut by pillar mining. Production from the secondary stopes has been successful and the high-strength backfill and conditions of the rock in the pillars has allowed the pillar mining to proceed with relatively minor problems. Figure 9 presents a photograph of backfill walls in stope D48, illustrating the stability of the backfill.

The single major failure of backfill that occurred in the mine was due to placement of a large block of backfill without cement. Upon developing the drilling level in the pillar, this fill flowed out into the drift over an approximately 15 m length. The zone was bulkheaded off and high-strength grout pumped in to stabilize the

area before extraction of the bench.

### 5.2 Field Measurements of Backfill Loading

Overburden loading of the backfill began to occur in the northern portion of the B-NORTH Zone roughly one year after the beginning of pillar removal operations. Instrumentation data, developed on a cross section at approximately 5,700 N in Figure 1, is presented to show the buildup of vertical stress in rock pillar D57 and then the transfer of loads to backfill pillars D50 and D55, as pillar D57 was overcut to relieve the stress. The relative locations of each stope are shown in Figure 1. Concentration of overburden weight in pillar D57 is illustrated in Figure 10. Load redistribution occurred slowly as pillars were removed throughout the B-NORTH, but then steeply accelerated 350 days after pillar production began when the remaining pillars around D57 were undercut. Destressing by overcutting D57 was initiated after 425 days because stresses were reaching levels that might make pillar development dangerous. Figure 11 shows the record of backfill stress in pillar D50 for the same period of time. The flatjack indicated four cycles of loading/unloading between day 150 and day 350 that are interpreted as overburden settling on the top of the backfill pillar, followed by failure of the point contacts at the interface zone which relaxed the loads. By day 350, the interface zone had compacted enough to mobilize the elastic response of the backfill pillar and a rapid buildup of backfill vertical stress occurred as pillars around D57 were undercut, and then as D57 was overcut. D50 was at the center of the undercut area between the west abutment and pillar D57 and was, therefore, the first backfill to support overburden. With the destressing of D57, the undercut width increased from 51 to 110 m, producing a change in width-to-depth ratio from 0.73 to 1.57. Backfill stress reached 2.28 MPa, or approximately 68 percent of the full overburden support (considering the area of pillars).

113

Figure 9. Photograph showing backfill walls in secondary stope D48.

The backfill in pillar D55 also developed high stresses in response to the removal of pillar D57. Figure 12 shows backfill vertical strain in pillar D55 and indicates a gradual compressive strain over the first 425 days, and then an abrupt change in rate that corresponds to the destressing of pillar D57. The gradual compression is interpreted as compaction settling because a corresponding backfill stress gage indicated no buildup in vertical stress during the compressive strain. The stress gage later malfunctioned, so there is no parallel record of stress buildup when the strain rate increased.

The strain record in Figure 12 indicates a compressive strain of 0.0023 mm/mm before the strain rate begins to stabilize. Since D55 was right next to pillar D57, it is clear that the rock pillar sheltered the backfill in D55 until the destressing. Assuming that D55 is carrying the same portion of overburden loads as D50, the change in stress would be 2.54 MPa and the resulting elastic modulus of the backfill would be 1103 MPa. This is 59 percent of the lab testing value.

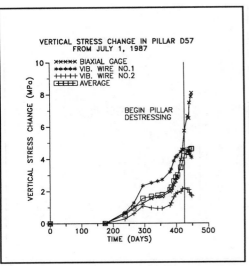

Figure 10. Vertical stress change in rock pillar D57 from the beginning of B-NORTH pillar mining.

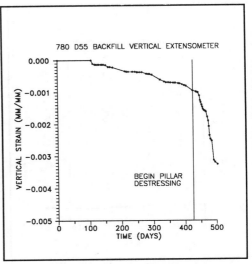

Figure 12. Vertical strain change in backfill pillar D55 from the beginning of B-NORTH pillar mining.

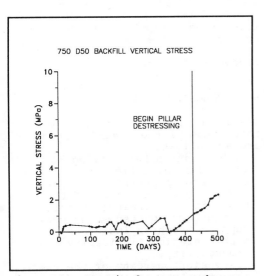

Figure 11. Vertical stress change in backfill pillar D50 from the beginning of B-NORTH pillar mining.

Stress and stain records for backfill pillar D45 are presented in Figures 13 and 14. D45 is one stope away from the western abutment, which has prevented the overburden from touching the backfill. D45 shows a substantial compressive strain (0.003) without a corresponding increase in stress. This behavior supports the inter-

pretation of long-term settling and compaction of the backfill assumed for D55.

Stress buildup in the western abutment zone is shown in Figure 15. Abutment stresses have increased gradually to the level of 2.21 MPa and have stabilized. This indicates that the overburden maintains the capability to transfer loads probably due to the limited vertical displacement allowed by the stiff backfill. These abutment loads are currently acting to limit backfill stresses to levels below full overburden.

5.3 Comparison of Design Approach to Field Results

Specification of the backfill strength based upon the assumption of full overburden support during pillar mining was a conservative approach, but has been justified by instrumentation data. The B-NORTH Zone is almost completely undercut, but many stope lifts remain to be extracted. Backfill pillar loads have reached significant proportions of the indicated strength. For example, the peak stress in the D55 backfill is currently estimated to be near 3.4 MPa, based upon instrumentation and self weight, or 60 percent of the average strength

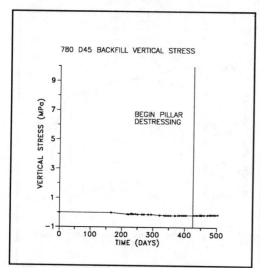

Figure 13. Vertical stress change in backfill pillar D45 from the beginning of B-NORTH pillar mining.

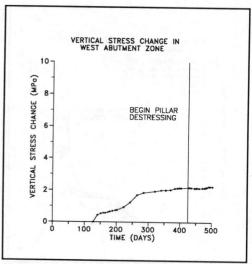

Figure 15. Change of vertical stress in the western abutment zone from the beginning of B-NORTH pillar mining.

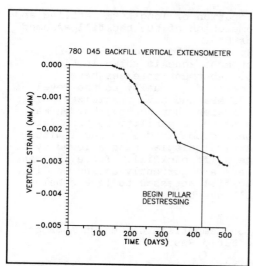

Figure 14. Vertical strain change in backfill pillar D45 from the beginning of B-NORTH pillar mining.

of the backfill cores. Portions of the stope to the south at greater depth will see higher overburden pressures. The overburden is currently transferring enough weight to the abutments to limit backfill stresses to 68 percent of the full overburden; however, further pillar removal to the east will increase the undercut width. The abutment

load transfer appears stable, but the overburden material is weak, and may allow increased backfill loading with time.

6 CONCLUSIONS

Major conclusions indicated at this stage of the B-NORTH mining are as follows.
• High-strength backfilling has allowed safe, efficient recovery of a large proportion of the B-NORTH ore. Benefits of the high strength include factors beyond the required structural support, such as increases in both operating flexibility and productivity.
• Sixty-eight percent of the overburden weight has been transferred into backfill pillars that are providing primary support during the continuing pillar recovery. Basing the design strength on full overburden support assures that the backfill pillars can carry this load during the ongoing pillar mining.
• Low-cost production of cemented backfills with the strength necessary for complete extraction of this zone is possible with proper materials selection. In-stope mass strength achieved in this

operation appears to be near the sample strengths.

ACKNOWLEDGEMENT

The authors wish to express their appreciation to Asamera Minerals (U.S.), Inc. for permission to publish this information. Some of the instrumentation used in this study was supplied by the U. S. Bureau of Mines, Spokane, Washington.

REFERENCES

Baz-Dresch, J. 1987. High strength backfilling at the Cannon mine. Proceedings, third western regional conference on precious metals, coal and environment. Rapid City, South Dakota: SME-AIME.

Brechtel, C. E., Agapito, J. F. T., Kelly, T. R., Karlson, R. C., and Gilbert, C. J. 1987. Mine design at the Cannon mine: Integration of operational planning and geomechanical design. First international conference on gold mining. Vancouver:SME.

Kelly, T. R. 1986. Stoping practices at the Cannon mine. Presented at the SME-AIME annual meeting. New Orleans, Lousiana:SME-AIME.

Koskela, V. A. 1983. Consolidated backfilling at Outokumpu Oy's Vihanti, Keretti and Vammala mines. Proceedings of the international symposium on mining with backfill. Rotterdam:Lulea-A. A. Balkema.

Oram, R. A. 1985. The Tara mines story. The mine operation, trans., Institution of Mining and Metallurgy. 94:A.

Ott, L. E., Groody, D., Follis, E. L. and Siems, P. L. 1986. Stratigraphy, structural geology, ore mineralogy and hydrothermal alteration at the Cannon mine, Chelan County, Washington, U.S.A. Toronto:Gold, 1986 Symposium - MacDonald, A. C., editor.

Starfield, A. M. and Crouch, S. L. 1972. Elastic analysis of single seam extraction, new horizons in rock mechanics. 14th U. S. Symposium on Rock Mechanics, Hardy, H. R. and Stefanko, R., editors.

*Innovations in Mining Backfill Technology, Hassani et al. (eds), © 1989 Balkema, Rotterdam. ISBN 90 6191 985 1*

# Instrumentation and modelling of the Cannon Mine's B-North ore body

D.R.Tesarik, J.B.Seymour & J.D.Vickery
*Spokane Research Center, US Bureau of Mines, Spokane, Wash., USA*

ABSTRACT: A cross section of the B-North ore body of the Cannon Mine was instrumented by the U.S. Bureau of Mines to monitor backfill and rock deformation during excavation. A two-dimensional, finite-element model was used to analyze the benching cut-and-fill mining method.

The instruments indicated that both the backfill pillars and rock abutments took load during secondary mining. The rate of loading in the backfill increased after one of the last secondary pillars was extracted.

A high correlation existed between relative displacements in the rock and relative displacements predicted by a two-dimensional, finite-element model, but there was no linear correlation between predicted relative displacements in the cemented fill and measured displacements in the fill.

## 1 INTRODUCTION

Located in central Washington in the eastern foothills of the Cascade Mountains, the Cannon Mine lies just outside the city limits of Wenatchee, an agricultural community with a population of about 45,000. Traditionally known for its fruit orchards, Wenatchee has recently become a major gold-producing region.

Although gold was first discovered in the Wenatchee district as early as 1885, the region's first successful mining operation did not begin until 1949. From 1949 until 1967, the Lovitt Mining Company produced over 12,767,800 g (410,500 oz) of gold and 19,464,300 g (625,800 oz) of silver (Baz-Dresch, 1987). Before the discovery of the new B-North ore zone 1.3 km (0.8 m) to the northwest, Asamera Minerals formed a joint venture with Breakwater Resources and constructed the facilities for the Cannon Mine. In July of 1985, they began processing ore from the B reef deposit. With the identification of potential ore reserves, the Cannon Mine is now regarded as one of the most significant recent gold discoveries in North America and is the second larg-

est underground gold mine in the United States (Argall, 1988).

## 2 GEOLOGY

In the Wenatchee district, gold occurs in a series of silicified arkosic sedimentary deposits known as reefs. The B reef complex of the Cannon Mine contains several distinct ore bodies that together form an elongate ore zone approximately 550 m (1,800 ft) in length. As the managing partner in the Cannon Mine operation, Asamera first began mining one of the larger, higher-grade ore bodies, the B-North. Containing 3,631,200 mt (4,000,000 st) of ore at an average grade of 7.47 g/mt (0.239 oz/st), this tabular ore body has a maximum width of 122 m (400 ft), a maximum length of 244 m (800 ft), an average thickness of 40 m (130 ft), and lies about 61-122 m (200-400 ft) beneath the surface.

Most of the B-North ore body lies within a repetitive sequence of interbedded feldspathic sandstone, siltstone, and claystone that has been folded, faulted, and intruded by a hornblende-andesite dike to the west and a biotite-rhyodacite porphyry stock to the east. A zone of extensively sheared sedimentary beds separate the ore body from the adjacent rhyodacite intrusion, while deformed claystone or mudstone beds delineate the ore body's upper and lower limits. Ground conditions within the ore body are generally good except in the vicinity of these sheared or deformed sediments. The rhyodacite intrusion served as a heat source and focusing mechanism for hydrothermal solutions that deposited gold and silver (primarily free gold, electrum, pyrargyrite, acanthite, and auriferous pyrite) throughout the fractured sandstone sediments.

Gold is disseminated within the silicified portions of the sandstone host rock in localized stocks, hydrothermal breccias, and widely spaced veins of quartz, adularia, and calcite. Although the gold is predominantly very fine-grained, visible coarse-grained gold can occasionally be found in high-grade veinlets. Silver accompanies the gold at a fairly constant 2:1 ratio throughout the ore body.

## 3 MINE FACILITIES

The mine's portal, ramp, shaft, and main access drifts were developed in the competent rhyodacite porphyry footwall to the immediate east of the ore body. A concrete-lined shaft, 5.5 m (18 ft) in diameter and 189 m (620 ft) in length, is used to hoist crushed ore out of the mine and serves as an exhaust path for the ventilation system. Inclined at a 15-pct slope, a 4.6- by 4.6-m (15- by 15-ft) ramp provides access to production levels and serves as the primary ventilation intake. Horizontal crosscuts lead from the main ramp to the ore body. Crushed ore from an underground primary crusher is transported by conveyor to a storage bin and loading pocket at the shaft, where it is hoisted to the surface in two 6.4-mt (7-st) skips by a drum hoist and then processed through a flotation mill. Currently operating at a 90-pct recovery rate, the mill processes 1,362 mt (1,500 st) of ore per day producing approximately 64 mt (70 st) of concentrate, which is shipped out for additional smelting and refining.

## 4 MINING METHOD

Because the B-North ore body lies only 61 m (200 ft) beneath the stables and

arena of a local riding club, a cut-and-fill mining method utilizing cemented backfill was chosen to minimize surface subsidence and yet enable a high percentage of the ore body to be recovered. As in slot-and-pillar mining, stopes are arranged in parallel panels across the strike of the ore body in 7.3-m (24-ft) widths and then mined and filled in an alternating sequence. Primary stopes are excavated and backfilled before secondary mining begins; therefore, no two adjacent stopes are mined at the same time. Stopes are excavated in 15-m (50-ft) vertical intervals by driving an upper and a lower sill cut 4.6 m (15 ft) high by 7.3 m (24 ft) wide from access drifts usually located in the rhyodacite footwall. After these sublevel headings are driven the length of the stope, a drop raise or slot is excavated at the end of the stope connecting the two levels. The resulting stope block is then benched toward the access drift on the upper level while blasted ore is removed on the lower level (fig. 1). Blasted ore is loaded onto 23.6 mt (26 st) diesel trucks using load-haul-dump (LHD) equipment. Remote-controlled loaders are used when mucking is beyond the brow of the stope.

After benching the entire stope block, fill fences are erected using either timber barricades or 5.1-cm (2-in) chain link fencing and cable slings to prevent the backfill from entering access drifts on the lower levels of the open stope. Another method of blocking off access to the stope after mining eliminates the use of fences. The fill is allowed to settle at its angle of repose, and the mucking access is cleared with a mucking bucket 6-8 hr after the fill is dumped. Cemented fill is dumped into the open stope from the upper heading using Dux DTD-30 Teledumper trucks. Equipped with a unique telescoping dump bed, the trucks can end dump their entire load from within the restricted 4.5-m (15-ft) height of the upper sill cut. The advancing fill pile is leveled with Caterpillar 930 loaders to establish a working platform from which the trucks dump backfill until the excavated stope is completely filled to the floor of the upper level. This sublevel then serves as the mucking level for the next vertical stoping interval, and the mining sequence is repeated. When the top of the ore block is reached, cemented fill is rammed tight to the back of the top sill cut using a 1.2-m$^2$ (4-ft$^2$) plate mounted on a beam bolted in the bucket of an LHD. Placing the ce-

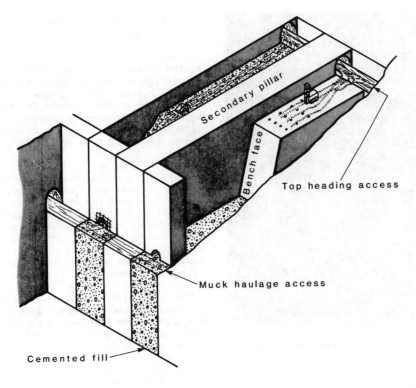

Figure 1.  Benching cut-and-fill mining method (after Brechtel et al., 1987).

mented fill tight to the back of the
top sill cut minimizes subsidence by
enabling the backfill pillar to provide
roof support as soon as secondary mining
commences.

After mining and backfilling the primary stopes in 15-m (50-ft) vertical
intervals, the secondary pillars are
extracted using a similar procedure,
except that blasted ore is removed from
between the cemented-fill pillars using
remote-controlled equipment.  Depending
on the ground conditions and mining
plans, the secondary stopes are backfilled with either cemented fill or waste
rock.  Completed stopes range in height
from 9-40 m (30-130 ft), depending on
their location in the ore zone.

Consisting of approximately 55-pct
coarse aggregate - minus-5.1-cm (2-in)
river gravel, 40-pct alluvial sand, 5-pct
cement, and a water-reducing agent, cemented fill for Asamera's overhand sublevel bench stopes is mixed underground
using an automatic batching system operated by a programmable controller.  The

proper amount of each component is blended together in a double-screwpug mill and
discharged into a loading bin from which
the backfill trucks are filled.

Laboratory tests performed with cylindrical backfill samples taken on a daily
basis yielded an unconfined compressive
strength of between 6.9-8.3 MPa (1,000-
1,200 psi).  Although some of the aggregate separated from the mix as the backfill was dumped into the open stope, only
very localized stability problems have
been observed.

5  INSTRUMENTATION

The mine implemented a geotechnical instrumentation program to monitor areas
that could have stability problems.
Digital vibrating-wire remote-reading
instruments were used because of the high
potential for cable cuts by large equipment.  Because the mine staff would be
monitoring all instruments, the Bureau
also chose to use vibrating wire gauges.

121

Figure 2. Vertical backfill extensometer.

Several types of instruments were selected to determine the response of backfill and mine rock throughout the mining
sequence. The response can be characterized by either displacement or stress.
Displacement information can be readily
achieved by the use of extensometers.
Several techniques have been developed to
measure changes in stress. All are derived from displacement measurements.

Commercially available joint meters
were modified and cast in place to measure transverse displacements of cemented
backfill pillars. Vertical extensometers
were constructed of steel pipe in the
stope during filling. The steel pipe was
encased in loosely joined plastic pipe to
prevent the pipe from adhering to the
backfill (fig. 2). A slip-joint allowed
the vibrating wire transducer to measure
displacements between the bottom and top
plate. Embedment strain gauges were
spaced along the height of the vertical
fill extensometers to determine if there
were any differential displacements.

Multiple-position borehole extensometers were used to measure horizontal
and vertical displacements in the secondary rock pillars and the vertical displacements occurring in the mine roof.

Hydraulic anchors were used to facilitate
installation.

Changes in rock stress and orientation
of the principal stresses in the plane
perpendicular to borehole axes were measured by biaxial stress gauges. Radial
deformation of the borehole is transferred to the instrument's three vibrating-
wire gauges, which span the diameter 60°
apart, through a thin annulus of grout.

Flatjack-type earth pressure cells were
cast in place within cemented backfill
pillars to monitor vertical stress. Some
were placed near the mine roof to provide
an indication as to when the backfill
pillar began taking overburden load.
These instruments were included mainly as
an indicator of changing conditions because there was some degree of skepticism
as to their accuracy.

The North 5650 transverse cross section
(fig. 3) was chosen for the location of
the instruments because the mining width/
depth ratio was greater here than at any
other location in the B-North ore body.
This area also had fewer development entries and shear structures than other
areas of the mine. Secondary rock pillars, cemented primary pillars, the back
and the west abutment were instrumented.
Table 1 summarizes instrument type, location, and anchor lengths when applicable.

6  DATA ACQUISITION AND RESULTS

The plots of the reduced data for selected instruments are included in figures 4
through 9. Displacements due to tension
and tensile stress are positive. A brief
summary of the behavior of selected instruments follows.

The biaxial stressmeter in the west
abutment showed increasing compressive
stress during excavation of secondary
headings at the 780 level in D43, D48,
and D53 (fig. 4). After this, overburden
stresses in the rock pillars had been
relieved from D41 through D55 and yet
compressive stresses recorded by the
instrument increased during benching.
Similarly, the stressmeter in pillar D57
responded to the mining of the headings
on the west side of the ore body (fig.
5). However, at day 222 stress increased
at a more rapid rate even though there
was no apparent associated mining activity. The stress increase from day 280 to
day 305 can be attributed to excavation

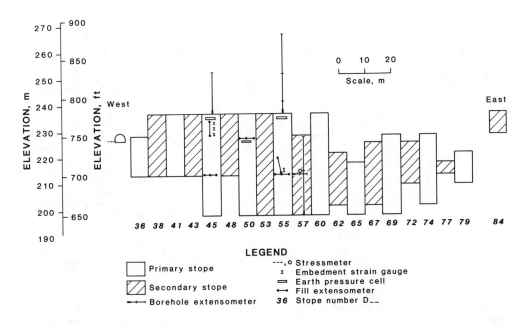

Figure 3.  Transverse cross section of the B–North ore body at North 5650.

Table 1.  Instrument description and location.

| Instrument | Location | | Dip, deg | Anchor depth, m (ft) |
| | Stope | Level | | |
|---|---|---|---|---|
| Stressmeter............ | West rib | 740 | 90 | 2.8 (9.3) |
| Do.................... | D57 | 700 | 90 | 16.8  (55) |
| Embedment strain gauge.. | D45 | 755 | 0 | NAp |
| Do.................... | D55 | 700 | 0 | NAp |
| Borehole extensometer... | D45 | 780 | 0 | 6.2 (20.2), 15.1 (49.7) |
| Do.................... | D55 | 780 | 0 | 4.4 (14.6), 10.5 (34.6), 19.7 (64.6) |
| Do.................... | D57 | 700 | 90 | 3.0 (10), 6.1 (20) |
| Do.................... | D57 | 650 | 0 | 16.5 (54), 30.4 (99.6) |
| Backfill extensometer... | D45 | 755 | 0 | 5.3 (17.4) |
| Do.................... | D45 | 700 | 90 | 5.1 (16.6) |
| Do.................... | D50 | 750 | 90 | 5.1 (16.6) |
| Do.................... | D55 | 700 | 90 | 5.1 (16.6) |
| Do.................... | D55 | 700 | 56 | 8.5 (27.8) |
| Earth pressure cell..... | D45 | 780 | 90 | NAp |
| Do.................... | D50 | 740 | 90 | NAp |
| Do.................... | D55 | 780 | 90 | NAp |

NAp  Not applicable.

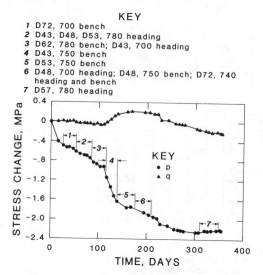

Figure 4. Plot of major (p) and minor (q) principal stresses in west abutment.

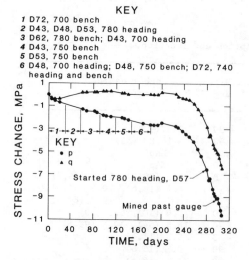

Figure 5. Plot of major (p) and minor (q) principal stresses in pillar D57.

of the 780 heading in pillar D57. The stressmeter in the west abutment did not record any stress change when this heading was mined. The directions of the principal stress planes for both stressmeters was 75–80° counterclockwise from the vertical.

Both horizontal fill extensometers in stope D45 showed initial displacements

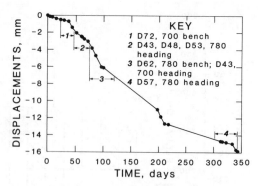

Figure 6. Plot for vertical backfill extensometer in stope D45.

of approximately 0.76 mm (0.03 in) in the compressive direction. These measurements are within the range that can be accounted for by cemented backfill shrinkage. The east extensometer recorded a displacement change of -0.46 mm (-0.018 in) when the 780 headings in pillars D43 and D53 were driven. An expansion of 0.66 mm (0.026 in) was recorded when the adjacent pillar (D48) was benched, removing the confinement on the east side of D45. No significant displacements were recorded after this point, even after the removal of the 780 heading of pillar D57.

The 5.3-m (17.4-ft) vertical fill extensometer on the 755 level, D45, showed a total displacement of -15.7 mm (-0.62 in) since installation (fig. 6). The rate of deformation was essentially constant. The maximum strain shown by the extensometer was in the range of the maximum strain measured by the three adjacent strain gauges (-700 to -1,100 microstrain). The range of the smaller strain gauges was exceeded after approximately 100 days. The lower and middle gauges showed a maximum strain of approximately -1,100 microstrain while the upper gauge showed a maximum strain of approximately -700 microstrain.

Both anchors of the two-point vertical borehole extensometer installed in the back of stope D45 recorded tensional displacements. The shorter anchor recorded a value of approximately 1.5 mm (0.06 in) after 328 days. The longer anchor had a value of 2.9 mm (0.115 in) after 191 days and then dropped to 0.89 mm (0.035 in). Because the displacements recorded by the shorter anchor continued to increase

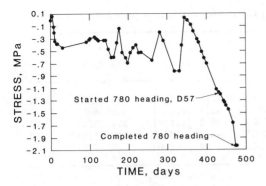

Figure 7. Plot for earth pressure cell in stope D50.

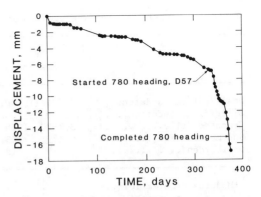

Figure 8. Plot for vertical backfill extensometer in stope D55.

after 191 days, the deeper anchor probably slipped in the borehole.

As shown in figure 7, the earth pressure cell in stope D50 had an increase in pressure of about -413.7 kPa (-60 psi) almost immediately after installation. This was most likely caused by the weight of the overlying cemented backfill. Readings oscillated from 0 to -827 kPa (0 to -120 psi) until 320 days after installation, after which the load increased to -2.1 MPa (-300 psi) 475 days following installation. Some of this load may have been caused by the removal of the 780 heading in pillar D57, which was one of the few remaining secondary pillars. The stress level at this time represented approximately 53 pct of full overburden load if it is assumed that the rock-filled stopes carried no load. The trend of the plot indicates that more load will be transferred to the cement pillars.

There was a noticeable correspondence between mining of the 780 sill cut of pillar D57 and the displacement recorded by the vertical backfill extensometer at the 700 level of stope D55 (fig. 8). The total displacement before removal of the 780 heading was -6.86 mm (-0.27 in), and the displacement measured during the period that the heading was removed was -10.4 mm (-0.41 in). An in situ backfill modulus of 1386 MPa (201,072 psi) can be calculated using stress changes from stope D50. The embedment strain gauge installed at the base of the vertical extensometer had a total microstrain of -1,000 while the larger extensometer had a microstrain change of approximately -200 during the same time period.

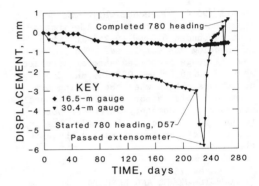

Figure 9. Plot for vertical borehole extensometer in pillar D57.

The earth pressure cell at the 780 level recorded approximately -69 kPa (-10 psi). This load reflected the 4.0-4.6 m (13-15 ft) of cemented fill above it and indicated that no additional load caused by mining was present. This instrument failed before the 780 heading in the adjacent pillar (D57) was extracted.

The three-point borehole extensometer in the back of stope D55 showed increasing displacements over time. The maximum displacement was 3.8 mm (0.15 in) and was recorded by the shortest anchor. The other two anchors were installed in unsilicified claystone and must have slipped down the borehole. This instrument quit working 210 days after installation.

Figure 9 shows the vertical borehole extensometer in D57 responding to cutting of the 780 heading with a displacement increase of -2.79 mm (-0.11 in) for the

Table 2. Material properties used in analysis based on average laboratory values.

| Material | $E^1$, MPa (psi) | $C^2$, MPa (psi) | $T^3$, MPa (psi) | $v^4$ |
|---|---|---|---|---|
| Unsilicified sandstone.. | 11,722 (1,700,000) | 25.4 (3,690) | 2.5 (369) | 0.25 |
| Silicified sandstone.... | 17,238 (2,500,000) | 45.6 (6,610) | 4.6 (661) | .27 |
| Cemented fill.......... | 3,792 (550,000) | 8.3 (1,200) | 2.1 (300) | .30 |
| Uncemented fill........ | 159 (23,000) | 4.1 (600) | 1.0 (150) | .30 |
| Overburden............. | 19 (2,818) | .7 (100) | .03 (5) | .30 |

[1]Modulus of elasticity.
[2]Uniaxial compressive strength.
[3]Tensile strength.
[4]Poisson's ratio.

30.4-m (99.6-ft) anchor while the heading approached the instrument. The instrument rebounded 6.5 mm (0.256 in) when the heading was completed. The 16.5-m (54-ft) gauge recorded a total displacement of −0.56 mm (−0.022 in) and did not show a change when the heading was blasted. This anchor and the horizontal borehole extensometer, which did not respond to the mining of the 780 heading either, were at approximately the same elevation.

7 COMPUTER MODEL AND RESULTS

The finite-element program used to model the cross section of the mine was UTAH2 (Pariseau, 1978). This program can be used for strictly elastic or elastic-perfectly-plastic analyses of plane strain or plane stress. The yield criterion is Drucker-Prager, where strength is dependent on all three principal stresses and associate flow rules are applied for determining strains in yielded elements.

The mining sequence modeled by UTAH2 represented actual mining and filling of the entire pillar or stope at cross section North 5650, where the instruments were located. Sixteen cuts and 14 filling steps were required to model the mining sequence from the driving of the first headings to October 24, 1988. At this point, only pillars D38 and D77 remained. All the other stopes except D57 and D67 had been filled.

The finite-element mesh had 4,443 elements and 4,286 nodes. The boundaries of the mesh were two ore body diameters away from the mine openings to eliminate the

influence of boundary conditions. The ore body section of the mesh had elements representing approximately 1.5 m (5 ft), providing five elements for the width of each pillar.

Laboratory material properties used in the computer model are listed in table 2. The rock and cemented backfill properties were obtained from uniaxial compression tests as reported by J. F. T. Agapito Associates, Inc. (Brechtel et al., 1985a, 1985b). The uncemented fill properties were obtained from one-dimensional compression tests on mine backfills (Nicholson, 1968). A unit weight of 2,082 $kg/m^3$ (130 $lb/ft^3$) was chosen for the overburden based on textbook values for soil (Sowers, 1979). Modulus values for the materials needed to be adjusted based on the slope of the measured displacements versus the displacements calculated by the finite-element code before the code could be used for prediction.

Figures 10 and 11 show measured versus predicted relative displacements for borehole extensometers placed in rock and fill, respectively. The linear correlation coefficients for the borehole extensometer data and backfill extensometer data are 0.81 ($r^2$ = 0.66) and 0.095 ($r^2$ = 0.009) respectively. One possible reason for the lack of correlation between the measured versus predicted fill displacements lies in the behavior of the horizontal fill extensometers which were deformed by the flow of backfill during installation. This could have caused some damage to the rod-transducer assembly. Another explanation may reflect the nature of the boundary between the backfill and rock which is repre-

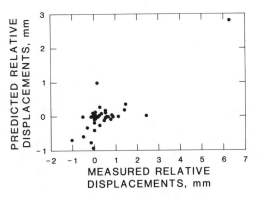

Figure 10. Measured vs predicted relative displacements for extensometers in rock.

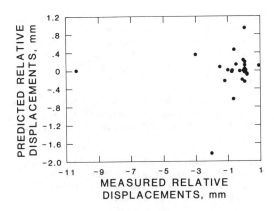

Figure 11. Measured vs predicted displacements for extensometers in backfill.

sented as a continuum by the finite-element method. Because the boundary has nodes shared by two elements (one with a material property of backfill and one with a material property of rock), the backfill elements are influenced by the modulus of the rock and vice versa. Some shear exists between the rock and the backfill in a mine, but this phenomenon may not be accurately represented by the finite-element method. A third possibility is that the mining sequence modeled may not represent the true mining sequence because the benches and headings were mined by a succession of small excavations not taken into account by the two-dimensional code. A three-dimensional model might improve the results.

## 8 CONCLUSIONS

The vertical backfill extensometers indicated that backfill stresses increased at a relatively constant rate during secondary mining. When one of the last secondary pillars was mined, the stress rate increased in an adjacent cemented pillar. The borehole extensometer in the secondary pillar showed compressive displacements as the heading advanced toward its location and showed tensional displacements when it was passed over during mining.

An earth pressure cell in the backfill showed evidence that the backfill received approximately 53 pct overburden load.

The value of $r^2$ for measured rock displacements versus rock displacements pre-

dicted by the finite-element code was 0.66. The measured versus predicted displacements for the instruments in the backfill resulted in an $r^2$ value of 0.009. This low value may be the result of malfunctioning instruments or unrealistic displacements predicted by the two-dimensional, finite-element method near the backfill-rock boundary. A three-dimensional model could possibly improve the results.

## ACKNOWLEDGMENTS

The authors wish to thank Jim Knowlson, chief engineer; Clark Gilbert, senior mine engineer; Mark Mudlin, mine engineer; John Baz-Dresch, mine engineer; and Chris Jacobsen, surveyor of Asamera Minerals (U.S.), for providing access to the B-North ore body, participating in instrumentation planning, providing mine maps and schedules, and assisting in instrument installation. The authors also wish to thank Carl Brechtel, associate, J. F. T. Agapito and Associates, for discussions regarding instrument placement and design; and William Pariseau, professor, University of Utah, for providing the finite-element code and assisting in the interpretation of results.

## REFERENCES

Argall, G.O. 1988. Cannon Takes Over Second Place Among U.S. Underground Gold Mines. Eng. and Min. J., pp. 34-39.

Baz-Dresch, J.J.  1987.  High Strength
Backfilling at the Cannon Mine.  Proceedings of the 3rd Western Conference
on Precious Metals, Coal and Environment. Soc. of Min. Eng. of AIME,
pp. 7-14.

Brechtel, C.E., et al.  1985a.  Mine
Layout and Stability Analysis - A
Report to Asamera Minerals (U.S.), Inc.
P. 24.

Brechtel, C.E., et al.  1985b.  Predicted
Subsidence Due to Mining the B-North
Orebody - A Report to Asamera Minerals
(U.S.), Inc.  Pp. 27-28.

Brechtel, C.E., et al.  1987.  Mine Design at the Cannon Mine:  Integration
of Operational Planning and Geomechanical Design.  First International Conference on Gold Mining.  Soc. of Min.
Eng. of AIME, Vancouver, BC. pp. 523-
536.

Nicholson, D.E., and R.A. Busch.  1968.
Earth Pressure at Rest and One-Dimen-
sional Compression in Mine Hydraulic
Backfills.  RI 7198, U.S. Bureau of
Mines.

Ott, L.E., D. Groody, E.L. Follis, and
P.L. Siems.  1986.  Stratigraphy,
Structural Geology, Ore Mineralogy and
Hydrothermal Alteration at the Cannon
Mine, Chelan County, Washington, U.S.A.
Proceedings of Gold '86 International
Symposium, Toronto, pp. 22-31.

Pariseau, W.G.  1978.  Interpretation of
Rock Mechanics Data (contract H0220077,
Dep. of Mining Engineering, Univ. of
Utah).  V. 2, 41 pp.

Sowers, G.F.  1979.  Introductory Soil
Mechanics: Geotechnical Engineering.
Macmillan, p. 34.

*Innovations in Mining Backfill Technology, Hassani et al. (eds), © 1989 Balkema, Rotterdam. ISBN 90 6191 985 1*

# Numerical modelling as a basis for evaluating the effectiveness of backfill as local support in deep mines

R.G.Gürtunca & I.H.Clark
*Chamber of Mines Research Organization, Johannesburg, RSA*

ABSTRACT: The application of numerical modelling is presented as a means to evaluate the effectiveness of backfill as local support along the gullies of filled panels. The results show that the most suitable support configuration is obtained by extending the uncemented backfill to the edge of the gully.

## 1 INTRODUCTION

One of the major reasons for using backfill in South African deep-level gold mines is to provide local support. In this capacity, the backfill is expected to improve face and gully conditions and to reduce rockfalls and rockburst damage.

It has been shown that high differential movements take place along the gullies of backfilled stopes due to closure variations occurring between points inside and outside the backfill (Gurtunca et al, 1989). These differential movements may induce tensile strains in the hangingwall across the gully, resulting in the fractured rock becoming loose and thus initiating a rockfall. It is believed that the number of rockfalls can be reduced in these areas by improving the support design at the shoulder of the gullies.

Three types of support are installed in the backfilled, longwall panels of South African gold mines. These are backfill, hydraulic props and packs. The integration of the installation of the supports with other mining activities presents some difficulties. The major problem is the installation of packs as this is both labour intensive and expensive.

In this paper the use of different support configurations along the gully will be evaluated by comparing the distribution and orientation of the induced principal stresses.

## 2 PRESENT SUPPORT LAYOUT IN BACKFILLED PANELS

A typical support layout showing the position of the packs and the backfill in a filled stope is illustrated in Figure 1. The composite- or mat-packs are used along the gullies with 2 m spacing. These packs are installed ahead of the backfill to provide support for the area in the vicinity of the advanced heading. The area between the backfill and the face is supported using a temporary support system such as hydraulic props. Backfill is hydraulically placed in paddocks or bags 5 to 6 m away from the face. As the face advances, new packs and hydraulic props are installed in the newly blasted areas after the face has been cleaned. Within the production cycle, the next backfill paddock is built and filled. This may require building a paddock after every second or third blast depending on the face advance per blast.

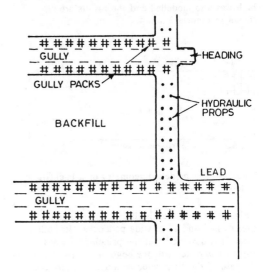

Figure 1 A typical support layout in a backfilled stope.

## 3 DEVELOPMENT OF STOPE MODEL

The modelling of a typical underground backfill stope is complex. The major parameters which need to be modelled are the backfill, the fractured rock mass, the gully packs, and finally the interaction of these elements. It is not practical to model all these factors simultaneously in one stope model. To simplify the modelling procedure it was decided to model backfill separately so that a suitable constitutive law could be selected to describe the in situ behaviour of the backfill. It was shown by Clark et al (1988), that the Cap model (Sandler & Rubin, 1987) incorporated into the program FLAC (Cundall, 1987) could accurately simulate the behaviour of backfill as measured in laboratory tests under $K_o$ (confined compression) conditions. FLAC is a two-dimensional finite difference program which allows large strain and non-linear material behaviour to be modelled.

In this study only the Mohr-Coulomb criterion will be used to model backfill as some software problems were encountered with the Cap model at the time that the paper was written. These have since been overcome by Dr. Cundall. The modelled geometry is shown in Figure 2.

The mesh consists of 50 elements along the horizontal axis and 5 elements along the vertical axis. The dimensions of the paddock are 10 m wide and 1 m high. The top and the bottom boundaries are fixed in the 'x' i.e. horizontal, and 'y' i.e. vertical, directions. Deformation was induced by displacing the boundaries with constant velocity. The principal stresses are presented in Figures 3 and 4. For comparison, a paddock 5 m wide and 1 m high was also modelled and the results are also shown in Figures 3 and 4.

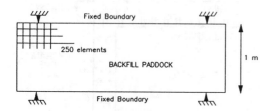

Figure 2  The modelled geometry of a backfill paddock.

Figure 3 shows the distribution of principal stresses in 5 and 10 m wide paddocks. In both cases, it is observed that the principal stresses approach zero towards the edge of the paddocks. In the case of the 10 m wide paddock, the stresses increase rapidly from the edge of the paddock and

become uniform towards the centre of the paddock. The 5 m wide paddock demonstrates a different behaviour. The extent of the low stress zone is greater and the stress increases only close to the centre of the paddock.

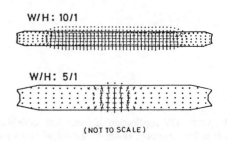

(NOT TO SCALE)

Figure 3  Distribution of principal stresses in 5 and 10 m wide paddocks.

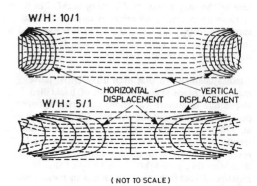

(NOT TO SCALE)

Figure 4  Distribution of horizontal and vertical displacements in the 10 and 5 m wide paddocks.

In Figure 4, the distribution of horizontal and vertical displacements in the 10 and 5 m wide paddocks are shown. The horizontal displacements' are restricted to the edges of the 10 m paddock whereas they extend further into the 5 m paddock. The displacement behaviour of the backfill observed in both cases corresponds with the principal stress distributions shown in Figure 3.

The model also demonstrates the effect of the change in width to height (w/h) ratio of the paddock. The result shows that the paddock with smaller w/h ratio is weaker in the sense that smaller stresses are generated in the paddock i.e. maximum 31 MPa in the centre of the 5:1 paddock compared to about 37 MPa maximum major principal stress in the 10:1 paddock. The area affected by horizontal displacement is also greater in the 5 m paddock as indicated by the flow of more backfill material towards the edge of the paddock.

This type of behaviour agrees well with observations and measurements made in both backfill paddocks underground and tests carried out in the laboratory. It is known that as the w/h ratio of a backfill paddock increases, the stress-strain response of the material becomes stiffer until at a w/h>11, the backfill response shows the same strain hardening behaviour as measured in confined ($K_o$) compression. Note also that the computer model simulates the bulging of the backfill paddock at the edges as is observed underground.

It is concluded that the selected model simulates the principal features of the expected material behaviour.

## 4 STOPE MODEL

In this section a typical stope gully geometry will be simulated to study the problems associated with design of support used along the gullies. The behaviour of the backfill will be modelled using the Mohr-Coulomb criterion. The rockmass will be treated as an homogeneous elastic continuum.

### 4.1 Description of the stope model

The modelled geometry is shown in Figure 5. Two axes of symmetry are used in the model to reduce the number of elements. These axes are taken through the centre of the gully and the centre of the panel. In this way, an infinite number of filled panels are generated in the model. Ten metres of hangingwall and footwall rock above and below the backfill are considered, and the stoping height is taken as 1 m. The distance between the centre of the gully and the panel centre line is 14 m. The edge of the backfill is located 3 m from the gully centre to simulate current practice.

The rockmass and backfill are modelled using 1410 elements. However, two interfaces along the boundaries between the backfill and the footwall and the hangingwall rock are used so that more elements can be assigned to the backfill. The

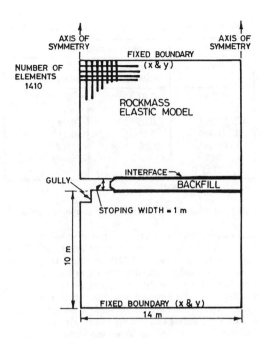

Figure 5   Geometry of modelled stope gully area

boundaries at the top and the bottom of the model are fixed in both the horizontal and vertical directions. The model is loaded by displacing both the upper and lower boundaries until 10% strain is measured in the backfill.

The rockmass around a stope is intensively fractured. However it is not possible to include all of these fractures in the model. It is assumed, therefore, that the fractured rock mass is a continuum with a low modulus. The fractured rockmass is attributed a Young's Modulus of 25 GPa rather than that of the intact quartzite of 70 GPa.

The input parameters for the Mohr-Coulomb criterion are listed in Table 1. These parameters represent a classified tailings material with 47% porosity. The normal and shear stiffness parameters used for the interfaces between the fill and the rock mass are derived by considering the relative stiffness of the backfill and rockmass.

Table 1: Input parameters used to model the different materials

| Criterion | Mohr-Coulomb | | Interface | Elastic | |
|---|---|---|---|---|---|
| Property | Uncemented Backfill | Cemented Backfill | Interface | Quartzite | Shale |
| c (MPa) | 0.01 | 0.5 | - | - | - |
| 0 (°) | 30 | 35 | - | - | - |
| G (GPa) | 0.05 | 0.08 | - | 10.4 | 1.0 |
| K (GPa) | 0.1 | 0.11 | - | 13.3 | 1.4 |
| d (kg/m³) | 2000 | 2500 | - | 2700 | 2700 |
| $k_n$ (GPa) | - | - | 6 | - | - |
| $k_s$ (GPa) | - | - | 6 | - | - |

### 4.2 Modelling of a conventionally supported panel

A conventionally supported panel was modelled as
Case 1 in order to compare it to the other backfill
layouts. Matpacks with 10 MPa in situ stiffness are
used in this model, spaced as shown in Figure 6.
The figure shows the distribution of principal
stresses in the hangingwall and footwall of the
panel. The stresses are quite uniform from about 2
m into the hangingwall and footwall. However,
because the packs become stress concentrators, the
principal stresses change direction around the interface between the rockmass and the packs. It can be
seen that high compressive stresses are generated
around the gully hangingwall and footwall.

These compressive stresses might be detrimental to
the sidewall of the gully in the footwall and cause
instability in this region. The model does not
display any tensile zone in the hangingwall around
the gully which could be interpreted as a favourable
condition for the gully hangingwall, particularly
from the point of view of the occurrence of rockfalls.

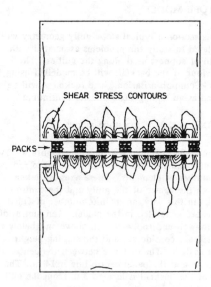

Figure 7 Modelled layout of conventionally supported panel showing the distribution of shear
stresses: [Case 1]

The shear stresses are concentrated around both
edges of each pack as expected (Figure 7) from the
distribution of the principal stresses. The shear
stresses do not extend far into the rockmass, but
unfavourable shear stress distributions are observed
around the gully especially in the footwall.

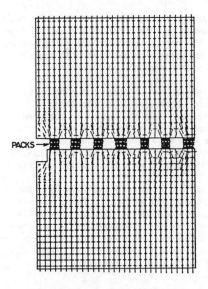

Figure 6 Modelled layout of conventionally supported panel showing the distribution of induced
principal stresses: [Case 1]

The vertical displacements are quite uniform across the panel simply because there is no difference in the stiffness of the support units used in the panel. This will be discussed further in Section 5.

### 4.3 Modelling of a backfilled panel with packs at the edge of the gully: Case 2

The configuration shown in Figure 8 is typical of the support layout used along the gullies in backfilled panels, in gold mines. A gully pack is installed at the shoulder of the gully and the backfill is located 3 m away from the centre of the gully.

The distribution of principal stresses in the rock mass surrounding the backfilled panel is shown in Figure 8 to be clearly different from the case of the panel supported by packs (Figure 6). The stresses in the rockmass are uniform towards the centre of the panel while they form an arch around the gully. It is also observed that a low stress zone i.e. a zone where the principal stresses are tensile or the compressive stresses are less than 1 MPa, is formed around the gully. The extent of this zone will be used in section 5 to compare the effect of different support configurations.

The model also shows that the pack at the edge of the gully does not influence the stress distribution of the rock mass as in the case of conventionally supported panels. The stiffness of the backfill is much higher than the pack and hence the backfill becomes the major stress concentrator in the rock mass, thus forming an arch around the gully and the pack. The results indicate that no high stresses will be generated at the sidewall of the gully in the footwall.

The maximum computed principal stress is at the point where the dilation of the backfill ends and the backfill starts demonstrating confined ($K_o$) compression behaviour. This is further illustrated in Figure 9 where the shear distribution in the rockmass is shown. Two shear stress lobes are formed in the hangingwall and the footwall and these lobes are concentrated at the same point as the computed maximum principal stress.

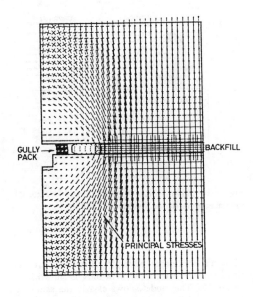

Figure 8 The distribution of principal stresses in a backfilled panel with timber packs at the edge of the gully: [Case 2]

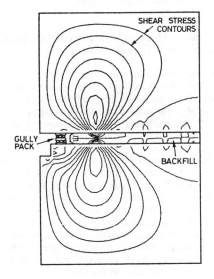

Figure 9 The distribution of shear stresses in a backfilled panel with timber packs at the edge of the gully.

133

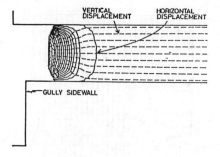

Figure 10 Distribution of vertical and horizontal displacements in the backfill: [Case 2]

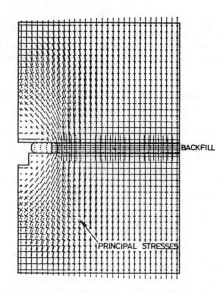

Figure 11 Distribution of principal stresses in a backfilled panel without packs at the gully edge. [Case 3]

The distribution of vertical and horizontal displacements within the backfill only is shown in Figure 10. The model shows clearly the same behaviour of the backfill as was observed in the modelling of the backfill alone (Figure 3). The horizontal displacements are concentrated at the edge of the backfill paddock resulting in lateral movement. The vertical displacements are uniform except at the edges of the backfill.

### 4.4 Modelling of a backfilled panel without packs at the edge of the gully: Case 3

As an alternative to the use of packs along the gullies, the backfill paddock could be extended to the edge of the gully. In this way, the backfill itself becomes the gully support. The results obtained from modelling such a configuration are shown in Figure 11.

The backfill is placed half a metre from the gully edge without the use of packs. The principal stress distribution in the rockmass is similar to that of the backfill and pack combination. However, the tensile zone is now smaller than in the previous case and again no high stresses are generated in the footwall around the gully edge. This is a favourable situation because there are no high stresses which may damage the footwall gully sidewall. Furthermore, the low stress zone in the form of an arch is reduced.

The distribution of the shear stresses are similarly distributed as those shown in Figure 9, i.e. in the form of lobes in both the hangingwall and the footwall.

### 4.5 Modelling a composite backfill as gully edge support: Case 4

The use of cemented backfill as gully edge support is modelled as shown in Figure 12. A 2 m wide cemented backfill is placed adjacent to the uncemented backfill to replace the packs normally used. The parameters used for the cemented backfill are listed in Table 1.

The results plotted in Figure 12 show that the low stress zone around the gully is reduced considerably compared to the cases discussed in Sections 4.3 and 4.4. However, high compressive stresses again appear in the footwall around the gully edge as in the case of the conventionally supported panel. The stresses become quite uniform towards the centre of the panel. Note that no dilation is observed along the edge of the backfill simply because of the high stiffness of the cemented backfill.

The shear stress lobes in the hangingwall and the footwall are also reduced considerably due to the presence of the cemented backfill as shown in Figure 13. However, the shear stress lobes move toward the gully and this may not be a desirable situation if the stability of the gully sidewall is to be maintained.

134

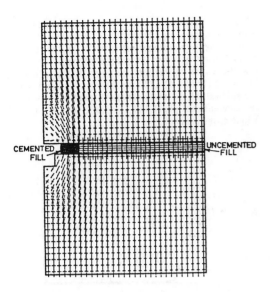

CEMENTED FILL    UNCEMENTED FILL

**Figure 12** Distribution of principal stresses for composite backfill panel: [Case 4]

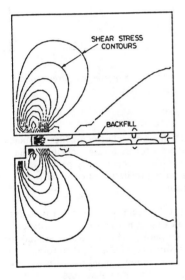

SHEAR STRESS CONTOURS

BACKFILL

**Figure 13** Distribution of shear stresses for composite backfill panel: [Case 4]

## 4.6 Modelling a shale layer in the hangingwall

Underground observations have shown that the existence of a shale layer in the hangingwall of a backfilled stope could be the cause of problems in the gullies. The hangingwall rock becomes unstable and rockfalls extending to the shale contact often occur in the gullies. Computer simulations were carried out firstly to study the influence of shale in the hangingwall on the stability of the gully and secondly, to compare the stress and displacement distributions in the rock mass for the configurations with and without the shale in the hangingwall, Cases 5 and 2, respectively. The properties of the shale are listed in Table 1.

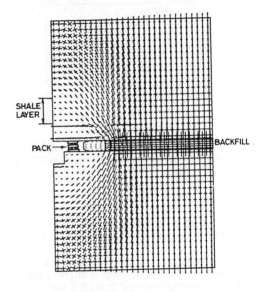

SHALE LAYER

PACK →    BACKFILL

**Figure 14** Distribution of principal stresses for rockmass containing a shale stratum: [Case 5]

The situation modelled for Case 5 comprises a shale layer of about 2 m thickness in the hangingwall only, with a 1,5 m thick layer of quartzite between the backfill and the shale. A pack is positioned next to the backfill as gully edge support.

The distribution of principal stresses in the rockmass is shown in Figure 14. The stresses in the shale and quartzite are horizontal at the bottom contact of the shale and the low stress zone is relatively large compared to the situation without shale.

Figure 15 shows that in addition to the usual shear lobe formed in the hangingwall, a smaller lobe is also observed in the quartzite layer between the backfill and the shale layer. Thus, slip may take place between the quartzite and the lower shale contact. The top contact of the shale with the quartzite, however, does not indicate any slip movement relative to the bottom contact. The displacement profile obtained across the panel will be discussed in section 5.

135

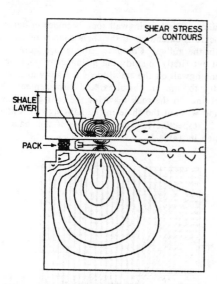

Figure 15 Distribution of shear stresses in rockmass containing a shale stratum in the hangingwall: [Case 5]

## 5 COMPARISON OF THE RESULTS OBTAINED FROM DIFFERENT GEOMETRIES MODELLED

Since it is rather difficult to draw conclusions from the computer modelling of the five different support configurations discussed in Section 4, two criteria were chosen to assess the relative benefits of the various layouts. The criteria are:
(i) The extent of the low stress zone around the gully, and
(ii) the closure profile across the panel.

The extent of the low stress zone (as defined in section 4.3) is plotted in Figure 16 for each case modelled. The figure shows that the smallest stress zone is obtained for Case 1, the conventionally supported panel (i.e. with packs) and the largest zone for Case 5 which has the shale layer in the hangingwall. The use of packs in conjunction with backfill (Case 2, representing current practice) yields the second largest low stress zone. The zone is quite small when a cemented backfill rib is placed next to the uncemented backfill. However, as discussed in sections 4.2 and 4.5, despite the formation of very small tensile zones in the hangingwall of the gully, high shear stresses which might be detrimental to the sidewall of the gully are computed for the conventional panel and composite backfill cases.

The most favourable situation is then obtained when the uncemented backfill is extended to the edge of the gully without using packs, so that the backfill itself becomes the gully edge support. The low stress zone in this instance is smaller than that in Case 2 (Figure 5) and also the stress levels around the sidewall of the gully are much smaller than in the conventional panel and composite backfill, Cases 1 and 4, respectively.

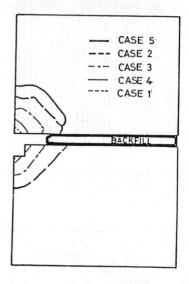

CASE 5
CASE 2
CASE 3
CASE 4
CASE 1

BACKFILL

Figure 16 The extent of the low stress zone for the different Cases modelled.

Figure 17 shows the closure profiles calculated across the panel for each case modelled. The conventional panel supported by packs displays a uniform closure profile because the stiffness of each pack is equal.

For Case 3 the backfill in the panel is stiffer than the packs in Case 2 but no support resistance is generated by the backfill in the area around the gully. The situation in Case 2, therefore, causes an increase in the closure around the gully as shown in Figure 17. Note however that Case 3 is not significantly better than Case 2. It is considered that the resultant change in the slope of the closure profile may indicate the onset of unstable behaviour of the gully hangingwall. The intensely fractured rock might become subjected to tensile strains due to the induced differential movements which might lead to rockfalls in the gullies.

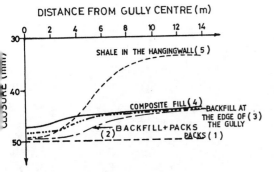

Figure 17 Closure profiles calculated across the panel for the various Cases modelled.

For Cases 3 and 4, where uncemented and composite backfills are used for gully edge support, respectively, the increase in closure over the gully is reduced by the cemented backfill. This reduction may increase stability of the hangingwall strata and thus decrease the likelihood of rockfalls in the gullies.

The existence of the shale layer in the hangingwall rockmass changes the shape of the closure profile considerably. The closure over the gully is similar to the other situations modelled. However, the closure in the panel is significantly less. The model indicates that the shale layer is compressed as a result of the high stresses generated in the backfill while no support resistance is provided over the gully. The ability of the model to demonstrate this type of behaviour of the rockmass is encouraging. The model also suggests an explanation for the difficult conditons experienced in the areas where shale layers exist in the hangingwall rock.

## 6 CONCLUSIONS

In this paper, numerical modelling has been used to provide a basis for evaluating the effectiveness of backfill as local support along the gullies of backfilled panels. The results show that uncemented backfill placed up to the edge of the gully is the most suitable support configuration of those considered. The following advantages can be obtained if the packs used along the gullies are replaced by the uncemented backfill:
(i)    the stability of the rockmass around the gully can be improved,
(ii)   the types of support elements used in a backfilled panel can be reduced from three to two; hence, the mining operation is simplified and becomes more efficient,
(iii)  the placement of the uncemented backfill is easier than the installation of the packs along the gullies,

(iv)   the gully support costs may be reduced by about 50% if the uncemented backfill is extended to the edge of the gully to replace the packs.

The use of cemented backfill as the gully edge support is not recommended as high stresses associated with the stiffer materials might damage the gully sidewall. The cemented backfill is also considerably more expensive than the uncemented backfill and the installation of cemented backfill next to uncemented backfill will only complicate the mining operation still further.

## 7 SUGGESTED SUPPORT LAYOUT FOR GULLIES IN BACKFILLED STOPES

Based on the conclusions drawn in this paper, the support layout shown in Figure 18 is suggested for backfilled panels.

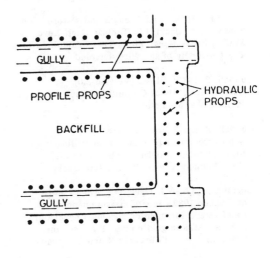

Figure 18 Proposed support layout for backfilled panels.

The backfill paddock can easily be extended to the edge of the gully. The suggested distance between the gully edge and the backfill paddock is 0,5 m. The area around the headings or leads ahead of the backfill can be supported by hydraulic props with headboards.

However, it is suggested that profile props should be installed 0,5 m inside the backfill, along the gully shoulders as shown in Figure 18. The function of the props is to provide initial stiffness to the support along the gully edge as it is known that the edge of the backfill does not provide enough support resistance during initial compression. The profile props will provide the necessary support resistance in this region until the backfill becomes stiffer.

## ACKNOWLEDGEMENT

The work described in this paper forms part of the research programme of the Chamber of Mines of South Africa Research Organization.

## REFERENCES

Clark, I.H., Gurtunca, R.G. and Piper, P.S. 1988. Predicting and Monitoring Stress and Deformation Behaviour of Backfill in Deep-Level Mining Excavations. Proc. Fifth Australia-New Zealand Conference of Geomechanics, Sydney, pp. 214-218.

Clark, I.H. 1988. The Strength and Deformation Behaviour of Backfill in Tabular Deep-Level Mining Excavations. Ph.D. Thesis submitted to The University of the Witwatersrand.

Cundall, P. 1987. FLAC (Version 2.0). User Manual. Itasca Consulting Group Inc., Minneapolis, Minnesota.

Cundall, P. and Board, M. 1988. A microcomputer program for modelling large-strain plasticity problems. 6th Int. Conf. Num. Methods Geomechanics, Innsbruck.

Gurtunca, R.G., A.J. Jager, D.J. Adams and M Gonlag, 1989. In situ measurements of backfills and surrounding rockmass behaviour in South African gold mines. Proc. 4th Int. Symp. Mining with Backfill, Ontario, Canada.

Sandler, I.S. and Rubin, D. 1987. Cap and critical state models: Short course notes. Second Int. Conf. on Constitutive Laws for Engineering Materials, Arizona.

*Innovations in Mining Backfill Technology, Hassani et al. (eds), © 1989 Balkema, Rotterdam. ISBN 90 6191 985 1*

# The use of models in sill mat design at Falconbridge

B.O'Hearn & G.Swan
*Falconbridge Limited, Sudbury Operations, Onaping, Ontario, Canada*

ABSTRACT: Through a research project sponsored by CANMET, Falconbridge Limited was able to: 1) measure the material properties required to numerically model fill, 2) instrument a backfilled stope and 3) calibrate a numerical model. Based on this initial work, an integrated design approach involving the stability analysis of sill mats was devised. Under static loading, voussoir compression arch analysis was found appropriate. For dynamic loading conditions the numerical model UDEC was used. Dynamic loads causing particle velocities, in a shear direction, of 0.3 - 0.2m/s were found to cause collapse of the sill mat and overlying fill. Increases in the cohesive strength and frictional resistance of the fill above the mat offered no significant resistance to shear failure. Improvements in stability can be obtained by following simple planning procedures.

## 1 INTRODUCTION

A cemented sill mat can be thought of as a thin, bedded plug situated below weaker backfill material and designed to support itself and the weight of overlying fill. Such a mat forms a cohesive back permitting the safe extraction of underlying ore. Sill mats can be designed from three different approaches:

1) experience,
2) classical solids mechanics theory,
3) and simulations (both physical and numerical).

At Falconbridge Limited, cemented sill mats are placed immediately above sill pillars to permit future sill recovery. They have also been used in undercut and fill stoping to routinely provide an artificial roof for each cut. Sill mats have been used over the course of several decades and this experience has tended to influence designs. Early mats consisted of massive timber construction, but with the advent of consolidated hydraulic fill, mat practice developed to a combination of a timber base immediately overlain with 1 to 2m of cemented backfill with tailings to cement ratios of less than 10:1. A major sill mat failure was recorded at Falconbridge and its failure was caused by a series of major rockbursts (Bharti and West, 1984).

In designs approached from solid mechanics theory, structural analogies can be made to: elastic beams, elastic plates and voussoir compression arches. Of these, the first two analogies are the most commonly used despite yielding extremely conservative designs. In light of some simplifying assumptions, stated later, and design procedures developed by Beer and Meek (1982), the voussoir beam model deserves wider application. Its use is demonstrated while examining the stability of a statically loaded sill mat.

Numerical simulations require the costs associated with acquiring hardware, software, specialized knowledge and calibrated models, and are therefore less appealing than the much simpler approach offered by the voussoir analysis. However, numerical simulations do offer a means of assessing the vulnerability of sill mats to failure by dynamic loads. Consequently, a series of numerical experiments designed to examine the stability of these structures when subjected to seismic events has been calibrated with measured data and reported in the following sections.

It is the objective of this paper to:

1) present a summary of a CANMET report completed by Falconbridge Limited listing in situ backfill material property data and report results of a calibrated numerical model simulating mining in the vicinity of a sill mat,

Table 1  a) Material properties of hydraulic tailings originating from a stope at Lockerby Mine.  b) Material properties of fill used in the numerical model.

|  | | Weakly Cemented (tailings/cement >25) | Strongly Cemented (tailings/cement <10) |
|---|---|---|---|
| Compr. Strength | MPa | 0.2 – 1.0 | 2 – 4 |
| Young's Modulus (E) | GPa | 0.05 – 0.1 | 0.3 – 0.5 |
| Cohesion (C) | MPa | 0.1 – 0.4 | 1.0 – 1.5 |
| Friction Angle ($\phi$) | Degrees | 35° – 50° | No reliable results obtained |
| Density (D) | kg/m$^3$ | 2100 | 2100 |

b)

|  | | 22:1 (tailings/cement) |
|---|---|---|
| E | GPa | 0.09 |
| Bulk Modulus (B) | GPa | 0.05 |
| Shear Modulus (G) | GPa | 0.037 |
| C | MPa | 0.2 |
| $\phi$ | Degrees | 35 |
| Tensile Strength (t) | Pa | 1000 |
| D | kg/m$^3$ | 2100 |

2) examine design aspects of a sill mat using elements of the three approaches listed previously and analyse the stability of the structures for both static and dynamic conditions, and

3) report on some general design criteria.

## 2 SUMMARY OF CANMET STUDY

Pertinent details from a report completed by Falconbridge Limited entitled, "In situ monitoring and computer modelling of a cemented sill mat and confines during a tertiary stage pillar recovery" (see reference list) are summarized here and form the ground work for this paper.

Unfortunately, Falconbridge's mining schedule precluded the availability of a suitable mat. Consequently, the exposed vertical wall of a backfilled stope, adjacent to a stope scheduled to be mined by vertical retreat, was monitored. The philosophy in choosing such a site was based on the premise that, if the fill associated with a tertiary pillar recovery could be modelled successfully, then a sill mat and confines could equally be simulated. The project involved efforts in two areas. In situ sampling and instrumentation was carried out to establish characteristic material properties of hydraulic fill in the test stope. Numerical modelling of the test

stope was done to increase confidence in our ability to use models to assist in the safe and economical extraction of tertiary pillars. Significant preliminary results of the modelling exercise include the key role of: gravity, zero boundary stresses around the fill, and the associated fill failure following the blast.

Table 1 lists data collected from Lockerby Mine by a geotechnical consultant. Additional data listing the numerical values used in the model, and compiled from the Lockerby data and other sources, are also given. Load cells, geophones and extensometers were used to collect additional data. Load cells installed in the backfill, before adjacent mining, recorded little change in load in both vertical and horizontal directions after mining took place. The fill therefore accepted very little load from adjacent boundaries. Blast vibration monitoring equipment was used with geophones to record key parameters describing the blast. Table 2 lists these parameters and measurements. Three extensometers were placed in the fill to record displacements near the unsupported fill wall. The values measured by these devices were plotted in contours of displacement and used to map out the sloughed failure surface which was observed subsequent to blasting.

Table 2  Data obtained from blast vibration monitoring equipment

| Peak Particle Velocity (PPV) * | 53 cm/sec. |
|---|---|
| Dominant Frequency | 30 Hz |
| Blast Duration | 15 millisec. |

* measured 15 m from the blast

Several numerical simulations were carried out in an attempt to reproduce the recorded fill failure. FLAC, a two-dimensional finite difference continuum code, was used in a scoping exercise to examine the kinematic potential for failure. The FLAC simulation produced a similar failure surface to that observed underground. However, the model was unable to match the timing of the fill failure to that noted underground. Extended analysis using MUDEC, a two-dimensional distinct element discontinuum code, allowed the analysis to simulate the dynamic impact associated with the production blasts. This analysis showed that the shock wave following each blast produced tensile forces causing fill to peel off the wall. In addition, the failure surface of the model was very close to that observed underground.

The numerical simulations were able to reproduce the mode and mechanism of the vertical fill wall failure adjacent to the pillar recovered by vertical retreat methods. Further modelling efforts demonstrating the susceptability of sill mats to failure over a range of conditions are reported in the remainder of this paper.

## 3 VOUSSOIR BEAM ANALYSIS

Cemented sill mats can be thought of as bedded deposits which, because of their confined situation between hangingwall and footwall, generate lateral stabilizing thrusts by self-weight. The voussoir beam analysis, described by Beer and Meek (1982), assumes a particular abutment load distribution and a parabolic-shaped thrust line, Figure 1. These assumptions are considered reasonable, though conservative, when it is considered that the beam is completely decoupled from the far-field stress state.

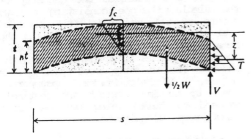

Figure 1   Voussoir Beam Geometry and
           Load Specification for Fill
           Sill Mat Analysis. (after
           Brady and Brown 1985).

For plane strain conditions and the beam geometry of Figure 1, the analysis yields a fourth-order equation for span s as a function of n, the load/depth ratio and $f_c$, the maximum compressive stress. This can be solved iteratively using a computer and requires the following fill properties as input: elastic deformation modulus, compressive strength, sill mat thickness, rock-fill interface friction angle and cohesion, and fill mass density. The theory for rectangular sill mats introduces two additional parameters: k, the ratio of short to long span, and the Poisson's ratio of the fill.

Three modes of failure are possible in voussoir fill beams. There is compressive failure in the arch, shear failure at the abutments and slough or tensile failure beneath the arch itself. In the first, the maximum longitudinal stress, $f_c$, is compared to the uniaxial compressive strength of the fill. A factor of safety can be defined from the ratio of the two. In the second, the shear stress is compared to a Mohr-Coulomb strength given by:

$$0.5 f_c \; nt \tan \phi + C$$

where $\phi$ and C are the interfacial friction angle and cohesion, respectively. Finally an estimate of the potential slough material (due to excessive vibration from blasting and/or rockbursts) may be obtained from a consideration of the material volume beneath the theoretical arch. This is given by:

$$0.667 \; t \; (1-n) s$$

From this, the gross weight of material to be supported in the event of sloughing (the arch otherwise stable) can be calculated. We suggest this is the load which a timber beam support system should be designed to carry.

### 3.1 Typical design calculation

In a typical Falconbridge design an 8:1 hydraulic fill is placed 1m thick. This is then overlain with 32:1 as the overhand cut-and-fill mining proceeds. Using the input data, Table 3, voussoir beam theory predicts a maximum allowable span of 12.8 m.

141

Table 3  Voussoir Beam, Fill Input Data

| Parameter | Value | Unit |
|---|---|---|
| Deformation Modulus | 0.21 | GPa |
| Compressive Strength | 1.00 | MPa |
| Thickness  (t) | 2 | m |
| Friction Angle | 35 | Deg. |
| Cohesion | 0.21 | MPa |
| Density | 1900 | kg/m$^3$ |
| Poisson's Ratio | 0.20 | - |
| k, short: long span | 0.074 | - |

For the example of an 8m span, the analysis produced factors of safety in compression and shear failure of 2.70 and 1.45, respectively. A further calculation shows that the 32:1 fill above the mat is essentially self-supporting, provided the thickness exceeds 6.1m with a 8m span. The potential slough volume for this mat is 2.45 m$^3$ per unit length orthogonal to its span.  Thus the maximum potential sloughing load will be in the order of 46 kN/m.  This is, in fact, the same order of magnitude as measured, by some load cells on timber caps, while extracting a sill pillar at Falconbridge's Strathcona Mine.

## 4  UDEC ANALYSIS

At Sudbury Operations, mining induced seismic activity is a significant factor. Consequently, it has become necessary to incorporate some sensitivity analysis of our designs to seismic loading.  The following discussion concerns an analysis using UDEC to simulate a stope filled with 32:1 hydraulic tailings: cement which is situated over a two metre thick layer of 8:1, referred to as a timberless sill mat.    Below the mat a four meter wide excavation has been created.  Upon coming to equilibrium under static loading the fill has been subjected to a range of dynamic loads.  The results of these numerical experiments are presented henceforth.   Note that no attempt has been made to model a timbered sill mat.

### 4.1  The UDEC model

UDEC is an explicit, time marching, distinct element code which solves the equations of motion in difference form, (Cundall, 1971 and 1974).  Unlike finite elements, distinct elements may interact with any of the other elements and can experience large scale rigid body translations and rotations. The geometry of distinct elements is defined by the spacing and orientation of joints or layers in the material mass being modelled, with each element corresponding to an individual block of material.

The code is based on force displacement relations which specify the forces between blocks, and a motion law which specifies the motion of each block due to unbalanced forces acting on the block. Numerical integration of Newton's Second Law is used to determine translation and rotation of each block about its centroid.

The distinct element method has three distinguishing features:

1) The rock mass is simulated as an assemblage of blocks which interact through corner and edge contacts.

2) Discontinuities are regarded as boundary interactions between blocks; joint behaviour is prescribed for these interactions.

3) The method utilizes an explicit timestepping algorithm which allows large displacements, rotations and general non-linear constitutive behaviour for both the mass and the joints.

In the simulation, the rock mass was modelled as a linearly elastic and isotropic material.    Backfill was modelled as an elastic/plastic material with a Mohr-Coulomb failure criterion. Joints situated within the fill were given the same material properties as the blocks of fill they encompassed.    In other words, no preferential weakness was defined.   However, when the assigned shear or tensile strengths were exceeded, the cohesion and tensile strengths in all subsequent calculations were assigned null values.  Table 4 lists the material properties of each material modelled.

Figure 2 illustrates the geometry of the simulated stope consisting of two meters of 8:1 hydraulic fill situated over the undercut with 32:1 filling the remainder of the stope.  In the model, the 32:1 fill is 20 m in height and the undercut has an unsupported span of four meters.   At Falconbridge, unsupported spans up to 5m in width are routinely exposed by bulk mining methods using remote mucking equipment.

No boundary stresses were assigned, therefore only gravity loading was simulated.    Viscous boundaries were assigned to the floor and the undercut side of the model.

The dynamic loading was simulated as a sinusoidal pulse 10 milliseconds in

Table 4  Material properties used in UDEC model

| Material Type | D | E GPa | B GPa | G GPa | C MPa | φ | t MPa | Dilation | |
|---|---|---|---|---|---|---|---|---|---|
| Rock Mass | 2800 | 40 | 22 | 17 | 15 | 50° | 2 | — | D = Density |
| 8:1 Fill | 2000 | 0.25 | 0.16 | 0.1 | 1 | 36 | 0 | — | E = Young's Modulus |
| 32:1 Fill | 2000 | 0.1 | 0.067 | 0.04 | 0.15 | 35 | 0 | — | B = Bulk Modulus |
| Rock/Fill Interface | — | — | — | — | 0.15 | 42 | 0 | 2.5° | G = Shear Modulus |
| 16:1 Fill | 2000 | 0.1 | 0.067 | 0.04 | 0.60 | 35 | 0 | — | C = Cohesion |
| Rock Fill | 2000 | 0.25 | 0.067 | 0.04 | 0.15 | 45 | 0 | — | φ = Friction Angle |

C = Cohesion
φ = Friction Angle
t = Tensile Strength

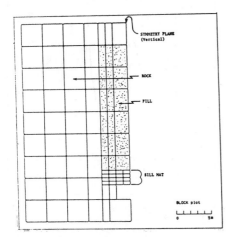

Figure 2   The Sill Mat Model

duration with a frequency of 100 Hertz. The source of the load was a 20 metre length along the left boundary.  Both pure compressive and pure shear waves were simulated.   In addition to the hydraulic fill a simulation replacing the 32:1 with 16:1 fill and cemented rock fill has also been investigated.  These results are reported in Table 5.

5  DISCUSSION

The results listed in Table 5 show collapse of the sill mat and overlying fill occurring at a maximum particle velocity, in the Z direction, of 0.3m/s, see Figure 3.  At a velocity of 0.2m/s the sill mat was damaged but the 32:1 fill remained stable, ie. stayed in place.
The dynamic loads created several vertical and horizontal joint planes,

predominantly in the fill, see Figure 4. A vertical shear plane, formed at the interface between the fill and the vertical walls of the undercut, when intersected by horizontal joint planes, formed a block which fell due to gravity.
Falconbridge Mine recorded a series of devastating rockbursts in June of 1984 causing the failure of a timbered sill mat.   Three very large seismic events occurred,   caused  by  a  fault  slip mechanism of failure, Bharti and West (1984).   The first was estimated to have a  magnitude  of  3.4  with  a  source location 20 m from the failed mat. (Note all  magnitudes  are  derived  from  the Nuttli  scale.   This  scale  is  used throughout eastern Canada and the eastern U.S.A.).  Using the following equation, Hedley (1988), a peak particle velocity of 2.1 m/s has been estimated.

$$V = 4000 \left(\frac{R}{10^{M/3}}\right)^{-1.6}$$

where V = vector sum peak particle velocity, mm/s
R = distance from source, m
M = rockburst magnitude using the Nuttli scale

The second event had a magnitude of 3.5 with the source located 50 m from the stope.   A peak particle velocity of 0.56 m/s was estimated.   Both these events caused significant damage, breaking timber and shaking down hundreds of tonnes of backfill.   A third event having magnitude 3.2 and being located 180 m away also occurred but produced no known damage.  The estimate of the peak particle velocity was 0.05 m/s.  It can therefore be deduced that the vibration at which collapse of the overlying fill occurred lies somewhere between 0.56 and 0.05 m/s.

Table 5  Numerical Model results of dynamic impact loading

| Wave Type | Dynamic Load | | Velocity Z Direction (m/s) | Fill Type | Comments |
|---|---|---|---|---|---|
| | Stress at Source (MPa) | Shear Stress at Rock/Fill Interface (MPa) | | | |
| Shear | 3 | 2.6 | 0.31 | 32:1 | Collapse |
| Shear | 2 | 1.8 | 0.2 | 32:1 | Damaged Mat, Stable Fill |
| Shear | 0.5 | 0.46 | 0.05 | 32:1 | Stable |
| Shear | 3 | 2.6 | 0.31 | 16:1 | Collapse |
| Shear | 3 | 2.6 | 0.31 | R/Fill | Collapse |
| Compressive | -14 | 1.2 | 0.2 | 32:1 | Damaged Mat, Stable Fill |

3a)

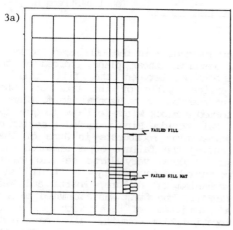

3c)

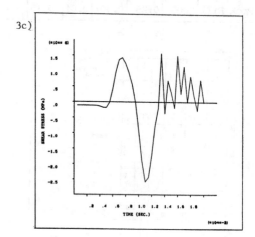

3b)

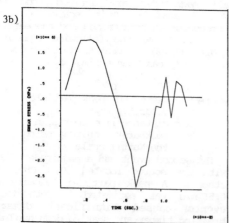

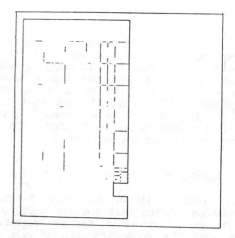

Figure 3a) Resulting collapse of sill mat and overlying fill when subjected to a dynamic load of 2.6 MPa at the rock/fill interface. 3b) The stress versus time graphs of the dynamic load as it passes a point near the source 3c) and at a point on the rock/fill interface.

Figure 4  Planes of Weakness created by Dynamic Loading

Comparing the peak particle velocities and associated damage of Table 5 with those occurring in the 1984 rockbursts demonstrate that the simulation results are consistent with recorded data. This gives a measure of credibility to the model. Using the combined data from the numerical models and Falconbridge Mine reveals that collapse of overlying fill into an undermined area is possible at peak particle velocities in the range of 0.3 - 0.2 m/s.

Thus a seismic event of magnitude 1.5 or greater located within 20 m of an exposed sill mat would endanger the stability of the fill above the mat. For the purpose of planning the location of future sill mats the following rules could be applied.

1) Where possible avoid locating sill mats within 100 m of a suspected seismically active fault or dyke. Results presented in this paper suggest fill stability could be maintained for events of 3.5 magnitude or smaller.

2) When the first rule is not possible plan to mine the stope below the sill mat by remote methods, such as longhole drilling from below the sill mat and using remote scoops for mucking, while keeping the mat as far away as possible from the discontinuity. However, the risk of heavy dilution exists if the overlying fill is shaken down.

What is also significant are the simulations in Table 5 which involve the substitution of both rock and 16:1 fill for the 32:1. Neither cemented rock fill nor 16:1 fill showed meaningfull resistance to shear failure over the 32:1. Therefore the increased resistance to shear offered by

1) the improved frictional resistance of rock fill and

2) improved cohesion of 16:1,

is not significant in improving the stability of sill mats to dynamic loads.

Shear waves were shown to cause damage at significantly lower amplitudes than compressive waves. (Note that all waves travelled from the left side of Figure 2 to the right side). This can be attributed to the fact that the shear wave is characterized by affecting particle motion in the shear direction (ie. in this case the $\pm Z$ direction), while the compressive wave characteristically has a much weaker affect on particle motion in the shear direction. Therefore, a significantly larger compressive wave is required to shear a vertical plane than a shear wave, when both waves travel in a direction normal to the vertical plane. Consequently, the orientation of the structure dictates which wave type to which it is most sensitive.

6 CONCLUSIONS

Calibrating numerical modelling results with a known sill mat failure has shown that the collapse of a sill mat and overlying fill into an undermined area is possible at peak particle velocities in the range of 0.3 - 0.2 m/s, measured at the rock/fill interface. In light of the inability of the heavily timbered sill mat at Falconbridge to stand up to the seismic activity which caused its failure, the use of timber must be questioned.

Since timber tends to crack under dynamic loads, the sill mat need only be designed with enough timber, or other support, to hold the portion of fill that is not supported by the arch, as predicted by a voussoir arch analysis.

The modelling also showed that cemented rock fill and 16:1 hydraulic fill offered no better resistance to shear failure by dynamic loading than 32:1 hydraulic fill. Therefore to change a backfill system from one type of fill to another, or to make significant increases in cement content in the hope that stability, with respect to shear on the interface, is being achieved does not appear to be the correct approach to the problem.

One approach that can be adopted lies in utilizing the following rules while in the planning stage.

1) Where possible avoid locating sill mats within 100 m of a suspected seismically active fault or dyke. Results presented in this paper suggest fill stability could be maintained for events of 3.5 magnitude or smaller.

2) When the first rule is not possible, plan to mine the stope below the sill mat by remote methods, such as longhole drilling from below the sill mat and using remote scoops for mucking, while keeping the mat as far away as possible from the discontinuity. However, the risk of heavy dilution exists if the overlying fill is shaken down.

The orientation of the structure dictates the wave type to which it is more sensitive. Consequently, a potential seismic hazard located in a direction normal to the vertical wall of a backfilled stope, where undermining is to take place, is more at risk than a stope facing a seismic hazard other than a 90° angle. Therefore, rule 1) can be tempered with this in mind.

ACKNOWLEDGMENTS

The authors would like to thank CANMET for providing funding to allow the ground roots portion of this project to occur. In addition, Falconbridge Limited is gratefully acknowledged for permitting this paper to be published.

REFERENCES

Beer, G. and Meek, J.L. "Design curves for roofs and hanging walls in bedded rock based on "voussoir" beam and plate solutions. Trans. Inst. Min. Met., 91, Jan. 1982.

Bharti, S. and West, D. "A Technical Assessment of the Rockbursts on June 20-23, 1984 at Falconbridge Mine" Report to the Ontario Ministry of Labour, Occupational Health and Safety Division, Mining Health and Safety Branch, Sudbury November 1984.

Brady, B.H.G. and Brown, E.T. Rock Mechanics for underground mining. George Allen & Unwin, London, 1985.

Cundall, P.A. "A Computer Model for Simulating Progressive Large Scale Movements in Blocky Rock Systems," Proceedings of the Sumposium of the International Society of Rock Mechanics (Nancy, France, 1971) vol. 1 paper II-8.

Cundall, P.A. "Rational Design of Tunnel Supports: A Computer Model for Rock Mass Behaviour Using Interactive Graphics for the Input and Output of Geometrical Data," Tech. Report MRD-2-74, Missouri River Division, U.S. Army Corps of Engineers, 1974.

Falconbridge Limited, "In Situ Monitoring and Computer Modelling of a Cemented Sill Mat and Confines During a Tertiary Stage Pillar Recovery." Canadian Centre for Mineral and Energy Technology Energy Mines and Resources CANMET Project number 5-9205, 1988.

Hedley, D.G.F., "Evaluation of Support Systems Subject to Rockbursts." Mining Research Laboratories, Division Report MRL88, Elliot Lake Laboratory, September 1988.

*Innovations in Mining Backfill Technology, Hassani et al. (eds), © 1989 Balkema, Rotterdam. ISBN 90 6191 985 1*

# Modelling of cut-and-fill mining systems – Näsliden revisited

W. Hustrulid
*Colorado School of Mines, Golden, Colo., USA*

Y. Qianyuan
*Jiang-Xi Institute of Metallurgy, Gan-Zhou Jiang-Xi, People's Republic of China*

N. Krauland
*Boliden Mineral AB, Boliden, Sweden*

ABSTRACT: Over the period 1975-1980 an extensive rock mechanics research and development program was conducted at the Näsliden Mine, Sweden by a team of investigators from the Boliden Company, the University of Luleå and the Royal Institute of Technology (Stockholm). The overall purpose of the program was to evaluate the degree to which rock mechanics calculation methods, mainly finite-element models, could be used as planning instruments in mining operations. Specifically a series of models (finite element and others) were used to predict the response of the ore body, the hanging and foot walls, and the fill during cut-and-fill mining at the Näsliden mine. Predictions were compared to field measurements and modifications were made until a calibrated model which could be used for simulation of future mining at Näsliden emerged. The major results from this work which was performed under sponsorship of the Swedish Board for Technical Development (STU) and the Swedish Rock Mechanics Foundation, (BeFo) were reported at the International Symposium of the Application of Rock Mechanics to Cut-and-Fill Mining (Luleå, Sweden) in June of 1980.

Although the values predicted by the final model were in general agreement with the measurements, some variation in both trend and amount of the fill pressures, wall convergence and pillar stresses were noted. One possible explanation was the difficulty of simulating the non-linear behavior of the fill using the finite element models selected. In this paper, the Näsliden data have been reevaluated using a technique that allows detailed modelling of the non-linear fill behavior and the stope pillars but greatly simplifies the reaction of the surrounding (wall) rocks. One-dimensional load-deformation curves for the fill determined using a split platen consolidometer have been incorporated into a filled slot analysis of the Näsliden cut-and-fill system. Actual and predicted fill pressures, wall convergence, and pillar stresses as a function of mining sequence are found to be in good agreement.

## 1 INTRODUCTION

### 1.1 The Näsliden Mine

The Näsliden Mine located in Northern Sweden was brought into production by the Boliden Company in 1970. Operations ceased in 1987 with the exhaustion of the complex sulfide ore. The ore body which was mined by modern cut-and-fill techniques dipped at about 70° to the west, had a length which varied between 110 and 220 m, and a maximum width of 25m (the average being 18 m). At the time the research program was initiated, mining was being conducted in four stopes located above the 460 m level (Figure 1). Each stope was planned to be about 100m high. Deslimed tailing sand transported hydraulically to the mining stopes was used as fill. The mining of 3.5 m high cuts was accomplished using large mobile equipment (drilling rigs, charging units, scaling units, loaders, trucks or dumpers, rockbolting rigs and mobile working platforms). Annual production averaged about 230,000 t with 48 mine employees.

### 1.2 An Overview of the Näsliden Project

This highly mechanized cut-and-fill mining system utilized
- few but large mining stopes with sometimes considerable roof spans

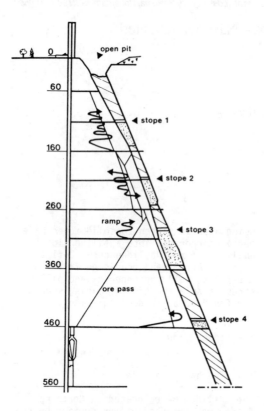

Figure 1.   Schematic Vertical Section of
            the Näsliden Mine (After
            Kolsrud (10).

o A large number of modern mining
  machines being concentrated in the
  stopes.

For the system to function properly it
was essential that mining be carried out
undisturbed by rock falls or other
stability problems in the stopes.  It
presented an ideal opportunity for
improving rock mechanics knowledge while
developing a prediction technique for the
Näsliden mine which could be used to
improve mine planning.  The Näsliden
research program included

  o theoretical studies
  o laboratory measurements
  o field observations and measurements
  o the development and application of
    finite element and other predictive
    models.

The extensive program of field
measurements included determining

  o in situ stress state
  o structural geologic setting
  o in situ fill properties

and evaluating

  o reactions of the backfilled
    excavation
  o reactions of the orebody and
    alteration zone above the stope
  o reactions around the stope

as a function of mining geometry.
   Laboratory studies were carried out to
provide the basis for estimating rock mass
and fill properties.  Tests on intact
cores taken from the different layers
yielded initial values of the elastic
(Youngs Modulus, Poissons Ratio) and
strength properties.  These values were
later modified using joint properties.
The rock mass properties eventually
incorporated into the final models were
"tuned" on the basis of field
measurements.  Several stages of finite
element modelling were performed including
the use of both elastic and 'joint'
elements.

2   THE NÄSLIDEN MODELLING APPROACH

For modelling purposes the geologic
situation at Näsliden was simplified to
that shown in Figure 2.

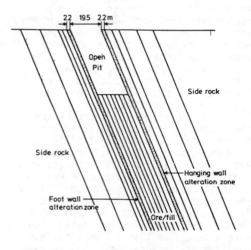

Figure 2.   Rock Types and geometries Used
            in the Näsliden Finite Element
            Modelling.  After Groth (11).

Six different rock materials were considered

- o Hangingwall rock
- o Hangingwall alteration zone
- o Ore
- o Fill
- o Footwall alteration zone
- o Footwall rock.

Although four different modelling stages were conducted, the results from stages 2 and 4 will only be described here.

Table 1. Elastic Material Properties of Rock

| Element | E-Modulus (GPa) | | $\nu$ | $\gamma$ ton/m$^3$ |
| | Stage 2 | Stage 4 | | |
| --- | --- | --- | --- | --- |
| Side rock | 45 | 30 | 0.25 | 2.6 |
| HW Alt. Zone | 6 | 10 | 0.25 | 2.6 |
| FW Alt. Zone | 15 | 10 | 0.25 | 2.6 |
| Ore | 110 | 150 | 0.30 | 3.9 |
| Fill | 0.032 | 0.032 | 0.32 | 2.1 |

In both stages a fully elastic simulation was performed (using the properties shown in Table 1) as well as one employing joint elements. In stage 2 modelling a single row of joint elements was inserted between the orebody/fill and each alteration zone. In stage 4 modelling, a row of joint elements was placed on the outer side of each alteration zone as well.

The element mesh for one of the global model used in Stage 4 is shown in Figure 3.

A comparison between the calculations and the in-situ measurements are given in Table 2. Trend refers to agreement between the shape of the curve for calculated results and that for in-situ measurements. The degree of agreement in the different comparisons was determined according to the scale.

    5 - excellent
    4 - good
    3 - fair
    2 - poor
    1 - no agreement

Comparison of magnitudes was accomplished by use of the relationship

$$\frac{\text{FEM} - \text{In-Situ}}{\text{In-Situ}} \times 100$$

for a number of points. The figure given represents the deviation from in-situ measurements in terms of the mean and standard deviation. As can be seen there are significant differences in the "goodness-of-fit" both with increasing modelling stage and with the incorporation of joints. In some cases the agreement gets better and in other cases worse. The stage 4 joint model provides an overall satisfactory fit to most of the measurements

## 3 ALTERNATIVE MODELLING APPROACHES

The selection and development of models always involves a series of trade offs. The finite element approach used at Näsliden provides coverage of a large region, and allows the introduction of a wide range of material behaviors, geometries and loading conditions. In this case joint as well as elastic elements were used. On the negative side it is very difficult to use some of this potential due to an inability, for example, to provide the required material properties. The description of joint geometries and load-deformation characteristics is a case in point. Furthermore, often the primary interest is in the near vicinity of the workings (behavior of the wall rocks, crown pillars, etc.). With usual finite element models, a large overall region must be described if the stresses and displacements in the region of interest are to be correct. This adds to model complexity with corresponding costs in preparation and running. A trade off made in the Näsliden study was to include some joint rows but to assume that the fill had linear elastic properties ($E_f = 32$ KPa, $\nu = 0.32$). The hanging and footwalls were assumed elastic and with the same properties. Narrow alteration zones between the ore and hanging/footwalls were included (both zones were assumed to have the same elastic properties. The ore zone had its own elastic properties.

An alternative modelling approach, which is the subject of this paper, is to focus on the behavior of the orebody/fill and the immediate hanging/footwalls. The model consists of a long slot in an elastic plate. The slot is filled with constant width elements which can be assigned properties representative of the ore, the fill, air, etc. Stresses within

Table 2. Comparison of measurements (trend: 5, excellent agreement, amount: deviation from in-situ) (7).

| Model type | Stage 2 | | | | Stage 4 | | | |
|---|---|---|---|---|---|---|---|---|
| | Elastic | | Joint | | Elastic | | Joint | |
| | Trend, 1-5 | Amount, % | Trend, 1-5 | Amount % | Trend 1-5 | Amount % | Trend 1-5 | Amount, % |
| Fill pressure | 2 | 37±24 | 4 | 18±8 | 2 | 54±27 | 5 | 6±10 |
| Convergence of backfilled excavation | 3 | 28±10 | | | 4 | 17±11 | 4 | 13±10 |
| Convergence at mid-height of backfilled excavation | 4 | 40±10 | | | 5 | 8±11 | 5 | 12±3 |
| Stresses in ore above stope 3 | 4 | 33±21 | | | 4 | 46±14 | 4 | 26±9 |
| Convergence of orebody and alteration zone | 3 | 57±45 | | | 4 | 27±19 | 4 | 58±38 |
| Convergence of orebody | 4 | 118±30 | | | 4 | 43±6 | 4 | 27±10 |
| Roof deflection | 2 | 30±34 | 3-4 | 24±25 | 2 | 68±12 | 4-5 | 60±16 |
| Displacement of hanging-wall | 3 | 40 | | | 3 | 40 | 4 | 17 |
| Open stope convergence | | 18 | | | | 6 | | 3 |
| Horizontal deformation of alteration zone | | 26 | | | | 14 | | 22 |

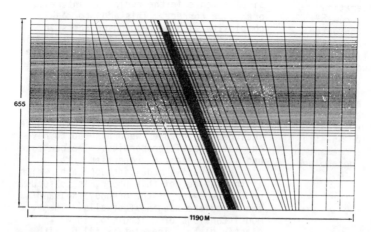

Figure 3. Finite Element Mesh Used in the Stage 4 Global Model (4004 elements). After Groth (11).

the slot normal to the walls can be determined as can wall convergence at points along the slot. Some of the tradeoffs involved in the use of such a model are

1. The hangingwall/footwall must have the same elastic properties.
2. The slot must be of constant thickness.
3. There is no opportunity to include the alteration zones.
4. There is no opportunity to include joint elements.
5. The non-linear behavior of the fill can be easily included.
6. The ore can be assigned an arbitrary load-deformation curve.
7. Modelling is done in the vicinity of the workings.
8. No information regarding behavior outside of the orebody is provided by the model.

In this case, the development and 'calibration' of such a model is accomplished by comparing predicted and measured quantities of

   o stope wall convergence
   o fill pressures
   o stresses in ore pillars

as a function of mining progress.

This paper deals with the development of such a model based upon the Näsliden data. This opportunity to explore alternative/complementary interpretations has only been possible due to the fact that the Näsliden research team published a significant amount of the data collected. The authors express their thanks and hope that other research teams will follow their lead.

The following sections include an examination of fill behavior, a brief discussion of the theory involved in the filled slot model, and the application of the model to the Näsliden data.

# 4 FILL STUDIES CONDUCTED FOR NÄSLIDEN

## 4.1 Stress-Strain Curves

The fill properties are particularly important to the present paper and the results of Börgesson (3) and Knutsson (4) will be discussed in some detail. Börgesson (3) carried out a series of triaxial and shear tests on the fill for the purpose of determining constitutive laws which could be used in the

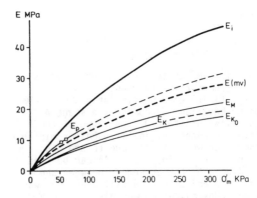

Figure 4. E Modulus Values Determined Using Five Different Test Methods. The Line E(mv) corresponds to the 'Average'. After Borgesson (3).

modelling. Figure 4 shows the results of five different testing methods. The values of

$E_i$ = the tangent modulus from the triaxial tests determined at the start of the tests.
$E_p$ = determined from pressure meter tests.
$E_m$ = determined from oedometer tests
$E_k$ = determined from isotropic consolidation tests
$E_{ko}$ = tangent modulus from the triaxial tests determined at the axial stress $\sigma_1 = \sigma_3/k_o$, where $k_o$ is the coefficient of earth pressure at rest ($k_o \approx 0.4$ for the fill).

are plotted as a function of the mean stress ($\sigma_m$)

$$\sigma_m = \frac{\sigma_1 + \sigma_2 + \sigma_3}{3} \qquad (1)$$

As can be seen the moduli are highly non-linear with respect to mean stress. The variation with the test is due to the fact that the samples sustain different shear stresses and they represent different stress paths. The 'average' curve represented by E(mv) on the figure can be expressed by

$$E = 490 \ \sigma_m^{0.7} \qquad (2)$$

151

where

$E$ = average modulus (KPa)
$\sigma_m$ = mean stress (KPa)

This is similar to curve $E_m$ obtained using the oedometer in the mine.

Two finite element simulations were run using equation (2) during the stage 4 modelling of stope 3. Only two of the five sequences originally used for mining the stope were studied due to the high computer costs involved. As can be seen in Figure 5, the agreement between the measured and predicted fill pressures was good. No indication was given as to how the use of the variable modulus curve affected either the stope convergence values or the stresses above the stope. The author suggested that to simulate the excavation properly (using a variable E modulus), the sequences should be smaller, preferably one sequence for every row of elements. This would eliminate irregularities in the stress curves.

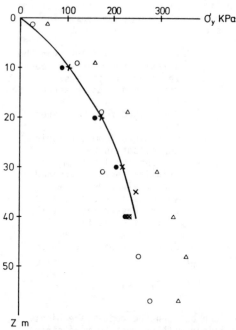

Figure 5.  Horizontal Fill Stresses as a Function of Depth Below the Fill Surface. [x = measured stress, o = stresses calculated from FEM calculations with constant $E_{modulus}$, and o = stresses from FEM calculations with variable E modulus]. After Borgesson (5).

## 4.2  Fill Pressure Prediction

Knutsson (4) performed a number of one-dimensional compaction tests on dry fill samples 254 mm in diameter and 75-85 mm in height (L). Sample shortening ($\Delta L$) was measured while increasing the applied pressure ($\sigma'$) stepwise to a maximum of 500 kPa (10, 20, 40, 80, 160, 300, 400 and 500 kPa). The deformation modulus (M) was calculated at each pressure level. A log-log plot of M versus $\sigma'$ yielded the values for constants m and $\beta$ in equation (3).

$$M = \frac{d\sigma'}{d\epsilon} = m \; \sigma_j' \; \left(\frac{\sigma'}{\sigma_j}r\right)^{1-\beta} \qquad (3)$$

where

$e_0$ = void ratio
$M$ = modulus of compressibility
$m$ = modulus number (a constant)
$\sigma_j'$ = atmospheric pressure expressed in the same units as M and $\sigma'$
= 100 kPa
$\beta$ = constant
$d\sigma'$ = pressure level
$d\epsilon$ = strain change calculated from zero load

Values of m and $\beta$ as a function of $e_0$ are given in Figures 6 and 7. Integrating equation (3) one obtains an expression which relates the increase in stress ($\sigma'$) due to an incremental strain of $\Delta\epsilon$ in the fill.

$$\sigma' = 100 \; \left[\Delta\epsilon \; m\beta + \left(\frac{\sigma_0'}{100}\right)^\beta\right]^{1/\beta} \qquad (4)$$

where

$\sigma_0'$ = the initial stress level in the fill (kPa)

Knutsson used this equation to estimate the increase in horizontal fill pressure due to measured side wall convergence during mining.

The other side wall component of fill pressure in cut-and-fill stopes is due to the dead weight loading of the backfill. To estimate this component, Knutsson (4) applied equations originally developed for the calculation of wall pressures in silos of rectangular cross-section.

The horizontal pressure against the wall is given by

$$P_s = \frac{g\rho A}{\tan \delta} \left[1-e^{\frac{-Zk_0}{A} \tan \delta}\right] \qquad (5)$$

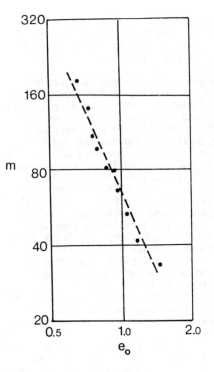

Figure 6. The Parameter m as a Function of the Void Ratio, $e_o$. After Knutsson (4).

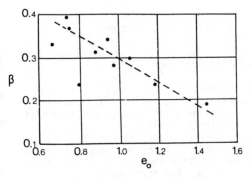

Figure 7. The Parameter β as a Function of Initial Void Ratio $e_o$. After Knutsson (4).

where

$A = \frac{a}{4}$ (pressure against short side of silo

$A = \frac{2ab - a^2}{4b}$ (pressure against long side of silo)

δ = angle of friction between the rock wall and the hydraulic back fill
z = depth below the surface of the backfill (m)
$K_o$ = relation between the horizontal and vertical stresses
ρ = bulk density $(t/m^3)$
a = short side of the silo (m)
b = long side of the silo (m)

For this particular case, he assumed

$K_o$ = $1 - \sin \phi' = 0.4$
$\phi'$ = Angle of internal friction for the fill = 36°
δ = 36° (same as for the fill)
ρ = 2.2 $t/m^3$

The appropriate horizontal dimensions (a, b) for stopes 1 - 3 are given in Table 3.

Table 3. Average Stope Dimensions

| Stope | Stope Dimensions | |
| | b (m) | a (m) |
|---|---|---|
| 1 | 200 | 17 |
| 2 | 160 | 14 |
| 3 | 150 | 15 |

Knutsson calculated the total side wall pressure due to the fill as the sum of the convergence (equation 4) and silo theory (equation 5) components.

Figure 8 is one example of the calculated and measured (Glötzl cell) results. In general, good agreement was observed. Knutsson suggested that the variations could be attributed to varying void ratios and backfill compression as well as to irregular room shapes.

He concluded that on the average 70-80% of the side wall pressure was produced by the dead weight of the backfill. Consequently only 20-30% of the total was due to wall convergence.

Unfortunately, the backfill pressure-convergence relationship developed by Knutsson is not very convenient for incorporation into finite element models.

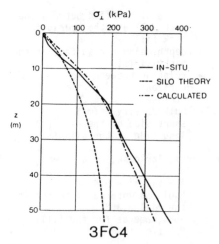

σ⊥ (kPa)

—— IN-SITU
--- SILO THEORY
-·-· CALCULATED

3FC4

Figure 8. Comparison Between Measured and
Calculated Fill Pressures
(Based on Silo Theory and
Convergence) for one Location
in Stope 3. After Knutsson
(4).

## 4.3 Finite Element Predictions vs Measurements

For most of the finite element modelling
the fill was assumed to have a constant
modulus ($E_f$) of

$$E_f = 32 \text{ kPa}$$

Other properties used in the modelling as
determined from laboratory testing
(Börgesson) are

| | | |
|---|---|---|
| $e_0$ | = | initial void ratio in the mine = 0.80 |
| ν | = | Poissons ratio = 0.32 |
| φ | = | friction angle (between 31 to 44° depending upon void ratio and stress level) |
| φ | = | 39° for the void ratio $e_0$ = 0.8 and the stress level 0-300 kPa applicable in the mine. |
| ρ | = | 2.2 t/m³ |

A comparison of measured and predicted
(Finite element) fill pressures for two
locations in stope 3 (13.5m and 32.5m
above the bottom of the stope) are given
as a function of stope height in figures 9
and 10.

As can be seen, the elastic models
predict a linear relationship between the
fill pressures and the height of the fill
above the measurement point. The
measurements reveal that this is true for
a small height change but then the

increase is much less than linear. With
the introduction of the joint elements one
obtains a result that better agrees with
the in-situ measurements both in trend and
amount.

As Börgesson (5) has pointed out, the
use of joint elements is a reasonably good
way to simulate the effect of slip between
the fill and the rock. This allows the
silo effect to come into play even when
using a constant E modulus for the fill.

A comparison of the measured and
predicted values of wall rock convergence
for two positions (15m and 57m above the
floor) in Stope 3 as a function of stope
height is presented in Figures 11 and 12.

The Stage 2 model predictions (both with
and without joints) differ considerably
from the actual behavior. The Stage 4
models agree somewhat better. The
introduction of joint elements along the
alteration zone was observed to have
little effect on the results.

Another indication of the "goodness" of
the models is a comparison of the
horizontal stress distribution in the roof
above stope 3.

This is plotted in Figure 13 for a roof
elevation of 295m and elements located
roughly 4m from the foot wall. As can be
seen, the predicted values are all higher
than those measured but have the same
trend. The Stage 4 joint model yields the
best agreement in amount.

## 5 THE FILLED INCLUSION MODEL

### 5.1 Introduction

An alternative way of modeling the cut-
and-fill sequence at the Näsliden Mine is
to approximate the vein by a filled slot
in an elastic medium.

The general procedure for the
determination of the boundary pressure and
closure distribution has been outlined by
Liberman and Khaimova-Mal'kova (12). The
extension of their approach to vein mining
systems in which pillars fill a portion of
the slot has been discussed by Hustrulid
and Moreno (8).

The approach is somewhat limited since
 o the medium (hanging and foot walls)
   must be assigned the same linear
   elastic properties (Youngs Modulus
   and Poissons ratio).
 o Only the stresses and convergences
   normal to the walls of the slot are
   determined.
 o The slot thickness and inclination
   must be constant.

154

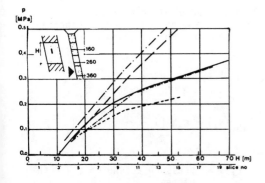

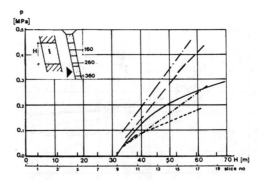

Figure 9.   Comparison Between Measured and
            Predicted Fill Pressure at
            Station 34, Stope 3.

Figure 10.  Comparison Between Measured
            and Predicted Fill Pressure at
            Station 84.

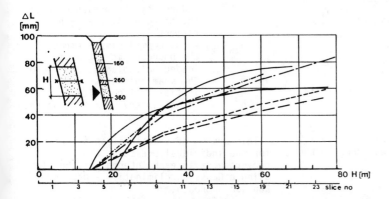

Figure 11. Comparison Between Measured and Predicted Convergence at Stations 43 and 64,
           Predicted Convergence Stope 3 (——, State 2: - · -  elastic, ---- joint,
           Stage 4: - · - elastic, ⁻ · ⁻ · ⁻ · joint).

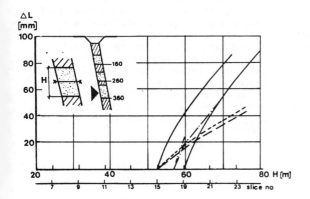

Figure 12. Comparison Between Measured and Predicted Convergence at Stations 154 and
           174, Stope 3. [— insitu, stage 2: — — elastic, ---- joint, stage 4:
           — · — elastic, - · - · - joint.]

155

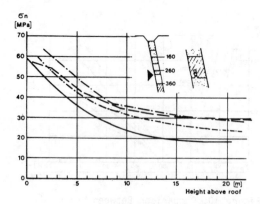

σn
[MPa]

Height above roof

Figure 13. Comparison Between Measured
and Prodict Stresses in the
Roof Above Stope 3. The Roof
Level in the Model is at 295 m
Elevation. [——— in situ,
stage 2: - - - elastic, stage
4, _._._ . ___ elastic, _._._.
joint]

o The problem must be two-dimensional
in nature.

These conditions do not allow for the
specific inclusion of layering in the
hanging and foot walls as was done in the
finite element models.

5.2  Liberman Approach

The approach taken by Liberman et al is to
approximate the slot by an ellipse. The
slot itself is then divided into a number
of intervals over which different boundary
pressures can be applied. The
relationship between the closure of the
slot and the applied pressure depends upon
the load-deformation relationship of the
material in the interval. It has been
found that the equation

$$\varepsilon = \varepsilon_0 \left[ 1 - e^{-\left(\frac{p}{p_0}\right)^N} \right] \qquad (6)$$

where
  $\varepsilon$ = strain
  $\varepsilon_0$ = initial porosity
  $p^0$ = Pressure
  N, $P_0$ = constants

represents the strain-stress behavior of
mine fill very well.
  The basic formula for determining the
boundary pressure and closure
distributions for a slot filled with a

nonlinear material in a gravity loaded
elastic medium as presented by Liberman
and Kraimova - Mal is given below

$$\sum_{k=1}^{n} P_k D_k (x_\ell)$$

$$= V_{\ell H} (x_\ell) - \varepsilon_0 h \left[ 1 - e^{-\left(\frac{P_\ell}{P_0}\right)^N} \right] \qquad (7)$$

where
  $P_\ell$ = pressure acting on the K-th
element
  $x_\ell$ = The coordinate of the mid-
point of the ℓ-th element
  $D_k(x_\ell)$ = The influence coefficient
such that the product of $P_k D_k (x_\ell)$
is the displacement
induced by the pressure $P_k$ at
the midpoint of the ℓ-th
element
  $V_{\ell H}(x_\ell)$ = the normal displacement of
the boundary in a gravity-
loaded half plane
  h = the seam thickness
  $\varepsilon_0$ = initial fill porosity
  n = no. of intervals making up the
slot.

This system of equations can be solved
for the individual pressures and
convergences in each interval of the slot
using a Gaussian elimination - Newton
Raphson approach.
  When using this equation to calculate
fill pressures in steeply dipping veins,
one finds that the calculated pressures
are smaller then the in-situ
measurements. The reason for this is that
the pressure in the hydraulic backfill
consists of two parts: one caused by the
dead weight of the backfill (silo effect)
and the second induced by the convergence
of the hanging wall and the footwall.
Application of the Liberman formula yields
only the pressure induced by
convergence. In the following section,
the modifications to Liberman's formula
required for calculating the total
pressure and convergence in the fill are
presented.

5.3  Modification For Gravity Loading of
     Fill

For the calculation of the normal pressure
in the slot, due to the weight of the
fill, the results from silo theory as
discussed by Knutsson (4) will be used.

156

The total stress (P) in the hydraulic backfill acting normal to the wall is the sum of that produced by the silo effect of the backfill ($P_s$) and that induced by the convergence ($P_\ell$) of the hanging-wall and the footwall.

$$P = P_s + P_\ell \tag{8}$$

where
$P_s$ = silo effect pressure
$P_\ell$ = convergence pressure

Equation (6) can be rewritten as

$$\varepsilon_{total} = \varepsilon_0 \left[ 1 - e^{-\left(\frac{P_s + P_\ell}{P_0}\right)^N} \right] \tag{9}$$

and

$$\varepsilon_{silo} = \varepsilon_0 \left[ 1 - e^{-\left(\frac{P_s}{P_0}\right)^N} \right] \tag{10}$$

Since

$$\varepsilon_{total} = \varepsilon_{silo} + \varepsilon_{conv}$$

then

$$\varepsilon_{conv} = \varepsilon_0 \left[ e^{-\left(\frac{P_s}{P_0}\right)^N} - e^{-\left(\frac{P_s + P_\ell}{P_0}\right)^N} \right] \tag{11}$$

Replacing the term

$$\varepsilon_0 \left( 1 - e^{-\left(\frac{P_\ell}{P_0}\right)^N} \right)$$

in equation (7) by

$$\varepsilon_0 \left[ e^{-\left(\frac{P_s}{P_0}\right)^N} - e^{-\left(\frac{P_s + P_\ell}{P_0}\right)^N} \right]$$

one obtains the more general equation

$$\sum_{k=1}^{n} P_k D_k (X_\ell) = V_{\ell H)} (X_\ell) - \varepsilon_0 h$$
$$\left| e^{-\left(\frac{P_s}{P_0}\right)^n} - e^{-\left(\frac{P_s + P_\ell}{P_0}\right)^H} \right| \tag{12}$$

This now includes both the effect of dead weight fill loading (silo effect) plus stope wall convergence.

## 5.4 Modification for Pillar (Solid) Intervals

In the original Russian article, the ellipse was used to bound only the fill zone. It is of course possible to include (a) the solid areas at the ends of the fill and (b) pillars internal to the total filled void within the overall ellipse representation. One needs only to provide the appropriate strain-stress relationship for the intervals involved. For the case when a linear elastic pillar is left in interval $\ell$, the term

$$\varepsilon_0 h \left| e^{-\left(\frac{P_s}{P_0}\right)^n} - e^{-\left(\frac{P_s + P_c}{P_c}\right)^n} \right|$$

in equation (12) is replaced by

$$\frac{P_\ell h}{E_\ell} \tag{13}$$

where

$E_\ell$ = elastic modulus of the material in interval $x_\ell$.

## 5.5 Numerical Solution of the Equations

The details for solving this set of non linear simultaneous equations has been discussed by Hustrulid et al (8), and only a brief summary of the steps will be repeated here.

The vein is divided into n elements and the modified Liberman's formula is written for each element. The resulting n simultaneous nonlinear equations can be solved by the following interation method

Step 1. Provide an initial guess for each $P_\ell$ and substitute these initial values into the right hand side of the modified Liberman's equation. One obtains n simultaneous linear equations. The required values for $P_s$ can be calculated directly using the silo equation.

Step 2. The Gaussian elimination technique is used to obtain a first approximation to the pressures in each interval.

Step 3. Because of the presence of the term

$$\varepsilon_0 h \left| e^{-\left(\frac{P_s + P_\ell}{P_0}\right)^N} \right|$$

the equation is ill-conditioned and in general one cannot get a convergent solution through the repeated use of Gaussian elimination. For the fill elements one uses the results from the Gaussian elimination as input for a Newton-Raphson evaluation of each equation. When solid elements are included several guesses for the pressures must normally be made due to the large modulus differences with fill values.

# 6 NÄSLIDEN MODELLED AS A FILLED SLOT

## 6.1 Properties of the Fill

Samples of the fill used at Näsliden were provided CSM by Boliden Mineral AB. A sieve analysis revealed that the size distribution was similar to that reported by Borgesson (3). Three one-dimensional load-deformation tests were performed on dry fill samples using a special split-platen consolidation device. For each test the initial void ratio ($e_0$) [defined as the ratio between void volume ($V_v$) and solid volume ($V-V_v$)]

$$e_0 = \frac{V_v}{V - V_v} \tag{14}$$

was calculated using solid and mass specific gravities. Corresponding initial porosity ($\varepsilon_0$) values [defined as the ratio between void volume and total volume (V)]

$$\varepsilon_0 = \frac{V_v}{V} \tag{15}$$

were calculated from the initial void ratios

$$\varepsilon_0 = \frac{e_0}{1 + e_0} \tag{16}$$

The fill properties for the pressure-strain curves shown in Figure 14 are given in Table 4. The curves can be well described by the Liberman et al (12) equation (6).

Table 4. Properties of the Fill

| Test | Initial Void Ratio ($e_0$) | Initial Void Volume Ratio ($\varepsilon_0$) |
|---|---|---|
| 1 | 0.942 | 0.485 |
| 2 | 0.835 | 0.455 |
| 3 | 0.754 | 0.43 |

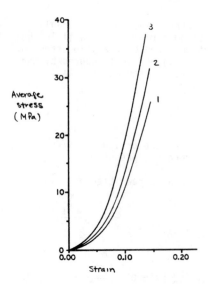

Figure 14. One Dimensional Stress-Strain Curves for Näsliden Fill Obtained Using the Split-Platen Consolidometer.

The values for the constants $\varepsilon_c$, N and $P_0$ are given in Table 5. In comparing the initial void ratios ($e_0$) with those suggested to be appropriate for Näsliden (3,4), the results of test 3 will be used for the simulations. In addition it will be assumed that
- o bulk density: $\rho = 2.2$ tons/m$^3$
- o angle of internal friction; $\emptyset = 36°$
- o friction angle between fill and wall: $\emptyset = 36°$

Table 5. Material Properties for Näsliden Fill

| Test | $\varepsilon_c$ | $P_0$ (MPa) | N |
|---|---|---|---|
| 1 | 0.485 | 126 | 0.54 |
| 2 | 0.455 | 345 | 0 43 |
| 3 | 0.430 | 256 | 0.49 |

## 6.2 Elastic Properties of the Näsliden Rock

In the initial simulations the same elastic properties for the hanging/footwalls and the orebody as used in the Stage 2 models were assumed. In the final runs, the hanging/footwall modulus was reduced to 27.6 GPa.

The elastic properties used in the final model are summarized Table 6.

Table 6.

| Youngs Modulii: | Footwall | 27.6 GPa |
| | Hanging wall | 27.6 GPa |
| | Orebody | 110 GPa |
| Poisson Ratio: | Footwall | 0.25 |
| | Hanging wall | 0.25 |

## 6.3  In Situ Stress Field

Leijon (2) found that the measured vertical field stress ($\sigma_v$) agreed well with that expected from gravity loading

$$\sigma_v = 0.026 \ H \qquad (17)$$

where
   H = depth below the surface (m).

The horizontal stress ($\sigma_H$) oriented perpendicular to the orebody was found to be

$$\sigma_H \ (MPa) = 6.1 + 0.045 \ H \qquad (18)$$

The nature of the Liberman formulation requires the use of a horizontal to vertical stress ratio ($\beta_0$) instead of absolute values.

$$\beta_0 = \frac{\sigma_H}{\sigma_v} \qquad (19)$$

The ratios for the mining zone (100-400 m below surface) are given in Table 7.

Table 7

| | Stress Ratio |
|---|---|
| H- depth (m) | $\sigma_H / \sigma_v$ |
| 100 | 4.1 |
| 200 | 2.9 |
| 300 | 2.5 |
| 400 | 2.5 |

For the purpose of this analysis, the ratio $\beta_0$ = 3 was selected.

## 6.4  Geometry Used For The Simulations

The model used to represent the mining of the orebody is shown diagrammatically in Figure 15. The "ellipse" is made up of 98 elements each having a length along the vein of 3.72 m (corresponding to the average cut of 3.5 m taken by the mine). The dip of the vein has been assumed to be constant at 70° and the vein thickness (normal to the dip) is 18.3 m. This geometry is the same as used by Groth (11).  As can be seen, stopes 1, 2, and 3 are represented by elements 68 -> 85, 40 -> 56, and 11 -> 33 respectively. Although one can "mine" the computer model in exactly the same cut-and-fill sequence as was actually done, for present purposes, the mining has been divided into the 10 sequences shown in Table 8.  The nomenclature indicates those elements which are solid and filled at any time. It is also possible to use unfilled (air) elements but this was not done.

## 6.5  Comparison of Predicted and Field Results

A great deal of fill pressure and stope wall convergence data were collected at Näsliden and the number of comparisons that could be made with model predictions is large.  In this paper comparisons are presented for the same approximate locations as discussed by Krauland et al (7).

Figures 16 and 17 for Stope 3 and Figure 19 for Stope 2 reveals good agreement between measured and predicted fill pressure values as a function of mining geometry.  A comparison of predicted and measured wall convergences for several positions in Stopes 2 and 3 are shown in Figures 19 through 22.  In general the agreement is good.  Another comparison which can be made is of the predicted and measured stresses in the roof of Stope 3.  As can be seen the agreement Figure 23 is also quite good.

It is noted that no particular effort has been made to select those field results considered 'best' for comparison, nor have the elastic properties for the ore and footwall/hanging wall rocks been further modified to achieve a better fit.  This of course could easily be done.

The relatively good agreement between predicted and measured values has been achieved without resorting to joint elements as was done in the FEM studies at Näsliden.

Table 8. Filled and Solid Elements for the Ten Mining Sequences

| Sequence | Solid element | Fill element | Sequence | Solid element | Fill element |
|---|---|---|---|---|---|
| 1 | 1-10 13-19 42-67 70-90 | 11-12 40-41 68-69 | 6 | 1-10 25-39 48-67 79-98 | 11-24 40-47 68-78 |
| 2 | 1-10 15-39 43-67 71-98 | 11-14 40-42 68-70 | 7 | 1-10 27-39 50-67 80-98 | 11-26 40-49 68-79 |
| 3 | 1-10 18-39 43-67 73-98 | 11-17 40-42 68-72 | 8 | 1-10 29-39 53-67 82-98 | 11-28 40-52 68-81 |
| 4 | 1-10 20-39 44-67 75-98 | 11-19 40-43 68-74 | 9 | 1-10 31-39 55-67 84-98 | 11-30 40-54 68-83 |
| 5 | 1-10 23-39 46-67 77-98 | 11-22 40-45 68-76 | 10 | 1-10 33-39 57-67 86-98 | 11-32 40-56 68-85 |

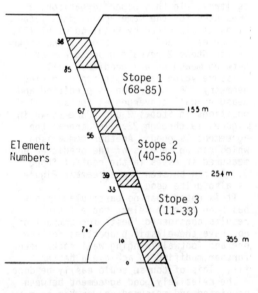

Figure 15. Mine Geometry Used for the Simulations

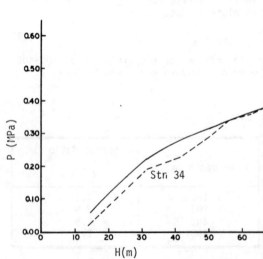

Figure 16. Comparison of the Predicted and Measured Fill Pressures in Stope 3 at Station 34.

160

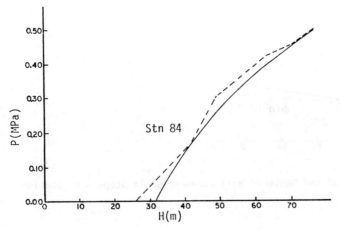

Figure 17. Comparison of the Predicted and Measured Fill Pressures in Stope 3 at Station 84.

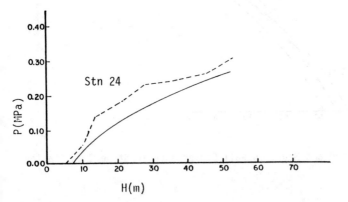

Figure 18. Comparison of Predicted and Measured Fill Pressure in Stope 2 at Station 24.

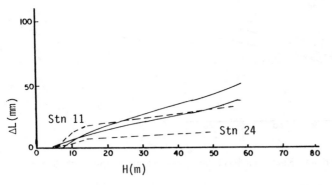

Figure 19. Comparison of Predicted and Measured Wall Convergence in Stope 2 at Stations 11 and 24.

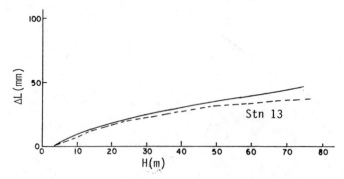

Figure 20. Comparison of Predicted and Measured Wall Convergence in Stope 3 at Station 13.

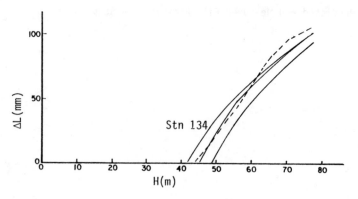

Figure 21. Comparison of Predicted and Measured Wall Convergence in Stope 3 at Station 134.

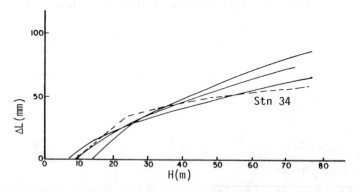

Figure 22. Comparison of Predicted and Measured Wall Convergence in Stope 3 at Station 34.

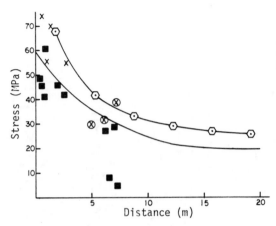

Figure 23. Comparison of Measured and Predicted Roof Stresses, Stope 3. [⊙ predicted in present models, __ fitted line in Näsliden Study, experimental points ■ , x, ❶]

## 6.6   Model Application

Although, as was noted earlier, the model does have limitations it can be used to make predictions regarding the outcomes of various possible mining sequences.

One application is to examine the build up of peak and average stresses in the pillars between stopes as the stopes are extended. A comparison of in-mine observations with the predictions may suggest geometries for which pillar crushing, bursting or other disruptive situations occur. Possibilities for avoiding these problems can be considered or preparations for handling the difficulties made beforehand. Pillar properties can be changed from one modelling sequence to the next thereby simulating the development of crushed/relaxed zones.

Hanging wall and foot wall stability problems are often related to the nature and amount of convergence. Wall convergence values can be obtained directly from this model and compared to mine stability conditions.

Another use of the model would be to provide a set of boundary conditions within the mined slot that could be applied to a more extensive finite element analysis. The non-linear properties of the fill could be evaluated in this simple model and the results applied as boundary conditions to the more elaborate model. This could prove to be a time and cost effective means of including the non-linear properties of the fill.

## 7   SUMMARY

A filled inclusion model has been developed for application to cut-and-fill mining systems. It allows for the simulation of gravity as well as wall convergence loading of non-linear fills. Predictions of wall rock convergence, fill pressures, and stresses in the pillars can be made.

Application of the model to data collected during the cut-and-fill mining of the Näsliden orebody has shown good agreement between measured and predicted data. It is felt that the model can provide some meaningful input to mine planning by itself as well as aiding in the development of more realistic finite element models.

## ACKNOWLEDGEMENTS

The CSM computer center provided the computer time necessary for conducting this work. The government of the People's Republic of China has provided the financial support for the study at CSM by Mr. Ye.

## REFERENCES

References 1 through 11 are from Application of Rock Mechanics to Cut and Fill Mining, (Ove Stephansson and Michael J. Jones, editors), The Institution of Mining and Metallurgy, 1981.

1. Krauland, N., "FEM Model of Näsliden Mine - Requirements and Limitations at start of Project", pp 141-144.

2. Leijon, B., Carlsson, H., and A. Myrvang, "Stress Measurements in Näsliden Mine", pp 162-168.

3. Börgesson, L., "Mechanical Properties of Hydraulic Backfill", pp 193-195.

4. Knutsson, S., "Stresses in the Hydraulic Backfill From Analytical Calculations and In-Situ Measurements", pp 261-268.

5. Börgesson, L., "Comparison of Hydraulic Backfill Measurements With Results From FEM calculations", pp 269-271.

6. Nilsson, G. and N. Krauland, "Rock Mechanics Observations and Measurements in Näsliden Mine", pp 233-249.

7. Krauland, N., Nilsson, G., and P. Jonasson, "Comparison of Rock Mechanics Observations and Measurements with FEM calculations", pp 250-260.

8. Hustrulid, W. and O. Moreno, "Support Capabilities of Fill - a non-linear Analysis", pp 107-118.

9. "The Nasliden Mine", pp 133-136.

10. Kolsrud, B., "The Näsliden Project - Why, What For and How", pp 137-140.

11. Groth, T., and P. Jonasson, "Application of the BEFEM Code to the Näsliden Mine Models", pp 226-232.

12. Liberman, Y. M., and R. I. Khaimova-Mal'kova, "Rock Pressure on Backfill with a Non-Linear Shrinkage Characteric", Soviet Min. Sci., 9, no. 2, March/April 1973, pp 109-112.

*Innovations in Mining Backfill Technology, Hassani et al. (eds), © 1989 Balkema, Rotterdam. ISBN 90 6191 985 1*

# Computer models for improved fill performance

R.Cowling
*Mount Isa Mines Limited, Mount Isa, Australia*

A.Voboril
*Mincom Pty. Limited, Brisbane, Australia*

L.T.Isaacs & J.L.Meek
*Department of Civil Engineering, University of Queensland, Australia*

G.Beer
*CSIRO Division of Geomechanics, Brisbane, Australia*

ABSTRACT: The constant drive to improve mining economics through changes to mining and processing methods requires associated changes in mine fill. With the large number of mining applications, changes to concentrating practice and the variety of fill types at Mount Isa Mines, it is essential to have techniques to assess the interaction of these as they impact on fill. The results of the application of a number of computer models are presented to demonstrate their role in a complex mining environment.

## 1 INTRODUCTION

Filling operations at Mount Isa Mines Limited are typified by a large number of fill types and applications, and high placement rates. In 1987/88 production of 5.4 million tonnes of copper ore and 4.5 million tonnes of lead/zinc/silver ore required the placement of 7.2 million tonnes of fill. Volumetrically, about 3.4 million cubic metres of void were created and 3.6 million cubic metres of void filled. The difference in void created and filled each year varies, depending on the status of mining but, generally, now there is a need to place more than is created. This arises from the volume created initially in open stoping operations before any fill can be placed. All of the copper production is by variations of open stoping and currently seventy percent of lead/zinc/silver, (hereafter referred to as lead), ore is from open stopes. The remaining production is by cut-and-fill methods.

Seven different fill types are used. This arises from the need to provide appropriate fills for different applications, to maximise the use of waste materials produced by processing

plants and to minimise costs. The ever-changing specifications of fill requirements resulting from changes to mining methods, and the changes in fill materials arising from improvements to processing plants require that there be methods to assess the consequences of these changes. In addition, because of the long time between placement and ultimate performance of some cemented fills, it is essential to be able to predict performance.

Although fill has been used for many years, there are still many aspects of its behaviour that are not clearly understood. For the above reasons, computer models are an essential tool for gaining improved understanding of fill behaviour, leading to improved performance and economics.

This paper presents the results of application of a number of computer models at Mount Isa Mines Limited.

## 2 MINING METHODS

Descriptions of mining methods have been presented by Alexander (1981) and Goddard (1981). Only brief details will be provided here, to enable an appreciation of the application of the computer models.

## 2.1 Copper Stoping

The 1100 orebody, at Isa Mine, is currently the source of most copper ore. Open stopes are typically 40m by 40m in plan and extend over the full height of the orebody, which ranges from 80m to over 300m, with an average height of 180m. As the orebody can be up to 500m wide, a number of stopes and pillars are required across the width of the orebody. To enable recovery of the pillars, the initial stopes, and the majority of the mined pillars,

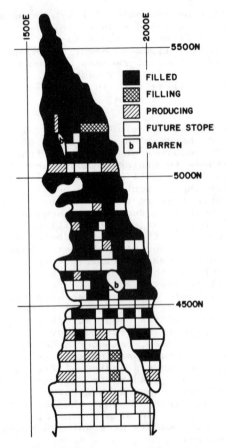

Figure 1. Idealised horizontal section of 1100 orebody showing status of mining and filling at June 1988.

have to be filled with a cement stabilised fill. The status of mining and filling at the end of

1987/88 financial year, is shown in Figure 1. With the dimensions and sequences implied in Figure 1, some filled primary stopes may have to be exposed on four sides. The main fill type is cemented rockfill and is described later.

## 2.2 Lead Stoping

Open stoping, and associated pillar recovery, is currently taking place in six orebodies at Isa Mine, and development has commenced at the nearby Hilton Mine for trial open stoping. The main differences between copper and lead areas are that the latter occurs in shale formations which dip at sixty-five degrees to the west, the plan areas of the stopes are smaller and, at most, only two sides of a filled stope are exposed by pillar recovery. The main fill type is cemented aggregate fill.

## 3 FILLING PRACTICE

A large number of materials are used in many different applications, as described by McKinstry (1989). Only the two main fill types will be described here. Cemented hydraulic fill is the common component of both fill types and is described first.

## 3.1 Cemented Hydraulic Fill

Cemented hydraulic fill is composed of de-slimed and, de-watered, concentrator tailings, (known locally as hydraulic fill), and cementing agents. Hydraulic fill is 100% minus 300 microns and 7% minus ten microns. To this is added Type A Portland Cement and copper reverberatory furnace slag in the ratio 3% Cement, 6% slag and 91% hydraulic fill. This mixture, cemented hydraulic fill, is reticulated at 69 weight percent solids through 150mm diameter pipelines and boreholes. Cemented hydraulic fill is seldom used on its own; instead it is a component of the two main fill types, cemented rockfill and cemented aggregate fill, which are used in copper and lead orebodies respectively.

## 3.2 Cemented Rockfill

Kennedy Siltstone is mined from a surface quarry, crushed and screened at 300mm and 25mm respectively, and conveyed to the top of two vertical passes in which it is choke fed to 13C sub-level and 15 level. These are respectively 590m and 730m below surface. There are 40 000t capacity pits at the top of each pass and both passes are kept full at all times. On 13C and 15 level, the rockfill is transported by conveyor belts to passes at the top of each stope to be filled. At these locations cemented hydraulic fill is introduced. The combined product is cemented rockfill.

## 3.3 Cemented Aggregate Fill

Lead ore is pre-concentrated by heavy medium separation. The reject is minus 15mm barren rock known as aggregate and when added to cemented hydraulic fill produces cemented aggregate fill. Aggregate is added at 20 to 25%, by weight. The product is reticulated through the same pipelines and boreholes as cemented hydraulic fill, at a pulp density of about 70 weight percent solids.

## 3.4 Fill Design

With the combination of different orebodies, variations in mining methods, number of fill types and inherent variation in fill products, it is essential to be able to predict performance under a wide range of varying conditions. Important factors include stope geometry, exposure stability, cost, drainage, bulkhead design, filling rates and pulp density.
The following sections will describe a number of models which have been developed and applied to obtain a better understanding of the many factors relating to fill design, performance and economics.

## 4 SIMULATION OF CEMENTED ROCKFILLING

Kennedy Siltstone and cemented hydraulic fill are introduced together at the top of the void to produce cemented rockfill. As shown in Figure 2, although the materials are introduced concurrently, there is large scale segregation, resulting in a cone of cemented rockfill and beach of cemented hydraulic fill. Experience has demonstrated that if the rockfill is permitted to build-up on walls which are to be exposed by subsequent pillar recovery, then there is a high probability that an inferior product will result. This leads to an increased probability of failure of the exposed fill wall. There is a need to limit the distribution of the rockfill within a stope, but at the same time to fill the stope as quickly as possible, by maximising the amount of rockfill. If all stopes were perfect prisms it would be a simple matter of introducing the components at the centre of the top of the stope and observing where the rockfill just touched the walls. In reality, stopes are not regular in shape and there is a cost involved in locating the pass to maximise rockfill usage and minimise filling time and total costs.

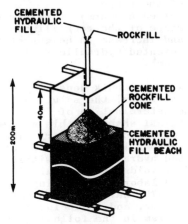

Figure 2. Idealised view of filling of copper open stope indicating zoning of materials.

Two programs have been developed to (i) simulate filling of previous stopes and (ii) predict the quantities for future stopes. The theory involved is identical with the exception that one program uses the known quantities

167

of materials used to fill a stope, whilst the other determines the materials for given pass locations and stope geometry.

## 4.1 FILDIS

FILDIS simulates the filling of previously filled stopes. Inputs include:
- three dimensional geometry of stope (pillar)
- location and orientation of fill pass(es) - (one or two passes can be considered)
- fill quantities - rockfill and cemented hydraulic fill - for each pour. (A pour is usually of less than 24 hours duration.)

Outputs include:
- three dimensional plots of the stope, showing where rockfill built-up against walls
- plan views showing the distribution of fill types at any level in the stope
- comparison of stope volume against fill volume.

In the simplest of terms the program completes the following steps:
1. Calculates trajectory of rockfill from top to bottom of the stope.
2. For each pour, determines the shape of the rockfill cone and increase in height of the cemented hydraulic fill beach.

Knowing the bulk densities of the fill types, the solution lies in the determination of geometrical shapes within a given three dimensional boundary. The only unknown is the quantity of cemented hydraulic fill within the rockfill voids. Large scale in-situ and laboratory testing (Dight,1979) have provided data on this factor, and is included in the program in the following fashion.

For a given pour, the quantities of rockfill and cemented hydraulic fill are given - RF and CHF tonnes respectively - and their ratio, R, = RF/CHF. The quantity of cemented hydraulic fill within the rockfill voids is equal to RF/F, where F is defined by the curve shown in Figure 3.

The lower limit of the curve is simply the case where all voids are filled by cemented hydraulic fill, and is derived as described below.

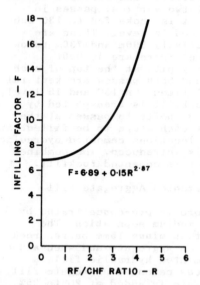

Figure 3. Relationship between rockfill / cemented hydraulic fill ratio, R, and proportion, F, of cemented hydraulic fill inside rockfill voids.

Rockfill has a dry solids density of $2.70t/m^3$ and a broken density of $2.10t/m^3$. Cemented hydraulic fill has a density of $1.37t/m^3$. For each cubic metre:

Volume of voids in rockfill
$= (1 - 2.10/2.70)$
$= 0.22m^3$

Mass of cemented hydraulic
fill in voids
$= 0.22 * 1.37$
$= 0.30t$

Mass of rockfill
$= 2.10t$

Lower limit of F
$= 2.10/0.30$
$= \underline{7.00}$

The remainder of the curve was established from results reported by Dight(1979). The implication of the form of the curve is that as the ratio R increases the volume of unfilled voids in the rockfill also increases, and results in a less well cemented product.

The model has been applied to a number of stopes and produces very good agreement between known volumes and fill distribution, and those derived by the program. Figure 4 shows the results of analyses to investigate the influence of ratio R on filling time, fill costs and volume of water to be drained from the fill.

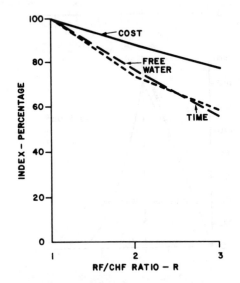

Figure 4. Cemented rockfill cost, filling time and free water as a function of ratio R.

The assumptions used in the analyses are:

Density
  Rockfill                    2.10t/m³
  Cemented                    1.37t/m³
    hydraulic fill

Slurry density                69%
(solids by weight)

Costs                         1987/88
                              statistics

Filling rates
  Rockfill                    R * 200tph
  Cemented                    200tph
    hydraulic fill

Filling time                  15 hours
                              per day

Using a ratio of R = 1 as an index equal to 100%, it can clearly be seen that changing from R = 1 to R = 3 reduces the costs by about 20% per cubic metre filled, reduces time (to fill the same volume) by over 40%, and also reduces free water by about 40%.

Having validated the model by back analysis, and demonstrated the potential benefits to be achieved by maximising ratio R, a variation of the program was developed to enable prediction of future fill ratios.

4.2 FILPAK

Program FILPAK predicts the quantities of rockfill and cemented hydraulic fill required for optimal filling of future stopes. One additional piece of information is required compared to FILDIS - namely the identification of whether, or not, a particular wall is to be exposed by future mining. Walls which are to be exposed can have the rockfill zone just touching them, whilst walls which are not to be exposed can have rockfill build-up against them. With this extra piece of information, the distribution of materials within the stope can be accurately determined.

In practice, mine planning engineers determine the optimal position for rockfill passes by a number of 'trial-and-error' runs of the program. Table 1 presents the results of the analyses of a particular stope for three different pass locations, and shows how sensitive the results are to small changes in pass geometry.

5 FILL DRAINAGE

The results shown in Figure 4 indicate how sensitive the quantity of water to be drained from filling stopes is to changes in ratio R. However, this is a total quantity and does not indicate at what rate the water will drain from the fill nor where it will report. There is also a need to determine the influence of different slurry pulp densities, to establish pore pressure levels within a filling stope and the effect of flushing water on total water balance.

Program ISAACS was developed for

Table 1. Influence of location of rockfill pass on quantities of rockfill and cemented hydraulic fill.

|  | Top of Pass RL = 2767, N = 5033, E = 1804 | | |
| --- | --- | --- | --- |
|  | Pass 1 | Pass 2 | Pass 3 |
| Dip | 90.0 | 87.5 | 85.0 |
| Bearing | 0.0 | 90.0 | 90.0 |
| Bottom of Pass   RL | 2729 | 2729 | 2729 |
| N | 5033 | 5033 | 5033 |
| E | 1804 | 1805 | 1807 |
| Length (m) | 38.0 | 38.0 | 38.1 |
| Tonnes | | | |
| Rockfill | 222779 | 272738 | 261024 |
| Cemented Hyd. Fill | 148916 | 113846 | 123769 |
| Ratio | 1.50 | 2.40 | 2.11 |
| Time (days) (200tph chf 15 hrs/day) | 50 | 38 | 41 |
| Free Water (tonnes) (69% density 35% moisture) | 14779 | 11302 | 12287 |
| Cost (pass 1 = 100) | 100 | 89 | 92 |

this purpose and has been reported previously by Isaacs (1983) and Cowling (1988). A three-dimensional version has been developed and is currently being validated. The results presented below report some of the uses of the original program.

5.1 Effect Of Pulp Density

Figure 5 shows the results of analysing the filling of stopes, of different size, with cemented hydraulic fill over a range of densities. All stopes are assumed to be filled for 8 hours per day, and to have drainage points on both sides at 40m intervals starting from the bottom. A number of observations can be made:

. Drainage performance in small plan area stopes ( Figure 5a) is much more sensitive to pulp

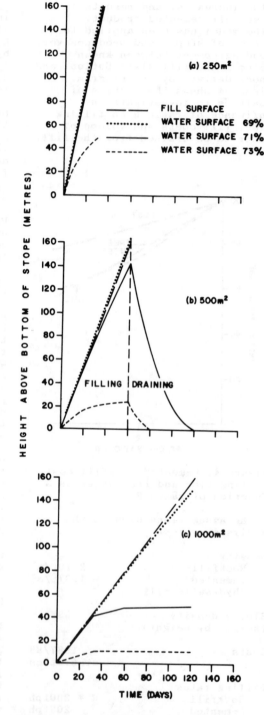

Figure 5. Fill and water heights during filling and draining of stopes of plan area of (a) 250m², (b) 500m² and (c) 1000m².

density than in large stopes
(Figure 5c).
. There is a large reduction in
drainage time, following the
completion of filling, for
small changes in slurry
density (Figure 5b).
. Large plan area stopes can be
filled by longer duration
pours than can be achieved in
small plan area stopes.

## 5.2 Pore Pressure

Results reported by Cowling (1988)
clearly indicate that pore
pressure is a function of the
stope size and existence and
location of blocked drains. Large
stopes experience higher pore
pressures in the bottom of the
stope than do small stopes, and
only blocked drains at the bottom
of stopes measurably effect pore
pressure distribution and
magnitude. Blocked drains above
the bottom of a stope have
negligible influence on pore
pressure or drainage performance.

## 5.3 Flushing Water

Normal practice at Mount Isa Mines
is to clean fill lines after each
pour with a known quantity of
flushing water. The quantity
varies as the distance between the
stope and the fill station.
    Figure 6 presents the results of
back analysis of a particular
stope. Observations during filling
recorded that there was a pond of
water on top of the fill for the
duration of the total filling of
the stope. Although the quantity
of flushing water used after each
pour is relatively small
(typically less than 50 tonnes),
the total is large when up to 50
pours can be used to fill a stope.
For the stope being analysed, a
total of 57 000 tonnes of cemented
hydraulic fill containing 11 000
tonnes of aggregate was placed.
About 24 000 tonnes of water was
used in the slurry and a further
2500 tonnes was used for flushing.
    Without flushing water the model
indicates that free water would
not pond on the fill surface,
whereas the use of flushing water
in the analyses resulted in 1 to 2
metres of water on top of the fill
at all times. This is in good

agreement with the quantities
observed. When this result is
combined with those discussed in
the section on pulp density, it is
apparent that any attempt  to
improve fill drainage by

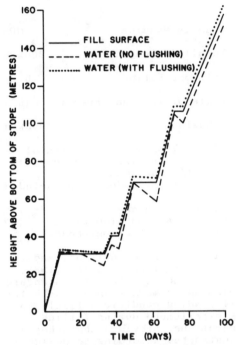

Figure 6. Influence of flushing
water on water height in lead open
stope during filling.

increasing pulp density has to
ensure that the improvement is not
lost by the use of excessive
quantities of flushing water. The
balance of pour time to total time
has also to be considered in
attempts to minimise free water.

## 6 BULKHEADS

To contain fill in stopes,
concrete brick walls - known
locally as bulkheads - are
constructed at all entrances. The
bulkheads have been described by
Cowling (1988) . In an exercise to
determine if the standard design
was applicable for all locations
and filling strategies, an
investigation of the modelling of
bulkhead stability was undertaken.
For this exercise a commercially
available, general purpose stress

171

analysis program, BEFE, was used, Beer (1986). Based on the results of large scale tests of bulkheads, as reported by Grice (1989), elastic and strength properties were derived and used in the analyses. They are summarised below.

| | |
|---|---|
| Elastic Modulus | 4GPa |
| Tensile Strength | 800kPa |
| Poissons Ratio | 0.2 |
| Angle of Internal Friction | 45° |
| Cohesion | 1000kPa |

Figure 7 shows the results of two of the analyses to compare alternative designs. Other investigations undertaken included

curved bulkheads
hitches at the side walls
ribs in front of bulkhead.

Based on these analyses, and the results reported by Grice (1989), it was concluded that the standard bulkhead is more than adequate for all loading conditions likely to be experienced during filling operations, provided that they are constructed according to design.


## 7 FILL EXPOSURE STABILITY

Cemented rockfill and cemented aggregate fill are eventually exposed by adjacent stope/pillar recovery. The fill exposures are usually of the same dimensions as the stopes and are typically 40m wide and 180m high for cemented rockfill in the 1100 copper orebody, and 20m wide and 160m high for cemented aggregate fill in the lead orebodies.
Program TVIS has been developed to enable simulation and prediction of fill exposure stability. As fill can eventually be exposed on four faces in the copper orebody, it is necessary that the model be capable of analysis of three dimensional geometries. TVIS is a finite element program with this capability. Its development and use have been reported previously

(Meek,1981). Only one application will be reported here.

A major requirement of most stress analyses is to obtain the correct initial and boundary conditions. With respect to stress analyses of fill, determination of the initial stress distribution is of vital importance.

Figure 8 shows the results of a number of analyses carried out to derive an appropriate model for the development of initial stresses in a filling stope. The models are described in Table 2.

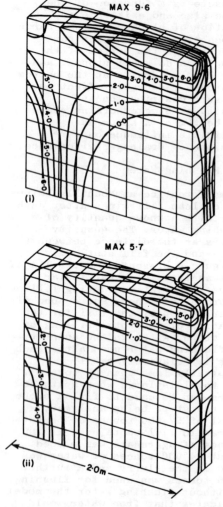

Figure 7. Principal stresses (MPa) on (i) straight bulkhead and (ii) straight bulkhead with strengthening rib. (Top left quadrant shown.)

An obvious conclusion to be drawn
from the field measurements
referred to in Figure 8, and the
results themselves, is that
initial stress levels are much
lower than depth stress. This is
presented graphically in Figure 9,
and is explained by the arching
developed during filling. The
support provided by arching is
critical to the performance of
fill exposures as is evidenced by
the fact that there is almost no
record of failure of fill in the
lead orebodies, where the fill can
only be exposed on two faces,
whereas there is increasing
evidence of failure in copper
stopes as the fill is
progressively exposed on 1, 2, 3
and 4 faces.

location on ratios and the
influence of these ratios on
economics, filling rates and
quantities of free water. All
stope designs for the 1100 orebody
are checked by program FILPAK for
filling strategies.

Table 2. Summary of models used to
investigate initial stresses in
filled stope. (After, Cowling
1983)

| Model No. | No. of Lifts | Modulus | Hydro-Static Option | Fill Type |
|---|---|---|---|---|
| 1 | 1 | Constant | – | CHF |
| 2 | 8 | Constant | – | CHF |
| 3 | 8 | Variable | – | CHF |
| 4 | 8 | Constant | Half | CHF |
| 5 | 8 | Constant | Full | CHF |
| 6 | 8 | Variable | Full | CRF |
| 7 | 8 | Constant | Full | CRF |

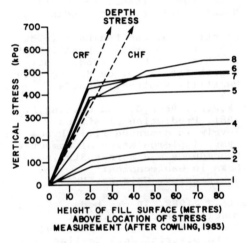

Figure 8. Initial vertical
stresses for models (1) to (7),
compared with measured values (8).

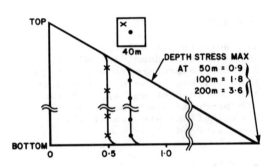

Figure 9. Effect of arching on
reduction of initial stresses
created during filling of open
stope.

8 CONCLUSIONS

The above sections demonstrate
that computer models have an
important role with respect to
understanding mechanisms of fill
behaviour and in designing
economic strategies. Based on the
results of application of these
models, a number of initiatives
are proceeding with the objective
of improving fill performance and
economics.
    Programs FILDIS and FILPAK have
highlighted the importance of pass

Based on results from the use of
program ISAACS, the plant in which
hydraulic fill is manufactured is
being partially re-designed to
enable a consistently higher
density fill (72 to 73 percent
solids) to be produced. Associated
work will ensure that the minimum
quantity of flushing water is
used.
    Investigation of bulkhead design
with program BEFE confirmed that
the standard design was adequate
for all conditions, and re-
emphasised the need for quality
control during construction and

173

routine inspections during filling.

Experience with cemented rockfill exposures has highlighted that initial exposures usually perform with zero or minimal failure, but that subsequent exposures progressively deteriorate in performance as the arching support, predicted by TVIS, is removed.

For the future it is envisaged that programs such as FILPAK and ISAACS (or its three dimensional equivalent, TRAVES) will be combined to produce a model which provides details of water drainage at the same time as ratios are derived. Developments are also planned to combine FILPAK with computer-based geology and mine planning software, such as CADMIN (Smith,1988). This will result in an ability to rapidly assess most aspects of alternative stope designs and their associated filling options.

## 9 ACKNOWLEDGEMENTS

The authors acknowledge the permission of Mount Isa Mines Limited to prepare this paper. Many colleagues were involved in the development and application of the programs discussed. Their contributions are gratefully acknowledged.

## REFERENCES

Alexander,E.G. and Fabjanczyk,M.W. 1981. Extraction design using open stoping for pillar recovery in the 1100 orebody at Mount Isa. Proc. Symp. Caving and Sublevel Stoping. Denver. USA.

Beer,G. 1986. Implementation of combined boundary element - finite element analysis with applications in geomechanics. Developments in Boundary Element Methods (Ed. P.K. Banerjee), Elsevier Appl. Science.

Cowling,R., Grice,A.G. and Isaacs,L.T. 1988. Simulation of hydraulic filling of large underground mining excavations. Proc. Sixth Int. Conf. Num.

Methods in Geomechanics. Innsbruck, Austria.

Dight,P.M. and Cowling,R. 1979. Determination of material parameters in cemented fill. Proc. Fourth ISRM Congress, Montreaux, Switzerland.

Goddard,I.A. 1981. The development of open stoping in lead orebodies at Mount Isa Mines Limited. Proc. Symp. Caving and Sublevel Stoping. Denver. USA.

Grice,A.G. 1989. Fill research at Mount Isa Mines. Proc. Fourth Int. Symp. on Mining with Backfill, Montreal, Canada.

Isaacs,L.T. and Carter,J.P. 1983. Theoretical study of pore water pressure in hydraulic fill in mine stopes. Trans. Instn. Min. Metall. (Sect.A:Min. Industry).

McKinstry,J.D. and Laukkanen,P.M. 1989. Fill operating practice at Isa Mine - 1983 to 1988. Proc. Fourth Int. Symp. on Mining with Backfill, Montreal, Canada.

Meek,J.L., Beer,G. and Cowling,R. 1981. Prediction of stress levels in cemented fill. Proc. Symp. Implementation of Computer Procedures and Stress Strain Laws in Geotechnical Engineering. Chicago, USA.

Smith,J. 1988. 3D ore body modelling and C.A.D.D. at Brunswick Mining and Smelting. Sixth District No.1 Meeting, Canadian Institute of Mining and Metallurgy.

# 4 Rockburst control

*Innovations in Mining Backfill Technology, Hassani et al. (eds), © 1989 Balkema, Rotterdam. ISBN 90 6191 985 1*

# Backfilling in deep level tabular stopes

A.J.Barrett, H.A.D.Kirsten & T.R.Stacey
*Steffen, Robertson & Kirsten, Johannesburg, RSA*

**ABSTRACT :** A four year programme of investigation into the backfilling of narrow tabular stopes at considerable depth using hydraulically placed deslimed tailings is described. The investigation dealt with stopes of shallow dip with a height of approximately 1 metre located at some 2000 metres below surface. Four main areas were dealt with: backfill material properties; interpretation of instrumentation results; interaction between rock and backfill; and support provided by backfill under seismic loads. The main conclusions are summarised as follows:

- Soft backfill can be used effectively for local support provided that the fill is placed close to the face and in contact with the hanging wall. These criteria may be more easily achieved if the fill material grading and backfill placement procedures are optimised.

- Elastic stress analysis techniques are not applicable for evaluating the stress distributions in the vicinity of stopes.

- Theoretical considerations indicate that backfill will perform well under dynamic conditions, and this is supported by practical observations.

## 1 INTRODUCTION

Backfilling has now been established as a common operation in many of the gold mines in South Africa. The investigations described in this paper were conducted during a pilot backfilling study at a major gold mine. Full plant tailings from the metallurgical plant was deslimed using a single stage cyclone at the pilot backfill plant. The deslimed tailings were then transported underground hydraulically via a specially formed borehole.

During the pilot project, filling was confined to a stope located at approximately 2000m below surface with an average stope height of approximately 1,2m. The dip of the reef in this area varies from 0° to 15°. The extensive faulting of the gold bearing sediments at the pilot site necessitates a scattered mining approach and the mining span usually does not exceed 100m. As a result, the primary objective of the backfilling was to provide local support to the hanging- and footwall rocks. The limited closures which occur in the working areas as a result of the mining geometry necessitate effective filling (ie intimate contact between the backfill and hanging wall) close to the face for reliable stope support. These requirements result in some unique backfilling problems at the mine. The authors were involved in the investigation of these problems over a period of approximately four years. Some of the more important conclusions of the investigation programme are discussed in this paper.

## 2 BACKFILL MATERIALS STUDY

A comprehensive materials testing programme was initiated some time before the commissioning of the pilot plant and continued throughout the pilot project.

This testing programme addressed two areas of testing, namely routine quality control testing and research-oriented testing.

Routine testing was carried out to maintain the quality of the backfill slurry and the placed backfill. For this purpose simple 'indicator' type procedures were favoured. The objective was to monitor material quality and to provide early warning of any parameters which vary significantly from the acceptable range. This testing was usually of an empirical nature and test results were not usually used for detailed analyses.

Research-oriented testing was aimed at measuring system properties and was usually more sophisticated in nature. In some cases, tests were conducted to obtain a relative comparison between the performance of alternative materials. In others, tests were conducted to obtain input parameters for numerical models.

The main objectives of the research-oriented testing programme carried out were as follows :

- to compare alternative backfill materials
- to investigate means of optimising material behaviour using additives, and by changing the material grading.

The results of testing programmes are discussed in detail by Barrett et al (1988) and the major conclusions of the study are summarised in the following paragraphs.

2.1 **Fines Content**

Reduction of the fines content of the backfill material has the following benefits.

- **Increased permeability** : the consolidation characteristics of any backfill material are closely related to that material's permeability. An increase in material permeability will result in more rapid consolidation during filling. The tendency of the material to move away from the hanging wall due to consolidation under its own weight shortly after

filling, generally referred to as shrinkage, will be reduced for fills of increased permeability. Conversely, an increase in permeability will allow a greater deposition rate for a similar amount of shrinkage.

In order to illustrate this point, consolidation times for freshly placed fills having a drainage path length of 2m were calculated. Certain assumptions were required in performing these calculations; the absolute times are therefore indicative only, with the trend being more relevant. The times presented are the calculated times for 95% consolidation.

- $k = 10^{-5}$ cm/sec, time = 13 hr
- $k = 10^{-7}$ cm/sec, time = 1 300hr
    = 54 days

(where k is the coefficient of permeability)

. **Increased placed density:** An increase in settled density with increased particle size was observed in the case of crushed rock materials. A series of simple sedimentation tests was conducted to determine whether the same trend is observed in the case of fine samples. Slurries with a range of particle sizes selected from a total tailings sample were agitated in a measuring cylinder and allowed to settle. The settled density was then determined, and confirmed that or pulp densities less than 1.7 a reduction of the fines in the slurry results in an increase in settled density. This relationship is illustrated in Figure 1.

. **Reduced passage of fines:** Although filters may be designed to prevent fine-grained materials from passing through, the degree of permeability of the filter cloth reduces in proportion to the entrained particle size. It is doubtful whether any cloth would be able to contain all colloidal-sized fines. Therefore, it has become essential to flocculate the classified tailings so as to enable the colloidal particles to bind together. Although this has been partially successful to date, a large percentage of fines still accumulates in the gullies. A reduced fines content in the backfill will greatly assist in overcoming this problem.

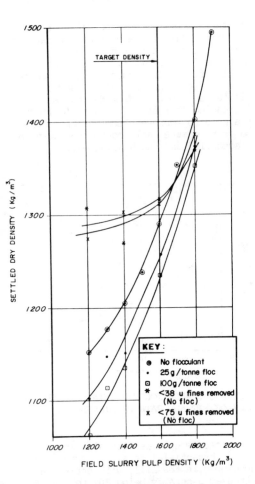

FIGURE 1 : RELATIONSHIP BETWEEN FEED
SLURRY DENSITY AND SLURRY DENSITY

## 2.2    Flocculent content

For the flocculents currently in use,
the ideal flocculent content for the
slurry lies in the range 10g/ton to
40g/ton.

Over- flocculation has the following
detrimental effects on backfill
properties:

.   At a high flocculent concentration,
    the backfill slurry tends to form a
    gel and the fines remain in
    suspension. This results in a
    reduction in placed density and an
    increase in retention of water.

.   At a high flocculent concentration,
    the backfill permeability reduces

significantly. This reduction will
tend to increase consolidation times
and hence increase shrinkage
movements.

## 2.3 Stress–strain characteristics

The stress-strain characteristics of
classified tailings with and without
aggregate addition are similar in the low
stress range, i.e. the stress-closure
behaviour will be similar for alternative
fill materials. Where stresses are in
excess of 500 kPa, however, the
aggregate-added fills are stiffer and
will provide a greater resistance to
closure. Consequently the aggregate-
added fills will become more attractive
than classified tailings fills only for
stresses of greater than 500 kPa. This
is illustrated in Figure 2.

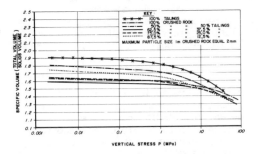

FIGURE 2 : CONFINED COMPRESSION
CHARACTERISTICS OF TAILINGS AND AGGREGATE-
ADDED BACKFILL MATERIALS

The mixing ratio for the tailings and
crushed rock which achieves the lowest
porosity appears to be 37.5% tailings :
62.5% crushed rock. This ratio is
dependent on the grading of the crushed
rock.

Barrett et al (1988) concluded that the
maximum potential closure resulting from
compression of the backfill in the working
areas at the mine was of the order of 7%
of the stope height. The potential
'shrinkage' resulting from sedimentation
of the slurry and subsequent self-weight
induced consolidation of the fill could be
many times this amount. The latter
phenomena are dependant not only on the
properties of the fill materials, but also
on the placement procedures used.

Analyses and optimisation of placement procedures are discussed in Section 4.

## 3  INTERPRETATION OF INSTRUMENTATION READINGS

Instrumentation was installed in a limited number of fill paddocks during the pilot project in order to assist with the evaluation of the fill performance. The instrumentation included pressure cells, closure meters, closure ride stations and wire extensometers. Visual observations of stope geometry and conditions were also recorded periodically. These observations provided an additional subjective assessment of the effects of backfilling.

Available instrumentation readings were analysed in detail after the pilot project had been underway for approximately one year. The main conclusions are summarised in the following sections :

### 3.1 Pressure Cells and Closure meters

Rosettes of hydraulic pressure cells were mounted perpendicular to the stope plane and in the dip and strike directions. Where possible, telescopic closure meters were installed close to the pressure cell locations.

Analyses of the pressure cell readings indicated that significant closure movements generally occurred before any stresses were monitored in the fill materials. This was attributed to 'lack of fit' resulting from time dependant shrinkage movements, stope convergence between instrument installation and filling, poor placement techniques, and difficulties in filling brows and cavities in the hanging wall. The 'lack of fit' was found to vary from 6% to 20% of the stope height with an average value of 14%. It was further evident from the analysis of the pressure cell readings that the time from filling to hanging wall/backfill contact varied from 20 days to 120 days. The corresponding face distance varied from 5m to 30m.

The slope of the stress-time plot for the pressure cells was typically approximately linear. The correlation was probably attributable to the mining cycle. Comparison of the slope of the

stress-time curves for the different pressure cells indicated that the rate of stress build up appeared to vary inversely with the 'lack of fit'. This relationship is illustrated in Figure 3. It is evident from Figure 3 that the efficiency of the fill as a support medium decreases with increasing lack of fit. This highlighted the need to improve placement procedures so as to ensure early contact between the placed fill and the hanging wall.

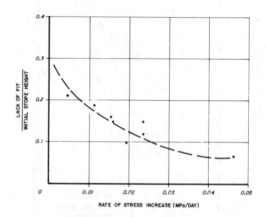

FIGURE 3 : RELATIONSHIP BETWEEN RELATIVE LACK-OF-FIT AND RATE OF STRESS INCREASE

The pressure cell results also indicated that substantial lateral pressures were developed in the fill, with strike and dip stresses generally being 30% to 80% of stresses perpendicular to the stope plane. These stresses are thought to originate from gravity effects, from lateral confinement of the backfill by the hanging- and footwalls, and from the fill surface not being parallel to the stope plane in some instances. For equilibrium, substantial shear forces must develop at the fill/hanging wall and the fill/footwall interface. These stresses must give rise to substantial clamping forces in the hanging and footwalls, helping to resist movement of potentially loose key stones. The potential adverse effect of these high shear stresses on the stability of gulley sidings was noted.

### 3.2 Closure Ride Stations

Closure ride stations were monitored in the gullies adjacent to fill paddocks. These stations generally consisted of three footwall points and one hanging wall point. Analyses of the readings taken at these stations indicated the following:

- Closure continues with time for a constant face distance. This is indicative of inelastic movement.

- Stations were generally installed at face distances ranging from 2m to 9m. It was probable that significant closure had occurred prior to installation of these stations.

- Ride movements appeared to be small. As the accuracy of the measurements was of the same order as the magnitude of the movements, no conclusions were drawn.

## 3.3 Extensometer Stations

Analyses of the extensometer results indicated the following :

- the major portion of bed separation was occurring before extensometer installation or above the top anchor (generally 3m).

- anchor slip appeared to have occurred in several cases, bringing into doubt the accuracy of the measurements.

The instrumentation measurements described gave an indication of lack of fit movements. The actual extent of this problem was larger than previously thought. Confirmation that the efficiency of the fill as a support medium was inversely proportional to lack of fit also highlighted the need to concentrate efforts on placement procedures.

## 4  BACKFILL PLACEMENT

As discussed in Sections 2 and 3, it became apparent early in the pilot project that the two most important aspects regarding the placement of backfill were:

- the fill should be placed regularly so as to minimize fill-to-face lag.
- the fill should be placed in intimate contact with the hanging wall, i.e. shrinkage should be minimised.

The former requirement necessitates a well disciplined mining and filling cycle. As the project progressed, it

became possible to ensure fill-to-face distances varying from 3m immediately after placement to a maximum of 6m after three further blasts. Fill was therefore placed in 3m paddocks every fourth day. This cycle was generally easily maintained provided that production personnel were timeous with sweeping of the footwall, and that problems with the backfill reticulation system did not delay filling.

## 4.1 Factors affecting backfill shrinkage

Minimising of shrinkage movements of the backfill proved to be complex. Tests were carried out in a laboratory flume (see Figure 4) to study the factors influencing backfill shrinkage.

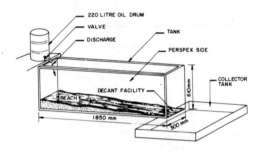

FIGURE 4 : LAYOUT OF LABORATORY FLUME

Standpipe piezometers were installed in the flume to monitor the development and dissipation of pore pressures during the filling process. Typical pore pressure observations during filling are illustrated in Figure 5.

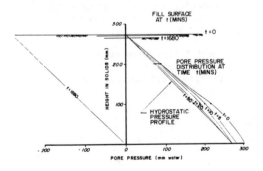

FIGURE 5 : PORE PRESSURE MEASUREMENTS FOR LABORATORY FLUME TESTS

The main conclusions of the study were:

. The expected shrinkage movement increased with:

    - increasing rate-of-rise of the backfill solids surface. This is illustrated in Figure 6.

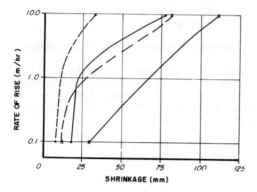

STOPE WIDTH = 1m

KEY

. CYCLONED TAILINGS (Cv = 600 mm²/min)
x CYCLONED TAILINGS (Cv = 6000 mm²/min)
—— IMPERMEABLE FOOTWALL
- - - PERMEABLE FOOTWALL

FIGURE 6 : RELATIONSHIP BETWEEN SHRINKAGE AND RATE OF RISE FOR 1m STOPE HEIGHT

    - with increasing fines content of the backfill.

    - as the stope height increases.

. The expected shrinkage reduces as the permeability of the footwall increases.

. No clear trends were evident regarding the influence on shrinkage of flocculents, or feed slurry density.

At the time of this study, backfill deposition in the stopes was taking place at a rate-of-rise of 0,5 to 1m/hr. Analyses to predict shrinkage movements, calibrated from the laboratory flume studies, indicated expected movements of 20mm to 70mm for a 1m stope height. These movements

appeared to be consistent with observations of actual shrinkage by mine personnel.

## 4.2 Placement techniques

Placement of backfill was initially carried out by operators manning backfill hoses in the fill paddocks. This technique was not successful and the large 'lack of fit' magnitudes discussed above were attributed to the placement techniques.

An innovative new technique referred to as 'pressure filling' was later developed, Hey (1988). The technique may be summarised as follows : filling takes place via multiple (typically 3) hoses attached to the hanging wall at third lengths of the paddock on dip. Backfill flow to these pipes may be controlled by means of valves. Filling takes place via these hoses, and on filling of the paddock near to a hose, slurry flow is transferred to the next up-dip hose using the valve. Special procedures were developed to prevent clogging of the hoses. After filling of the paddock through the uppermost hose, backfill flow was routed to the hose on the down-dip side of the paddock and second stage filling commenced. This was continued until back pressure developed, when the flow was diverted to the next hose.

Mine personnel had hoped that 'pressure filling' could be carried out so that excess water would be driven out of the paddock and replaced by fill solids. In this way, a 'prestress' might be developed in the backfill. A theoretical analysis was carried out to evaluate the feasibility of this using mass balance techniques and indicated :

. the lack of fit after primary filling of a 1.2m high stope could be up to 0.25m due to the presence of excess water in the paddock and shrinkage.

. the more water removed by decantation, the shorter the required duration of pressure filling. Decantation losses could be maximised by selection of a permeable filter fabric for paddock construction, in combination with effective wetting agents and/or flocculents. In addition, the deposition technique should be optimised so as to maximise the fabric area available for decantation of excess water.

. pressure filling could work provided the flow was maintained at sufficient pressure for sufficient time to drive off excess water.

. for an unfavourable combination of material parameters, the pressures required to achieve effective filling may be too high. It is therefore essential that the permeability of potential fill materials be maximised by removal of fines.

Following on from the above study, alternative filling methods were evaluated as follows :

. bottom up filling : the deposition nozzle is gradually retracted as the paddock fills until complete filling is achieved.

. top down filling : the deposition point remains at the top of the paddock until paddock filling is completed.

. multiple entry, staggered filling : similar to the pressure filling layout described above.

. multiple entry, simultaneous filling : the overall slurry feed is split by means of a manifold and simultaneous placement is carried out at multiple points.

. perforated pipe : a pipe, perforated with holes along its entire length is attached to the hanging wall along the full length of the stope. Flow through the pipe is from the top of the paddock.

In comparing the above methods, the following criteria were used :

. rate of rise of the backfill solids should be minimised.

. the area of barricade available for decantation of surface water should be maximised.

. the pressures occurring over the supply piping and barricade should

not exceed design values.

Comparison of the above alternative filling methods was carried out for stopes of moderate dip. This showed the 'top down' filling procedure to be most preferable, followed by 'multiple entry, simultaneous', 'perforated pipe', 'multiple entry, staggered' and 'bottom up' methods in order of decreasing preference.

The 'top down' filling technique was successfully applied at that point in time. It was found that, provided sufficient attention to detail was given to placement, reasonable and consistent results could be achieved.

## 5 STRESS ANALYSES OF BACKFILL PERFORMANCE

Analyses to assess the potential effectiveness of backfill were carried out some years ago by Ryder and Wagner (1978). These analyses were of the elastic mining simulation type commonly used for rock mechanics planning purposes in the deep level gold mines. Such analyses had indicated that backfill would be unlikely to be effective unless spans were large, and that stiffer backfills would be far more effective. This was contrary to the observed effectiveness of soft backfill in stopes of limited span in a scattered mining environment. In an attempt to resolve this discrepancy, a programme of stress analyses was carried out. This involved the following:

. three-dimensional elastic displacement discontinuity analyses of actual and idealised scattered mining stope layouts

. two-dimensional elastic displacement discontinuity analyses

. two-dimensional non-linear finite element analyses

The results of these analyses are dealt with in detail by Kirsten and Stacey (1989). The mining simulation analyses confirmed the previous predictions that backfill would provide very little benefit. Considerable attention was therefore given to the non-linear analyses.

The finite element solution scheme used was a continuum approach described by Stacey (1973), and further developed by Diering (1982). Two joint or fracture sets may be assigned to each element, and orientation, spacing, continuity, cohesive, frictional and tensile strengths specified for each set. Excessive shear and tensile stresses which develop on the joints are redistributed to unfailed areas using an iterative process.

The results of these analyses indicated a stress regime around the stope which was completely different from that predicted by the elastic analyses. In particular, vertical stresses in the region of the stope face were about half the elastic magnitude, and, instead of biaxial tension, a significant compressive stress in the immediate hanging wall of the stope. Further, deformations were approximately twice those calculated in the elastic analysis, and were in reasonable agreement with measured closures in scattered mining stopes.

Analyses were carried out using a single step for the mining and filling operation. To check on the validity of this, a multi-step mining and filling operation was simulated. This showed that the single step analyses underestimate the closures and backfill pressures by about 10%, and overestimate the vertical stresses immediately ahead of the stope face. Figures 7 and 8 show comparisons of stope closure distributions, and vertical stress distributions immediately ahead of the face. It was concluded that single step analyses are adequate for the purposes of comparative analyses. Using this approach, the effectiveness of backfill at different depths, greater spans and various horizontal to vertical stress ratios was investigated. From the results of all the finite element analyses, it was possible to draw the following conclusions:

- elastic analyses are inapplicable for evaluating the stresses and deformations around a stope in discontinuous rock. Consequently, they cannot be used to determine, reliably, the effectiveness of backfill.

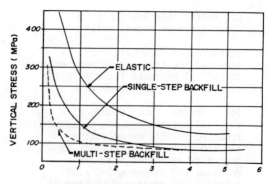

FIGURE 7 : VERTICAL STRESS DISTRIBUTION AHEAD OF FACE

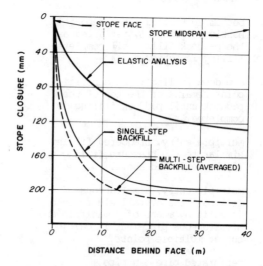

FIGURE 8 : HANGING WALL DEFLECTION FOR ALTERNATIVE ANALYSIS TECHNIQUES

- in a scattered mining environment, soft fills are as effective as stiff fills. The initial closure which occurs before the fill is placed, and the shrinkage of the fill which delays the point of initial loading of the fill, have a much greater effect than the difference between 'soft' and 'aggregate-added' fills.

- the most important attributes of backfill are the continuity of support that it provides, and the exponential increase in stiffness

184

that develops as the fill is compressed.

. at greater mining depths and spans, backfill stabilises the stope excavations successfully. The stress magnitudes adjacent to the stope are affected by the stiffness of the backfill - the vertical stresses increase and the horizontal compressive stresses decrease with an increase in fill stiffness.

. the rock in the immediate vicinity of the stope is shielded from the effects of depth, span and horizontal stress by the arching action in the super-hanging wall and footwall.

# 6 EFFECTIVENESS OF BACKFILL UNDER DYNAMIC LOADING

Under dynamic loading conditions, the rock adjacent to stope excavations is subjected to accelerations as a result of the passage of various waves and resulting vibrations. Backfill is effective in reducing the effects of such dynamic loading (Kirsten and Stacey, 1988):

i) transmission of seismic waves across the fill: in a filled stope, free surfaces are not present at which incident compressive shock waves may reflect completely to form tensile waves. The latter cause the ejection of rock blocks from hanging- and footwalls. Owing to the difference in density of the rock and backfill, however, there is some reflection, and only part of the wave energy is transmitted. For typical tailings backfill materials under significant compression behind the face, about 40% of the wave energy will be transmitted.

ii) resistance to ejection of rock blocks: the partial reflection indicated in i) above, and the effects of vibrations cause the ejection of rock blocks from hanging and footwalls. Newton's laws of motion and the load-stiffness relationship for backfill were used to calculate the depth to which ejected blocks would penetrate the fill. It was shown (Kirsten and Stacey, 1988) that even soft, unconsolidated fills are adequate to contain blocks, and maintain the integrity of the hanging- and footwall strata.

iii) damping of vibrations: according to Spottiswoode and Churcher (1988) backfill is likely to be beneficial for stability by damping vibrations of the hanging wall in particular. The presence of fill reduces the span of the detached strata and increases their resonant frequencies. Since high frequency vibrations are susceptible to attenuation by fractured rock, less energy is therefore available for excitation at the resonant frequency.

# 7 FIELD EVIDENCE OF THE BENEFICIAL SUPPORT ACTION OF BACKFILL

The effect of backfill as a support medium has been reported by several authors for both scattered and longwall mining environments. The following is a summary of some of these reports:

. during the implementation of an hydraulically-placed backfill operation, placement was interrupted on three occasions (Kirsten, 1987). On each occasion the stoping width increased substantially and the condition of the hanging- and footwall surfaces deteriorated substantially. This behaviour was reversed when the placement of fill was resumed. This case indicated the adequacy of soft backfill as a local support medium.

. Close and Klokow (1986) reported that in backfilled areas, an increase in face advance of 25% was achieved over a three month period. Seismic events caused significant damage only in those areas where fill was inadequately placed, or not at all. After a severe rockburst, it took three weeks to reopen an unfilled stope, whereas filled panels were accessible after three days.

. during filling of a panel, a rockburst occurred, but had no significant effect, other than to cause bulging of the geofabric walls of the backfill paddock (Gay et al,

1988). These authors also reported that gulley conditions were better in filled stopes, and that damage in filled stopes subjected to seismic events was generally less than in conventionally supported stopes.

in stopes with poor shale in the hanging wall, backfill improved the conditions markedly from those under conventional support conditions (Faure, 1988), demonstrating the local support effect. The number and magnitude of damaging seismic events appeared to be reduced, as was the associated damage. Falls of ground in gullies were eliminated almost completely.

## 8  CONCLUSIONS

The most important conclusions of the investigations described above into the use of 'soft' backfill in a scattered mining environment are as follows:
- 'Soft' backfills can be effectively used to provide local support provided that the fill is placed close to the face and in contact with the hanging wall (ie shrinkage should be minimised). These objectives may be more easily achieved if the grading of the fill material and the placement procedures are optimised.

- Elastic stress analysis techniques are not applicable for evaluating stress distributions in the vicinity of stopes.

- Theoretical considerations indicate that backfill will perform well under dynamic conditions, and this is supported by practical observations.

## REFERENCES

BARRETT, A.J., WRENCH, B.P. and BLAIR HOOK, D. 1988. An overview of materials testing relevant to backfilling at Vaal Reefs Gold Mine. Backfill in South African Mines, SAIMM, Johannesburg 1988, pp 61 - 75.

CLOSE, A.J. and KLOKOW, J.W. 1986. The development of the West Driefontein Tailings Backfill Project, South African Mining World, 5:4, pp 49 - 67.

DIERING, J.A.C. 1982. Mining simulation of tabular excavations using finite elements, Proc. 4th Int. Conf. on Num. Methods in Geomechanics, Nagoya (Japan) pp 1549 - 1555.

FAURE, M. 1988. Experimental backfilling at Harmony Gold Mine in Backfill in South African Mines, SAIMM, Johannesburg 1988, pp 405-428.

GAY, N.C., JAGER, A J and PIPER, P.S. 1988. Quantitative evaluation of fill performance in South African Gold Mines, in Backfill in South African Mines, SAIMM, Johannesburg 1988, pp 167-202.

HEY, C. 1988. Backfill placement technique on Vaal Reefs. Backfill in South African Mines, SAIMM, Johannesburg 1988, pp 347 -352.

KIRSTEN, H.A.D. 1987. Reliability of backfill as local support in narrow tabular stopes - state of the art with particular reference to Vaal Reefs, Unpublished Report CI4052/15, Steffen, Robertson and Kirsten, 79p.

KIRSTEN, H.A.D. and STACEY, T.R. 1988. Hanging wall behaviour in tabular stopes subjected to seismic events, J. S Afr. Inst. Min. Metall. 88:5, pp 163-172.

KIRSTEN, H.A.D. and STACEY, T.R. 1989. Stress displacement behaviour of the fractured rock around a deep tabular stop of limited span, J. S. Afr. Inst. Min. Metall., 89:1.

RYDER, J.A. and WAGNER, H. 1978. 2D Analysis of backfill as a means of reducing energy release rates at depth, Unpublished Report, Chamber of Mines of South Africa, 1978.

SPOTTISWOODE, S.M. and CHURCHER, J.M. 1988. The effect of backfill on the transmission of seismic energy, Backfill in South African Mines, SAIMM, Johannesburg, pp 203-217.

STACEY, T.R. 1973. Stability of Rock Slopes in Mining and Civil Engineering Situations, Doctral Thesis, University of Pretoria, 217p.

## ACKNOWLEDGEMENTS

The permission of Vaal Reefs Exploration and Mining Co. Ltd to publish the material contained in this paper is gratefully acknowledged.

*Innovations in Mining Backfill Technology, Hassani et al. (eds), © 1989 Balkema, Rotterdam. ISBN 90 6191 985 1*

# The in situ behaviour of backfill materials and the surrounding rockmass in South African gold mines

R.G.Gürtunca, A.J.Jager, D.J.Adams & M.Gonlag
*Chamber of Mines Research Organization, Johannesburg, RSA*

ABSTRACT: A backfill rock mechanics monitoring programme to determine the in situ behaviour of backfills and rockmass has been carried out at six South African gold mine for more than three years. The results obtained from this monitoring programme are presented in this paper.

## 1 INTRODUCTION

Backfilling is an important mining strategy to alleviate rock pressure problems in South African gold mines. The use of various types of backfill has increased considerably in recent years and at present 11 of the 32 major gold mines are placing backfill underground. It is planned that the rate of backfilling will reach over a million tons per month within the next five years.

Backfill is used for both regional and local support in gold mines. As regional support, the backfill is required to reduce the volumetric convergence in the mined out areas, and the high stresses at the face so that the occurrence of seismic events can be reduced by achieving lower energy release rates (ERR) and excess shear stresses (ESS). The percentage extraction of the reef is also increased when backfill is used as regional support in place of or to supplement stabilizing pillars.

As local support backfill is required to maintain the integrity of the fractured rockmass; to improve face and gully conditions; to reduce rockburst damage and to provide better stope width control. In order to quantify the potential of backfill to achieve these objectives it was necessary to determine the in situ behaviour of the backfills and their influence on rockmass behaviour. An additional benefit of the monitoring programme has been that it has given further insight into fractured rockmass behaviour.

A monitoring programme to determine the in situ behaviour of backfills and rockmass has been carried out at six underground sites for more than three years. The results obtained from this monitoring are presented and a conceptual model of the behaviour of a backfilled stope is derived from these results.

## 2 BACKFILL TYPES USED ON THE GOLD MINES

Metallurgical plant tailings and waste rock from the development of off-reef tunnels are the two main sources of backfill materials. Four types of backfill, namely classified tailings, dewatered full plant tailings, comminuted waste and cemented tailings are used on the gold mines. The first three types are used as uncemented backfills. At present approximately 90 per cent of the backfill placed underground in narrow stopes is classified tailings, 7 per cent is dewatered tailings, 2 per cent is comminuted waste and 1 per cent is cemented tailings. In addition to the small amount of cemented tailings placed in narrow stopes, a large tonnage is used in conjunction with massive mining methods in stopes with widths varying from 3 to 12 m. At present the amount of cemented backfill placed in high stopes accounts for 25 per cent of all backfill used in the gold mines.

The particle size distribution of the backfill materials is shown in Figure 1. To produce classified tailings the −38 μm particle size fraction is reduced from 50 per cent to approximately 10 per cent as a result of the classification process. The relative density (RD) of the classified tailings slurry is about 1,6 to 1,65 and the optimum and placement porosities are 40 and 52 per cent respectively. The optimum porosity is the minimum porosity attainable with a particular material and is related to the water content of the backfill.

Dewatered tailings are produced from vacuum filters or centrifuges positioned

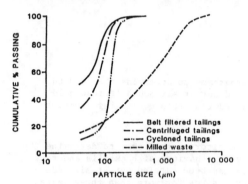

Figure 1 Cumulative particle size particle distribution of various backfill materials (After Stewart et al., (1986).

close to the stope. The product, of RD 1,9, is pumped to the stopes. The optimum and the placement porosities are 35 and 47 per cent respectively. The comminuted waste is obtained from development waste rock which is crushed underground to provide a material of good grading with a maximum size of 5 millimetres and with about 15 per cent material less than 38 micron. The fill which is pumped to the stopes at an RD of 1,9 has optimum and placed porosities of 22 per cent and 42 per cent respectively. The cemented tailings are simply obtained by adding 5 to 10 per cent binder to dewatered tailings. To facilitate pumping, a lesser amount of water is removed and the fill is placed at a density of 1,7 to 1,8. Because the initial stiffness of cemented tailings is determined by its binder content the porosity is of less consequence.

The uncemented backfills are usually placed in narrow, extensive and relatively flat dipping stopes where high stope closure rates and consequently high stresses in the fill, are generated. The cemented backfills are used currently at shallow depth (600 - 1 000 m) where the stope closures are relatively small and hence stiffer backfill is required. The cemented fill is placed primarily to

support the middlings in multi-reef mining situations and the strata falling within the tensile zone above wide reef, relatively shallow mining areas.

The uncemented fills are placed in geotextile material bags or in paddocks built with timber and geotextiles. The average width of these bags or paddocks is 2 - 4 metres depending on the face advance and the mining and filling

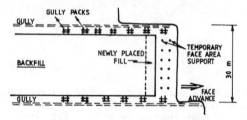

Figure 2 A typical layout showing the position of the backfill and the other support units in a panel.

cycle arrangements. A typical layout showing the position of backfill ribs in a panel is displayed in Figure 2. The concurrent mining and backfilling operation should be carried out so as to keep the backfill close to the face. This means that the backfill bags or paddocks should be built and filled as the face advances every 2 - 3 metres.

## 3 IN SITU BEHAVIOUR OF BACKFILLS

In situ performance of three backfill materials has been monitored at six mines for more than three years. The main objectives of the monitoring were:
 i) to obtain in situ confined compression stress-strain behaviour of various backfill materials,
 ii) to obtain stress-strain behaviour across complete backfill ribs, and
 iii) to determine in situ response of backfills to rockbursts.

### 3.1 Instrumentation for in situ monitoring of backfill

Two types of instruments have been developed to determine the in situ behaviour of backfills. These are;
 i) hydraulic stress meters, and
 ii) mechanical closure meters.
 These instruments which are able to reliably measure stress and closure in the aggressive environment in the backfill were developed at the Chamber of Mines Research Organization of South

Africa and are able to measure stresses up to 100 MPa and deformations of 0,5 m in the backfill.

A typical layout of instruments in a backfill paddock is shown in Figure 3. The hydraulic stress meters are placed to measure stress in three mutually perpendicular directions at each station. The one orientation, referred to as the 'vertical stress' is actually measured normal to the reef plane on the axis of maximum closure. A mechanical closure meter which measures closure normal to the reef plane is installed at each measuring station.

## 3.2 Results of in situ measurements

Two types of backfill behaviour, namely confined compression and complete backfill rib have been monitored.

The confined compression behaviour of backfill is usually measured in the centre of a paddock where the maximum confinement is likely to occur. Additional, further stress and closure meters are installed across the paddock towards the gully to establish the change of backfill behaviour at different points in the paddock.

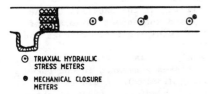

TRIAXIAL HYDRAULIC STRESS METERS

MECHANICAL CLOSURE METERS

Figure 3. The position of the backfill instruments in a paddock.

### 3.2.1 Confined compression behavior of backfills.

In this section, the in situ confined compression behaviour of the backfill types studied is presented.

#### Dewatered Tailings

Figure 4 shows the in situ confined compression behaviour of dewatered tailings backfill up to 45 MPa vertical stress at about 19 per cent strain. The measured strike and dip stresses are also included in Figure 4. The figure shows that the ratio of horizontal dip and strike stresses to the vertical stress varies from about 0,3 at low stresses to 0,5 at higher stresses. This stress

ratio is known as $K_O$ (Clark et al., 1988) and many other measurements show the $K_O$ ratio to vary between 0,3 to 0,6. Therefore, in order to simplify the subsequent figures, the horizontal stresses will not always be included.

Also shown in Figure 4 is the good agreement between the in situ behaviour and the standard confined compression laboratory test result. The laboratory test curve used in the figure is for a backfill sample with an initial porosity of 40 per cent, very close to the measured porosity of the fill in the paddock in which the instruments were installed.

It is important to note that all of the stresses that have been measured are significantly higher than those that are predicted by elastic stress analysis programs such as MINSIM-D using conventional input parameters. For example the simulation for the situation shown in Figure 4 predicts a vertical stress of only 23 MPa for the position where the measured stress was 45 MPa. This discrepancy is accounted for by the difference between the theoretical convergence and the measured closure. The reasons and implications of this will be discussed in Section 4.1.1.

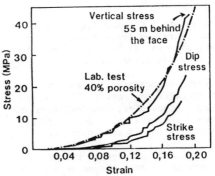

Figure 4 The in situ confined compression behaviour of dewatered tailings

#### Classified Tailings

The confined compression behaviour of classified tailings with different porosities has been measured at three sites. The initial porosity of the placed fill is of paramount importance to its stress-strain performance (Jager et al., 1987). It is known that as the porosity of backfills reduces, the backfills display stiffer stress-strain response.

The first result was obtained from a mine where classified tailings backfill

with about 52 per cent porosity was
used. The measured confined compression
stress-strain curve is shown in Figure 5.

Since the scattered mining method is
used on this mine, the closures were
relatively low. Hence the maximum
vertical stress measured was only 6 MPa
for 14 per cent strain at this station
when it was 70 m behind the face. The

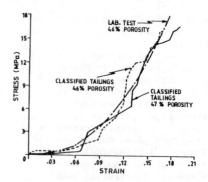

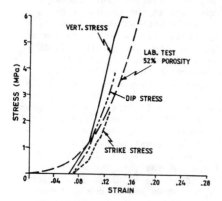

Figure 5 The in situ confined compression
behaviour of classified tailings
backfill with 52 % porosity.

$K_0$ ratio is about 0,6 and 0,4 for dip
and strike stresses respectively and the
agreement between the laboratory and the
in situ vertical stress curves is reasonable.

The measured stress-strain behaviour of
classified tailings backfill with 46 and
47 per cent porosities, obtained from the
other two mines, are shown in Figure 6.
A laboratory stress-strain curve for
material with an initial porosity of 46
per cent is also included in the graph.
It demonstrates a good agreement with the
measured data. The $K_0$ ratios for this
type of fill also varied from 0,4 to 0,6.

The results in Figure 6 indicate that
vertical stresses of 16 and 17 MPa were
measured for 17 and 19 per cent strains
for the classified tailings with 46 and
47 per cent porosities respectively.

As for the dewatered fills, the measured stresses at all three mines discussed
above are also considerably higher than
the calculated theoretical stresses.

Comminuted Waste

The in situ confined compression behaviour of the comminuted waste type of
backfill has been established up to
25 MPa vertical stress as shown in

Figure 6 The in situ confined compression
behaviour of two classified
tailings backfills with 46 and
47 % porosities.

Figure 7. The corresponding strain
value is about 14 per cent. The dip and
strike stresses are 10 and 11,5 MPa for
the same strain giving a $K_0$ ratio of
between 0,4 and 0,45.

Also shown in Figure 7 is the good
correlation between the laboratory curve
for 27 per cent initial porosity
material and the measured stress-strain
curve. The discrepancy in the initial
part of the curves is mainly due to in-
complete filling of the backfill to the
hangingwall and subsequent shrinkage of
the material resulting from drainage and
consolidation.

A back analysis was carried out to
simulate stresses generated in the backfill by using MINSIM-D. Again it was
found that the predicted stress is about
half of that measured in the backfill.

3.2.1.1 Comparison of confined compression behaviour of backfills

Figure 8 shows the comparison between the
in situ behaviour of dewatered tailings,
comminuted waste and classified tailings
types of backfill. Firstly, the figure
clearly shows that the comminuted waste
backfill is the stiffest of the
materials. Secondly, the backfill with
46 per cent porosity is stiffer than the
equivalent tailings with 52 per cent
porosity, as would be expected.

3.2.2 Complete backfill rib behaviour

The behaviour of complete backfill ribs
of comminuted waste, classified and
dewatered tailings is being monitored at
a number of sites in the gold mines. As

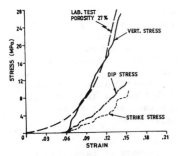

Figure 7 The in situ confined compression behaviour of comminuted waste backfill with 27 % porosity.

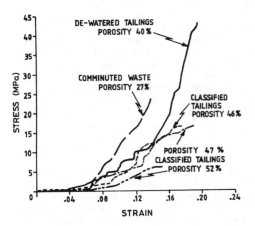

Figure 8 Comparison of in situ confined compression behaviour of different backfills.

each station was plotted from data obtained about 9 months after installation, when the face had advanced 34 m from the instruments. The vertical stress measured at station C3 at different stages of closure development is also indicated.

Figure 9 clearly shows that the closure measured inside the fill is 40-50 per cent less than that measured at the closure station outside the fill adjacent to the gully. However, it can be seen from the figure that the initial closure at all stations develops more or less

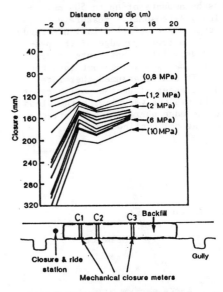

Figure 9 The distribution of closure across a backfill paddock.

similar results are being obtained for all the different types of fill, only the results from a dewatered tailings site will be discussed in this section.

Figure 9 shows the variation of closure along the dip direction, of a filled panel as the face advances. Three triaxial stations C1, C2 and C3 were located 3 m, 5 m and 11 m away from the edge of the paddock respectively. In addition, a closure station was also installed in the gully outside the backfill paddock.

The curves shown in the figure indicate the closures measured at each closure meter on the same day. The upper curve shows the measurements taken about 2 weeks after the installation of the meters, when the face was 8,5 m from the instruments. The lowest curve which displays the maximum closures measured at

equally up to the point where the vertical stress in the backfill reaches 1 MPa. Subsequently the closure inside the panel is restricted while the station in the gully continues to be displaced at a relatively high rate until a stress of 3 MPa is developed in the fill. The closure differential between the gully and the panel mainly develops during this period. With higher backfill stresses, the amount of closure developing inside and outside the fill is fairly similar. This type of behaviour of the rockmass in response to the stress development in the fill indicates that large inelastic movements (i.e. bed separations and dilations) may take place only when the stress in the backfill is less than about 3 MPa. At higher stresses the strata become clamped, and the inelastic deformation is significantly reduced (see also

191

Figure 11). The clamped beam bridges across the gully and closes almost uniformly over the filled panel and unfilled gully area and converges into the excavation as a single structure.

A consequence of the high initial differential movements between points inside and outside the fill is that rockfalls in gullies are more likely to occur in the period before the rockmass is clamped and when the fractured hangingwall becomes subjected to tensile strains induced by differential movements.

It is observed underground that the edges of a backfill bag or paddock bulge outwards by amounts of up to 500 mm. Preliminary extensometer measurements indicate that the backfill dilation extends only 2 to 3 m from the edge of the paddock, and very little horizontal movement of the fill takes place deeper into the paddock. It can be assumed therefore that the stresses along the periphery of a backfill rib are very low, being almost zero at the edge where there is effectively no lateral confinement.

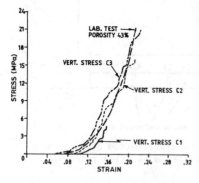

Figure 10 The vertical backfill stresses measured at different points in a backfill paddock. (seeFig. 9)

However, the central section of the rib demonstrates a different picture. The vertical stress-strain curves for Stations C1, C2 and C3 are plotted in Figure 10. A laboratory stress-strain curve for 43 per cent porosity material is included in the figure. The stress-strain response at stations C2 and C3 are very similar and show a good agreement with the laboratory curve. The stresses at C1 are somewhat lower indicating the influence of lower confinement towards the edge of the paddock. This means that the stresses at different points in a fill, outside the dilating zones, are developed according to the confined compression behaviour

of the material. However, the induced strains are larger at the narrow stoping width sections of the stope and hence the corresponding stresses are also larger at those sections. If it is assumed that the stoping width is the same across the bag, a uniform stress distribution is expected in the central sections of the bag while the stresses approach zero towards the edges of the bag.

### 3.2.3 In situ response of backfills to rockbursts

Jager et al. (1987), studied the relative ability of backfill and other support systems to counteract the kinetic energy imparted to the rock by seismic ground motion. They calculated the average work done by four support systems between the stope face and 15 m back, during rapid convergence of 300 mm. It was concluded that the backfill was superior to the other support types in absorbing seismically generated energy.

Since the publication of that paper other rockbursts have occurred in the backfill monitoring sites and the stress-strain response of various backfills to rockburst induced closure have been measured. The results are listed in Table 1.

Table 1 clearly shows that the average work done by the different backfills during a rockburst is considerably higher than can be achieved by any other conventional support system such as timber packs, timber props, hydraulic props etc. It was calculated by Jager et al., (1987) that the average work done by a timber pack system close to the stope face is about 43 kJ/m$^2$ for 300 mm rapid convergence and 70 - 100 kJ/m$^2$ for hydraulic prop systems. The figures listed in Table 1 indicate that the maximum work done by the backfill during the rockbursts was 260 kJ/m$^2$ and averaged 117 kJ/m$^2$ for the six rockbursts for which results were obtained. These measurements are unique and have very important implications as to the design of support systems for rockburst conditions. However, more measurements of this nature need to be made and interpreted before final conclusions can be drawn. It should be noted also, that compared to filled stopes, excessive closure can occur in conventionally supported stopes during seismic events due to the relatively low support resistance supplied by most of these support systems. The fact that the closure in backfill stopes is considerably less

Table 1  The stress-strain response of backfill to rockbursts in the backfill sites
* Two events occurred on the same day

| Backfill Type | Distance to Face at the time of the event (m) | Closure Increase measured (mm) | Vert. Stress before the event (MPa) | Stress Increase measured (MPa) | Work done (kJ/m²) | Magnitude of the event M_L |
|---|---|---|---|---|---|---|
| Classified Tailings 45 % Porosity | 11 | Stn 1   48 Stn 2   46 Stn 3   46 | 1,3 2,5 0,16 | 5,2 6,3 0,11 | 180 260 10 | 2,8 |
| Classified Tailings 52 % Porosity | 12 | Stn 1   37 Stn 2   62 Stn 3   35 | 0,2 0,0 0,045 | 3,2 '2,8 0,3 | 67 87 7 | 2,8 |
| Classified Tailing 46 % Porosity | 25 | Stn 1    6 Stn 3   7,5 | 15 16 | 1 0,5 | 93 122 | 1,7 |
| Classified Tailings 46 % Porosity | Panel 1  27 | 30 | 4,5 | 3,5 | 188 | 2,1 |
| | Panel 2  9,5 | 25 | 1,6 | 0,9 | 103 | |
| Dewatered Tailings 40 % Porosity | 18 | 20 | 0,6 | 2,2 | 44 | 2,1 |
| Comminuted Waste 27 % Porosity | 52 | 10 | 23 | 2 | 240 | *1,45 1,85 |

during seismic events than in conventionally supported panels, partly explains
the reduced damage reported in backfilled
stopes subjected to rockbursts.

## 4 IN SITU ROCKMASS BEHAVIOUR

Certain changes in the rockmass behaviour
can be expected with the introduction of
backfill in the stopes.  It is important
to identify and quantify these changes
with a view to understanding the behaviour of the rockmass surrounding a
filled stope so that computer programs to
model the interaction between the backfill and rockmass can be developed, or
existing models modified appropriately.
One of the important roles of backfill is
to provide adequate local support at the
face and in the gullies by reducing rockfalls induced by gravity or rockbursts.
If the interaction between the rockmass
and the backfill is well understood and
the influence of the backfill on the
fractured rockmass behaviour is determined, then the effectiveness of the backfill as local support can be maximized.

To achieve these objectives, closure
and extensometer measurements have been
carried out in filled and equivalent
unfilled stopes at a number of mines and
the results of this monitoring programme
are presented in the following sections.

The equivalent unfilled stopes were
chosen so that the depth, the geology,
and the mining related factors such as
stope geometry are as similar as possible
to the filled stopes.

### 4.1 Closure Measurements

Closure measurements have been carried

out along gullies and inside panels at
the monitoring sites.  The objective was
to determine the influence of different
types of backfill on closure by comparing
results obtained in filled and equivalent
unfilled stopes.  The difference between
elastic convergence and the closure in
filled stopes was also investigated as it
was thought that the high stresses
measured in backfill should reduce the
difference between the elastic and the
measured closures.

A summary of the closure rates
determined on three different mines for
the three types of backfill discussed in
this paper is given in Table 2.  The results reveal that the closure rates expressed in terms of closure per metre of
face advanced and per day are significantly less in filled panels than in
unfilled panels, though the difference is
less marked in certain of the gullies.

It is also interesting to examine the
differences in closure recorded in
gullies and in panels.  In filled panels,
the closure rates in the gullies are

Table 2 Summary of closure measurements

| Sites | Number of Closure Stations | FILLED | | UNFILLED | |
|---|---|---|---|---|---|
| | | Gullies | In Panels | Gullies | In Panels |
| Comminuted Waste | 156 | 20-35mm/m 1,5-5mm/day | 6-15mm/m 0,7-1,3mm/day | 10-40mm/m 4-12mm/day | 50-65mm/m 15-30mm/day |
| De-watered Tailings | 107 | 30-70mm/m 1,2-10mm/day | 4-40mm/m 6,6-25mm/day | 10-60mm/m 1,2-7mm/day | Not Available |
| Classified Tailings | 55 | 5-10mm/m 0,6-3,3mm/day | 3-7mm/m 0,5-1mm/day | 17-39mm/m 3,6-16mm/day | 50-60mm/m 15-25mm/day |

193

twice those measured inside the panels
(Section 3.2.2). However, in unfilled
panels the opposite is the case, the
closure is higher inside the panels than
in the gullies. This can be explained as
follows. In both filled and unfilled
panels, packs are used currently along
the gullies. The difference in stiffness
lies in the panel support where packs or
elongates can be used in unfilled
panels. The support resistance generated
by backfill is considerably greater than
for systems based on the other support
units. Therefore, in filled panels,
since the backfill is a stiffer support
system than the packs, the backfill inside the panels reduces closure more than
the packs along the gullies. Moreover,
in the unfilled panels, these packs develop higher support resistances than the
pipe sticks after a certain amount of
closure, hence the closure rates reduce
along the gullies. This means that the
closure, particularly the inelastic components of the closure, are reduced by
different amounts by the different types
of support used in stopes.

Composite diagrams showing the difference in closure measured in the filled
and the unfilled panels are shown in
Figure 11. The elastic convergence profiles as computed by MINSIM-D, where a
Young's Modulus of 70 GPa was used as an
input parameter, are plotted for comparison with the filled and the unfilled closure profiles. It is seen that the component of inelastic closure can be several times the elastic convergence and
that the backfills reduce the closure
considerably compared to that in the unfilled panels. However, the amount of
closure is not reduced to that computed
by elastic analysis, inferring that some
permanent inelastic deformation is retained in the fractured rock.

The function of local support is to
maintain the integrity of the hangingwall
beam. This might be achieved by restricting closure and hence the differential
movements of key blocks in the rockmass.
If the key blocks do not move with respect to each other, it is unlikely that
they would lose the cohesion between
their common surfaces and there should be
a reduction in rockfalls in the filled
stopes due to the observed reduction in
closure compared to the unfilled stopes.
Work is still continuing at underground
sites to quantify any reduction in rockfalls in filled stopes.

It is also interesting to observe for
all the cases in Figure 11 that there is
a significant difference between the com-

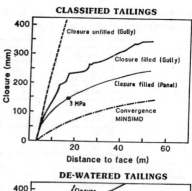

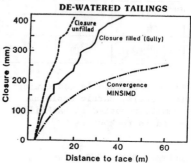

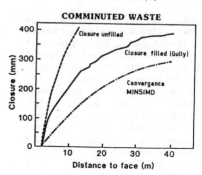

Figure 11 Composite diagrams showing the
difference in closure measured
in the filled and the unfilled
panels.

puted elastic convergence and the filled
panel closure profiles. This difference
in closure is the reason for the discrepancy between the computed and the
measured vertical backfill stresses as
indicated in Section 3.2.1; MINSIM-D uses
constitutive equations which approximate
the stress-strain behaviour of backfill
(Jager et al. 1987) and calculate the
backfill stresses according to the
induced closure and hence strain at that
point. If the convergence estimated by
the program is lower than the measured
closure then the corresponding stress
value would also be lower than the
measured stress in the fill.

194

The closure measured underground has three components. These are elastic convergence, inelastic bed separations and closure due to dilation. The elastic convergence is the elastic relaxation of the rockmass in response to mining. The inelastic bed separations take place very close to the face where the first few layers in the hangingwall displace downwards due to their own mass. Dilation in the rock mass ahead of the face, due to shear movements and extension fracturing also induces closure as a result of the generation of horizontal compressive forces which cause the strata to deform into the stope (Brummer, 1987). It is also found that some vertical dilation takes place on fractures in the plane of bedding and in response to horizontal shearing movements which cause the strata to ride up irregularities on bedding surfaces and the formation of gouge, which forces the strata apart and thus down into the excavation (Section 4.2)

MINSIM-D assumes the rockmass to be a homogeneous and isotropic medium and therefore can simulate only the elastic component of the closure. In addition to the differences observed between the measured and the theoretical closure profiles which reflect the inability of MINSIM-D to allow for inelastic closure, the joints and different layers in the rockmass also contribute more convergence than the program predicts.

The closure profiles measured both inside the filled paddock and outside the fill on the gully edge are shown in the top graph of Figure 11. These curves confirm the findings discussed in Section 3.2.2 that the closure is less in the filled section of the stope than in the unfilled area adjacent to the gullies. It is also observed that the change in slope of both of these curves takes place at the same distance behind the face, 15 m and where the closure in the panel is 140 mm. This converts from the stress-strain curve for classified tailings to a backfill stress of 3 MPa. It is also of significance that from this point onwards not only are the two measured closure curves approximately parallel but that they are also almost parallel to the curve depicting the elastic convergence. This implies that the strata have been clamped by a support resistance of about 3 MN/m$^2$ to form a rigid beam spanning the gully. The rockmass then displaces into the excavation as a single body at a rate approximating that of elastic convergence. It would appear therefore that vertical inelastic deformation over the stope is largely

inhibited by stresses of the order of 3 MPa.

4.2 Extensometer Measurements

To gain further understanding of the effect of backfill on the deformation and to quantify the extent and amount of vertical dilation in the hangingwall of the stope several borehole wire

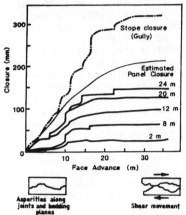

Figure 12 The vertical displacements measured at different horizons measured along a borehole in a filled stope.

extensometers were installed in the hangingwall of filled stopes. King et al. (1988) reported vertical dilation of over 300 mm extending at least 15 m into the hangingwall over the centre dip gully of an unfilled stope. The dilation occurred continuously during the period when the stope span was mined from 12 m to 90 m. Smaller amounts of dilation were measured between 2 and 10 m from the face.

Great difficulty is experienced in obtaining these measurements as the boreholes are often displaced by horizontal shearing along bedding planes, cutting off or jamming the extensometer wires. A higher proportion of successful measurements can be gained in backfilled stopes if the extensometer boreholes are drilled close to the backfill. However where boreholes are drilled from positions some metres behind the face the early history of strata deformation is missed.

Figure 12 shows the results obtained from a borehole drilled into the hangingwall of a classified tailings backfill stope. The borehole was about 30 m long and was drilled 1,5 m ahead of the backfill bag and about 4 m behind the face. The anchors were installed at 2, 8, 12, 20 and 24 m along the borehole. A clo-

195

sure station was also installed next to the borehole collar to quantify the stope closure at that point.

From this figure it appears that the bed separation takes place within almost all the strata along the borehole for about 15 m of face advance. After this the displacement profiles level off and become parallel to each other. The stope closure also follows the same trend. These results also demonstrate that gaps occur between the layers which do not close, even when subjected to high stresses generated in the fill.

It was originally thought that the gaps between layers should close due to the reaction of the fill and that the displacement profiles for each horizon should tend to approach each other instead of becoming parallel. As shown in Figure 12, the overriding of asperities and irregularities on bedding planes during horizontal shearing can cause the vertical dilation. In addition the contact between high points on opposing surfaces would prevent the closing of the opened bedding planes when subjected to compression.

It is interesting to note that the displacement profiles and the estimated closure profile inside the panel become parallel to each other, again at 3-4 MPa vertical backfill stress.

## 5 CONCEPTUAL MODEL OF A BACKFILLED PANEL

The conceptual model of a backfilled panel is derived from the results obtained from the monitoring program and is shown as closure and stress profiles superimposed on a diagram of a stope in Figures 13 and 14. Figure 13 depicts the distribution of closure in a filled panel. The figure shows that the minimum closure occurs in the centre of the panel while the closure becomes excessive in the gullies. It is emphasized again that the closures measured even in the centre of the filled stopes were much higher than originally expected from MINSIM-D modelling.

Following from the closure distribution shown in Figure 13, the stress distribution in a filled panel based on in situ stress-strain measurements is illustrated in Figure 14. In this figure the zero backfill stress line is drawn a certain distance back from the edge of the first backfill paddock because stresses in the fill are not generated immediately on placement of the fill as the consolidation of the fill due to drainage and

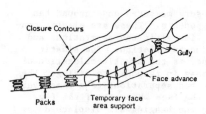

Figure 13 The distribution of closure in a filled panel.

shrinkage results in a gap between the fill and the hangingwall.

It follows that the stress in the backfill only develops some distance back from fill front depending on the rate of closure and amount of shrinkage. The zero stress line can therefore be up to 10 m behind the newly placed fill.

Figure 14 also shows that the backfill stresses approach zero towards the edges of the paddock, parallel to the gullies on both sides of the panel and that the stresses become uniform and relatively high in the centre of the paddock.

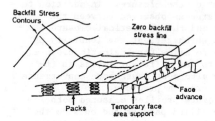

Figure 14 The distribution of stress in a filled panel.

## 6 IMPLICATIONS OF LARGE CLOSURE AND HIGH STRESSES MEASURED IN BACKFILL FOR NUMERICAL MODELLING

As discussed in Section 3.2, the stresses measured in backfill underground are considerably higher than would be predicted by MINSIM-D.

It has also been shown that the closure within 20 m of the stope face measured in backfilled stopes is 40 to 150 per cent greater, than would be predicted by MINSIM-D using conventional input parameters. This discrepancy in closure is reduced to between 20 and 80 per cent at about 40 m behind the face since the amount of inelastic deformation which occurs initially does not increase significantly further behind the face. However the rate of closure by this stage correlates closely to the predicted elastic rate of closure. Nevertheless it

196

is clear that the stresses measured in the backfill are greater than those predicted by MINSIM-D modelling, due to the discrepancy between the predicted and actual convergence. This has important implications in the assessment of the effectiveness of backfill, particularly as a regional support. Although it is highly probable that the effectiveness of backfill as a regional support is greater than predicted by elastic theory, further work is necessary to quantify this benefit of backfill.

# 7 CONCLUSIONS

The results of a rock mechanics monitoring program to determine the in situ behaviour of backfills and the rockmass surrounding a filled stope are presented in this paper. The most pertinent conclusions reached from the results are as follows:

## In situ behaviour of backfill

(i) the confined compression stress-strain behaviour of the various uncemented backfills used in South African gold mines has been established. The comminuted waste type of backfill has been confirmed to be the stiffest type of backfill currently used;

(ii) good correlation has been obtained between in situ and standard laboratory measurements;

(iii) although further measurements are required, the complete backfill rib behaviour has been established and it has been found that the maximum and minimum closure takes place in the gullies and near the centre of the filled panel respectively. The results also show that the minimum stress occurs at the edge of the backfill paddock and the stresses become higher and more uniform towards the centre of the paddock;

(iv) the reduction of convergence during rockbursts in filled stopes is considerable compared to conventionally supported stopes. Closure and stress changes in backfill resulting from rockbursts indicate that the energy imparted to the rock by seismic ground motions and which has to be absorbed by the supports can be much higher than previously estimated. This could have significant implications regarding the design of stope support for rockburst conditions;

## In situ behaviour of rockmass

(v) in unfilled stopes high rates of closure persist 30 m or more behind the stope face;

(vi) in filled stopes, the high rates of inelastic closure take place only until a vertical stress of 3-4 MPa is generated in the fill. Above 3-4 MPa, the rockmass becomes clamped and displaces into the excavation as a single body. It appears that several mechanisms account for the inelastic deformation and that these generate high forces which translate into a vertical stress component of 3-4 MPa;

(vii) the closure rates are significantly lower in filled stopes than in unfilled stopes, and

(viii) the absolute closure measured in both filled and unfilled stopes is much higher than predicted by MINSIM-D.

ACKNOWLEDGEMENT

The work described in this paper forms part of the research programme of the Chamber of Mines Research Organization of South Africa. The co-operation of management and staff of the mines using backfill, and their permission to publish the paper, are gratefully acknowleged.

REFERENCES

BRUMMER, R.K. (1987). Fracturing and deformation at the edges of tabular gold mining excavations and the development of a numerical model describing such phenomena. Ph.D. Thesis Rand Afrikaans Univ., Johannesburg.

CLARK, I.H., GüRTUNCA, R.G. and PIPER, P.S. (1988). Predicting and monitoring stress and deformation behaviour of backfill in deep-level mining excavations. Proc. 5th Aust. New Zealand Conference on Geomechanics, p 214-218.

JAGER, A.J., PIPER, P.S. and GAY, N.C. (1987). Rock Mechanics aspects of backfill in deep South African gold mines. Proc. 6th Int. Congress on Rock Mech., International Society for Rock Mechanics Montreal, Canada.

KING, R.G., ROBERTS, M.K.C. and TURNER, P.A. (1988). Rock Mechanics aspects of stoping without back-area support. Chamber of Mines Research Organization, Research Report. In preparation.

STEWART, J.M., CLARK, I.H. and MORRIS, A.N. (1986). Assessment of fill as a basis for selecting and developing optional backfill systems for South African gold mines. GOLD 100 Proc. Int. conf. on gold, Vol. 1, Johannesburg, p.255-270.

*Innovations in Mining Backfill Technology, Hassani et al. (eds), © 1989 Balkema, Rotterdam. ISBN 90 6191 985 1*

# Assessment of a new mine layout incorporating concrete pillars as regional support

D.J.Adams, R.G.Gürtunca, A.J.Jager & N.C.Gay
*Chamber of Mines Research Organization, Johannesburg, RSA*

ABSTRACT: Reef stabilizing pillars are commonly used as regional support on deep level South African gold mines where longwall mining layouts are adopted. A mine layout which utilizes concrete pillars in place of the reef pillars has been considered from a rock mechanics, a cost and an implementation point of view. Energy Release Rates and Average Pillar Stresses were used to assess the effectiveness of the mining and support layouts. The paper concludes that a mining scheme using concrete pillars is feasible but recognizes that changes to the mining system will be necessary.

## 1 INTRODUCTION

At present, deep level South African gold mines use a system of stabilizing reef pillars to reduce elastic convergence and therefore the stored energy along the mining faces. This is measured by the Energy Release Rate (ERR). It has been recognized that there is a direct relationship between ERR and the incidence of rockbursts on mines using longwalling mining methods (Hodgson & Joughin, 1967) and thus there is a commitment to keeping the ERR values as low as possible.

The regional support benefits of stabilizing reef pillars is unequivocal, but on mines where they are used, only about 85 per cent extraction of the reef occurs, with the remaining 15 per cent of the reef being locked up in the stabilizing pillars. As an example, one mine has implemented 40 m wide reef stabilizing pillars spaced 240 m apart, measured from the pillar's inside edges.

The extraction ratio will have to decrease as mining depths increase as more of the reef must be left unmined in order to maintain current or acceptable levels of ERR's and Average Pillar Stresses (APS). It is desirable from the point of view of profitability and the life of the mine, to achieve 100 per cent extraction or at least to maintain the present extraction ratios at greater depths without increasing ERR or APS values. A scheme whereby total extract-ion is achieved while not increasing the seismic risk or reducing the safety of the work force would be extremely attractive. Such a scheme will require a departure from the current mine layouts for deep level mines where longwall mining is now generally used with reef stability pillars. The idea put forward in this paper is to introduce concrete into the stopes to replace the reef stabilizing pillars. The use of large quantities of concrete underground will however, also necessitate innovations for conveying and pumping of concrete into the face area.

One of the critical considerations for the success of a system where concrete replaces reef pillars is the placement of the concrete in panels designed to have low rates of closure in order that the concrete can cure before being subjected to large strains. If concrete were to be placed behind a normal advancing set of longwall panels the closure rate would be too rapid, and hence, the stiffness of the concrete would be reduced due to insufficient curing of the material before crushing occurred (Fulton 1977). The possible benefits which could be derived from the concrete would thus be lost.

The strategy presented in this paper is to mine 50 m wide panels which are spaced at about 200 m on dip for 400 to 500 m on strike, as a first mining phase. These panels would be filled with 40 m wide concrete pillars which would replace the presently used stabilizing reef pillars. They would thus become the regional support for the mine. The second phase of mining would then be to mine the ground between the stabilizing concrete pillars.

Three major aspects have been considered in assessing the viability of such a mining method. These are:

(i) the schemes effectiveness as regional support - an investigation of the rock mechanics implications of the proposed mining method and layout has been carried out,

(ii) a preliminary cost estimate - an initial cost exercise has been done to see if such a scheme is financially viable, and

(iii) the implementation of the scheme - a preliminary study has been undertaken to explore the feasibility of the idea from a mining point of view to find out if any major obstacles are likely to prevent the project from ever being successfully implemented.

## 2 ROCK MECHANICS CONSIDERATIONS

The use of concrete to replace reef stabilizing pillars will only be acceptable if it can be shown to be comparable or better, from a rock mechanics point of view, than the present situation with stabilizing reef pillars. Two criteria, namely ERR and APS, were used to assess the rock mechanics effectiveness of the scheme.

### 2.1 Determination of in situ modulus of the rockmass

The field observations in backfilled stopes have shown that the assumption of linear elasticity is inappropriate for modelling the in situ deformation of the rockmass (Gürtunca et al., 1989), in that the field measurements do not agree with the values predicted using the MINSIM-D program. In particular, the model underestimates the effect which backfill materials have in modifying the stress environment underground and is not able to simulate the high closures which have been recorded underground, unless much reduced values of Young's Modulus are assumed (Gürtunca et al, 1989).

The MINSIM-D model assumes that the rock around the stope is homogeneous and elastic. It can thus only approximate the detailed behaviour of the inhomogen-

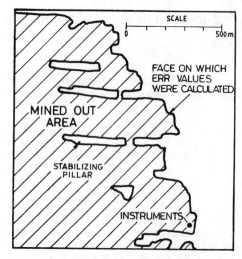

Figure 1. The layout of a deep level mine where reef stabilizing pillars are used.

eous, layered and jointed nature of the rock mass, which is typical of the Witwatersrand sedimentary basin rocks. Conventionally, the Young's Modulus of intact quartzite, 70 GPa, is used as an input parameter for MINSIM modelling.

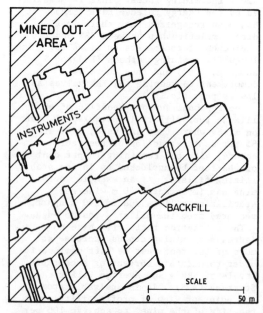

Figure 2. The backfilled stope where instrumentation was installed and where a back analysis of stress-strain profiles was conducted.

The fact that larger closures are observed in situ, than those predicted implies that the effective modulus must be much lower than 70 GPa.

To confirm this layers of jointed rock of different stiffnesses were modelled using FLAC to simulate 1 000 m of rock overburden above a stope. The results showed that, compared with a homogeneous elastic model the closures were 17 per cent higher. (Gürtunca & Adams, 1989).

Thus it was decided that the Young's Modulus E in the MINSIM-D model should be reduced to obtain a more realistic simulation of the stope closure which is a critical factor in determining the reaction of the backfill. In order to arrive at a value to which the modulus E should be reduced for the area being modelled, a back analysis of a backfill site where instruments had been install-ed and which was close to the stabiliz-ing pillars was conducted (Figures 1 and 2). The readings of stress and closure at the instrumentation site are considered to be reliable.

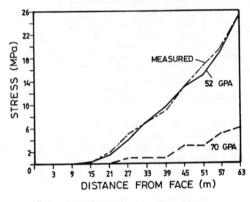

Figure 3. The stress profile along a strike section in a backfilled stope.

Figure 3 shows the measured stress in a backfill paddock plotted against the distance from the face. Also shown are theoretical curves computed by MINSIM-D assuming values of Young's Modulus of 52 and 70 GPa. From this figure it is clear that there is good agreement between the stresses measured at the underground backfill site and the model which used 52 GPa, as a Young's Modulus for the rockmass.

The parameters used in the back analysis model are shown in Table 1. The fill parameters used were establish-ed from laboratory confined compression curves of the placed backfill materials

and the measurements obtained at the underground instrumentation site. To arrive at a suitable value of E, all the known parameters from the field site were modelled as accurately as possible. It is important to model the mined stoping width and the fill height at which the fill becomes active accurately. Using the stoping width at the time of the backfill placement and the closure ride measurements recorded before backfill placement, plus the closure which occurred before the closure/ride instruments were installed (this value was estimated from computer modelling), it was possible to arrive at a mined stoping width of 1,05 m. From stress-strain curves which were derived from instrumentation in the backfill it was possible to determine that 4 - 5 per cent shrinkage occurred before the back-fill became active. Allowing for this also the effective height of the back-fill was established as 0,814 m.

Table 1. Parameters used in back analysis models.

| | |
|---|---|
| Fine grid size | = 6 m |
| Poisson's ratio | = 0,2 |
| Young's modulus | = 40, 52, 70 GPa |
| Depth | = 2 800 m |
| Dip | = 20 degrees |
| Stoping width | = 1,05 m |
| *Fill width | = 0,814 m |
| Fill parameters a = 16 MPa | b = 0,30 |

| *Determination of fill width | |
|---|---|
| Stope width | 1,050 m |
| Closure before measurements | −0,064 m |
| Closure before filling | −0,120 m |
| Shrinkage | −0,052 m |
| Effective fill width | 0,814 m |

## 2.2 Computer modelling of reef and concrete stabilizing pillars

A schematic diagram (Figure 4) shows the present mining layout with positions along the pillar and face where Average Pillar Stresses and Energy Release Rates were calculated respectively. Figure 5 shows the situation with the present reef stabilizing pillars replaced by concrete and extending ahead of the faces. Figure 6 has an additional 40 m wide concrete pillar between the existing pillar layout.

Table 2 contains information which is relevant to the models which were run to examine stabilizing pillars, and the quality of the concrete which was assumed in the simulations.

A number of computer runs have been carried out by using MINSIM-D to compare the effectiveness of using concrete pillars in place of stabilizing reef pillars. As indicated previously, ERR and APS were used as criteria to evaluate the scheme. As determined from the back analysis described in Section 2.1, an in situ Young's Modulus of 52 GPa was used in all the computer runs.

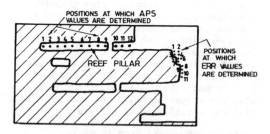

Figure 4. Schematic plan of the present layout showing the positions along a pillar and a face where APS and ERR values were established respectively.

Table 2. Parameters used in concrete pillar models.

| | | |
|---|---|---|
| Coarse grid size | = | 20 m |
| Poisson's ratio | = | 0,2 |
| Young's modulus | = | 52 GPa |
| Depth | = | 2 800 m and 3 800 m |
| Stoping width 1 m | | |
| Fill width = 0,95 m | | |
| Concrete stiffness = 20 GPa | | |

2.2.1 Mechanical properties of concrete used in computer modelling

A concrete material with about 20 GPa stiffness and 20 MPa compressive strength was used in this study. It was also required that this concrete would be able to maintain its properties at high stresses (i.e. 300 - 400 MPa) when used underground, to replace reef pillars.

A concrete specimen with 8:1 width to height ratio was prepared in the laboratory and tested for ten day strength. The results are shown in Figure 7. The figure demonstrates that the stiffness of the concrete was about 20 GPa while the maximum stress carried was 80 MPa, which was the limit of the press for the sample size.

To test the concrete samples at higher stresses, up to 400 MPa, tests were carried out in confined compression using an oedometer of diameter 120 mm.

The average aggregate size for the concrete used in the 8:1 tests was about

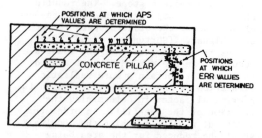

Figure 5. Schematic plan with concrete pillars replacing reef pillars. Positions at which APS and ERR values were established are indicated.

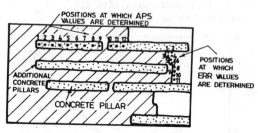

Figure 6. Schematic plan with concrete pillars plus an additional concrete pillar. Positions at which APS and ERR values were established are indicated.

20 mm. Obviously, if the same material were to be tested in the oedometer, one would expect to have inaccuracy due to large aggregate with respect to the size of the oedometer. Therefore, a finer aggregate with about 5 mm particle size was used to produce the concrete for the confined compression tests. The results obtained from the 7 day test showed that the concrete material with finer aggregate could carry stresses up to 440 MPa when the test was terminated due to limitations of the press (Figure 8). However, the 28 day characteristics of the concrete are to be used in the design of the pillars, and the strength of the concrete is expected to have increased in this period. The stiffness at that time is expected to be close to 20 GPa.

It is planned to place 40 m wide

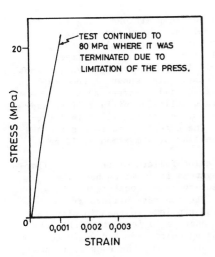

Figure 7. The stress-strain curve for
20 MPa concrete tested after 10 days.

concrete pillars underground. Pillars
of this width will exhibit confined
compression behaviour towards the centre
while the edges of the pillar will
dilate but still provide the necessary
confinement for the inner section of the
pillar.

This argument is further supported
by the triaxial stress increase in
strength shown in Figure 9 (Price,
1951). The figure clearly shows that
the strength of the concrete improves

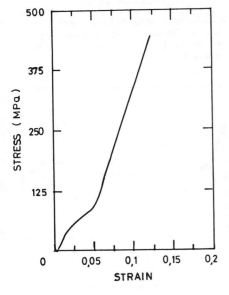

Figure 8. The stress-strain curve from
an oedometer test after 7 days.

considerably with the higher confinement
stresses.

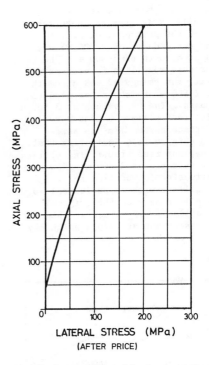

Figure 9. Relationship between lateral
restraint and axial compressive strength
for concrete.

## 2.2.2 Comparison of energy release rates

A number of faces between two stabilizing pillars were selected to compare the
ERR's in the present situation with reef
stabilizing pillars, with the ERR values
which would exist if concrete pillars
replaced the reef pillars. The results
are shown in Figure 10. The values of
ERR are plotted for the designated
positions along a mini-longwall between
the two stabilizing pillars as in
Figures 4, 5 and 6. The present situation, i.e. reef stabilizing pillars (Fig
4), is compared with the stabilizing
pillars replaced with concrete (Figs 5
and 6). The ERR values for concrete
pillars are slightly higher than the
present values but are not unacceptably
high. If it were necessary to reduce
the ERR values, an additional 40 m wide
concrete pillar between the original
concrete stabilizing pillar positions
could be introduced, which would reduce

ERR values to levels approximately 30 per cent below those experienced at present.

To demonstrate the benefits of concrete pillars at greater depth the geometry of the present reef pillar situation (at a depth of 2 800 m) was modelled at a depth of 3 800 m. The result of this modelling is presented in Figure 11. The values of ERR along the faces at 2 800 m are shown for comparison purposes. The ERR values at 3 800 m depth with reef pillars become very high, approaching 50 MJ/m$^2$ in some instances (Figure 11), indicating that this layout for reef stabilizing pillars at is not suitable at 3 800 m. However, if the situation is modelled for concrete pillars with an additional 40 m concrete pillar between the original concrete pillars, ERR values are higher than at present but at levels which may still be acceptable.

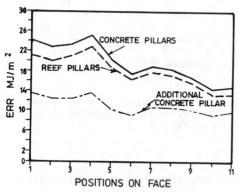

Figure 10. The Energy Release Rates along a mini longwall with either concrete or reef stabilizing pillars.

2.3 Comparison of average pillar stress

A stabilizing pillar, whether reef or concrete, adjacent to and up dip of the panels which were assessed for ERR's, was examined to determine average pillar stress.

The stress which pillars carry is important to the siting of off-reef excavations and it would be unacceptable if models showed that the concrete pillars were required to carry higher stresses than that of which they were capable. The results of the runs for reef and equivalent concrete pillars are shown in Figure 12. In this figure average pillar stresses are plotted for

the designated positions along the stabilizing pillar. From the simulations it appears that concrete pillars carry slightly less stress than the reef pillars. The difference however, is not significant. Also shown on the graph is the average pillar stress on the concrete stabilizing pillars if an additional 40 m concrete pillar is introduced. The effect is to reduce the APS in the pillar by a maximum of 12 per cent.

Foundation failure of the reef pillar-stope systems is known to have the potential to cause rockbursts (Hagan, 1986). The concrete pillars are probably less problematic than in situ rock as the concrete is less stiff than the surrounding quartzites and would tend to fail before causing the rock surrounding

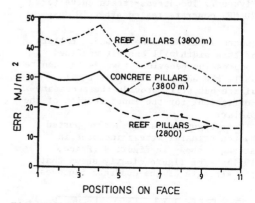

Figure 11. The Energy Release Rates along a mini longwall with either concrete or reef stabilizing pillars at 3 800 m depth.

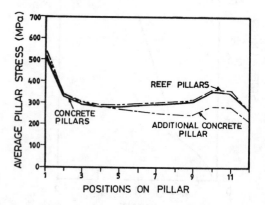

Figure 12. The Average Pillar Stress on concrete reef stabilizing pillars.

it to fail. It is also unlikely that there would ever be a decrease in stress on the pillars but rather a continuous increase in stress with increased mining.

# 3 COST ANALYSIS

An exercise was carried out to establish whether the cost of installing concrete in the stopes would change overall profits by significant levels.

The results show that the income generated from mining the reef which would normally be left as stabilizing pillars, far exceeds the cost of the additional mining and the concrete replacing them. Table 3 shows the profit for two different grades of ore, 5 g/tonne and 15 g/tonne for 220 m of reef face advanced 1 m. The three cases considered are reef stabilizing pillars (180 m of panel mined), concrete pillars replacing reef stabilizing pillars (220 m of panel mined) and concrete stabilizing pillars plus an additional 40 m concrete pillar between them (220 m of panel mined). The table shows the scenario of a mine with high mining costs and one with low mining costs under the various circumstances. The analysis does not include any costs related to the piping, pumping or conveying of the concrete from surface to underground. However, even with the introduction of an additional concrete pillar the profits obtained by placement of concrete stabilizing pillars in place of reef stabilizing pillars increase by 10 - 20 per cent. Such profits should be sufficient to pay for the infrastructure for the backfill system necessary, including pipe columns, pumps, crushers and conveyor belts.

# 4 IMPLEMENTATION

The proposed scheme will be successful if it is possible to design a feasible mine layout incorporating concrete pillars. As has already been emphasized, it is important to be able to place concrete at stoping widths close to the initial stoping width. Also the environment must be such that closures are not initially rapid. Therefore the layout requires advance panels to be mined before the main stoping operation begins.

A proposed mining layout is shown in Figure 13. The layout requires that

Table 3. Profit comparison.

| | Low Cost | High Cost | Low Cost | High Cost |
| --- | --- | --- | --- | --- |
| | 5 g/ton | | 15 g/ton | |
| Reef Pillars | R56 000 | R27 000 | R181 000 | R153 000 |
| Concrete Pillars | R64 000 | R30 000 | R219 000 | R185 000 |
| Additional Concrete Pillar (40 m) | R62 000 | R28 000 | R216 000 | R182 000 |

cross-cuts are developed at vertical intervals of 75 - 90 m to intersect the reef along the same dip line. These cross-cuts should be about 1 000 m apart on strike (Figure 13). Raising and ledging on reef between the various levels is completed before mining of the ± 50 m panels begin. One or perhaps two such panels will be mined on both sides of the raises between any two adjacent levels. The lower panel between any two levels should be above the elevation of the lower cross-cut in order to provide boxhole storage capacity (Figure 14).

The 50 m panels should be mined on strike to a position midway between two consecutive raise lines, a distance of about 500 m. Only once these panels have holed and the 40 m wide concrete pillars have been placed in two adjacent ribs should mining between the concrete pillars proceed. Each set of panels between the concrete pillars should be mined as a mini-longwall.

Safety is an important consideration in the proposed scheme because of the long accessways on reef to the mining faces, particularly when the initial 50 m panels are approaching holing. The gullies and travelling ways will have to be well supported. The possibility of seismicity in the area while workers are excavating the advance panels should also be minimized. This can be achieved by not mining between the advance panels until these are complete. This will ensure that low ERR values (3 MJ/m$^2$) are experienced in the faces of the advance panels and in the general mining area.

The advance panels should have two entrances and dip gullies connecting the top and bottom strike gullies at regular intervals (Figure 15). The area between the top and bottom gullies will be fill-

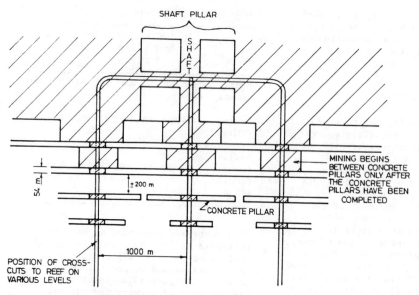

Figure 13. The proposed mine layout using concrete stabilizing pillars (plan).

ed with concrete except for the conveyance ways adjacent to the concrete and a line of gully edge support. The conveyance ways could be above and below the top and bottom strike gullies respectively. The gullies will be used for travelling ways or for pipe columns or rope transporters. Continuous scrapers or conveyor belts could be located in the conveyance ways.

The long distances from the cross-cuts to the advancing faces, particularly close to holing positions will pose problems from a ventilation, travelling, cleaning and material supply, point of view. However, the placement of concrete should limit the area to be ventilated and minimize the need for timber support, thus reducing material handling problems. Pipe lines and/or conveyor belts will bring materials into the stope for the construction of the concrete ribs.

It is envisaged that the faces will be mined on rapid yielding hydraulic prop support in spite of the minimal seismic risks and that the concrete will be placed as close to the face as possible (i.e. approximately 5 m).

The panels should be blasted every working day so that more than 20 m of advance is achieved per month. Therefore a complete rib between two raise lines should be completed in 24 months. The mini-longwalls between concrete pillars should also utilize continuous scrapers for cleaning, to minimize any

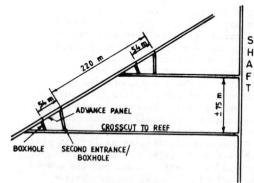

Figure 14. A section through a raise line of the mine layout shown in Figure 13.

additional footwall development. The support for the secondary mining operation should consist of hydraulic face support and uncemented backfill as the permanent support.

5 DISCUSSION

The use of concrete in place of reef stabilizing pillars appears to have potential from the points of view of rock mechanics effectiveness, cost and practical implementation. Certainly at greater depths the extraction ratio need not decline from the present situation if concrete is used in conjunction with reef stabilizing pillars. More positively, extraction ratios of 100 per cent may be achieved, even at greater depth than has been modelled, by using concrete pillars as regional support. This would require the closer spacing of concrete pillars than the present reef pillars, which themselves would have to be spaced more closely at greater depth. However, the cost of concrete is far less than the revenue generated from the mining of the reef even at grades as low as 5 g/tonne.

It is recognized that a change in mining layout and method is required, but given the benefits, it is probable that solutions can be found to address these problems. The envisaged mine layout offers the possibility of reducing the amount of off-reef development, which would also contribute to increased profitability, while the method of mining advance headings would be a means of obtaining knowledge of geological structures and reef grade distribution before the major stoping operations begin.

Only one layout has been presented and the authors realise that there are many more options which will have to be examined and improvements which can be made.

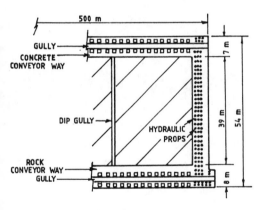

Figure 15. A detailed plan of an advance 50 m panel in which concrete will be placed.

6 CONCLUSIONS

The work on concrete pillars to date has examined, with the aid of computer models, the possibility of replacing reef stabilizing pillars, with concrete pillars. Additional concrete pillars were also modelled in between the present pillar spacing in order to assess any benefits from this changed layout. The conclusions which emerged from the modelling study undertaken are as follows:

(i) the use of concrete pillars as an alternative regional support, to reef stabilizing pillars, is possible from a rock mechanics point of view. The ERR values along working faces and APS's along concrete stabilizing pillars can be maintained at almost the present levels associated with reef stabilizing pillars. With the introduction of additional concrete pillars, ERR's and APS's can be reduced to below present values experienced,

(ii) the replacement of reef pillars with concrete will result in greater profits for the mines. These profits may be used to construct additional concrete pillars which could have significant safety benefits,

(iii) the extraction ratio of the mine could be increased, thus extending the life of the mine even at present mining depths. The use of concrete may make viable very deep mines which would not be profitable at lower extraction ratios required for reef stabilizing pillar as regional support,

(iv) the success of concrete regional stabilizing pillars will depend on a change in the present mining method and layout. Concrete will have to be placed in low convergence areas and also be allowed to cure before accepting high loads,

(v) a concrete of reasonable quality will be required close to the stope face of the advance panels. This means that a complete backfill system to handle concrete conveying and placement will need to be developed,

(vi) it will be necessary to carry out far more detailed cost investigations before such a scheme is implemented. A major cost in this scheme, namely pumping, piping and conveying of concrete, has not been considered in the costs presented in the paper. Once a system to convey concrete of the quality desired to the face has been devised further cost exercises will have to be undertaken, and

(vii) the layout adopted is not unique as a solution and it is probable that other mine layouts may be found which will have greater benefits. Therefore further computer modelling is necessary to design and investigate these alternatives.

# 7 ACKNOWLEDGEMENTS

The work described in this paper forms part of the research programme of the Chamber of Mines Research Organization of South Africa.

# 8 REFERENCES

Fulton, F.S. Concrete Technology. A South African Handbook, 1977. The Portland Cement Institute, Johannesburg.

Gürtunca, R.G., Jager, A.J., Adams, D.J. & Gonlag, M. 1989. In situ measurements of backfills and surrounding rockmass behaviour in South African gold mines. Proc. 4th International Symposium of Mining with Backfill, Ontario, Canada, October.

Gürtunca, R.G. & Adams, D.J. 1989. Determination of in situ modulus of rockmass by using backfill measurements. Unpublished Report in preparation. Chamber of Mines Research Organization.

Hagan, T.O. (1988). Mine design strategies to combat rockbursting at a deep South African gold mine. In P. Cundall et al. (eds.), Key Questions in Rock Mechanics, pp.249-260. Rotterdam: Balkema.

Hodgson, K. & Joughin, N.C. 1967. The relationship between energy release rate, damage and seismicity in deep mines, Proc. Eighth Symposium on Rock Mechanics, University of Minnesota, 1966, in Failure and Breakage of Rock, C. Fairhurst (ed.), 1967, p.194-203.

Price, W.H. (1951). Factors influencing concrete strength, Jour. of American Concrete Inst., V.22, No. 6.

*Innovations in Mining Backfill Technology, Hassani et al. (eds), © 1989 Balkema, Rotterdam. ISBN 90 6191 985 1*

# Examination of the support potential of cemented fills for rock burst control

J.K.Whyatt & T.J.Williams
*Bureau of Mines, Spokane Research Center, Spokane, Wash., USA*

M.P.Board
*Itasca Consulting Group, Inc., Minneapolis, Minn., USA*

ABSTRACT: Recent interest in switching from the traditional overhand cut-and-fill mining method to the underhand longwall cut-and-fill method as a step towards controlling the rock burst problem in the Coeur d'Alene mining district has focused attention on underhand backfilling practices. The underhand method requires the backfill to function as a stable roof during mining and makes filling a cut tight to the sandfill back difficult. The gap increases the stope closure required to load the fill significantly and thus increases the effective span between mining face and backfill abutment. The influence of changes in backfill density and placement gap on the rock burst hazard were examined with an energy release rate (ERR) analysis. South African studies suggest that ERR is related to rock bursts triggered by pillar and abutment crushing, but not by discontinuity stick-slip. Both triggering mechanisms play an important role in the rock burst problems in the Coeur d'Alene District. The analysis was based on the assumption of an elastic rock mass, which was checked by a elastoplastic parametric study. The ERR analysis showed that changing from present practice to one in which backfill is omitted altogether increases ERR by 42%, while "ideal" backfill reduces ERR an additional 28%. The most promising avenue for further reduction in ERR is elimination of the gap, which accounts for half of the ERR difference between present practice and the ideal case. The ideal backfill case also requires placing backfill at maximum density. While improvements in backfill density have been attained by vibrating the backfill, these improvements have fallen well short of maximum density. Increasing the cement content of lean cemented backfills does not appear to affect ERR.

## 1 INTRODUCTION

Rock bursts have become a major concern in the Coeur d'Alene Mining District of northern Idaho and threaten the future of many district mines. The Bureau of Mines, Spokane Research Center (SRC), has long been involved in developing alternatives to the overhand cut-and-fill stoping method that would reduce the rock burst hazard and be amenable to mechanization. One of these methods, underhand longwall cut-and-fill, was recently tested at the Lucky Friday Mine under a cooperative agreement between SRC, Hecla Mining Co., and the University of Idaho.

The principle of a single advancing face, which is central to the underhand longwall method, is not new. In fact, the South African High-Level Committee on Rock Bursts and Rockfalls recommended longwalling as a means of reducing the rock burst hazard associated with mining remnants (or sill pillars) as early as 1924. Longwalling is now standard practice in South Africa and has greatly reduced the rock burst hazard. The theoretical underpinnings of this recommendation were developed further in the energy release rate (ERR) method. Both longwalling recommendation and ERR methods primarily address crushing-type rock bursts in pillars or at the edges of excavated zones, which are common in South Africa. Other mechanisms of rock bursting, especially slip along discontinuities, are not properly addressed by an ERR analysis, but such mechanisms are common in the Coeur d'Alene District (Salamon, 1988).

Present backfilling practice in the underhand method leaves a gap of about 0.5 m in each 3.3 m cut. A number of researchers, including Salamon (1988), have expressed concern that this gap may seriously increase the rock burst hazard. This paper examines the influence of backfill quality and placement procedure on rock burst potential, as measured by ERR, to evaluate the contribution of present backfill practice and the potential for further contributions towards controlling rock burst potential in underhand longwall mining. To this end, force-displacement curves for a range of backfill quality and placement procedures were developed. The relative influence of various backfills on rock bursting was estimated by ERR calculations based on numerical models. Further studies are envisioned to examine the feasibility of adding stabilizing pillars, as advocated in South Africa, to the underhand longwall method.

## 2  BACKFILL BEHAVIOR

The behavior of the backfill typically used in the Coeur d'Alene District has been tested at SRC in a number of field and laboratory investigations. Patchet (1983) identifies three major parameters that control backfill behavior: the particle size distribution, cement content, and the density or void ratio at placement. Since particle size distribution is largely a function of the milling process, this parameter is generally taken as a constant, except for removal of slimes. The range of uniaxial compressive behavior for fine-to-coarse Coeur d'Alene fills with 6% cement are illustrated in figure 1. For a more detailed discussion of Coeur d'Alene backfill characteristics, see Boldt et al. (1989).

The primary effect of adding cement is that it increases the initial strength of the backfill, which is useful primarily for providing a stable back in the underhand method or providing a slusher floor in the overhand method. The initial cuts of the experimental Lucky Friday underhand longwall (LFUL) stope were backfilled in two stages; the bottom half of the fill was fortified with cement. Stope closure/fill pressure instrument pairs were placed in both the cemented and uncemented portions of the fill to obtain in situ compression measurements. The results (fig. 2) show that the initial strength developed by the cemented fill was fractured by the 5 cm closure during excavation of the following stope cut. This rapid fracturing of the cemented fill prevented the fill from sustaining a

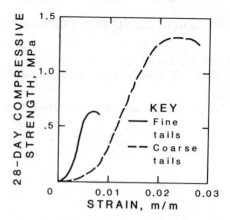

Figure 1.  Unconfined compression stress-strain curves for 6% cement fill using fine and coarse mill tailings.

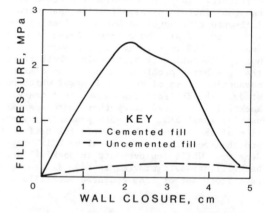

Figure 2.  Measured stope closure/fill pressure curves for cemented and uncemented fill in the Lucky Friday experimental stope.

significant support load until confinement was attained. Laboratory confined-compression tests on Coeur d'Alene backfill also showed very little influence of lean cement mixtures (Boldt, 1987).

The "closable" void ratio of backfill at placement has been measured in situ by Corson and Wayment (1971) at approximately 15% axial strain in confined compression. That is, the uncemented fill does not pick up appreciable loads under confined compression until roughly 15% strain. In the overhand method, each cut is totally filled. But when cemented sandfill is placed in the underhand stope

210

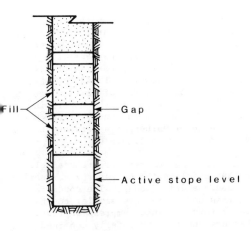

Figure 3. Backfill placement in a typical underhand stope.

using current technology, it is virtually impossible to fill tight to the back. At best, the stope is filled to within 0.5 m of the back, leaving about 15% of the cut as void (fig. 3). This void or gap should effectively increase the strain required for development of appreciable support loads under confined compression to about 30%. Better placements have been obtained in South Africa with paste fills that have very low water contents and can stand at high angles. Research has also been conducted on ways to place backfill at closer-to-maximum density. Corson (1971) managed to increase density by 6% to within about 9% of maximum by using vibrators during the backfilling operation.

## 3 NUMERICAL ANALYSIS OF THE EFFECTS OF GAPS IN THE FILL

A numerical investigation of observed cemented backfill behavior was undertaken with an experimental strain-softening constitutive law implemented in the finite-difference program FLAC (Cundall and Board, 1988). Standard elastic parameters and a Mohr-Coulomb criterion with a linear plastic-strain cohesion-softening relationship (table 1) were defined to match the results from typical laboratory tests (fig. 4). Earlier work (Whyatt and Board, 1988) has shown that such a procedure will produce sample shape effects similar to those observed in laboratory tests. However, the constitutive law must be considered experimental because of mesh dependence during softening.

Table 1. Backfill properties used in numerical simulations.

| Property | Value |
| --- | --- |
| Shear modulus (MPa) | 22 |
| Bulk modulus (MPa) | 64 |
| Cohesion[1,2] (MPa) | 0.21 |
| Friction angle[2] (deg.) | 37 |
| Density (kd/m[3]) | 2,000 |

[1]Cohesion varies linearly from 0.21 MPa at 0% plastic strain to 0 cohesion at 4% plastic strain.
[2]Estimated values taken from backfill of similar composition (Board and Crouch, 1984).

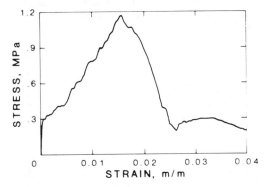

Figure 4. Numerical model simulation of an unconfined compression test with backfill properties.

A FLAC grid was defined with a number of fill levels and intervening gaps (fig. 5A) to examine variations in the stope closure/fill pressure relationship when the unfilled volume was changed. Program logic automatically accounted for large strains and detected contact between fill pillars as gaps closed. Once contact was detected, normal and shear forces were transmitted across the interface. Thus, the model followed the transition from unconfined to fully confined compression. A series of grid plots for the 90% fill case shows this progression (fig. 5B).

Complete stress-strain curves (fig. 6) were developed for this progression using backfill gaps of 0%, 10%, and 20%. In the numerical model, the plotted stresses were calculated for the center of the fill pillars, to compare with field measurements. Thus, these stresses were slightly larger than the average support

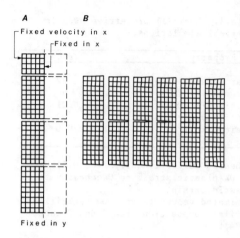

Figure 5. Problem mesh (A) and deforming sandfill pillars (B) during a numerical model simulation of stope closure.

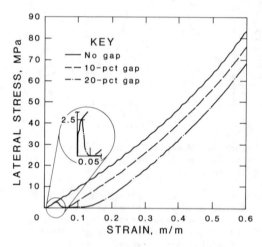

Figure 6. Stress-strain response at the center of a model backfill pillar.

pressure. The 10% and 20% gap curves peaked initially at approximately 2.5 MPa. The model pillar, with a 1:1-aspect ratio, was appropriately stronger than the reference 2:1-aspect ratio simulated laboratory test and compared well to field measurements of 2.5 MPa peak pressure. The 3% strain shown by the model at peak load is somewhat greater than the 0.7%-1.2% measured in the stope (20-37 mm for a 3-m-wide stope) and the 1.5% strain-to-failure for the model laboratory test. These discrepancies were caused in part by the laboratory and field fills having slightly different

materials and in part by mesh dependency problems in the experimental constitutive law. A further check on these curves was provided by Corson and Wayment (1967), who reported lateral pressures of 2.4 to 3.5 MPa after closures of 14% to 18% in sandfill placed at 85% of maximum density (a 15% effective gap in the model) in overhand stopes at the Lucky Friday Mine. This range of values lies between the 10% and 20% curves in figure 6.

At the other extreme from the single-pillar model is the no-gap, or perfectly placed, sandfill model. The conditions in this model are very similar to the laboratory confined compression test. Because the fill is totally confined, loss of cohesion with plastic strain is overwhelmed by increases in frictional resistance. In the intermediate models, which have gaps of 10% and 20%, there is an intermediate region between the initial peak strength and the onset of confined compression as partial contact occurs between fill pillars.

## 4 STOPE MODEL PARAMETRIC STUDY

Backfill is placed in deep mines to provide a stressed abutment behind mining and reduce stress concentrations at the mining face. The location of this stress abutment, and hence the span between abutment and mining face, are strongly influenced by the backfill properties discussed in the previous section. They are also influenced by the characteristics of the rock mass being mined. Since rock mass properties are difficult, if not impossible, to define exactly, a probable range of properties was examined for influence on backfill behavior. The mesh shown in figure 7 was developed for this purpose and halved through symmetry arguments. The perfect backfill force-displacement curve from figure 6 was approximated by a piecewise linear spring across the vein, a feature developed for the Chamber of Mines of South Africa (Cundall, 1987).

Initially, a 400-m-long and 3-m-wide portion of the mesh was mined and filled to represent prior mining. The model was then used to follow the development of stress concentrations at the face and loading of the fill. Since shear components of the in situ stress field have little influence on these factors, the stope section was assumed to be oriented with the principal stress directions.

212

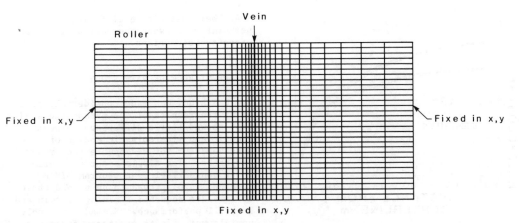

Figure 7.  Finite-difference mesh for underhand stope model.

Principal stresses of 32 MPa vertical and 64 MPa horizontal were assumed on the basis of a survey of overcore measurements (Whyatt, 1986).

The model properties were chosen in an attempt to cover the range of probable rock mass properties for formations in the Coeur d'Alene Mining District (table 2). A Mohr-Coulomb failure criterion was used for runs allowing failure. A further simplification was made by considering only the extreme cases of 100% (no gap) and no-fill models.

Table 2.  Rock mass properties used in large-scale model.

| Property | Value |
|---|---|
| Shear modulus (GPa) | 8.62 |
| Bulk modulus (GPa) | 11.50 |
| Cohesion (MPa) | 10-20 |
| Friction angle (deg.) | 30-40 |

The following assumptions were made for the six model runs:

1. Elastoplastic rock mass with cohesion of 10 MPa, friction angle of 40°, and no fill (these are considered to be realistic rock mass properties),

2. Elastoplastic rock mass with cohesion of 10 MPa, friction angle of 30°, and fill (this checks fill performance for assumption of a weak rock mass),

3. Elastic rock mass, fill,

4. Elastic rock mass, no fill,

5. Elastoplastic rock mass with cohesion of 10 MPa, friction angle of 40°, and fill (model 1 with fill), and

6. Elastoplastic rock mass with cohesion of 20 MPa, friction angle of 40°, and fill (this model checks fill performance when the strongest rock mass is assumed).

Figures 8 and 9 show the half-closure (i.e., closure/2) of the stope as a function of distance from stope midspan and horizontal stress from the stope face for all six models. The effective ground support pressure generated by the fill as a function of distance from the face is shown in figure 10. The effective ground support pressure was calculated by dividing spring force by bearing area (approximately 3.3 m²). The following observations can be made from these figures.

1. There was little difference between elastic and elastoplastic models in stope closure. The yield zone surrounding the stope was limited and resulted in only slightly greater closures.

2. The closure of the stope was, however, affected by the presence of the fill. Closure at midspan was reduced by roughly 20%. Fill strain was approximately 35% at midspan, decaying only near the mining face.

3. The backfill reduced rock mass stress by roughly 10% immediately ahead of the face, with virtually no difference 20 m from the face.

213

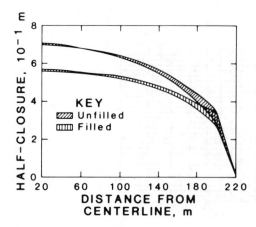

Figure 8. Half-closure of the stope for various models.

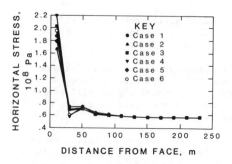

Figure 9. Horizontal stress at the stope face for various models.

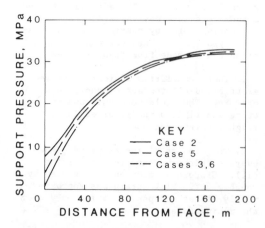

Figure 10. Effective fill support pressure.

4. There was little difference in fill performance between the various rock mass property assumptions.

5 ENERGY RELEASE RATE STUDY

The ERR study was made considerably easier by the results of the parametric study that showed little difference between elastic and elastoplastic models of rock mass behavior for reasonable rock mass strengths. The displacement discontinuity boundary-element program MINAP (Crouch, 1976) proved to be a good choice to simulate sequential mining of vein and backfill performance. Second-order polynomials were fit to the three force-displacement curves of figure 6 and an extrapolated 30% gap curve. The ERR was calculated as the kinetic energy accompanying the mining of a cut and was approximated by one-half the product of pre-existing force across the cut and change in closure. This method assumes a linear reduction of load with closure.

MINAP models were run for the underhand mining geometry with no fill, no gap, 10% gap, 20% gap, and extrapolated 30% gap cases. The energy release rates for these cases are presented in percentage changes from the present practice case (30% gap including densification) since no correlation between the absolute value of energy release rate and pillar-crushing rock bursts has been established for the Coeur d'Alene Mining District. The results showed that:

1. Underhand mining without backfilling increased ERR by 42% from present practice. Thus, the present backfill practice provides an important contribution to reducing the rock burst hazard.

2. Reducing the stope closure required for confined compression conditions by filling the gap or densifying the backfill reduces ERR about 1% for each 1% reduction. Ideal backfill practice (no gaps and maximum density) would reduce ERR an additional 28% over present practice.

6 DISCUSSION AND CONCLUSIONS

An analysis of backfill properties and practice typical of the Coeur d'Alene Mining District showed that the backfill does contribute substantially to reducing the rock burst hazard in underhand stopes. Removing backfill increases ERR by 42%. Furthermore, the backfill can be modified to further reduce the ERR. The first step should be to reduce the unfilled void between cuts. Complete elimination of this space could reduce ERR by 14%. Further gains will require placing backfill closer to its maximum density. Bureau of Mines attempts at densification were only partially successful, eliminating about 5% of the closure required for confinement. Finally, the addition of cement in lean backfill mixtures does not appear to affect ERR.

In applying these results to any particular mine, it should be noted that the two-dimensional numerical models used in this analysis assume a very long vein, capable of attaining complete closure for a significant length without end effects. Also, the ERR method is aimed primarily at rock bursts caused by pillar and abutment crushing rock burst mechanisms, but does not apply directly to discontinuity slip rock bursts.

REFERENCES

Board, M.P., and S.L. Crouch. 1984. The Physical, Mechanical and Hydrologic Properties of Backfills from the West Driefontein Mines. Dep. of Civil and Min. Eng., Univ. MI, Minneapolis, MI.

Boldt, L. 1987. Personal communication.

Boldt, L., P.C. McWilliams, and L.A. Atkins. 1989. Backfill Properties of Total Tailings. U.S. Bureau of Mines RI, in press.

Corson, D.R. 1971. Field Evaluation of Hydraulic Backfill Compaction at the Lucky Friday Mine, Mullan, Idaho. U.S. Bureau of Mines RI 7546.

Corson, D.R., and W.R. Wayment. 1967. Load-Displacement Measurements in a Backfilled Stope of a Deep Vein Mine. U.S. Bureau of Mines RI 7038.

Crouch, S.L. 1976. Analysis of Stresses and Displacements Around Underground Excavations: An Application of the Displacement Discontinuity Method. Univ. MI Geomechanics Report, Minneapolis, MI.

Cundall, P., and M.P. Board. 1988. A Microcomputer Program for Modelling Large Strain Plasticity Problems.

Proceedings of the 6th International Conference on Numerical Methods in Geomechanics, Innsbruck, Austria.

Hedley, D.G.F. 1987. Rock Burst Mechanics. CANMET Min. Div. Rep., Elliot Lake Laboratory, MRL 87-118(TR).

Patchet, S.J. The Use of Fill for Ground Control Purposes. Chap. 11 in Rock Mechanics in Mining, pp. 241-255. Practice, S. Budavari, ed. The South African Institute of Mining and Metallurgy Monograph Series, No. 5, Johannesburg, S. Africa.

Salamon, M.D.G. 1988. Personal communication.

Whyatt, J.K. 1986. Geomechanics of the Caladay Shaft. M.S. thesis, Univ. ID, Moscow, ID.

Whyatt, J.K., and M.P. Board. 1988. A Strain-Softening Model for Representing Shear Fracture in Continuous Rock Masses. 2nd International Symposium on Rock Bursts and Seismicity in Mines, Univ. MI, Minneapolis, MI.

*Innovations in Mining Backfill Technology, Hassani et al. (eds), © 1989 Balkema, Rotterdam. ISBN 90 6191 985 1*

# The assessment of cemented rockfill for regional and local support in a rockburst environment, LAC Minerals Ltd, Macassa Division

W.J.F.Quesnel & H.de Ruiter
*LAC Minerals Ltd, Operations Management Group, Kirkland Lake, Ontario, Canada*

A.Pervik
*LAC Minerals Ltd, Macassa Division, Kirkland Lake, Ontario, Canada*

ABSTRACT: The use of cemented rockfill for rockburst control has proven successful for both local and regional support in the mine. Post burst reconditioning of drifts and stopes has been reduced significantly. An apparent reduction in the rockburst frequency has occurred with the implementation of cemented rockfill.

## 1 INTRODUCTION

The LAC Minerals Ltd., Macassa Division, located in Kirkland Lake, Ontario, has been in continuous operation for approximately 55 years, and has produced in excess of 2.7 million ounces of gold. The narrow ore veins have been primarily recovered with overhand cut and fill mining methods incorporating recycled waste development as backfill.

The main mining horizons are now located 1200 to 2150 meters below surface, and the ore zone is open below the 2200 meter elevation. The orebody is on average 2.5 m wide and has a fairly uniform dip of 75° S.

The wall rocks are generally stronger than the ore zone and have compressive strengths as high as 345 MPa, with average strengths in the order of 200 MPa. The average ore strengths are 170 MPa. The modulus of elasticity for the ore and walls rocks can range from 35 to 80 GPa (Quesnel and Hong 1987).

## 2 ROCKBURST HISTORY/MECHANICS

The mine has experienced, both, strain bursts during development, and pillar bursts during the mining of stope crown pillars (Cook and Bruce 1983). Arjang and Nemcsok (1986) have indicated that from 1935 to 1985, a total of over 400 rockbursts ranging from strain to pillar bursts have been reported at the mine. Approximately 70% of these bursts occurred in the ore zone or stopes and 10% of these events were classed as heavy bursts with displacements of more than 50 tonnes of rock.

In general, the larger bursts have occurred in cut and fill crown pillars with a nominal thickness of 15 m. The largest burst magnitude, recorded by the National Seismograph Network, has been 3.1 $M_N$ with a rock displacement of more than 1000 tonnes. The mine's microseismic network usually locates the epicenter of these bursts in the plane of the crown pillar. The bursts generally occur with or shortly after production blasting in the stope.

In these types of pillar bursts, the largest source of damaging seismic waves are derived from the release of potential energy due to the convergence of the sidewalls. Underground closure stations (tape extensometer) at the mine have clearly indicated that rapid or sudden differential convergence of stope sidewalls was directly related to rockbursting. Incremental closure magnitudes of 25 to 35 mm have been measured after minor rock bursts in stopes utilizing unconsolidated waste rock as backfill (Quesnel and Hong 1986).

One of the major operational problems after large rockbursts, was the rehabilitation of collapsed timber support of overlying backfilled stopes. The failure of the timber support resulted in unconsolidated rockfill completely filling ore access drifts. This resulted in significantly increasing the cost and time for the reconditioning of the stopes.

## 3 BACKFILL ALTERNATIVES

In 1986, a new No.3 Shaft was completed to a depth of 2202 m and is now used as the main access to the ore zone for handling of men and materials. Previous to this, the mine was serviced by two shafts and two internal winzes.

Prior to 1986, the mine was restricted to using unconsolidated waste development rock as backfill due to the limited hoisting capacity. In addition, sufficient mill tailings could not be reclassified for backfill due to the fine grind of the material.

The development of the No.3 shaft, permitted a review of the mine's backfill system with respect to the handling of alternative backfill materials. In addition, the No.3 shaft intersected the ore zone at a depth of 1615 m. At this elevation a shaft pillar was required.

N-Fold numerical modelling, conducted by Golder Associates (1985), for dimensioning of the shaft pillar, clearly illustrated that future mining would result in the overstressing of critical elements located along some sections of the shaft pillar. In addition, modelling of various mining sequences at depth, outlined several critical ore and waste remnants in close proximity to the shaft pillar. The stresses approached 145 MPa which exceeded the estimated unconfined in-situ yield strength of the rock mass (Hanson, Quesnel and Hong 1987).

These factors led to the conclusion that to maintain the integrity of the shaft pillar during mining, a consolidated stiff backfill would be required. It was decided to test concrete as a backfill material. Concrete was initially selected because placement techniques had been developed during the mining of a surface crown pillar at LAC's Lake Shore Mine Division.

A test stope was selected at Macassa for experimenting with underhand cut and fill with concrete. This stope was located at depth of 1400 m in area known as the Central Pillar. This section of the mine had been temporarily closed due to frequent rockbursting during mining. The selected test stope was an ore remnant located in the eastern portion of the Central Pillar.

This stope was mined from 1984 to 1986 and a detailed account of the handling system was given by De Ruiter and Hong (1986).

The specified strength requirements for the concrete was 28 day strengths of 20.7 MPa. This strength was achieved for the most part, as 75% of the concrete placed had an actual strength of 20.4 MPa. The remaining 25% of the concrete achieved only 50% of the design strength or 11.0 MPa.

The nature of the underhand cut and fill

method used, only allowed for 50% of the individual cuts to be filled. This resulted in alternating 1.4 m slabs of concrete separated by 1.4 m of open ground. As the stope advanced, convergence pins were installed between the concrete layers and were monitored on a regular basis.

During the mining of this stope five minor rockbursts occurred with the largest burst displacing approximately 10 tonnes. The bursts were not recorded by the National Seismograph Network which indicated all of these events had a magnitude of less than 2.0 $M_N$. The bursts occurred in the footwall of the stope and were associated with a narrow band of tuff. Historically, this rock unit at the mine has proven to be the most prone to violent failure or bursting (LeBel, Quesnel and Glover 1987).

Review of the convergence data after each burst, showed no change in the trend or rate of wall closure indicating that sudden or "instantaneous" closure of the sidewalls had not occurred. In comparison, significant changes in the rate of wall closure with an incremental or stepped convergence profile have been measured after pillar type bursts at the mine.

The above analysis led to the conclusion, that the bursts experienced during mining of the test stope were not pillar type bursts, which would have been expected in this area, but were strain bursts due to the stress readjustment during mining. Hedley (1987) has indicated that certain strain bursts can be the result of the released stored strain energy in the rock mass and this energy release occurred when the rock suddenly changed from a triaxial stress condition to either a biaxial or uniaxial stress state. The energy release or seismic efficiency of a strain burst is much lower than a pillar type burst. Hedley (1987) has indicated that strain bursts have a seismic efficiency of 30-60% while pillar bursts are in the range of 70-90%.

Figure No.1 is a plot of the estimated convergence (see Golder's 1985) versus the actual measured convergence in the test stope. The modulus of elasticity for the concrete was assigned a value of 6.9 GPa which was believed to be a realistic estimate of the in-situ modulus. The modulus used for the rock was 69 GPa which gave a rock:fill modulus ratio of 10. Golder Associates (1985) found that for estimating stope convergence, that modelling (N-Fold) of alternate slabs of concrete and voids was equivalent to modelling of a stope completely backfilled with material having a modulus equivalent to 50% of that of concrete.

The maximum measured convergence in the test stope at mid span was approximately 35 mm. This compared to observed closure of 75 to 150 mm in stopes utilizing unconsolidated rockfill. The rate of closure measured in the test stope was directly related to the extraction ratio.

The measured convergence in the stope clearly illustrated that the concrete backfill placed a limiting constraint on the sidewall convergence. This reduced the potential of a pillar type burst by minimizing the volumetric closure, which in turn, reduced the mining stress concentration on the stope sill pillar.

Although this test proved successful as a rockburst control measure, the major disadvantage was the marked increase in the average stoping costs. The use of concrete as backfill increased the stoping costs by 52%. Approximately 75% of the cost increase was the cost of concrete while the remaining 25% was due to the preparation and placement costs. This would limit the use of concrete to a very few number of select stopes.

This limitation led to the investigation of cemented rockfill (C.R.F.) as a replacement for concrete. The major advantage of the C.R.F. would be the possible incorporation of the existing backfill handling system.

## 4 CEMENTED ROCKFILL

In 1986, a testing program was initiated to evaluate if the gradation of the waste development would be suitable for use as cemented rockfill. Grain size analysis of the material gave a gradation of approximately 85% minus 150 mm (coarse fraction) and 15% minus 10 mm (fines). The ideal gradation determined at LAC's other operations using C.R.F. was 75% coarse and 25% fines.

The variation of the C.R.F. strength as a function of the fines content is shown on figure 2. This curve indicated that less than optimum strengths could be expected for the C.R.F. because of the coarse gradation of the waste development.

The initial test work consisted of 150 mm x 300 mm cylinders that were cast with minus 50 mm waste rock and various normal portland cement contents by weight of aggregate. It was concluded that a minimum 28 day strength of 3.5 MPa could be achieved with the addition of 5% cement to the waste development. It was felt that this strength could be significantly upgraded because of the gain in fines due to the natural attrition of the waste in the fill raises.

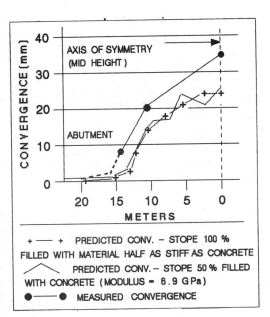

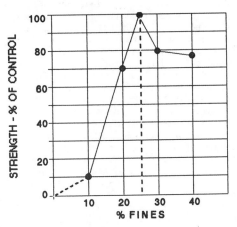

Figure 1. Stope convergence profile - modelled v.s. actual. Axis of symmetry divides the stope at mid height. Convergence profile is along the mid span of the stope.

Figure 2. C.R.F. strength as a function of the fines content. Control samples had a gradation of 75% coarse/25% fines. Minus 150 mm rockfill aggregate utilizing 450 mm diameter cylinders with 5% cement content by weight of aggregate.

219

Yu and Counter (1983), reviewed the in-situ compressive strength of C.R.F. with a designed average of 5% cement and found the strengths ranged from 1.3 to 11.0 MPa. The highest strength occurred directly below the fill dumpoint into the 25 m wide stope. This strength range compared to their average laboratory strengths of 5 MPa using 150 mm x 300 mm cylinders with 5% cement. The ore widths at Macassa of 2.5 m wide and mining methods would minimize this segregation problem and methods of placement could be developed to insure a uniform graded backfill with minimal downgrading of the strength.

Based on figure 2, it was extrapolated that the range of in-situ compressive strengths of C.R.F. at 5% could be from 3.5 MPa to a theoretical 9.0 MPa. This strength range is 17% to 44% of that of concrete.

The above results suggested the C.R.F. would have similar limiting convergence properties as the concrete test stope if
1. 100% of the stope would be back-filled.
2. The in-situ strengths would approach 9 MPa.

The application of C.R.F. in the conventional overhand cut and fill stopes would result in 100% of the stope being filled. The strength of 9 MPa was dependant upon the degree of confinement in the stope and the fines content and this strength could only be verified by in-situ testing.

Review of the costs of producing a cement slurry on site and combining it with the existing waste handling system indicated significant cost savings could be realized in comparison to the use of concrete. The use of 100% C.R.F. would increase the average stoping costs by 15% to 20%. This cost increase for C.R.F. in the overhand cut and fill stopes, would be fully offset by:
1. Distinct contact between ore and backfill during slushing. This would reduce dilution and improve recovery in comparison to unconsolidated waste fill. An increase of only 4% in the average recovered grade would more than offset the additional costs for C.R.F.
2. Maintaining the mining widths in blocky ground and therefore dilution would be reduced during mining.
3. Reducing the damage as a result of bursting. Blast vibration monitoring at LAC's other operations illustrated that the peak particle velocity could be attenuated by 30 - 50% when blasting adjacent to C.R.F. This suggested that the seismic waves released during a burst would be partially absorbed by the C.R.F., resulting in lower rehabilitation costs by minimizing the amount of scabbing failure along the drifts and the dynamic overloading of installed support.

It was concluded that C.R.F. would not only reduce the frequency of bursting but could also minimize the amount of damage if a rockburst occurred. The cost of C.R.F. would be more than offset by gains in stoping productivity, and it was therefore decided in 1987, with the completion of No.3 Shaft, to introduce C.R.F. to several test stopes.

### 4.1 C.R.F. handling system

For the initial test stopes, the cement packaged in 40 kilogram bags, was transported on standard pallets. The pallets would be moved through the No.3 shaft and trammed on the level by flatcar to the stope to be backfilled. The cement was then mixed manually with water in a ore car to give a cement slurry with a pulp density of 55%. To coat the waste rock with slurry, several mixing methods were tested:
1. Mixing method. Pumped the slurry from the mixing car to the waste car and sprayed the slurry onto the waste. The C.R.F. was then dumped directly down the stope raise and was then reslushed into the stope.
2. Percolation method. In this method the dry waste rock would be dumped via the raise into the stope. The waste would be reslushed in the stope. The slurry would then be gravity feed, using a 12 mm hose, down the raise and into the stope. The slurry would then be sprayed and injected into the rockfill.

Method 1 gave the best results for consistent fill quality and was also, the most productive. The re-slushing of the C.R.F. in the stope gave a relatively uniform graded backfill and insured all of the aggregate was coated with slurry.

After this initial testwork proved that the placement of C.R.F. was feasible, various options were investigated for improving the handling of the slurry. The strike length and depth of the mine workings with multiple levels being mined simultaneously, precluded the use of a central slurry system located on surface. In addition, the mining sequence required the placement of C.R.F. along the boundary of the No.3 shaft pillar by 1988.

To satisfy these constraints, it was concluded that a portable slurry mixing car would be required. After reviewing available slurry mixing cars on the market, it was determined that for the narrow drifts (2.4 m x 2.4 m) at Macassa, specialized mixing cars would have to be constructed. In 1987, the first slurry mixing car was constructed in-house and consisted of v-shaped holding tank, a horizontal agitator with vertical rakes and a slurry pump mounted on the front of the mixing car.

The simple design of the mixing cars proved to be quite successful due to the high availability and mobility. Todate, the mine has a total of three mixing cars. The present C.R.F. handling system is shown schematically on Figure 3. The numbers in this figure are referenced to the following description:

1. Pallets with 40 kilogram bags of cement (Type 10) are conveyed through No.3 shaft to the appropriate level. Presently 1.35 tonne tote bags are being tested as an alternative to the 40 kilogram bags.

2. The pallet is conveyed by flatcar from the stations to the stope.

3. The waste rock is obtained from either development (3A) or if required a waste stope (3B).

4. The waste can be dumped directly down a waste pass (4A). Development progressing on lower levels can be recycled through a mid shaft dump (4B) and then through a waste pass.

5. The waste is then trammed from the waste pass to the stope using 3.6 tonne cars.

6. The cement is mixed in a slurry car to a pulp density of 55%. The slurry is then pumped to the waste car and then the mixed C.R.F. is dumped down the stope raise.

7. The C.R.F. is then reslushed in the stope to make a working floor for the next lift.

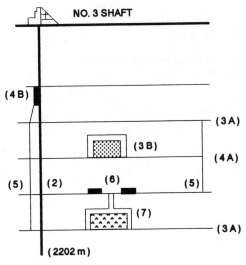

Figure 3. C.R.F. handling system

## 4.2 Local support

The introduction of C.R.F. has resulted in significant improvements in ground conditions in the stopes, as well as minimizing the damage after a burst. This is demonstrated with the following examples.

A major advantage of C.R.F. was the elimination of runs of unconsolidated backfill, after a burst, into the main ore access drifts. In one instance, the 60-40 stope, the first cut had been blasted and the initial 1.2 m thick C.R.F. cap was placed on the timber flats. A large burst occurred shortly after placing the fill. A total of 95 tonnes of rock was displaced with 18 tonnes being displaced directly below the C.R.F. cap. Although the timber caps and posts supporting the C.R.F. were damaged, there was no run of fill into the drift.

The two largest bursts, both having magnitudes of 3.1 $M_N$, have occurred in the same area of the mine. There was a marked diference in the rock displacement between these two bursts, and, this was felt to be attributed to the use of C.R.F. A comparison of these bursts is given below.

The first burst, in 1982, was located in 61-38 stope at an elevation of 1875 m below surface. This burst caused significant damage in the stope and displaced more than 1000 tonnes of rock which resulted in the temporary closure of the area. It should be noted that this stope was only being developed at the time of the burst and was located directly adjacent to a stope backfilled with unconsolidated waste.

The second burst, in 1989, occurred in 60-38 stope located directly above the 61-38 stope. The total displacement was only 130 tonnes. This stope using C.R.F. had an extraction ratio of approximately 40% at the time of the burst. In addition, this stope was also located adjacent to a stope filled with unconsolidated waste rock. Table 1 illustrates the significant reduction in rock displacement as a result of C.R.F. in the 60-38 stope.

Table 1. Comparison of rock displacement between C.R.F. and unconsolidated rockfill (R.F.).

| Stope | Fill Type | Ext. (%) | Displacement (Tonnes) | Rehab |
|---|---|---|---|---|
| 61-38 | Waste | 15 | 1000 | 5 months |
| 60-38 | C.R.F. | 40 | 130 | 1 month |

Figure 4 is a longitudinal section of the 60-38 stope. The cross hatched areas define the location of the displaced material as a result of the rockburst. The location of displacement appears to be defined by the contact between C.R.F. and unconsolidated backfill. The majority of the damage occurred directly above the stope with no C.R.F.

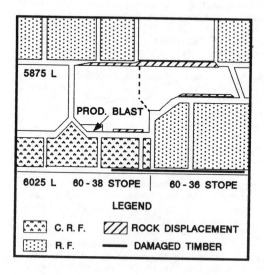

Figure 4. Longitudinal section of 60-38 stope.

C.R.F. has also been used in reconditioning raises after a burst. In several cases the stope access raises have sloughed, as a result of bursting, to approximately twice their original size. In these cases, a 1.1 m diameter culvert with 1.8 m long sections, would be lowered into the raise. The outside annulus between the culvert and raise walls would then be filled with C.R.F. This technique has proved quite successful and has improved the safety and working conditions during the rehabilitation of the actual raise.

4.3 Regional support

Figure 5 is a histogram plot of the historical frequency of bursting at Macassa. For the period 1935 to 1989, the average frequency per year was calculated and is shown on figure 5. For the period from 1985 to March 1989, the total displacement of all the bursts/year and the displacement and magnitude of the largest burst in the year were plotted on figure 6. It should be noted that for this analysis, a

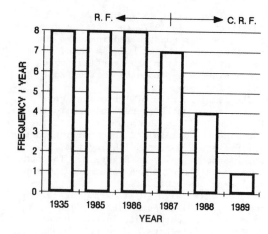

Figure 5. Rockburst frequency. The frequency shown for the year 1935 is actually the average frequency/year for the period 1935-1984.

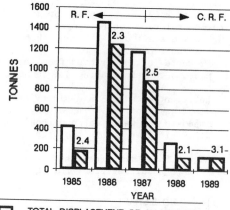

☐ TOTAL DISPLACEMENT OF ALL BURSTS PER YEAR

▨ DISPLACEMENT AND MAGNITUDE OF THE 2.1 LARGEST BURST PER YEAR

Figure 6. Rock displacement caused by rockbursts.

rockburst was defined as any seismic event which resulted in more than 5 tonnes of displacement.

C.R.F. was introduced to the mine in mid 1987 and by 1988 over 70% of the active stopes were using C.R.F. From figures 5 and 6, it appears that a major reduction in both the frequency of bursting and the total displacement coincides with the implementation of C.R.F.

222

## 4.4 Mining methods

The success of the C.R.F. in reducing rockburst damage, prompted the mine department to re-evaluate the crown pillar recovery methods. The observations that the attenuation characteristics of the C.R.F. reduced the seismic wave propagation led to the idea of a C.R.F. mat above the crown pillar. This sill mat would act as a low velocity buffer zone and therefore could reduce the track heave after a burst.

A stope has been developed to test this approach and figure 7 is a schematic illustrating the various stages of mining this stope with a C.R.F. mat. An initial 5 m deep trench would be mined below the overcut and filled with C.R.F. Culverts would be placed at approximately 10 m intervals to act as future fill raises for the recovery of the crown pillar. The stope would then be mined from the undercut with standard cut and fill methods using C.R.F. The crown pillar would then be recovered using a uphole or benching, longitudinal retreat method, with C.R.F. introduced through the previously installed culverts.

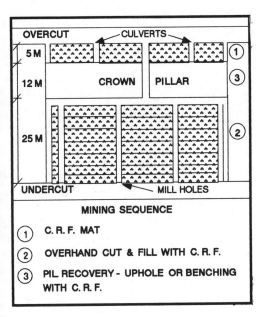

Figure 7. A schematic of a stope with a C.R.F. mat for absorbing rockburst generated seismic waves.

Another mining method is presently being tested for future deep mining as well as for mining with 100% C.R.F. along the No.3 shaft pillar. This method coined central rill cut and fill (C.R.C.F.) stoping is illustrated in figure 8, parts A and B.

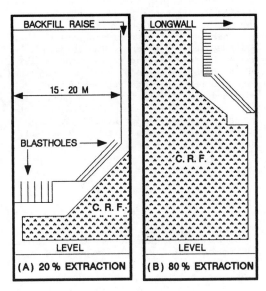

Figure 8. Central rill cut and fill mining method. The backfill raise is located at the center of the stope and the stope is retreated from the raise in both directions. (A) and (B) represent only half of the stope and (B) is the mirror image of (A).

The major advantage of this method for rockburst control is the placement of high strength C.R.F. at the mid span of the stope. This results in the rapid, strategic placement of C.R.F. at the point of maximum convergence or potential energy release during a burst. If the results are positive, the C.R.F. protective sill mat will be incorporated into the C.R.C.F. stoping method.

## 5 CONCLUSIONS

The use of C.R.F. for alleviation of rockbursts has exceeded all of the initial expectations. The use of a semi stiff backfill has resulted in minimizing the damage caused by rockbursting. In addition, an improvement in the overall mine stability has been achieved as a result of a apparent reduction in the rockburst frequency. The additional cost of C.R.F. has been outweighed by the improvements in overall ore recovery and reduced post burst reconditioning costs. This has been translated into overall improved working conditions.

A major research project, justified by the experience gained todate, has been initiated. This project will center around the avenues for improving the C.R.F. stiffness as well as modifying the handling system for cement. The initial areas of investigation will be the review of the possible methods for increasing the fines content, partial replacement of portland cement with flyash and quantification of the in-situ physical and attenuation properties of C.R.F. It is hoped that this optimization program will give us sufficient data to further develop an engineered backfill for rockburst control.

REFERENCES

Arjang, B. & Nemcsok, G. 1986. Review of rockburst incidents at the Macassa mine, Kirkland Lake. Division report MRL 87-21(TR): Canmet.

Cook, J.F. & Bruce, D. 1983. Rockburst control through destressing - a case example. Rockbursts: prediction and control, p.81-91. London, England: The Institution of Mining and Metallurgy.

De Ruiter, H. & Hong, R. 1986. The use of underhand cut and fill with concrete support, LAC Minerals - Macassa Division Paper presented at the mine backfill design seminar. Montreal: McGill University.

Golder Associates 1985. Numerical modelling for no.3 shaft pillar and backfill requirements - Macassa mine. Internal report 851-1064.

Hanson, D., Quesnel, W.J.F. & Hong, R. 1987. Destressing a rockburst prone crown pillar - Macassa mine. Division report MRL 87-82(TR): Canmet.

Hedley, D. 1987. Rockburst mechanics. Division report MRL 87-118: Canmet.

LeBel, G.R., Quesnel W. & Glover, W. 1987. An analysis of rockburst events during sinking of Macassa no.3 shaft. Presented at 89th annual general meeting - CIM, Toronto. Montreal: Canadian Institute of Mining and Metallurgy.

Quesnel, W.J.F. & Hong, R. 1987. Mining induced seismicity: monitoring and interpretation, p.227-237. Fred Leighton memorial workshop on mining induced seismicity. Montreal: Canadian Institute of Mining and Metallurgy.

Quesnel, W.J.F. & Hong, R. 1986. The use of rock mass conditioning (destressing) for the safe recovery of rockburst prone crown pillars, Macassa Division, LAC Minerals Ltd. Ground Control Symposium, Haileybury School of Mines.

Yu, T.R. & Counter, D.B. 1983. Backfill practice and technology at Kidd Creek Mines. CIM Bulletin, Vol. 76, No.856: 56-66.

*Innovations in Mining Backfill Technology, Hassani et al. (eds), © 1989 Balkema, Rotterdam. ISBN 90 6191 985 1*

# Spiral slot-and-pillar mining with backfill

J.D.Dixon
*Spokane Research Center, US Bureau of Mines, Spokane, Wash., USA*

ABSTRACT: An innovative mining method termed "spiral slot-and-chamber mining with backfill" has been conceptualized as a possible alternative to room-and-pillar mining. It is a total extraction method applied to nearly horizontal, strata-bound ore deposits. Conventional methods of mining tabular deposits extract ore from panels laid out in a rectilinear pattern. It may be possible to transform this layout into a curvalinear one so that one continuous panel, along a spiral, is mined. There are several benefits of the spiral pattern of mining. This paper discusses the novel mining method and a feasibility study proposed to the Bureau of Mines.

## 1 INTRODUCTION

The U.S. Bureau of Mines is exploring alternate mining methods to replace room-and-pillar methods for extracting tabular, nearly horizontal seams. Room-and-pillar methods have many disadvantages. They are wasteful of resources, inherently dangerous, and prone to cause surface subsidence.

One method for the total extraction of tabular ore bodies, known as "concrete pillar stoping," has been developed in Finland. Ore is extracted from a series of alternating parallel excavations, here termed slots and chambers (see fig. 1). Subsequently, the slots are filled with cemented backfill and the chambers with sand. This method has been applied where the country rock is strong and competent, but has not been applied where ore deposits occur in soft rock, such as potash, trona, or coal. A disadvantage of the method is the cost of the cement that must be added to the backfill to make pillars strong enough to withstand self-weight and overburden pressures.

It may be possible to transform this layout to a curvalinear one so that one continuous spiral can be mined. In a geometrical sense, the concept, called the "spiral slot-and-chamber mining with backfill method, represents a radical departure from conventional mining practices.

It offers several potential benefits, including better ground control, productivity, and economy, and is a candidate for soft rock mining.

## 2 CONCRETE PILLAR STOPING

Concrete pillar stoping was developed at the Keretti Mine in Finland (Heiskanen 1980; Lappalainen et al. 1980; Matikainen 1982) and has been in use since 1954. The Keretti Mine produces copper, iron, zinc, sulphur, and cobalt. The rock is very strong and consists of unfractured schist, gneiss, dolomite, and quartzite.

Mining begins by excavating a series of parallel slots. After the slots are filled with cemented backfill, the intervening rock pillars are excavated. Finally, the empty pillar voids (chambers) are filled with sand. Both the slots and chambers are excavated in two sequences. First, a top heading is excavated by drilling and blasting. Second, the depth of the slot is increased to the full thickness of the seam by excavating a bench, following the top heading. These operations are indicated schematically in fig. 2.

## 3 FUNCTIONS OF BACKFILL

A common practice in concrete pillar stoping is to use backfill to fill the

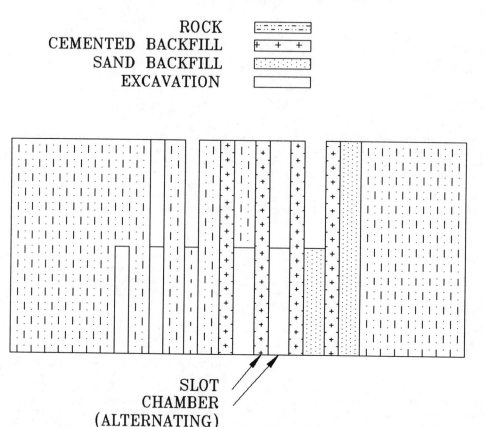

**PLAN VIEW**

**CONCRETE PILLAR STOPING**

ROCK
CEMENTED BACKFILL
SAND BACKFILL
EXCAVATION

SLOT
CHAMBER
(ALTERNATING)

Figure 1

voids left by excavation. The backfill prevents instability failures, and roof falls are reduced because less area of roof rock is exposed. Surface subsidence is greatly reduced. The use of backfill also reduces problems associated with surface mine waste disposal since the primary constituents of backfill are mine refuse and tailings.

In the slot-and-chamber method, backfill has, in general, two functions. First, it provides overall support of the mine back when successive slot-and-chamber excavations create large roof spans. Second, backfill provides localized support adjacent to unfilled chambers. That is, the

cemented backfill pillars provide stable sidewalls for excavation of the intervening rock pillars and bulkheads for the subsequent filling of the empty chambers with sand. Thus, whereas overall support is provided by both cement and sandfill components, localized support can only be provided by the cemented backfill pillars.

The cemented backfill pillars also sustain loads that may be caused by deformation from the mine back. Therefore, the strength of the cemented pillars must at least be sufficient to withstand their own self-weight plus the induced loads.

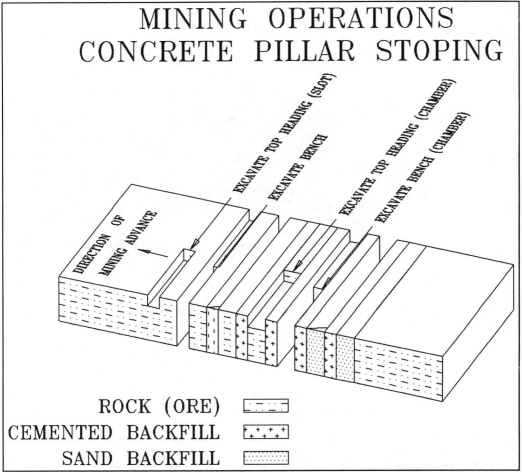

# MINING OPERATIONS
# CONCRETE PILLAR STOPING

EXCAVATE TOP HEADING (SLOT)

EXCAVATE BENCH

EXCAVATE TOP HEADING (CHAMBER)

EXCAVATE BENCH (CHAMBER)

DIRECTION OF MINING ADVANCE

ROCK (ORE)

CEMENTED BACKFILL

SAND BACKFILL

Figure 2

From an economic standpoint, it is desirable that induced loads do not develop on the cemented backfill pillars and thus increase the strength requirement. In strong, competent rock with relatively narrow chamber widths, such as exist at the Keretti Mine, it appears that induced loads are very small or nonexistent. Indeed, the strength of the cemented backfill pillars at the Keretti Mine is not much more than that required to sustain self-weight. This may not be the case for less competent or soft rock conditions, where rock deformability is much greater.

One important measure that can be taken to minimize mine back deformability (and decrease induced loads) is in the planning of the mine layout. In this respect, it is prudent to be aware of eventual roof spans that will be developed during the

mining sequence. The mine layout can be selected to minimize roof spans for the volumes of ore extracted. The idea that led to development of the spiral method was that of providing a central support in the panel to be mined. For a uniformly loaded, simple beam, a center support reduces the maximum beam deflection by 94 to 97 pct (depending on beam end constraint conditions) over that of a similar beam without a center support.

## 4 SPIRAL MINING

### 4.1 Method

There is no precedent for spiral mining. However, in a variation of longwall coal mining, panels have been "turned" radially to permit the continuous mining of succes-

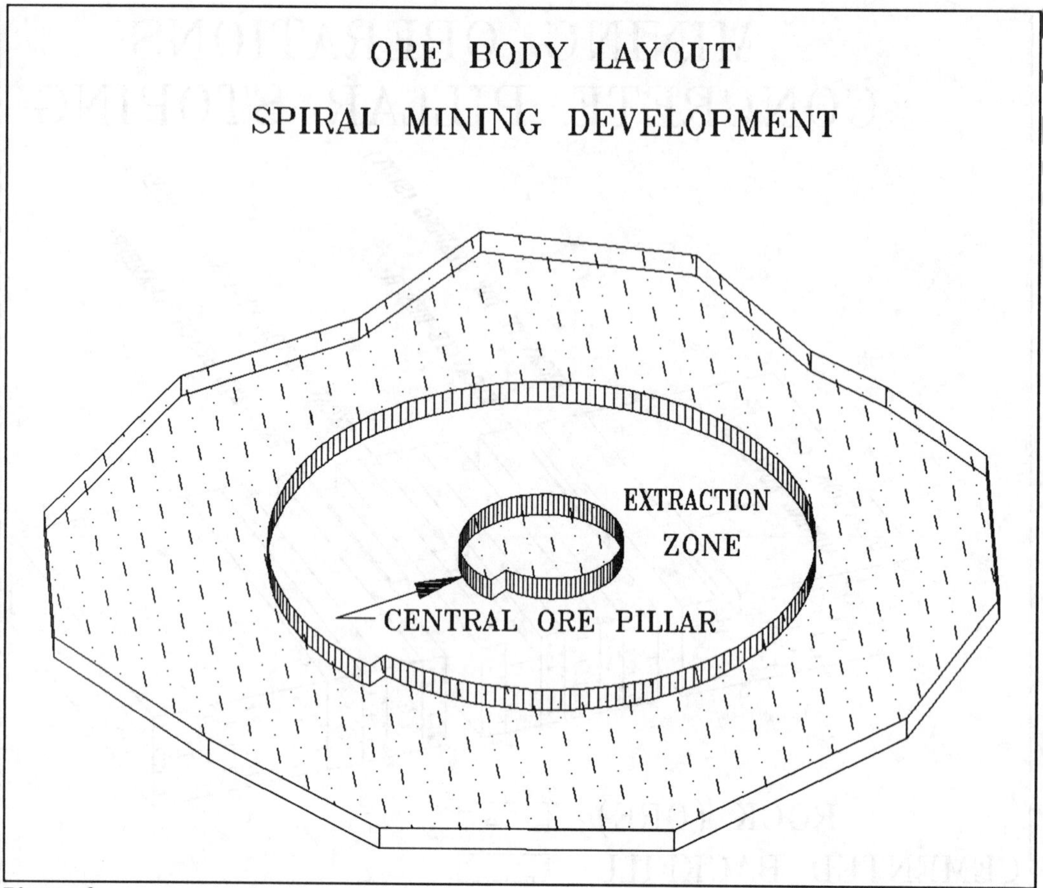

# ORE BODY LAYOUT
# SPIRAL MINING DEVELOPMENT

EXTRACTION ZONE

CENTRAL ORE PILLAR

Figure 3

sive panels. This has the advantage of maintaining a continuous operation without removing and reinstalling face equipment. The spiral method has similar advantages.

The spiral mining method consists of mining seams along a continuous flat spiral. The spiral need not be circular, and may be adapted to the shape of the ore body. In this discussion, a concentric spiral pattern, that increases uniformly each revolution, is used. The planview of the extraction zone is that of a donut, with a central ore pillar left for support (see fig. 3).

Mining the spiral begins after shaft construction and excavation of mine access openings and ancillary mine facilities. Access to the initial spiral is made through perpendicular crosscuts from the shaft. Depending on the thickness of the ore body, crosscuts are provided on one or two levels of the mine. For thin ore

bodies that may be excavated in one pass, crosscuts are necessary on only one level. For thicker ore bodies, it is more economical to excavate a top heading in one pass, followed by a benching excavation to the full depth of the seam. In this case, two sets of crosscuts are used, one for servicing each level (see fig. 4).

Mining the slot spiral is done in three operations: top heading, benching, and backfilling. Drill-and-blast methods may be used to excavate the top heading. After the top heading is advanced sufficiently, the benching operation deepens the excavation to full seam thickness. The fully excavated slot is filled with cemented backfill. Each operation is isolated from the others to prevent conflicts and to assure free access to the shaft (see fig. 5).

Mining of the chamber spiral is begun after the slot spiral has been backfilled

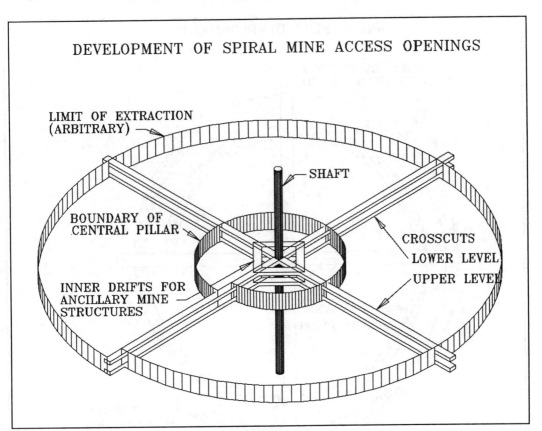

# DEVELOPMENT OF SPIRAL MINE ACCESS OPENINGS

LIMIT OF EXTRACTION (ARBITRARY)

SHAFT

BOUNDARY OF CENTRAL PILLAR

CROSSCUTS
LOWER LEVEL
UPPER LEVEL

INNER DRIFTS FOR ANCILLARY MINE STRUCTURES

Figure 4

and the cemented backfill has cured. The chamber operations are identical to the slot operations with the exception that sand is used instead of cemented backfill (see fig. 6).

Establishing the dimensions of the mine depends on many factors and requires rigorous analysis and integration of other considerations. The dimensions used in this discussion are arbitrary, but they approximate the dimensions used in the Keretti Mine. Thus the initial spiral begins at a radius of 46 m (150 ft) from the center of the shaft. The slot width is 6 m (20 ft) and the chamber width is 9 m (30 ft). The thickness of the ore body is 15 m (50 ft). The top headings are excavated to a depth of 6 m (20 ft) and the benching excavations to 9 m (30 ft). The mining operations are illustrated in a three-dimensional view in fig. 7.

4.2 Benefits

There are several potential benefits of the spiral pattern. Ground pressures should be evenly spread out, minimizing localized and random stress concentrations and reducing ground control problems. Stress distributions in the rock and backfill are likely to be more favorable because an ore pillar left in the center of an ore body reduces the effective roof span for equal volumes of ore as compared to roof spans in conventional mine layouts. With less stress, the amount of cement used, and therefore costs could be reduced. Productivity should be increased because the face of mining is continuous and there would be less time lost because of equipment transfers. A shaft located through a central pillar can be used to service the mining face, thus decreasing the number of entries needed to access the production panel and shortening haulage and tramming distances. With a central shaft, ventilation circuits could be shortened, thus

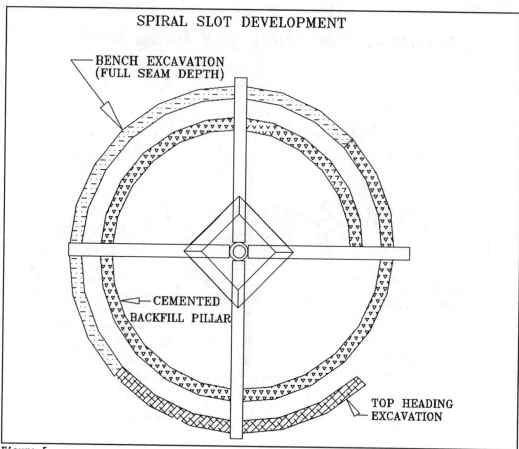

**SPIRAL SLOT DEVELOPMENT**

BENCH EXCAVATION
(FULL SEAM DEPTH)

CEMENTED
BACKFILL PILLAR

TOP HEADING
EXCAVATION

Figure 5

ventilation systems. Other potential
benefits are summarized below.

### 4.2.1 Rock bursts

As the spiral layout creates lower stress
levels than the conventional slot-and-pillar layout, rock bursts should be reduced.
Energy release during excavation is an
indicator of rock burst potential. Rock
bursts are reduced if the stiffness of the
overall mine system is greater than that
of individual elements, i.e., pillars.
The spiral system is likely to develop
less deformation, and therefore release
energy more slowly or build up less stored
energy than the conventional slot-and-pillar method, given equal volumes of ore.
Therefore, the rock burst potential of a
spiral system is hypothesized to be less
than that of a conventional system.

### 4.2.2 Mine roof presupport

Whereas backfill provides overall ground
control, it does not contribute significantly to ground support in those areas
where mining operations are going on.
Conventional roof bolting is usually required for this purpose. There may be
significant advantages to the use of presupport as a substitute for or replacement
to conventional bolting.

Presupport is a relatively new method of
ground control. Prior to excavation of
the seam below, long holes are drilled
into the roof strata at shallow angles,
and bolts or resin (or both) are installed. By installing presupport, the
strength of the roof strata is preserved
and immediate collapse of the roof after
excavation is prevented. This is a promising method of roof control, especially
in poor or loose rock conditions.

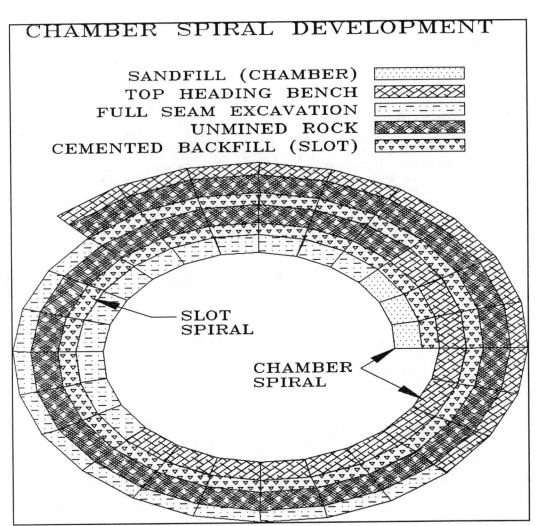

# CHAMBER SPIRAL DEVELOPMENT

SANDFILL (CHAMBER)
TOP HEADING BENCH
FULL SEAM EXCAVATION
UNMINED ROCK
CEMENTED BACKFILL (SLOT)

SLOT
SPIRAL

CHAMBER
SPIRAL

Figure 6

The adaption of presupport concepts to spiral mining complements the mining method. During excavation of the slots, presupport could be installed in the adjacent ribs, up and over the pillar row. Later, when extracting the pillars (which now would have cemented backfill on both sides), the stability of the mine back would be maintained by the presupport. With the benefit of presupport, it may be possible to use more sandfill and less cemented backfill. This might be accomplished by increasing the chamber-slot width ratio. The technical problem is to find what new patterns of mining are possible by using presupport.

### 4.2.3 Automation

The spiral layout appears more adaptable to automated mining techniques than does conventional systems, which are fixed with irregular geometries and stop-and-go operational practices. There are six independent mining operations in progress at any given time, and within each operation, several subsidiary tasks are being done, i.e., drilling, blasting, mucking, and bolting. In the conventional method, all six operations are interrupted when one panel is completed and labor and machinery are transferred to the next panel. Due to the proximity of operations and shared mine access openings, these operations are dependent on one other. In the spiral pattern, the panels continually advance and each operation can be independent. These two aspects--continuous but independent operations--are fundamental to automation in other industries.

## 4.3 Mining considerations

### 4.3.1 Layout variations

Mine layout geometry is influenced by many factors. Flexibility is necessary to achieve a compromise in the mine layout.

tral ore pillar must be determined from analysis, and total mining extraction may be dependent on the central ore pillar's stress and strength. These conditions would change as mining progresses outward.

Since the spiral layout introduces radically different geometrical aspects, it is probable that dimensions taken from existing mining are not optimal. One important consideration is the amount of cement used in cemented backfill pillars, a costly aspect of the mining method. This amount could be reduced by increasing the chamber-slot width ratio.

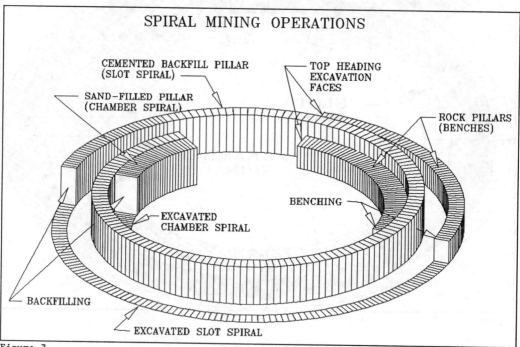

SPIRAL MINING OPERATIONS

CEMENTED BACKFILL PILLAR (SLOT SPIRAL)

SAND-FILLED PILLAR (CHAMBER SPIRAL)

TOP HEADING EXCAVATION FACES

ROCK PILLARS (BENCHES)

BENCHING

EXCAVATED CHAMBER SPIRAL

BACKFILLING

EXCAVATED SLOT SPIRAL

Figure 7

Production of a mine, which is controlled by supply and demand, needs to be adjusted from time to time. This flexibility can be served by spiral patterns having multiple mining faces.

The direction of a spiral can be from inward out or from outward in. If it is from outward in, the central ore pillar is initially very large, but becomes smaller, and stresses on the pillar increase. In this case, mining could continue inward until signs of ore plug instability are noted. If the mining direction is from inward out, the initial size of the cen-

### 4.3.2 Access to mining operations

Access to spiral mining operations may be from a center shaft or from outside the spiral. If access is from a center shaft, it can be from below the seam in the footwall or from the surface.

Access to the mining faces could be made either through crosscuts developed in the footwall or through the crosscuts as illustrated (fig. 4). The latter is preferred to avoid extra mining costs. However, in this case, crosscut routes become blocked after the slots and chambers are backfilled. In this respect, the mining operations should be planned to minimize face crosscut blockages and conflicts. Three methods of gaining access to blocked mining faces may be considered.

In lower level crosscuts, face access can be provided by installing enclosed structural liners along the crosscuts passing through slots and chambers before they are backfilled. After filling, this would provide a passageway through the backfill. Loading on the liners should be light and predictable.

In upper level crosscuts, face access can be provided by excavating passageways through filled slots and chambers. Excavation through cemented backfill pillars should be easy and should require minimal or no support. Excavation through sand-fill would require forepoling and spiling placed to prevent intrusions of sand. Through proper planning, access to mining faces over empty slots and chambers (at the upper level) can be avoided. Otherwise, it would be necessary to bridge across the slot and chamber chasms. In this case, it may be possible to develop adjustable, movable bridges.

### 4.3.3 Recovery of central ore pillar

After mining of a spiral panel is completed, unmined ore remains in the central pillar. Recovery of this pillar may be possible, but access may be limited and ground control may become difficult. It may be possible to gain access to the central ore pillar from drifts outside the spiral mined zone, either through drifts constructed through the backfill or excavated in the footwall. Assuming that the ore pillar can be accessed, other problems still exist concerning how the pillar can be excavated and how much can be mined.

### 4.3.4 Underground plants

The shaft could also be used to service an underground ore crusher and backfill preparation plant, located in the footwall, to minimize ore and backfill transport distances. If the ore crusher and backfill preparation plants were located underground, perhaps in the footwall, and serviced from a central shaft, the cost of material transport might be reduced as compared with surface facilities.

### 4.4 Problems and hazards

A list of potential mining problems includes the stability of the central shaft, catastrophic mine failure, air percussions caused by blasting, mine water disposal, heading alignment, and avoidance of seam anomalies.

### 4.4.1 Shaft stability

With the spiral method, a central ore pillar offers a likely location in which to excavate a service shaft, which could be used for ventilation, mine access, and ore transportation. However, because of the shaft's proximity to mining, its stability is of concern. Two factors should be considered: the stress levels on the shaft during mining and the influence of subsidence on shaft stability.

### 4.4.2 Catastrophic failure

Catastrophic mine failure is a type of dynamic pillar collapse that, once begun, spreads throughout the mine with a domino effect. This type of failure has occurred in a South African coal mine that lacked sufficient barrier pillars to localize the failure zone.

Also, as mining progresses, stress levels in the rock and backfill gradually increase and may eventually lead to structural instabilities in the mine. This condition might be avoided by leaving spiral barrier pillars for ground support at radial intervals.

### 4.4.3 Blasting percussions

In a spiral layout, blasting the top heading and the bench and backfilling would occur along one continuous panel. Each of these operations should be isolated from the other to eliminate damage from air percussions and flyrock.

### 4.4.4 Water disposal

Drainage in a spiral layout may pose problems if there are no provisions for water control. In most mines, water drainage is aided by orientation of the mine entries, which are either aligned downdip or updip. In such alignments, water drains downdip to a point where it is collected. In a spiral layout, the alignment of the panel would be changing continuously and there would be no point along the spiral where water could be trapped.

## 4.4.5 Heading alignment

Because the spiral mining layout is a radical departure from the rectilinear layout of conventional mining patterns, special methods of underground surveying may be needed.

## 4.4.6 Seam anomalies

Because of adverse geological conditions or low-grade ore, parts of an ore body may not be minable, and it may be desirable to redirect mining around such areas.

## 5 CONCLUSIONS

It is clear that the feasibility of the spiral mining concept cannot be assessed other than by rigorous evaluation of the many factors presented above. This evaluation includes a comprehensive review of the rock mechanics issues, productivity, types and extent of ore reserves, ventilation, material transport, and mining economics.

The spiral method has potential advantages and is an interesting alternative for mining near-horizontal, tabular seams. It is hoped that presentation of this concept will elicit useful comments in this regard.

## REFERENCES

Heiskanen, R. 1980. Planning and Developing Small Mines. Paper presented at First International Mine Planning and Development Symposium, Beidaihe, China, Sept.

Lappalainen, P., R. Matikainen, P. Sarkka, and P. Kupias. 1980. Application of Rock Mechanics to Cut and Fill Mining at Outokumpa Oy, Finland. Paper presented at the Lulea Cut and Fill Symposium, Sweden, June.

Matikainen, R., and P. Sarkka. 1982. Cut and Fill Stoping as Practiced at Outokumpa Oy. Chapter 9 in Underground Mining Methods Handbook, W. A. Hustrulid, editor, Society of Mining Engineers, AIME, New York, NY.

# 5 General topics

*Innovations in Mining Backfill Technology, Hassani et al. (eds), © 1989 Balkema, Rotterdam. ISBN 90 6191 985 1*

# Stability of classified tailings backfills containing reinforcements

R.J.Mitchell
*Queen's University, Kingston, Canada*

ABSTRACT: Recent centrifugal model studies (Mitchell, 1988) have clearly shown that cement can be used more efficiently, in mine stopes where the height to exposed fill length ratio, H/L, is greater than unity, by creating layered fill systems composed of thin layers with high cement content between thicker layers of low cement content fill. In cases where L ≥ H, however, this technique is not found to be effective and it is suggested that structural fill reinforcements might be used to increase the stability of relatively long stopes. The results of centrifuge model tests carried out to study reinforced cemented mine backfills are reported in this paper.

It has been previously shown from unconfined laboratory test samples (Li and Mitchell, 1988) that geogrids, mesh elements and anchored fibre reinforcements can all contribute substantially to the strength of a weakly cemented tailings sand. These reinforcements do not rely on bonding with the cement paste but develop strength in a sand material by resisting tensile strains and the dilation required for grain slippage. Three series of centrifuge model tests on backfills containing these three types of reinforcements are reported in the paper. The mechanics of failures which developed in each of the three types of backfill models and the benefits of reinforcing backfills are discussed in the paper. It is concluded that the use of reinforcements can produce overall economies under certain operating conditions.

## INTRODUCTION

Waste tailings are commonly classified for use as backfill in large underground openings in metalliferous mines. These waste tailings must be cemented to provide for ore pillar removal. Increasing costs of cementing agents have given impetus, in recent years, to considerable research on methods of improving backfill economics. Most research on fill stability has concentrated on the three dimensional nature of fill blocks and the beneficial effects of arching (see, for example, Smith et al, 1983). The most costly backfills, however, are associated with relatively long fill exposures (fill exposure strike length, L, greater than exposed fill height, H) in room and pillar mining of flat-lying or laterally extensive ore bodies. In such cases, relatively greater economic gains are to be realized by the reduction in cement usage through improvements in backfill characteristics. Improvements in fill characteristics can be achieved by altering the grading of the fill materials, increasing the placement density, reducing the delivery water content or by addition of reinforcing elements. All of these methods, of course, have associated costs but will generally produce overall savings if these costs are not in excess of two to three dollars per tonne of ore mined.

The technologies associated with these four methods are at various stages of development and it is currently considered viable to continue exploring all four. Indeed, combinations of two or more of the methods may eventually produce a most economical cemented sand backfill. This paper outlines recent centrifugal model studies carried out to determine the beneficial effects of using various types of reinforcing elements for stabilizing cemented sand fills.

### Reinforced centrifuge models

Li and Mitchell (1988) report on the results of plane strain tests on cemented fill samples containing one of five different types of reinforcement inclusions at different variations and concentrations. They found that mesh elements and anchored fibre reinforcements more that doubled the plane strain shear strength of a 33:1 T:C mixture. The other types of reinforcements, which included deformed plastic wires and straight steel wires, were not as effective but all inclusions improved the durability of the backfill. Following completion of this work in early 1987, Falconbridge Ltd. supported several centrifuge model tests incorporating placed geogrid reinforcements as well as model mesh elements which were mixed and poured with the backfill models. Further centrifuge model testing was carried out during the summer of 1988 to extend these complementary studies and all of the centrifuge data are combined in this paper into three series as follows:

1. Mesh element reinforced models: Netlon mesh elements, commonly called 'snippets' are commercially available from the Tensar Corporation. Shown on Figure 1, these elements are composed of 0.5mm polypropylene strands of 3.5 kpa tensile strength. This test series includes models in which the snippets were placed as directional reinforcing elements and models in which snippets were mixed and poured as part of the backfill.

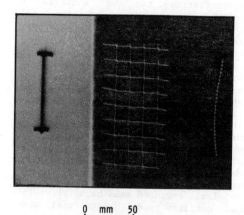

Figure 1   Anchored wire, snippet and deformed plastic wire reinforcements

2. Geogrid reinforced models: Geogrid SS-1 reinforcing material is available from Tensar Corporation in large rolls. It is a post-tensioned polypropylene webbed material with a tensile strength of 12.5 kN/m in the transverse axis, as used in the model tests. The tests were carried out with horizontal layers of geogrid at various spacings in both cemented and uncemented fills. The locations of reinforcement layers in the 0.3 m high models are shown on Figure 2.

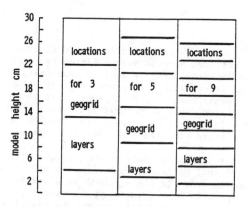

Figure 2   Geogrid reinforcement locations

3. Lineal inclusion models: Lineal inclusions were fabricated from fencing wire (to form anchored wire fibers) and from commercial 'Secur-a-tie' fasteners (to form 1mm diameter plastic fibers with 2mm deformations). These inclusions, intended to produce a backfill strength increase by providing increased interlocking in the material, are shown on Figure 1.

Centrifuge models were prepared by pouring the design mixes into form-ply molds. After curing for 28 days, the cemented models were placed in appropriate centrifuge strong boxes for testing. To maintain a two dimensional plane stress condition, these models are fitted into the strong box with a small edge clearance of about 0.5 mm so that no significant boundary friction could develop during the model tests. Boundary friction was eliminated in uncemented fill models by using silicone grease and silicone membranes between the metal sides of the box and the model.

Model testing was carried out by conditioning the samples for a few minutes at a centrifuge speed below 80 RPM (equivalent to about 20 gravities) and then increasing the simulated gravitational stress in increments until failure occurred. The centrifuge speed at model failure is used to calculate the equivalent prototype height at failure based on the proven principle that "prototype stresses are produced in a λ scale identical model by simulating λ gravities in centrifuge flight". A schematic of Queen's geotechnical centrifuge is shown on Figure 3.

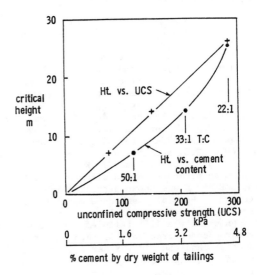

Figure 4    Critical heights of backfills

## Mesh element reinforced model results

Table 1 contains data from models reinforced with mesh elements. The

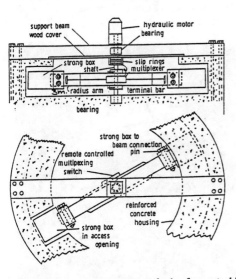

Figure 3    Queen's geotechnical centrifuge

Unconfined compression specimens of each model pour are tested in order to compare any improvements with the stability of an unreinforced model. Figure 4 shows the relationship between critical height and unconfined compressive strength (UCS) as established from centrifuge model tests on plane cemented backfills with frictionless boundary conditions. The relationship is linear with a stability number of 3.7, compared to the commonly used value of 3.8 for vertical slopes in cohesive soils.

Probably the greater effect of tensile crack propagation in this brittle cemented fill explains this minor difference. The form of the height vs. cement content curve on Figure 4 is typical of cemented classified failings sands.

TABLE 1 - MESH ELEMENT REINFORCED MODEL RESULTS

| TEST NO. | SAMPLE DESCRIPTION | UCS OF CONTROL CYLINDERS kPa | CENTRIFUGE RPM AT FAILURE | EQUIVALENT PROTOTYPE HEIGHT m | IMPROVEMENT FACTOR |
|---|---|---|---|---|---|
| D | 50:1 T:C 1 kg/m³ snippets | 74 | 108 | 11 | 1.5 |
| J | 33:1 T:C 1 kg/m³ snippets | 166 | 157 | 22 | 1.5 |
| E | 50:1 T:C 2 kg/m³ snippets | 78 | 128 | 15 | 2.1 |
| K | 33:1 T:C 2 kg/m³ snippets | 100 | 138 | 17 | 2.0 |
| F | 50:1 T:C 3 kg/m³ snippets | 78 | 160 | 23 | 3.3 |
| L | 33:1 T:C 3 kg/m³ snippets | 100 | 175 | 28 | 3.2 |
| OS/3 | 40:1 T:C overlapped snippets at 3 cm spacings | 110 | 230 | 48 | 4.8 |
| OS/5 | 40:1 T:C overlapped snippets at 5 cm spacings | 108 | 190 | 33 | 3.3 |
| SR/5 | 40:1 T:C joined snippet reinforcements at 5 cm spacings | 103 | 183 | 32 | 3.2 |

improvement factor is calculated as the ratio of the corrected equivalent prototype height divided by the unreinforced prototype height (7m for 50:1 T:C, 10m for 40:1 T:C and 14m for 33:1 T:C fills). The corrected equivalent prototype height is the calculated prototype stable height from the model test multiplied by the ratio of average UCS to model UCS for that T:C ratio. The average UCS are 70kPa for 50:1 T:C, 100kPa for 40:1 T:C and 150kPa for 33:1 T:C fills. Models D, E, F and J, K, L contain model mesh elements poured with the fill. Models E and F are shown after failure on Figure 5. Partially embedded model mesh elements (snippets cut to 50mm x 10m) can be seen after the fill failures. Although greater concentrations of model mesh elements resulted in higher prototype heights and reduced volumes of failed material, the largest ratio of improvement factor to mesh element concentration is at a concentration of 1 kg of snippets per m³ of fill. An equivalent 50% height increase using cement would require an additional 0.7% cement or about 12 kg of cement for one cubic meter of fill. Either way the improvement cost is about a dollar a cubic meter of fill for a 50% improvement in prototype height. The main advantage of using the snippets is that the fill is made less brittle and failures are likely to be less extensive.

Models OS/3 and OS/5 contained snippets overlapped to form continuous horizontal reinforcing layers. In the overlapped arrangements, these models contained approximately 2.0 and 1.2 kg of snippets per meter of fill respectively, and the improvement factors are much greater than realized by the same concentration in a randomly orientated (poured) arrangement. This agrees with the findings of Li and Mitchell (1988) that randomly oriented reinforcements generally produced less improvement than horizontally oriented reinforcements. Figure 6 shows failure of

Figure 6    Models OS/5 and OS/3

Figure 5    Snippet reinforced models

the models with horizontally overlapped snippet layers. Failures resulted due to slippage and pullout of the reinforcing elements. It is concluded, however, that fill reinforcement using snippets would be efficient and effective if a method could be developed to ensure that these reinforcements would be mainly horizontally aligned in the fill. Perhaps the use of larger sized elements in fill pours would result in sub-horizontal alignment.

The snippets in model SR/5 were wired together to form a continuous reinforcing layer and prevent pullout. This model failed on a continuous shear plane through the fill and reinforcements as shown on Figure 7. The improvement factor in this case was 3.2, equivalent to the highest concentration of randomly poured snippets with about one-tenth of the mass of reinforcing material. It is obviously more cost effective to employ continuous horizontal reinforcing layers if in-stope placement of reinforcements is possible. Model SR/5 would have reached an even larger equivalent prototype height had the reinforcements been stronger or had the layers been placed closer together. To study this type of fill reinforcement in greater detail, geogrid reinforced models were cast and tested.

## Geogrid reinforced model results

Table 2 lists relevant data from geogrid reinforced samples and the failures in samples A, B and C are shown on Figure 8. At the widest spacing between geogrid layers in cemented fill, shear failure developed after some facial spalling and subsequent edge rotation in the exposed geogrid. With the closest geogrid spacing, failures were limited to facial spalling. Although the improvement factor was about 7 with the nine geogrid layers, the ratio of improvement factor to numbers of geogrid layers was greatest with the three layers. This improvement factor of about 3 would require an increased cement usage of about 2.2% or about 38 kg of cement per cubic meter of fill. The cost of this extra cement would, in most cases, run better than $2/tonne. The placement of geogrid reinforcements appears to have economic potential. Closer examination of the cemented fill data on Table 2 shows that all of the models reached a failure condition when the equivalent prototype geogrid layer spacing (equivalent prototype height divided by the number of

Figure 7    Failure through snippet reinforcements

TABLE 2 - GEOGRID REINFORCED MODEL RESULTS

| TEST NO. | SAMPLE DESCRIPTION | UCS OF CONTROL CYLINDERS kPa | CENTRIFUGE RPM AT FAILURE | EQUIVALENT PROTOTYPE HEIGHT m | IMPROVEMENT FACTOR |
|---|---|---|---|---|---|
| A | 50:1 T:C 3 layers geogrid | 78 | 160 | 23 | 3.3 |
| G | 33:1 T:C 3 layers geogrid | 101 | 162 | 24 | 2.7 |
| B | 50:1 T:C 5 layers geogrid | 78 | 185 | 31 | 4.4 |
| H | 33:1 T:C 5 layers geogrid | 100 | 218 | 43 | 5.0 |
| C | 50:1 T:C 9 layers geogrid | 74 | 240 | 52 | 7.4 |
| I | 33:1 T:C 9 layers geogrid | 118 | 260 | 61 | 6.0 |
| U/10 | uncemented 10 layers geogrid | 0 | 133 | 16 | 11 |
| U/15 | uncemented 15 layers geogrid | 0 | 160 | 23 | 15 |

separated fill layers) was about 5.8 m for the 50:1 T:C and 7.0 m for the 33:1 T:C. This result sponsored the two tests without cement, since it is obvious that the greatest cost savings could be realized by eliminating both the capital and the operating costs associated with cement addition.

Uncemented fill will stand vertically to a height of about one third of the height of capillary rise in the fill. For a typical gravity drained classified backfill sand, it was expected that facial sloughing would develop in the uncemented models when the distance between geogrid reinforcing levels was about 1.5 meter -

Figure 8    Geogrid reinforced models

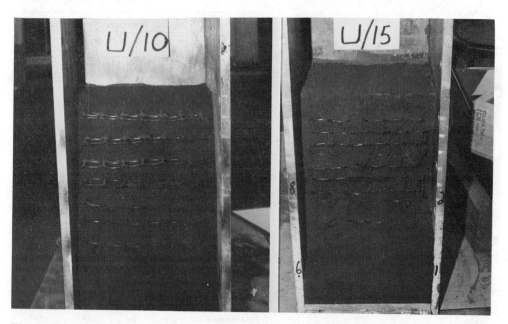

Figure 9    Models U/10 and U/15 after testing to 23 m equivalent prototype height

that is when model U/10 was about 15 m in equivalent prototype height and model U/15 was about 23 m in equivalent prototype height. This expectation was closely realized in the two uncemented tests reported on Table 2. The sloughing in test U/10 produced a substantial amount of uncemented fill dilution while model U/15 was subject only to minor facial sloughing at the 23 m equivalent prototype height. Photographs of the models after centrifuging at 160 RPM are shown on Figure 9. These data indicate that the drained uncemented tailing used in these models (about 5% water content after drainage) could be stabilized (consolidated) by using geogrid reinforcements at about 1.2 meter spacings. Some dilution would be expected, however, unless the geogrid was wrapped around the open face to prevent sloughing. With a wrap-around utilization, greater geogrid spacings might be anticipated. Further economy might be gained by using a spacing which varied with fill depth. Certainly, further model testing of reinforced uncemented fills is warranted.

Figure 10 shows the relation between the required prototype geogrid spacing for a stable fill and the cement content of the fill. The greatest rate of improvement is found at low cement concentrations and it would be interesting to explore the early

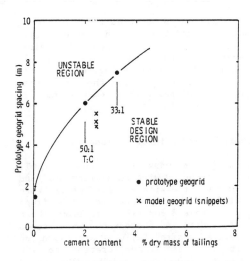

Figure 10   Stability chart for geogrid reinforced cemented sandfills

shape of the relation. Perhaps the most economical usage would be to simply cement grout the face of reinforced uncemented fill to prevent sloughing. Post placement grouting could be accomplished by drilling from the fill surface. It is clear that more research is needed to investigate these alternatives before the efficiency and economics of geogrid reinforced fills can be determined.

Figure 11   Models PW30 and AW30

It should also be noted that the model studies do not indicate the actual size requirement for the geogrid since stresses, and not forces, are correctly modeled. It is interesting, however, to note that the results from the models with layers of snippet reinforcements do lie below the geogrid data on Figure 10.

## Anchored fiber reinforced models

Following the work of Li and Mitchell (1988), anchored wire reinforcements and deformed plastic wire reinforcements were selected for placement in centrifuge models. Table 3 contains data from these model studies and photographs of Models PW30 and AW30 after failures are shown on Figure 11. After acceleration to 200 gravities (260 RPM), model AW30 showed only facial slough back to a distance of about 1 cm.

TABLE 3 - ANCHORED WIRE REINFORCED MODEL RESULTS

| TEST NO. | SAMPLE DESCRIPTION | UCS OF CONTROL CYLINDERS kPa | CENTRIFUGE RPM AT FAILURE | EQUIVALENT PROTOTYPE HEIGHT m | IMPROVEMENT FACTOR |
|---|---|---|---|---|---|
| PW30 | 40:1 T:C 10 layers of 30 deformed plastic wires per layer | 120 | 156 | 20 | 2.0 |
| PW15 | 40:1 T:C 10 layers of 15 deformed plastic wires per layer | 110 | 130 | 15 | 1.5 |
| AW30 | 40:1 T:C 10 layers of 30 anchored steel wires per layer | 130 | 260 | 52 | 5.2 |
| AW15 | 40:1 T:C 10 layers of 15 anchored wires per layer | 118 | 220 | 41 | 4.1 |

It is a little difficult to predict exactly what would happen to the anchored fiber fill at prototype scale. If the linear scaling rule is used, the facial slough of 1 cm is equal to 2 meters and the 1 mm wire thickness is equal to 0.2 m. The prototype anchored wire reinforcements would then be steel sections of about 10 m length with 0.5 m long anchors. On the other hand, the improvement factors of 5.2 and 4.1 can be compared with sample improvement factors of approximately 3 and 2.5, respectively, derived from Li and Mitchell (1988) for 0.2 m high plane strain samples of equivalent anchored wire concentrations. The question is one of size effect - do the anchored wires provide an integral improvement in the bulk characteristics (strength, stiffness, durability) of the backfill or do they act as independent reinforcing elements to which excess forces are transferred on a mass scale - the answer will be clear only when further modelling is carried out. Both physical modelling of models and numerical modelling of forces and displacements should be conducted. It is clear that the deformed plastic wire reinforcements are not as good as the anchored wires, a conclusion in agreement with the plane strain testing by Li and Mitchell. This may be due either to the low bending stiffness in the plastic wires on to their inability to develop adequate anchorage. If the answer to this were known, the basic requirements, (hence the modelling) requirements, for anchored fibers would be known. Any consideration of the economics of anchored fiber reinforcements, beyond the simple statement that the potential improvement factor of 4 could save up to $3/tonne in cement usage, would be premature.

## Economic considerations for reinforced fills

Tailing backfills derive their greatest economic advantage from the fact that they can be delivered hydraulically. Excess delivery water is, on the other hand, an extreme disadvantage in cement consolidation. Either the excess water or the cement should be eliminated at delivery and methods to do both are being explored. In the meantime, any cost savings by reduced cement content can be beneficial to operating mines, particularly where long fill exposures require relatively high cement contents for fill stability. Figure 12 provides a summary of the data obtained from centrifuge testing. It is immediately obvious from Figure 12 that operations where H > L have a geometric advantage due to arching and stress transfer. Layered fills can provide some reduction in overall cement usage but are of greater benefit for high fills (H > L) as well. The economies of using reinforcements are not easily determined and must be considered for individual fill heights. Many other factors are also site specific.

Two examples will be considered:

Example 1: low height fill, long exposure

Exposed to 25 m height, our long homogeneous fill would require 4.5% cement content. Alternatives might be 2 kg/m³ of snippets poured into a 3% cement content fill (33:1 T:C) or geogrids at 1.5 m spacings (16 layers) placed with uncemented fill. The geogrid reinforcing has potential benefits as indicated on Table 4 but would depend mainly on the costs associated with stops access and geogrid placement.

Example 2: moderate height fill, long exposure

Exposed to 50 m, our long homogeneous fill would require 6.5% cement content. Alternatives are anchored wire reinforcements or 7 layers of geogrid with a 2% cement content fill (40:1 T:C). The benefits of anchored wire reinforcements are uncertain, to be beneficial these

elements must achieve a sub-horizontal alignment when poured with the fill and must act to improve the bulk characteristics of the fill. The geogrid reinforcing alternative looks very attractive in this case. If the volume was in the order of 0.1 M m³, the cement savings would be in the order of $1 M which would probably provide for associated costs in operations with good quality stable roof and pillar rocks. A general benefit of reinforced fills is improved ductility, hence improved resistance to blast damage and reduced dilution.

TABLE 4 - BACKFILL ALTERNATIVES

| FILL HEIGHT | FILL CEMENT CONTENT | ALTERNATIVES AND APPROX. MATERIAL COSTS | CEMENT COST SAVINGS AT $100 PER TONNE | OTHER MAJOR ASSOCIATED COSTS AND EXPECTED BENEFIT |
|---|---|---|---|---|
| 25m | 4.5% | (1)3% cement plus 2 kg/m³ of snippets at estimated cost of $1.3/kg. | $3/m³ | Storage and mixing of snippets. Marginal benefits. |
| | | (2)16 levels of geogrid reinforcements at $2/m³. | $9/m³ | Geogrid storage and placement in stope. Could be beneficial if stope access already provided. |
| 50m | 6.5% | (1)7 levels of geogrid reinforcements at $0.5/m³. | $9/m³ | Same as above but some additional savings available for provision of stope access. |
| | | (2)10k anchored wires /m³ at estimated cost of $5/m³. | $9/m³ | Storage and aligned delivery of anchored wires with fill. Benefits not fully predictable but likely marginal. |

Similitude of reinforced models

In centrifuge modelling, the in-flight radial accelerations are used to simulate gravitational stresses so that the stress distribution in the model will be identical to that in a prototype having the same boundary conditions. By using the prototype material in the model, similarity of strains is ensured and the patterns of deformation and failure (if it occurs) will be similar to prototype expectations.

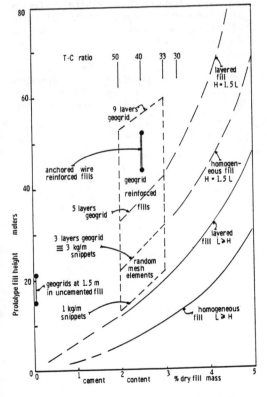

Figure 12   Summary comparison of model data

Areas in the model remain constant during flight so that model forces develop at a scale of $\lambda^2$ (where $\lambda$ is the linear scaling factor). In homogeneous models, it is sufficient, for similitude, to ensure that the model is sufficiently large, with respect to the material particle size, to create a statistical number of contacts within the model. Smooth lateral boundaries provide the plane strain conditions used in the models described in this paper. When reinforcing inclusions are used in models, however, similitude requires that these inclusions be modelled (reduced area or reduced strength) so that either pullout or rupture of the reinforcing elements can occur. The reinforced models described in this paper do not satisfy that condition of similitude and, thus, do not examine the potential for these modes of failure. Observed failures generally involved flow of the fill with associated bending of the reinforcing elements. This mode of failure would appear to be critical and the models do provide similitude in respect of such failures.

## Conclusions

Three series of centrifuge physical model studies designed as a preliminary investigation into the benefits of reinforcing classified backfills with geogrid layers and with various inclusions are reported. The major conclusions are that inclusions designed to be poured with a cemented backfill (snippets, anchored wires, etc.) are likely to be of marginal benefit but that geogrid type reinforcing layers are capable of producing significant overall cost benefits. The most important site specific cost factor in using geogrid type reinforcements is the cost of making stopes safe for entry by men and/or vehicles used in placing the reinforcements. In cases where stopes have been secured for entry during mucking operations, geogrid reinforcing is likely to be cost-effective with cement savings in the order of $10/m^3$ and geogrid material costs between $0.5 and $2 per $m^3$. Methods of local face support (wrap-arounds or face grouting) should be explored to reduce geogrid usage with uncemented fill. The use of reinforced uncemented fill would eliminate the major incompatibility between hydraulic delivery and cement consolidation. Further model studies and prototype trials using geogrid reinforcements are considered worthwhile research endeavors.

## Acknowledgments

The financial support of the Natural Sciences and Engineering Research Council (NSERC) for continued centrifuge operation is greatly appreciated. Falconbridge Ltd. supported much of the model testing reported in this paper and particular credits go to Bill Hedderson, Stan Bharti and Phil Hopkins for letting us become involved in Falconbridge research projects. Tensar Corporation provided the geogrid and snippet materials used in the model testing. Sincere appreciation is also extended to Jim Roettger, Research Assistant, whose many contributions made it possible to complete several models per week during the summer of 1987.

## REFERENCES

Mitchell, R.J. (1988) Centrifuge Model Studies on Layered Backfill Systems. In Centrifuge 88, Proc. Int. Conf. on Geotech. Centrifuge Modelling, Paris, April, 1988. Ed. J.F. Corte pp. 225-232.

Li, L. and Mitchell, R.J. (1988) Effects of Reinforcing Elements on the Behaviour of Weakly Cemented Sands. Canadian Geotechnical Journal.

Smith, J.D., DeJongh, C.L. and Mitchell, R.J. (1983) Large Scale Model Tests to Determine Backfill Strength Requirements. Proc. Int. Symposium on Mining With Backfill, Lulea, Sweden, June, 1083, p. 413-423.

*Innovations in Mining Backfill Technology, Hassani et al. (eds), © 1989 Balkema, Rotterdam. ISBN 90 6191 985 1*

# Field studies by full displacement pressuremeter in mine backfills

L.Piciacchia
*Trow Ontario Ltd, Sudbury, Ontario, Canada*

M.Scoble
*Department of Mining & Metallurgical Engineering, McGill University, Montreal, Québec, Canada*

J.-M.Robert
*Centre de Recherches minérales, Ministère de l'Energie et des Ressources, Sainte-Foy, Québec, Canada*

ABSTRACT: This paper examines the use of pressuremeters, in particular the full displacement pressuremeter technique (FDPMT) for determining the in situ modulus of deformation and limit pressure of mine backfill materials. These parameters represent potential input for numerical models as well as criteria for quality control. Data is reported from recent underground testing in 7 Canadian mines. This experience is also the basis for discussion of underground equipment design, testing procedures and limitations, as well as accuracy and precision for material properties determination.

## 1. INTRODUCTION

Increased interest in the in situ measurement of backfill properties has arisen from the need to:

... define property variability resulting either from segregation associated with transport and placement or from inadequate distribution quality control;

... improve the accuracy and precision of properties data input into numerical design models, accounting for actual stope environmental factors;

... avoid the limitations in core sampling of stope backfill.

The adoption by the mining industry of developments with in situ instrumentation from geotechnical engineering has not been widespread. The U.S.B.M. adapted a radioactive density and moisture probe system, aimed to yield a measure of the amount of densification caused by vein-wall closure, and an indication of the movement of water [1]. The vane shear test, traditionally employed in clays, has been used for comparitive studies in hydraulic sandfill [2] and cemented aggregate backfill [3]. The plate bearing test, originated to determine the bearing capacity of homogeneous soils, was used at the Mt. Isa Mine to measure modulus of deformation of cemented sandfill and rockfill [4]. Similar studies were

reported at the Kidd Creek Mine [5], and the Cobar Mine [6]. Unconfined compression and direct-shear tests were also undertaken in consolidated backfill at Mt. Isa [4].

Experimentation with standard penetration, dynamic penetration and static cone penetrometer techniques has been limited. Recent studies were completed at the Dome Mine by McGill using a new electronic, piezo-cone penetrometer system in studies of liquefaction potential in paste backfill [7].

This paper is based on underground studies with a Pencel pressuremeter in a range of mine backfills. 91 pressuremeter tests were conducted at 17 stope locations in 7 Quebec and Ontario mines. The test data is summarized and related to conclusions regarding precision and applicability to backfill testing in stope environments.

## 2. PRESSUREMETER DESIGN

### 2.1 Classification

The pressuremeter is an expandable probe, placed in the material and expanded under controlled conditions. It offers potential for determining the deformational and strength characteristics of backfill materials [8]. Types may be classified according to:

.... _probe characteristics_ ..
probes are available in a variety of diameters, selection being dependant on the material grading and method of insertion. It consists of single or triple expandable cells. The tricellular variety uses two guard cells to ensure that the applied deformation can be estimated as a cylindrical cavity. These have proven to be more difficult for field use as the differential pressure between the guard cells and the test cell must be monitored at all times.

.... _method of inflation_ .. this may be by strain or pressure control. Although this should have no effect on the quality of data and analysis, it has been found in practice that the strain control instruments have experienced less operational problems. The main problem encountered in stiffer backfills with the pressure control instruments has been a high incidence of probe bursting.

and .... _insertion conditions_ .. three basic insertion methods are presently used: the Prebored Pressuremeter Test (PMT), the Self-Boring Pressuremeter Test (SBPMT) and the Full Displacement Pressuremeter Test (FDPMT). The amount of material disturbance would appear to vary significantly with each method. In the PMT and SBPMT methods where boring of test holes is mandatory then this becomes operator dependant. Boring techniques may represent a potential source of imprecision. The FDPMT, when push insertion is used, creates the highest degree of material disturbance, although this has been considered to become repeatable and operator independent.

The applicability of the pressuremeter may thus depend on the instrument type. In the commonest type, the PMT probe (based on the Menard system with a tricellular probe), data is presented as volumetric strain at one minute after the pressure increment has been applied, versus the applied pressure. The PMT has been used to determine the modulus of deformation and the limit pressure directly from primary data. Empirical correlations have also been derived to estimate the undrained shear strength and angle of internal friction.

A _take-off_ pressure can be determined and has been used as a crude estimate of the pre-existing lateral stress in the test material, because of the disturbance created by pre-boring. Efforts to minimize this disturbance gave rise to the SBPMT, which is essentially a thick-walled probe with a hollow core, through which material is displaced to accommodate the probe. The problem of borehole wall compaction and disturbance are thus reduced. The SBPMT is more costly and the amount of borehole disturbance is operator dependent. It can be used to determine the material's strength and deformational characteristics with the same accuracy as the PMT. The ideally less disturbed borehole conditions makes the SBPMT the most appropriate of the pressuremeter type tests for the determination of in situ stresses. The SBPMT insertion procedure has not as yet been standardized and the range of procedures in use are generally slow and costly.

In an effort to reduce the costs associated with insertion for the SBPMT and to a lesser degree the PMT, the FDPMT was developed. Disturbance during probe installation is large but considered to be repeatable. The costs involved with FDPMT insertion and operator training are substantially reduced. It is generally of smaller diameter than the PMT or SBPMT in an effort to reduce probe wall friction during insertion. The data types obtainable for material strength and stiffness are the same as for the PMT and SBPMT, and includes the angle of internal friction, shear modulus and undrained shear strength. Correlations with in situ stress are at best rough estimates due to insertion disturbance. The shear modulus for the FDPMT are calculated at 45% lateral displacement, compared to 15% for the PMT method [9].

## 2.2 Prior mine studies

Backfill pressuremeter testing, reported to date, appears to be limited to some PMT work in Canada by CANMET [10,11,12] and SBPMT work in Sweden [13].

A tricellular, Menard pressuremeter, was developed by the University of Alberta for CANMET, designed to be used in a pre-drilled borehole. The probe was 115 mm diameter and 1 m long in an attempt to cater for scale effects, particularly associated with cemented rockfill. The CANMET pressuremeter was used at the Selbaie and Lockerby Mines to measure modulus of deformation, [12]. At Selbaie, 4 tests were reported in a weakly-cemented rockfill (2:1 rock:tailing, 1.3 % cement by volume); the mean modulus at the probe's pressure limit of 2 MPa was 32.7 ± 20.1 MPa, (assumed Poisson's ratio was 0.2). Borehole T.V. was used to evaluate excessive borehole damage related to drilling in weak and friable zones. Equivalent laboratory data for modulus was 30% in excess of the pressuremeter data. At Lockerby, 29 tests were reported in a cemented sandfill, giving moduli ranging from 160 to 1200 MPa. The wide range was attributed to variability in density, observed in drill core. The pressuremeter was evaluated to be a useful and cost effective tool, with one advantage being related to its ability to be used to monitor backfill at later dates without resort to drilling.

The Camcometer, a SBPMT, was used at the Näsliden Mine to measure modulus of deformation in uncemented, hydraulic backfill (a silty sand tailings). Moduli of 9.0 and 10.5 MPa were reported at 3.3 and 3.8 m depths below a fill floor in a cut and fill stope [13].

## 3. THE PENCEL PRESSUREMETER

### 3.1 Insertion system

The Pencel pressuremeter was developed by civil engineers for acquiring foundation design parameters where probe push-insertion is feasible with large surface drilling equipment. This is not readily available in underground mines, where available space is often restrictive. An alternate insertion system for the Pencel was thus developed for the stope studies [21]. This comprised a heavy duty frame, 10 tonne hydraulic ram with a 1.1 m stroke and an air powered hydraulics control unit. The Pencel pressuremeter was selected for its economy, simplicity and ruggedness for underground use. The frame disassembled into components no larger than 0.8 x 0.8 m for passage through spaces such as raises. The air-powered hydraulic system was chosen due to the ready availability of compressed air and the absence of electrical power in many underground mine locations. The maximum operating pressure of 35 MPa is achieved with .85 cu m/min. of compressed air at 415 kPa gauge pressure. This insertion system proved to be inexpensive and versatile in use, both on surface and in several different underground situations.

Four insertion frame anchoring systems were developed. The first and simplest anchoring system for both surface and underground locations, where the roof was nonexistent or inaccessible, was to load two large plates on either side of the frame with waste rock or backfill.

Secondly, when LHD vehicles were not available it was also possible to anchor the frame with double-flighted, .15 m diameter, anchoring augers drilled to 1m depth with a hand-held, rotary drill. Thirdly, where the roof was easily accessible with upto 3 m of available height, the top of the frame was propped to the roof with timber cribbing. This system provided the quickest means of providing a rigid and stable frame assembly to counteract the stresses associated with forcing the pressuremeter probe and insertion rods upto 10 m into the various backfill types tested. Lastly, where the only access to the material to be tested is via a bulkhead, the frame was mounted horizontally and supported by 25 mm diameter cables. These insertion methods have proved effective to date in backfill of up to 65 MPa modulus and 10 m depth, beyond which pre-drilled holes have been required.

### 3.2 Data interpretation

The Pencel pressuremeter, used in these studies, is a 32 mm diameter, strain controlled, monocellular, full displacement probe instrument [14]. The operating range is 0 to 90 cc volume displacement at a maximum working pressure of 2.5 MPa. The testing procedure comprises 18 increments of 5 cc in volume, V. The pressure, P, is noted at 30 sec intervals after each increment. Studies have shown that data from monocellular probes with length to diameter ratios over 6.5 compares well with that from tricellular probes [15]. Typical pressuremeter data from tests in 3 mines representing different backfill classes is presented in Figure 1(a). This has been processed only for corrections due to dilation in the pressurized lines and the inertia of the pressuremeter probe sleeve. It shows the good reproducibility in test data experienced in uniform backfill masses.

The derivation of the <u>shear modulus</u>, $G_m$, is based on the Lamé equation for the expansion of a cavity in an infinitely elastic medium [16] and is interpreted from the linear portion of the P-V curve, see Figure 1(b), as:

$$G_m = V_m \frac{\Delta P}{\Delta V} \qquad \ldots 1$$

$V_m$ = Volume at mid-height of P-V curve linear portion

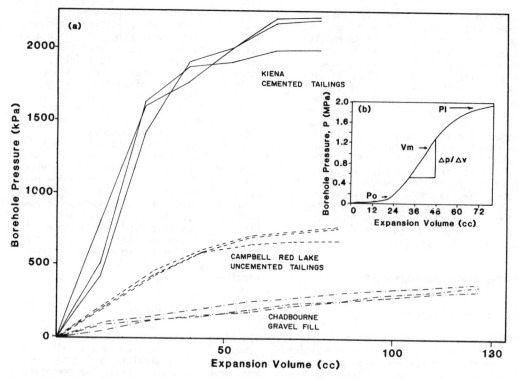

Figure 1  Typical FPDMT Test Data from 3 mine test holes [3].

The <u>deformation modulus</u>, $E_m$, is subsequently related to the shear modulus by the following:

$$E_m = 2(1+\nu)G_m \qquad \ldots 2$$

$\Delta P$ = incremental pressure change
$\Delta V$ = incremental volume change
$\nu$ = Poisson's Ratio (assumed = .33)

The <u>in situ lateral stress</u> may be estimated, [8], by the *take off* pressure, $P_o$, see Figure 1(b). Use of the FDPMT yields a borehole with compacted walls, and a redistributed stress field. Thus, the data achieved can only serve as a crude estimate of lateral stress. Furthermore, the FDPMT data in many instances does not exhibit a *take off* pressure, i.e. when the probe remains in close contact with the borehole walls. The lateral stress has also been related to the effective angle of internal friction. The coefficient of earth pressure at rest, $K_o$, represents the ratio of

lateral stress to vertical stress due to self weight, and can be estimated as:

$$K_o = 1 - \sin (\phi') \qquad \ldots 3$$

The <u>limit pressure</u>, $P_L$ is defined as the pressure at double the initial volume of the pressuremeter. The operating volume of the Pencel is a maximum of 90 cc, with the initial volume of the pressuremeter being 230 cc. A method for the experimental data to the theoretical definition of limit pressure is thus required. Four extrapolation methods are:
  1. Manual or Visual Method,
  2. Log - Log Method,
  3. Relative Volume,
  4. Inverse Volume Method.
In methods 2,3 and 4 the objective is to linearize the volume displacement curve and then extrapolate the curve to twice the initial probe volume to obtain the limit pressure. In the manual method, the point where the characteristic curve of

Figure 1 flattens within the operating range of the test instrument is used to determine a visual asymptote, which is then used to estimate the limit pressure. It has been found that regardless of the method used the estimated limit pressure remains consistent. The inverse volume method was used to estimate the limit pressure in this study, involving the development of a linear relation between the cell pressure, P, and the inverse of the volume. The relation is then extrapolated to 230 cc of injected volume, thus yielding the theoretical value for limit pressure.

Several approaches exist to obtain the undrained shear strength, $S_u$, from pressuremeter data. The major theories are based on elastic - plastic assumptions. Three theoretical solutions have been reviewed by Baguelin *et al.* [8]. Setting Poissons' ratio to .5 (undrained loading), these equations all reduce to:

$$P_{lc} = \frac{S_u\,[\,1 + \ln\,(\,E_m)\,]}{3\,.\,S_u} \qquad \ldots\ 4$$

$P_{lc}$ = limit pressure (corrected for in situ stress, $P_l - P_o$)

Each of these equations require the pre-existing horizontal stress, estimated from $P_o$. The evaluation of $S_u$ is thus prone to the same source of error as the estimate of in situ lateral stress. The effective angle of internal friction, $\phi'$, has also been related empirically to the limit pressure, Gambin [17]:

$$P_{lc} = b\,.\,2^{\,[(\,\phi'-\,24\,)\,/4]} \qquad \ldots\ 5$$

where: b is a material-dependant constant (1.8 for homogeneous, wet soil ; 3.5 for heterogeneous, dry soil ; 2.5 on average).

Baguelin *et al* [8] indicated that no theoretical method existed to convert pressuremeter test data to effective stress strength parameters, c' and $\phi'$ for free draining soil, due mainly to the effects of borehole unloading, disturbance, deformation and volume changes. Equation 5 requires that the in situ lateral stress be known; so it is prone to the same limitations as the evaluation of $S_u$. Calhoon [18], proposed an empirical method to derive $\phi'$, graphically evaluated from $P_{lc}$ and $E_m$.

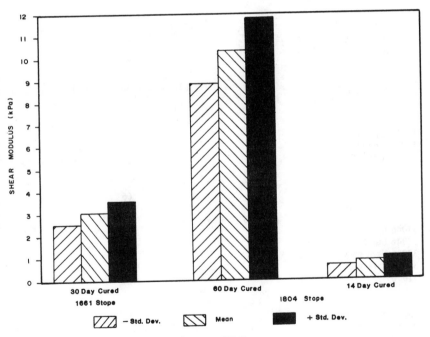

Figure 2. Shear Modulus Test Data - CRLM

## 3.3 Data variability

Two Pencel test series enabled an assessment of data variability within uniform backfill test sections and the apparent FPDMT sensitivity to material property variation. The first series of three testhole insertions, .25 m apart in plan, were conducted in the 1804 stope of the Campbell Red Lake Mine (CRLM). 11 tests were conducted through its 20:1 cemented tailings backfill, to a depth of 3.25 m. Two distinct materials were present in this section of the stope's backfill: a 14 day-cured material within the first 2.25 m, overlying a 60 day-cured material. The test data above and below 2.25 m were grouped separately; each data set exhibited a standard deviation from the mean ($S_D$) of ± 12% for $G_m$, Figure 2.

(Also shown is data from tests in CRLM's 1661 stope 30 day-cured 20:1 cemented tailings).Throughout the overall testing program the Pencel demonstrated a similar ability to perform consistently within uniform material and to distinguish between backfills with wide ranging material properties, such as those arising here from drainage and curing. A second test series was conducted in the 37036, 37038, and 37046 stopes in consolidated sandfill at the Denison Mine. Tests in two segregation layers, (accessed in backfill faces exposed as the rib pillar between 37036 and 37038 was recovered), indicated a 42 % difference in mean $G_m$, although $S_D$ for each layer was only ± 4 %. Based on full test program experience it is considered that data scatter arising from the test procedure itself can be expected up to ± 15 % $S_D$.

Table 1.  Summary of In Situ Test Data

| Mine | Stope | Number of Tests | Material Grading | | | mean $P_l$ (MPa) | mean $E_m$ (MPa) |
|------|-------|------|------|------|------|------|------|
| | | | Cu | Cc | $D_{50}$ (μm) | | |
| Lac Matagami | | | 6.88 | 2.00 | 100.5 | | |
| | surface | 9 | | | | 1.00 | 11.97 |
| Chadbourne | | | 1.57 | 1.21 | 17370.0 | | |
| | surface | 8 | | | | .41 | 2.29 |
| | 4a stope | 7 | | | | .27 | 1.06 |
| | 4c stope | 7 | | | | .24 | 1.96 |
| Remnor | | 5 | 2.81 | 1.29 | 1972.4 | .86 | 20.60 |
| Campbell Red Lake | | | 2.50 | 1.05 | 90.0 | | |
| | 1661 stope | 6 | | | | .75 | 8.00 |
| | 1804 stope (14 day) | 6 | | | | .21 | 2.40 |
| | 1804 stope (60 day) | 5 | | | | 2.50 | 27.93 |
| Sigma | | 8 | 11.96 | 6.34 | 153.1 | | |
| | | | | | | .66 | 9.97 |
| Denison | | | 3.55 | 1.29 | 477.6 | | |
| | 37046 u/c | 5 | | | | 1.97 | 43.9 |
| | 37046 bulkhead | 3 | | | | -- | 65.2 |
| | 37036 light grey | 2 | | | | .91 | 18.5 |
| | 37036 dark grey | 2 | | | | 1.19 | 10.9 |
| | 37038 | 1 | | | | 1.10 | 11.0 |
| Kiena | | | 8.91 | 1.62 | 70.8 | | |
| | non flocculated | 6 | | | | 2.11 | 15.3 |
| | flocculated #1 | 7 | | | | 2.07 | 19.9 |
| | flocculated #2 | 4 | | | | 1.85 | 22.2 |

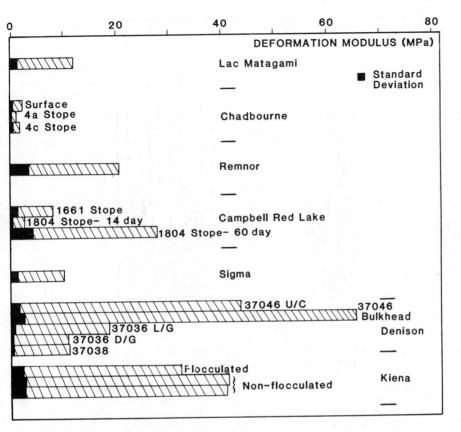

Figure 3. Deformation modulus data, all mines

The Pencel's sensitivity would appear to be such that, provided insertion procedures are carefully reproduced, then test data differences beyond this limit relate to actual variation in material properties.

## 3.4 Backfill deformation and shear moduli

Backfill mechanical properties were extracted from the corrected pressure-meter data curves as three basic parameters: the shear modulus, $G_m$ the limit pressure, $P_l$, and with a lesser degree of confidence the pre-existing lateral stress. $P_l$ is important as it has been empirically correlated to undrained shear strength and angle of internal friction. Measured values of $E_m$ in the 7 mine test series ranged upto 65 MPa, beyond which severe limitations were imposed by the capacity of the insertion system and the probe pressure. Table 1 and Figure 3 indicate how $G_m$ and $E_m$ were observed to vary significantly with cementing agent content, curing period and flocculants. The addition of cementing agents, for instance, was observed to stiffen uncemented sandfills from a 2-12 MPa $E_m$ range to a 20-40 MPa $E_m$ range for typical 20:1 cemented sandfills.

## 3.5 Backfill strength and classification

The pressuremeter curves, Figure 1(a), were also used to extract $P_l$. Table 1 illustrates that a $P_l$ value of 1 MPa consistently distinguished between uncemented and cured, cemented materials.

253

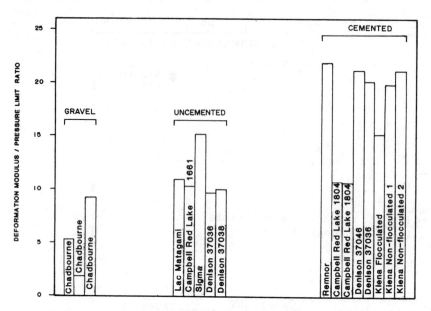

Figure 4. $E_m / P_l$ ratio classification, all mines

The uncemented materials could further be subdivided by $P_l$: coarse grained ($P_l$ = 0.2 - 0.4) and fine grained ($P_l$ = 0.66 - 1.00). The $E_m / P_l$ ratio, used to develop soil profiles, also proved to be useful for material's classification. Three backfill groups were apparent, see Figure 4:

.. the coarse grained loose gravel at Chadbourne Mine with $E_m / P_l$ ratios in the order of 4 to 8.

... the uncemented fine grained sandfills at the Lac Matagami, Sigma and Denison Mines. The $E_m / P_l$ ratio for these materials is in the range of 10 to 15.

.. the cemented sandfills used at the Red Lake, Denison and Kiena Mines and the Remnor Mine sulphide-cemented smelter slag indicated $E_m / P_l$ ratios between 15 and 24, typically in excess of 20.

### 3.6 Undrained shear strength and angle of internal friction

The undrained shear strength, $S_u$, defined as one half the compressive strength of an undrained soil sample, was empirically correlated to the $P_l$ data derived from the pressuremeter test program as:

$$S_u = .141 \, P_l + 12.5 \ (kPa) \qquad \ldots 6$$

The empirical relationships proposed between $\phi'$ and $P_l$, $E_m$ by Gambin (8), equation 4, and Calhoon (9) were applied to the pressuremeter test data. Calhoon's graphical method consistently under-estimated $\phi'$ when compared to Gambin's equation. In many instances, Calhoon's equation predicted low angles of $\phi'$ in the order of 25°. However, the largest limitation of Calhoon's method is the narrow range of applicability. Although the method appears applicable to $E_m$ values of .05-40 MPa, and $P_l$ values of 75-2000 kPa, combinations of high $P_l$ and low $E_m$ or vice versa fall outside the bounds of his graphical method. Gambin's equation, however, was applicable to the entire $P_l$ range encountered, predicting $\phi'$ in the range of 30-37°.

### 3.7 Lateral stress

The total in situ lateral stress in a mine backfill is comprised of stresses due to self-weight and stresses which are absorbed or transmitted through the back-fill mass by the redistribution of stresses around the mine opening. Although the stress redistribution component cannot be measured directly by the pressuremeter, an indirect estimate can be made. The stresses generated by the backfill's self-weight are related to $\phi'$ by $K_o$, equation 3, which can be used with the material's bulk density to estimate the lateral stress due to self-weight.

254

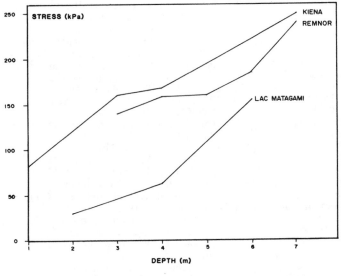

Figure 5  Absorbed stress estimation with the FDPMT

Total lateral stress has been measured by PMT and SBPMT tests. In the mine test program it could be estimated in some instances with the Pencel pressuremeter. The data from the Remnor, Lac Matagami and Kiena mines allowed total lateral stress estimation, since a $P_o$ was consistently distinguishable.The combination of the $K_o$ estimator and the measurement of total lateral stress enabled the estimation of absorbed stresses. These values were comparable, though lower than those reported by others, which ranged from 200 to 700 kPa, e.g. Corson and Wayment [19], Zahary *et al* [20]. The comparison of the absorbed lateral stresses, Figure 5, for the  Lac Matagami (uncemented) and Kiena (cemented 20:1) sandfills indicated an enhanced support capability from cementation, although this should be qualified since the materials at Lac Matagami and Kiena exist in different stress environments and data was limited.  From this study program the FDPMT offers not only a versatile tool to monitor the strength and modulus within a backfill mass over time, but also the potential to monitor stress changes. It is felt that this prospect justifies further work in a more controlled environment, with the ability to reference pressuremeter data with pressure cell data and stress change predictions from analytical modelling.

## CONCLUSIONS

The developed Pencel FDPMT system proved to meet the requirements for adaptability, ruggedness and economy in  underground stope environments. As many as ten tests were conducted daily at stope test sites. In a quality control context, it demonstrated an ability to provide consistent data within a uniform material, whilst responding to significant changes in material properties. The cost effectiveness and observed data repeatability, makes this technique appropriate for material's quality classification. This paper has illustrated the depth of data analysis potentially available with a pressuremeter, the most important parameters being the modulus of deformation and the limit pressure. It is considered that future work should investigate further the potential to derive undrained shear strength and angle of internal friction data to assess failure criteria, together with estimates of lateral pressure. The applicability and precision of this and other pressuremeter types and testing techniques would appear to warrant further study.

255

## ACKNOWLEDGEMENTS

The author's wish to acknowledge the assistance of the following mines in the studies reported: Lac Matagami, Chadbourne, Remnor, Kiena, Sigma, Campbell Red Lake and Denison. This work was facilitated by support provided by the Centre de recherches minérales, Québec, and the Natural Science and Engineering Research Council of Canada.

## REFERENCES

[1] Corson, D.R., 1977. Field Evaluation of Hydraulic Compaction at the Lucky Friday Mine, Mullan, Idaho. U.S. Bur. Mines, Report of Investigations. 7546.

[2] U. S. Bur. Mines, 1983. Placement and Evaluation of High-Modulus Backfill. Open File Report 151-84, U.S. Dept. Interior.

[3] Knissel, W. and W. Helms, 1983. Strength of Cemented Rockfill from Washery Refuse, Results from Laboratory Investigations. Proc. Int. Symp. Mining with Backfill, Lulea, pp 31-37.

[4] Gonano, L.P. and R.W. Kirkby, 1977. In Situ Investigations of Cemented Rockfill in the 1100 Orebody, Mt. Isa Mine, Queensland. Tech. Report 47, Div. Appl. Geomechanics, Comm. Sci. and Ind. Res. Org., Australia.

[5] Yu, T.R., 1983. Ground Support with Consolidated Rockfill. Proc. Symp. Underground Support Systems, Sudbury, Can. Inst. Min. Metall., pp 85-91.

[6] Richards, B.G., 1975. The Determination of Experimetally Based Load-Deformation Properties of Mine Fill. Proc. 2nd Aust.- N.Z. Conf. on Geomechanics, Brisbane, pp. 56-62.

[7] Aref, K., 1988. A Study of the Geotechnical Characteristics and Liquefaction Potential of Paste Backfill. Unpublished Ph.D. Thesis, McGill University, Montreal.

[8] Baguelin F., Jezequel J.F. and D.H. Shields, 1978. The Pressuremeter and Foundation Engineering. Trans Tech Public., Series on Rock and Soil Mechanics. Clausthal, Germany.

[9] Withers N.J., Schaap L.H.J. and C.P. Dalton, 1986. The Development of a Full Displacement Pressuremeter. The Pressuremeter and Its Marine Applications: Proc.2nd Inter. Symp. A.S.T.M. STP 950, pp 38-55.

[10] Swan, G., 1983. Compressibility Characteristics of a Cemented Rockfill. Proc. Symp. Underground Support Systems, Sudbury, Can. Inst. Min. Metall., pp 93-97.

[11] Herget, G., 1981. Borehole Dilatometer for Backfill Studies. CANMET, Mining Research Laboratories, Division Report MRP/MRL 82-2 (TR).

[12] Swan, G. and G. Vaillancourt, 1983. In Situ Backfill Testing with a Borehole Dilatometer. CANMET, Mining Research Laboratories, Division Report MRP/MRL 83-64 (TR).

[13] Borgesson, L., 1981. Mechanical Properties of Hydraulic Backfill. Proc. Conf. Appl. Rock Mech. to Cut-and-Fill Mining, Lulea, Inst. Min. Metall., pp. 193-195.

[14] Anon, 1986. Pencel Pressuremeter : Instruction Manual. Roctest Limited, St. Lambert, Quebec.

[15] Capelle J.F., 1983. New and Simplified Pressuremeter Apparatus. Proc. Int. Symp. on Recent Developments in Laboratory and Field Tests and Analysis of Geotechnical Problems, Bangkok, Asian Inst. Tech., pp 159-164.

[16] Lamé G., 1852. Lecons Sur la Théorie Mathématique de L'Elasticité des Corps Solides. Bachelier, Paris, France.

[17] Gambin M., 1977. Le Pressiometre et la Determination de L'angle de Frottement et de la Cohesion d'un Sol. Internal Report, Geoprojekt, Paris, France, 1977, pp 1-21.

[18] Calhoon M., 1970. Field Testing With the Pressuremeter. Lecture given at the University of Kansas, U.S.A. (unpublished).

[19] Corson, D.R. and W.R. Wayment, 1967. Load-Displacement Measurement in a Backfilled Stope of a Deep Vein Mine, U.S.Bur.Mines, Report of Investigations 7038.

[20] Zahary, G., Zorychta, G. and S. Zaidi, 1972. Results of Fill Pressure Measurements in a Cut-and-Fill Stope. Mines Branch, Dept. Energy, Mines and Resources, Internal Report 72/119.

[21] Scoble, M.J., Piciacchia, L. and J-M. Robert, 1987. In Situ Testing in Underground Backfilled Stopes. Bulletin Can. Inst. Mining and Metallurgy, 80, 903, pp. 33-38.

*Innovations in Mining Backfill Technology, Hassani et al. (eds), © 1989 Balkema, Rotterdam. ISBN 90 6191 985 1*

# Delayed consolidation of slag backfill for pillar recovery

L.A.Beauchamp
*Mines Accident Prevention Association of Ontario, North Bay, Ontario, Canada*

M.J.Scoble & H.S.Mitri
*Department of Mining and Metallurgical Engineering, McGill University, Montreal, Canada*

ABSTRACT : This paper reports on preliminary research into delayed backfill consolidation of HBM&S slag backfill, a well graded gravel, using pressure grouting with
particulate grouts. A physical and mechanical testing program is reported, which aimed
to establish the slag properties in natural and grouted forms. This also considered the
influence of slag gradation, compaction and grout mix type on mechanical properties.
Preliminary analysis of consolidated wall stability is reported from numerical
modelling.

## 1. INTRODUCTION

Flin Flon smelter slag has been used
underground as backfill for over 50 years
by the Hudson Bay Mining and Smelting Co.
Ltd. (HBM&S), at its Flin Flon and Snow
Lake mines in Canada. Significant
tonnages and often attractive grades of
pillar reserves are now located adjacent
to such loose, cohesionless backfill.
Principal options for pillar recovery
against unconsolidated backfill are cut-
and-fill stoping or open stoping (with
remnant pillars or consolidated backfill
walls, see Figure 1). Mitchell [1],
reviewing remnant pillar design, noted
that these may represent 10-15% of recoverable mine ore reserves. Alternate techniques to create a consolidated wall,
Figure 2, include (a) temporary backfill
freezing (b) sheet piling, slurry trenching in backfill; (c) cemented cut-and-
fill stoping on pillar boundaries; (d)
pressure grouting; (e) percolation of
cement slurry through backfill; and (f)
sequential drawdown and replacement by
cemented fill.

U – Unconsolidated Backfill          C – Consolidated Backfill Zone

P – Rib Pillar                              R – Remnant Pillar

Figure 1. Schematic of open stoping pillar recovery options.

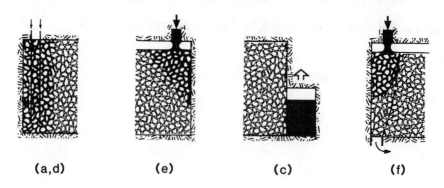

|  (a,d)  |  (e)  |  (c)  |  (f)  |

Figure 2. Potential backfill wall consolidation techniques.

Consolidation design requires establishing by analysis the **thickness** and **mechanical properties** of the wall, required for stability during complete pillar extraction. Backfill consolidation results in increased brittleness, with more susceptibility to cracking under blast loading [2]. The **Kidd Creek mine**, for example, uses amongst other backfill types, a 20:1 cemented rockfill, designed for a uniaxial compressive strength (UCS) of 7 MPa [3]. Considering self-weight alone, the fill was defined to require a UCS of 2.8 MPa after 28 days curing, for stability of an exposed face at least 120 m high and 70 m long. A 2.5 safety factor was used to account for blast loading and variation in strength due to inadequate mixing.

Size distribution and porosity, as seen in the current slag consolidation studies, exert significant control over groutability and mechanical properties. Little has been reported on segregation arising from gravity placement of coarse backfill against pillar walls, a potential consolidated wall locality. In dry rockfill placement, the natural formation of rill slopes under gravity was observed to result in segregation [4]. Segregation due to mass flow mechanisms during cemented rockfill placement has also been studied at **Kidd Creek** and **Mt. Isa mines** [3,5]. Dry rockfill placement, forming rill slopes with segregation, potentially poses two problems: the formation of weaker planes inclined at over 40° down towards the pillar face and the concentration of coarse fill at the pillar face. These features remain to be clearly defined in dry slag backfilled stopes.

## 2. PRIOR DELAYED CONSOLIDATION STUDIES

A unique filling system was developed at **Noranda's Geco mine** in Ontario from 1957, [6]. Rockfill was introduced into the voids above broken ore piles in open stopes, added continuously to prevent weak walls from collapsing. Broken ore was drawn until all boxholes showed rockfill, constantly added to keep voids full at all times. Cemented hydraulic fill, (3% cement by weight), was introduced to **percolate** through the rockfill. The initial rockfill role was to maintain the stope walls as ore was withdrawn, but the addition of hydraulic fill also minimized the dilution of broken pillar ore when adjacent pillars were mined. Stable exposures of cemented rockfill faces of up to 60 m are reported.

In 1972, at **INCO's Birchtree mine**, Manitoba, uncemented rockfill was placed in a test stope and then a 6:1 sand to cement hydraulic fill placed to **percolate** along the pillar face in an attempt to consolidate the rockfill [7].

Substantial ore reserves at the **Broken Hill mine**, Australia, were locked between uncemented fill, mainly hydraulic classified tailings, with some development waste [8]. In 1976, research concluded that **sheet piling** and **slurry trenching** to create a high-strength diaphragm wall were impractical and excessively costly. **Grouting** with particulate materials was rejected because the permeability of the fill was too low. All grouting systems for consolidation were ultimately abandoned because of concerns over *windows* of unstabilized backfill, together with excessive materials and placement costs.

Underground trials were conducted at Cominco's Sullivan mine, British Columbia [9]. Cement slurry was sprayed onto the upper surfaces of a sized float backfill to percolate down both sides of a pillar. Subsequent pillar recovery was good, although trials were suspended. Later studies injected grout into boxes of float backfill to study groutability. Laboratory and in situ tests were also conducted concerning the introduction of a cement and sand slurry into stopes to percolate into rockfill at the Kidd Creek mine, Ontario [10]. The cemented sand was aimed to act as a void-filler, thereby minimizing the effect of segregation and blasting vibration on stability.

In 1981, methods for stabilizing uncemented alluvial sand or mill tailings, were investigated by INCO, Ontario [11]. Laboratory studies, including physical modelling, were conducted to consider cement-based or solution grouts.

Routine freezing for delayed consolidation of rockfill to recover pillars between open stopes was reported at Outokumpu Oy's Pyhasalmi mine, Finland [12]. Laboratory tests of frozen fill at -5° C gave a compressive strength of 4 MPa, corresponding to the normal strength of the mine's cemented rockfill.

Research was initiated in 1987 when excessive dilution was encountered in a longhole stope at the HBM&S Chisel Lake mine, which had been filled with uncemented Flin Flon smelter slag [13]. Assessment was made of the flowability and cementation characteristics of neat slurries, with water: cement ratios from 0.7 to 2.0 by weight. It was concluded that the optimum ratio was 0.7, at which UCS values were 8.8 MPa (7 days) and 10.8 MPa (28 days). This ratio was then used in underground injection trials.

## 3. HBM&S SLAG BACKFILL CHARACTERISTICS

### 3.1 Physical characteristics

700 kg of Flin Flon smelter slag was used to determine its grain size distribution, Table 1. This also shows the characteristics of 3 other mine rockfills for comparison, tested recently at McGill. The bulk specific gravity of the sample gradation was 3.21. The slag may be classified as a well-graded gravel, with a coefficient of uniformity of 4.7. It may be grouped with both the rockfills of the Bousquet and Selbaie mines, as well as the more uniformly graded slag backfills of the Chadbourne and Remnor mines.

The HBM&S samples were obtained from surface and the extent of size reduction in transit underground has yet to be fully resolved. The amount of degradation is seen as an important factor, e.g. dry rockfill degradation was observed at Brunswick Mining & Smelting's 12 mine to reduce the maximum particle size by 50 % per 350 m of vertical travel [14].

### 3.2 Uniaxial compressive strength (UCS)

Point load tests were performed on coarse lump samples of HBM&S smelter slag, sawn into blocks, according to ISRM recommendations [15]. Sample dimensions and point load values were used to obtain size-corrected point load strength indices, accounting for sample anisotropy. UCS was calculated according to Bieniawski [16]. Mean $I_{s(50)}$ values were 99.8 MPa and 63.6 MPa, perpendicular and parallel to the plane of weakness respectively, defined either by sandwiches of contrasting porosity, i.e. either highly vesicular and porous or massive with very small bubbles, or by parallel alignment of flattened bubbles in the slag.

### 3.3 Shear strength

Leps [17] assembled and analyzed dam data published between 1948 and 1966 for large scale triaxial tests. Materials tested ranged from sand specimens to 15 rockfills with maximum particle size up to 20 cm. This showed how rockfill friction angle decreases significantly with normal stress, corresponding to a nonlinear Mohr failure envelope. This is considered due to material crushing on grain contacts and a decrease of the dilatation component of shear resistance [18]. Leps noted that:
... at a given normal pressure, increasing the unit weight results in an increased friction angle;
... improving the gradation of rockfill increases the friction angle;
... a rockfill of strong particles has a greater friction angle than one of weak particles;
... the more angular the particles, then the greater is the friction angle.

Friction angles ranged widely from 30° (low density, poorly graded, weak particles under high pressures), to 60° (high density, well graded, strong particles under low pressures).

Barton and Kjaernsli [19] noted the similarity in rock joint and rockfill behaviour, i.e. a log-linear trend in normal stress vs friction angle data. They proposed an empirical peak shear strength equation for rockfill, extending from Barton's original empirical method to estimate rock joint peak shear strength, which reflected reduced strength with increased particle size:

$$\tau = \sigma_n \cdot \tan [ R \cdot \log ( S / \sigma_n) + \phi_b]$$

where: ... (1)

$\tau$ = peak shear strength
$\sigma_n$ = effective normal stress
$R$ = equivalent roughness
$S$ = equivalent compression strength
$\phi_b$ = basic friction angle

S is size dependent and calculated from UCS and $d_{50}$. R is dependent upon the origin, particle roundedness and smoothness, and porosity of the rockfill. R can either be estimated from charts or back-calculated from tilt tests. A tilt box was constructed at McGill to test HBM&S slag and back-calculate R for ranges of gradation and compaction. The box, (40 x 40 x 45 cm) comprised a separate top and bottom, joined by plates which were removed just prior to testing, thereby exposing a potential shear failure plane. Testing was conducted as follows:
... the box was filled with a pre-weighed quantity of slag (usually > 100 kg) of the required gradation;
... when required, the slag was compacted using a portable vibrator for 15, 30 or 60 seconds;
... the box was slowly tilted until shear failure occurred, when the tilt height was measured and the tilt angle calculated.
... tests at a given gradation and compaction time were repeated five times to calculate a mean tilt angle, $\alpha_m$.
... R was back-calculated using $\alpha_m$ and the slag basic friction angle, $\phi_b$, estimated by simple tilt tests on sawn surfaces of smelter slag: the mean $\phi_b$ was 25°. S was determined using the $d_{50}$ particle size and the UCS.
The effect of slag gradation on shear strength was studied by preparing several gradations:

+18.8 mm ($d_{50}$ = 30 mm);
+13.3 mm -18.8 mm ($d_{50}$ = 16 mm);
in situ gradation ($d_{50}$ = 11 mm);
wide gradation ($d_{50}$ = 10 mm).

The effect of porosity on shear strength was considered over a range of compactions for each above gradation: no compaction (light tamping only); 15, 30 and 60 seconds. Longer vibration was found to have negligible effect on porosity. Decreased porosity reduced significantly the peak drained friction angle, e.g. Figure 3 shows the reduction for the in situ gradation over a porosity range 38-47 %. The effect of decreased porosity was more pronounced at higher porosity values, 1° in peak drained friction angle per 1% porosity decrease. The laboratory porosities ranged from 37-52%, whereas those present in rockfill dams can undoubtedly be lower, due to compaction. Leps [17] did not specify porosities in his literature review, but Marachi [20] reports values of 20-25%. Failure to attain low porosities in the laboratory is thought to explain the lower shear strength values. Figure 4 shows a comparison with Lep's review dam data. Note that the HBM&S +18.8 mm fraction, when uncompacted, has very low friction angles (12-25° lower than Leps' weakest rockfill), whereas the wide HBM&S gradation, when compacted to a 37% porosity has friction angles almost identical to Leps' low density, poorly graded rockfill.

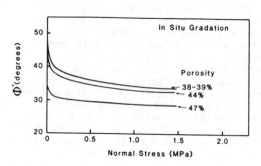

Figure 3. Sample tilt test data

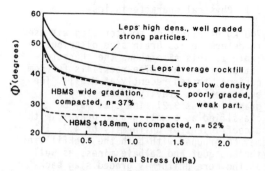

Figure 4. Tilt test vs dam data

Tilt tests were conducted on the other coarse mine backfills of Table 1: Figure 5 shows peak drained friction angle data from the tilt tests for uncompacted samples. The strength, controlled by particle interlocking i.e. frictional resistance to failure (in terms of the friction angle), appeared to be governed primarily by porosity, but also related to grading and particle angularity.

Table 1. Rockfill physical properties

| Mine | Uncompacted Porosity % | Uniformity Coefficient | Aspect Ratio |
|---|---|---|---|
| HBM&S | 47 | 4.7 | 11.3 |
| Bousquet | 53 | 1.8 | 5.9 |
| Chadbourne | 47 | 1.6 | 7.8 |
| Selbaie | 34 | 83.4 | 9.5 |

Equation (1) indicates that the slag angles of internal friction range from $26°$ (+18.8 mm fraction, porosity = 52%, normal stress = 1 MPa) to over $50°$ (wide gradation, porosity = 37%, normal stress < 0.05 MPa). Underground behaviour indicates the slag backfill to be cohesionless, with an angle of repose of $40$-$45°$. Particles can be described as very sharp, angular, and very rough.

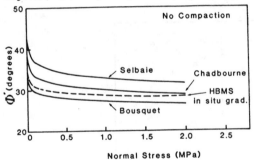

Figure 5. Uncompacted tilt test data

## 4. HBM&S GROUTED SLAG CHARACTERISTICS

### 4.1 Grout mixture studies

Grout mixtures comprised water:cement ratios between 1 and 2. ASTM Type III, High-Early Portland cement was used, as its finer gradation ensures better penetration in injection tests. The effect of adding sodium bentonite to maintain cement in suspension was evaluated (ASTM C940-81).

Bleeding tests to determine the optimum proportion of bentonite in the grout mix were conducted at water:cement ratios (w:c) of 1, 1.5, and 2.0. The grout mix was placed in a 1 l cylinder and the percentage of clear water on top of the mix, related to the total volume of the sample, was determined at 10 min intervals. The optimum proportion of bentonite in the grout mix was the smallest quantity which resulted in less than 5% sedimentation after 90 min settling (2% bentonite by weight of water for a 1:1 w:c grout, 3 % bentonite at 1.5 w:c, 4% bentonite at 2.0 w:c). Viscosity tests were also performed at various w:c ratios and rates of shear, Table 2.

Table 2. Grout mix properties

| w:c by weight | 2:1 | 1.5:1 | 1:1 |
|---|---|---|---|
| sodium bentonite | 4% | 3% | 2% |
| density (mg/m$^3$) | 1.28 | 1.35 | 1.47 |
| viscosity (cp) | 10.0 | | 20.0 |
| UCS (MPa) | 0.52 | 2.11 | 6.67 |
| E (GPa) | 0.09 | 0.33 | 1.08 |

### 4.2 Grouted slag UCS testing

Laboratory UCS tests, after 28 days curing, were conducted on a range of slag gradations (+18.8 mm, -18.8 mm to +13.3 mm, -13.3 mm to +4.7 mm, and -4.7 mm) with 3 ASTM type III Portland cement grouts (1:1 w:c by weight, 2% sodium bentonite by weight of water; 1.5 w:c, 3% bentonite; and 2:1 w:c, 4% bentonite). 15 cm diameter, 30.5 cm long cylinders were tested at a strain rate of 5 mm/min on a RDP servo-controlled rig (ASTM C39-83b). Particles over 2.5 cm were scalped to maintain the sample diameter:maximum particle size ratio above 5:1. Modulus of deformation (E) was calculated as the tangent at 50% UCS.

Figure 6 by example shows mean stress-strain data for 1:1 and 1.5:1 w:c ratios for the gradations and respective grouts (numbers in parenthesis indicate the number of tests). These show the marked effect on UCS and E exerted by w:c ratio, see Table 3.

The influence of gradation is also seen to be marked. Failure was clearly evident to be confined to the grout. The higher degree of interlocking in the finer gradations (-13.3 mm), related to reduced porosity, was also seen to increase the strength and modulus above those values of the respective grouts. Figure 7 shows further the influence of gradation and grout mix on strength and modulus.

Table 3. Grouted slag mechanical properties

| w:c ratio | bentonite ( % ) | UCS ( MPa ) | E ( GPa ) |
|---|---|---|---|
| 1.0 | 2 | 3.5 - 6.3 | 0.41 - 2.13 |
| 1.5 | 3 | 0.89 - 3.6 | 0.09 - 0.51 |
| 2.0 | 4 | 0.39 - 0.59 | 0.03 - 0.08 |

### 4.3 Laboratory injection trials

Initial injection trials were conducted in the laboratory using a Warren Rupp

except where material finer than 4 mesh was present. Attempts at washing out such fines by initial pumping of water were largely unsuccessful. Bleeding, the loss of cement from suspension, was acute where bentonite was not included in the grout mix.

## 5. NUMERICAL MODELLING OF EXPOSED FILL

Numerical modelling was undertaken to determine the maximum allowable free-standing vertical exposure of a wall of grouted slag for a given thickness and strength of the grouted mass. All modelling used FLAC, a 2-D explicit finite difference code developed by

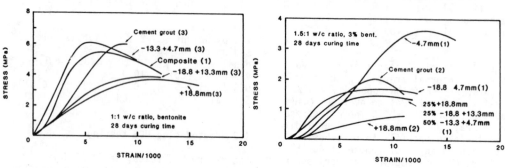

Figure 6. Mean stress-strain data for grouted slag gradations

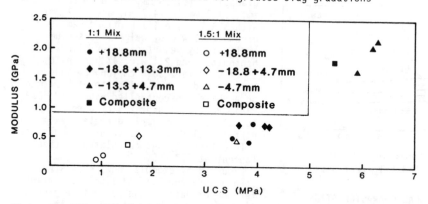

Figure 7. UCS related to E with variation in gradation and grout mix

air-powered double diaphragm pump. 1:1 w:c grouts were injected into a range of uncompacted slag gradations in 8 and 23 cm diameter, 1.8 m long perspex pipes, at pressures up to 700 kPa. ASTM Type III cement was used throughout, and grout mixes included 0 % or 2 % sodium bentonite. Injection in all gradations was successful to distances over 1 m,

Itasca Consulting Group Inc. Properties of the HBM&S rock mass were based on published data [21], while properties of the grouted slag were based on the laboratory prior work. Input data are summarized in Table 4; a Mohr-Coulomb plasticity model was assumed for the grouted slag. The interface between pillar and grouted slag was assigned a low friction angle and zero cohesion.

The modelled region was assumed to be an ore pillar, 60 m high by 20 m wide, against which a wall of grouted slag of uniform thickness of 2 m to 8 m was created. The resulting mesh comprised 900 elements, Figure 8. The vertical stress gradient was assumed to be 0.026 MPa/m, and the top of the modelled region 250 m below surface. Lateral pressure was exerted by uncemented slag against the grouted slag. Gravity was applied to all elements.

Models were run with w:c ratios of 1.0, 1.5, and 2.0. The pillar was excavated in 5 m high slices. Equilibrium was assumed to be reached when the maximum out-of-balance force approached zero and displacements approached a constant value. Failure was deemed to have occurred when grouted slag elements had yielded. Results are shown in Figure 9; a 35 m exposure, for example, requires a 4 m thick wall at 1.0:1 w:c, or a 7 m wall at 1.5:1 w:c.

Table 4. Material properties assumed in FLAC modelling

| | Grouted slag | | | Rock mass |
|---|---|---|---|---|
| | 2.0:1 w:c | 1.5:1 w:c | 1.0:1 w:c | |
| E (GPa) | 0.08 | 0.4 | 1.0 | 55.0 |
| Poisson's ratio | 0.25 | 0.25 | 0.25 | 0.22 |
| UCS (MPa) | 0.6 | 2.0 | 5.0 | |
| Angle Int. Friction | 40 | 40 | 40 | |
| Density ( t /m$^3$) | 2.0 | 2.0 | 2.0 | 2.7 |

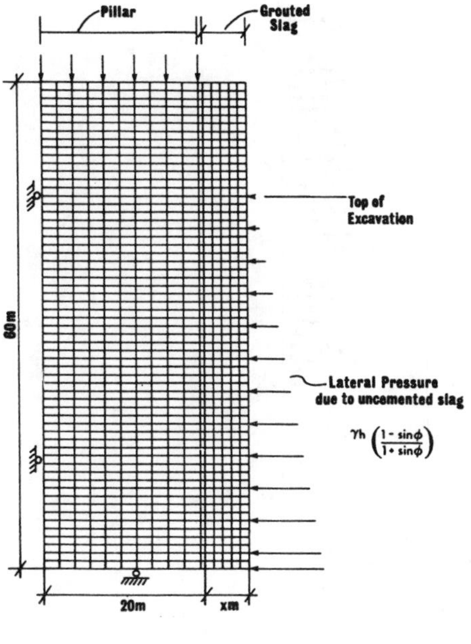

Figure 8. Finite difference model

## CONCLUSION

Despite several Canadian attempts to develop delayed backfill consolidation techniques, these still remain to find mine implementation. Field and laboratory observations indicate that both porosity and size gradation exerts a significant influence on HBM&S slag backfill mechanical properties, both in original and grouted states. Porosity and gradation, particularly the proportion of fines, have also been seen to strongly control slag groutability. There has been a marked lack of in situ testing of the physical and mechanical properties of both unconsolidated and grouted backfills. Attempts at correlation of any laboratory studies with field data have been scant.

Future field studies should investigate the potential modification of size distribution through degradation during gravity fall underground. Segregation on rill slopes during placement is also considered important for future underground study, in order to define gradation variation in the vicinity of pillar faces and the formation of potential shear failure planes in the backfill mass.

Future field studies should investigate the potential modification of size distribution through degradation during gravity fall underground. Segregation on

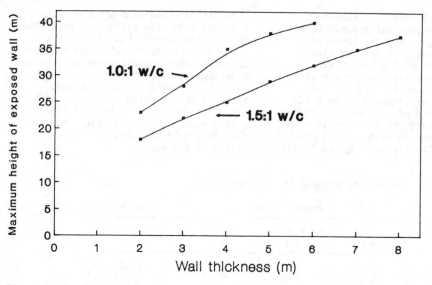

Figure 9a. Predicted maximum vertical exposure vs thickness of grouted wall

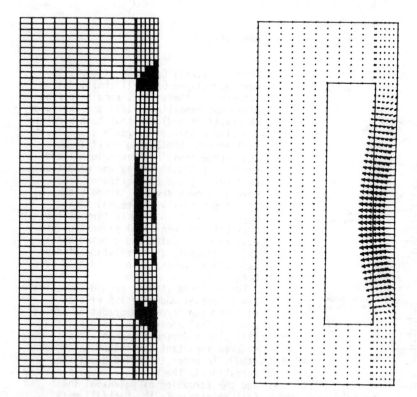

b  Yield state of grouted slag elements    c  Displacement vectors

Figure 9.   Analysis results of a 40 m vertical exposure of a 4 m thick wall
            at 1.5:1 W:C

rill slopes during placement is also considered important for future underground study, in order to define gradation variation in the vicinity of pillar faces and the formation of potential shear failure planes in the backfill mass. Although preliminary numerical modelling indicates that stability can be maintained by consolidation, studies have yet to evaluate either economics or field performance of pumping and grout types.

## ACKNOWLEDGEMENTS

The authors wish to express their thanks to the personnel of HBM&S who cooperated with this work. Acknowledgement is also due to Energy Mines and Resources for their support of this research. The views expressed are entirely those of the authors.

## REFERENCES

[1] Mitchell,R.J., "Earth Structures Engineering". Allen and Unwin Inc., Winchester, Mass., 1983.

[2] Mitchell, R.J., Smith, J.D., "Mine Backfill Design and Testing." Trans. Can. Inst. Min. Metall., 72, 801, 1979, pp 82-88.

[3] Yu, T., Counter, D.,"Backfill Practice at Kidd Creek Mines". Trans. Can. Inst. Min. Metall., 1983, pp 56-65.

[4] Nichols, R.S.,"Rock Segregation in Waste Dumps". Ann. Gen. Meeting, Can. Inst. Min. Metall., Montreal, 1986.

[5] Barrett, J.R., Cowling, R.,"Investigations of Cemented Fill Stability in 1100 Orebody, Mt. Isa Mines, Ltd.". Trans.Inst. Min. Metall., 89, 1980, pp A118-128.

[6] Schwartz, A., "Pillar Recoveries using Consolidated Fill at Noranda Mines Ltd., Manitouwadge, Ontario", Mining with Backfill, CIM Special Volume 19, 1978, pp. 56-68.

[7] Kerr, R.W., "The Fill System for 108 Blasthole Stope at Inco's Birchtree Mine", Mining with Backfill, CIM Special Volume 19, 1978, pp. 16-20.

[8] Askew, J.E., McCarthy, P.L., and Fitzgerald, D.J., "Backfill Research for Pillar Extraction at ZC/NBHC", Mining with Backfill, CIM Special Volume 19, 1978, pp. 100-110.

[9] Baptie, M., Kimberley, Personal Communication, April 1988.

[10] Yu, T., Kidd Creek Mines Ltd., Personal communication, Jan. 1987.

[11] Kelly, J., Inco Ltd., Sudbury, Personal communication, Mar. 1987.

[12] Rajalahti, M., "Ground Freezing at the Pyhasalmi Mine, Finland". Proc. Int. Symp. Large Rock Caverns, I.T.A./I.S.R.M., Helsinki, Finland, 1986, pp 1577-1582.

[13] Haapamaki, S., "Chisel Lake Mine - 611 VPF Stope Slag Cementation", Internal Report, H.B.M.&S., May 1987.

[14] Gignac, L., "Filling Practices at Brunswick Mining - No. 12 Mine": in Mining with Backfill: Proc. 12th. Can. Rock Mech. Symp., Sudbury, Can. Inst. Min. Metall., pp. 30-36.

[15] International Society Rock Mechanics, "Suggested Method for Determining the Point Load Strength Index", Int. J. Rock Mech. Min. Sci. & Geomech. Abstr., Vol. 22, No. 2, 1985, pp. 51-60.

[16] Bieniawski, Z.T., "Estimating the Strength of Rock Materials", J. S. Afr. Inst. Min. Metal., 74, 1984, pp. 213-320.

[17] Leps, T.M., "Review of Shearing Strength of Rockfill", Int.J. Rock Mech. Min. Sci. & Geomech. Abstr., Vol. 22, No. 2, 1985, pp. 51-60.

[18] Sarac, D. and Popovic, M., "Shear Strength of Rockfill and Slope Stability", Proc. 11th. Int. Conf. on Soil Mech. and Found. Eng., A.S.C.E., 1985, pp. 641-645.

[19] Barton, N., and Kjaernsli, B., "Shear Strength of Rockfill", J. Soil Mech. Found. Div., A.S.C.E., July 1981, pp. 873-891.

[20] Marachi, N.D., Chan, C.K., Seed, H.B., "Evaluation of Properties of Rockfill Materials", J. Soil Mech. Found. Div., A.S.C.E., Jan. 1972, pp. 95-115.

[21] Hanson, D.S.G., "Geotechnical Analysis to Aid Recovery of North Main Shaft Pillar - Flin Flon Mine", Stability in Underground Mining, Society of Mining Engineers, 1983, p. 874-901.

*Innovations in Mining Backfill Technology, Hassani et al. (eds), © 1989 Balkema, Rotterdam. ISBN 90 6191 985 1*

# In situ determination of high density alluvial sand fill

D.Thibodeau
*Mines Research, Ontario Division, Inco Ltd, Copper Clif, Ontario, Canada*

ABSTRACT: In 1986 a contract was awarded to INCO through the Canada-Ontario Mineral Development Agreement, in collaboration with CANMET, for the in situ determination of high density alluvial sand fill properties in Ontario Mines. The fill is a pumpable mix of well graded, alluvial pit run sand, portland cement, water and -325 mesh tailings slimes. It was mixed in a surface pilot plant and pumped underground. This project attempted to outline and develop a quantitative understanding of the in situ support and fill properties of high density alluvial sand fill. It was found that the fill could withstand relatively high compressive loads, from 1.38 MPa to 6.90 MPa and still act as a very stable support. The mechanical behaviour of the fill under load was studied. It was found that arching and the elastic beam behaviour could not be applied since the principal stresses in the fill were much larger than the maximum fiber stresses calculated. The principal stresses were believed to be caused by pillar convergence.

## 1 INTRODUCTION

One of the major concerns of the mining industry is to improve production while maintaining high safety standards. To achieve such a goal, Canadian mining companies in association with the federal and provincial governments are currently involved in extensive technological research. Mine backfill is one major area of this joint venture. Research is directed at producing a superior product at the lowest possible cost.

As a result of INCO's initial test work with high density fill, a contract was awarded to Inco through the Canada-Ontario Mineral Development Agreement in collaboration with CANMET to determine the mechanical properties of in situ high density alluvial sand fill. INCO high density fill was successfully placed underground on 700 level at Levack Mine. The fill, a pumpable mix of well graded, alluvial pit run sand, portland cement, water and -325 mesh tailings slimes was mixed in a surface pilot plant and pumped underground. A standard dual piston pump was used to transport the fill product through a pipeline at a pulp density of 80 to 85 weight percent solids.

The project objectives were as follows:

1. To outline and develop a quantitative understanding of the in situ support and mechanical properties of a high density sand paste fill.

2. To relate laboratory tested fill properties to those observed in situ.

This report describes the equipment, the procedures used, and the results obtained.

The results are divided into three categories:

1. Laboratory results from test cylinders.

2. Results from monitoring in situ instrumentation.

3. Results obtained from in situ sampling of the fill.

It was assumed that the in situ behaviour of the fill underload would correspond to arching. Using the elastic beam theory, the data obtained from the monitoring program were analyzed and evaluated to see if they fit this chosen model.

The in situ results were also compared to the laboratory results to determine if the compressive testing of cylindrical samples is an effective means of evaluating the mechanical properties of a cemented backfill material.

The in situ samples were analyzed. for cement content to determine if the 60% reduction in binder, over conventional slurry backfill had been achieved, in conjunction with a reduction in water content.

The results obtained from this investigations will be used to increase our general understanding of how backfill behaves mechanically in situ. This will allow the definition of fill requirements for various mining methods.

## 2 EQUIPMENT AND PROCEDURE

The pilot plant at Levack Mine consisted of the following equipment: (fig. 1)
1. Two slimes holding tanks (maintained at a pulp density of 60% by weight solids).
2. One SRL pump.
3. One 25 tons cement silo with screw feeder.
4. One sand hopper with belt feed.
5. One Schwing BP250 concrete pump.
6. One front end loader.
7. One screw mixer conveyor.

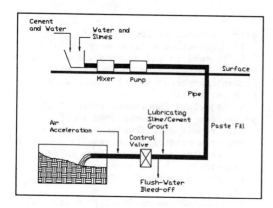

Figure 1:  Prototype high-density paste fill system.

High density fill was prepared by mixing alluvial sand, portland cement, slimes and water in the screw mixer conveyor. It was then pumped underground by a Schwing pump with a maximum capacity of forty tons per hour (1).

An undercut and fill stope at INCO's Levack Mine on 900 level was selected as a test site (located 45 km N.-W. of Sud-

bury). The first step was to mine out slice 2WA and fill it (fig.2). Prior to filling the stope instrumentation was placed in the mined out area. Filling started in September 1986 and was completed in April 1987; filling had to be delayed on numerous occasion due to delays in instrumentation delivery.

The in situ instrumentation consisted of 36 electronic total earth pressure cells (see fig.2), 3 electronic piezometers, 2 Young's modulus cells, 4 thermosistors and 7 convergence monitors. Boxes for in situ samples recovery were placed in the hanging wall sector.

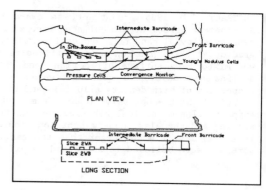

Figure 2:  9600 Block - 900 level Levack mine.

Samples of the fill product were taken in both 15.24 cm and 7.62 cm diameter cylinders on surface and at the pour point. Laboratory analyses were made to determine:

1. Size distribution.
2. Chemical composition.
3. Portland cement content.
4. Moisture content.
5. Bulk density.
6. Compressibility modulus.
7. Uniaxial compressive strength.

Similar tests were performed on in situ samples after the undercut was mined out.

All the instrumentation, with the exception of the Young's modulus cells, was connected to three dataloggers to allow for continuous monitoring of the cells, while the filled slice 2WA was undercut.

The data from the dataloggers were down loaded on a regular basis during the mining of 2WB to identify stress changes in the self-supporting fill. A timberless undercut and fill mining method was used for this experiment to show that mechanized undercut and fill could be used with a suitable backfill. The use of high density fill in timberless undercut and fill results in increased productivity while providing safer mining conditions. The mining of undercut 2WB was completed on November 26, 1987.

Shortly after the completion of mining slice 2WB, Trow Ltd of Sudbury was contracted to recover in situ samples of 15.24 cm and 7.62 cm diameter cores. This met with limited success because coring of the fill proved to be impractical due to pieces of gravel which ground away the samples. Only three samples were recovered and tested. These included one 15.24 cm and two 7.62 cm diameter samples. Three boxes, that had been placed in the fill for such an eventuality, were recovered and sent to Trow for testing. Attempts were made to drill 7.62 cm diameter core out of the boxes. Only one of six attempts proved successful. As a last resort, seven 5.08 cm side cubes were cut out of the boxes and tested for uniaxial compressive strength.

# 3 RESULTS AND DISCUSSION

## 3.1 Laboratory results

All the cylinders poured were sent to Trow Ltd of Sudbury. A series of tests were conducted to determine the mechanical properties of the paste fill.

Uniaxial compressive tests were performed on fill cylinders after 7, 28, 162 and 183 days curing. Following each compressive test the wet and dry specific gravity of each cylinder was determined as well as moisture content. Results are summarized in table 1.

Table 1a. Laboratory test results

| Miscellaneous data | (% by wt of solids) |
|---|---|
| Average cement content: | 4.6% (15 samples) |
| Average moisture content of the mix: | 12.33% (63 samples) |
| Average moisture content of the sand: | 4.1% (4 samples) |
| Average wet specific gravity: | 2.2 (63 samples) |
| Average dry specific gravity: | 1.96 (63 samples) |

Table 1 continued:

Table 1b. The average confined compressive strength (MPa) of the paste fill cylinders.

| Confinement pressure: | 0 | 0.17 | 0.34 | 0.69 |
|---|---|---|---|---|
| 7 days | 1.28 | 2.48 | 2.80 | 3.92 |
| 28 days | 1.63 | 2.45 | 4.11 | 4.55 |
| 168 days | - | 3.30 | 3.43 | 3.52 |
| 183 days | - | 3.28 | 4.83 | 5.00 |

Table 1c. The average Young's Modulus (GPa) of the paste fill test cylinders.

| Confinement pressure (MPa): | 0 | 0.17 | 0.34 | 0.69 |
|---|---|---|---|---|
| 7 days | 0.47 | 0.26 | 0.22 | 0.28 |
| 28 days | 1.12 | 0.25 | 0.37 | 0.39 |
| 168 days | - | 0.30 | 0.19 | 0.25 |
| 183 days | - | 0.36 | 0.57 | 0.43 |

## 3.2 In situ monitoring

Data were collected from the stope on an hourly basis and the daily average was compiled from July 3, 1987 until June 1988.

The results obtained from the in situ monitoring program was used to determine if the elastic beam theory could be applied to the behaviour of the fill mass.

If a simple beam with uniformly distributed load is assumed, then:

$$Mx = \frac{wx}{2}(1-x),$$

where w is the load applied on the beam, l is the span of the beam, x is any given position along the span of the beam and M is the applied moment of the beam. The span (l) of the beam will be equal to the width of the stope (e.g. 4.27 m). The load (w) is equal to the unit weight of the fill mass multiplied by the height of the stope and the 0.3048 m section (e.g. 21.55 KN/m$^3$ x 3.6575 m x 0.3048 m = 24.02 KN/m).

From the flexure theory, assuming that the neutral axis of the beam is situated at mid-height of the fill mass,

$$\sigma = \frac{2Fc}{1.83m \times 0.3048m} \quad , \quad M = 2x2/3(1.83m)Fc$$

and Fc = 0.28 m$^2$ x $\sigma$ therefore

$M = 0.68 \text{ m}^3 \times \sigma$

where $\sigma$ is the fiber stress, Fc is the applied force; thus

$Mx = 0.68 \text{ m}^3 \ \sigma_x = \dfrac{wx(1-x)}{2}$ and

$\sigma_x = \dfrac{wx(1-x)}{1.36 \text{ m}^3}$

for $x = 2.13$ m, M is maximum and so is $\sigma$, therefore

$\sigma_{max} = \dfrac{24.02 \text{ KN/m} \times 2.13 \text{ m}(4.26 \text{ m} - 2.13 \text{ m})}{1.36 \text{ m}^3}$

$\sigma_{max} = 0.08$ MPa.

Figures 3 and 4 show the pressures measured by the total earth pressure cells during mining.

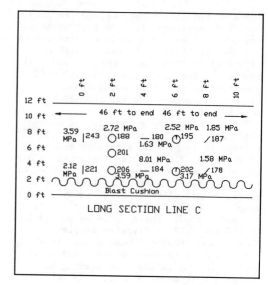

Figure 4: Pressure Distribution

Using the following equations, the principal stresses $\sigma_1$, $\sigma_2$ and $\sigma_3$ were calculated:

$\sigma_n = \sigma_x \cos^2\theta + \sigma_y \sin^2\theta + 2\tau_{xy}\sin\theta\cos\theta,$

$\sigma_{1,2} = \dfrac{\sigma_x + \sigma_y}{2} \pm \sqrt{((\sigma_x - \sigma_y) + \tau_{xy}^2)}$

$\tan 2\theta = \dfrac{2\tau_{xy}}{(\sigma_x - \sigma_y)}$

Table 2 gives a summary of the results and figure 5 display the results at their location in the stope.

Table 2. Principal stresses within the fill mass.

| Reference Line | Height (m) | $\sigma_1$ (MPa) | $\sigma_2$ (MPa) | $\sigma_3$ (MPa) | $\theta$ |
|---|---|---|---|---|---|
| A | 2.74 | 6.77 | 2.43 | 7.30 | -10,2° |
| A | 1.52 | 3.00 | 2.26 | 7.49 | 34.9° |
| B | 1.82 | 5.39 | 2.76 | 5.42 | - 8.5° |
| C | 0.91 | 9.23 | 2.37 | 9.22 | 25.0° |

The results show that the load carried by the fill is much greater than can be accounted for by beam theory. Hence the

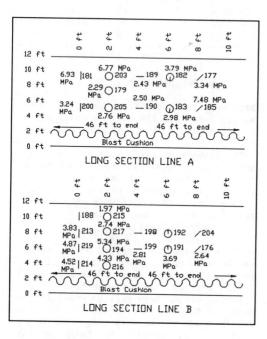

Figure 3: Pressure distribution.

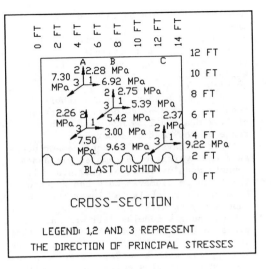

CROSS-SECTION

LEGEND: 1,2 AND 3 REPRESENT
THE DIRECTION OF PRINCIPAL STRESSES

Figure 5: Principal stresses.

stresses due to arching within the beam
are negligible compared to the principal
stresses measured in the fill.

It is suspected that as the slice 2WB was
mined, loads which were supported by the
ore were redistributed partly to the rock
mass enclosing the fill and partly to the
fill mass itself. Figure 6 schematically
illustrates this phenomena; wall closure
of the stope, induced shear within the
fill mass, and an increase of the lateral
stress ($\sigma_1$) carried by the fill mass. From
a Poisson effect, the vertical stress will
also increase.

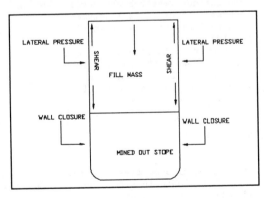

Figure 6: Schematic diagram of stresses
induced by wall closure.

From the test cylinder results, the angle
of internal friction and the cohesion of
the fill were calculated and yielded
respective
values of $\phi = 30°$ and $C = 0.44$ MPa. Using
these values Co and q can be calculated
with the following formulas:

$$Co = 2c\tan(45° + \phi/2) = 1.52 \text{ MPa}$$

and

$$q = [\tan(45° + \phi/2)]^2 = 3.$$

A plot of a $\sigma_3$ versus $\sigma_1$ (fig. 7) shows
that the fill mass is very near failure.
This means that the lateral stresses are
so high that it squeezed the fill into
place and crushed it, up to a point where
failure occurred.

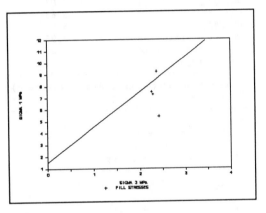

Figure 7: Fill failure envelope

3.3 In situ sampling

The in situ sampling program was not very
successful and only a few samples were
recovered. Boxes placed in the fill were
taken out and sent to Trow for testing,
but taking core samples from these boxes
turned out to be too difficult. Thus,
seven 5.08 mm cubes were tested for uniaxial compressive strength. Table 3 summarizes the results obtained from the in
situ sampling program.
The boxes were made of plywood and open at
the top end. They were placed at the
bottom of slice 2WA so that they would be
filled at the same time as the stope.

271

Table 3. In situ sampling test results.

| Miscellaneous data   (% by wt of solids) | |
| --- | --- |
| Average moisture content: | 10.6% |
| Average wet specific gravity: | 2.17 |
| Average dry specific gravity: | 1.95 |
| Cement content: | 6.42% |

| Uniaxial compressive strength (MPa) | |
| --- | --- |
| 15.24 cm core from stope | 3.00 |
| 7.62 cm core #1 from stope | 1.57 |
| 7.62 cm core #2 from stope | 1.62 |
| 7.06 cm core from box | 2.79 |
| Cube #1 | 3.59 |
| Cube #2 | 5.65* |
| Cube #3 | 3.97 |
| Cube #4 | 1.72 |
| Cube #5 | 2.07 |
| Cube #6 | 3.63 |
| Cube #7 | 3.08 |
| Overall average | 2.95 |
| 7.62 cm core average | 2.00 |
| Cube average | 3.39 |

| Young's modulus (no confinement pressure) (GPa) | |
| --- | --- |
| Cube average | 0.07 |
| 15.24 cm core from stope | 0.52 |
| 7.62 cm core #1 from stope | 0.08 |
| 7.62 cm core #2 from stope | 0.18 |

* Large stone present.

The average in situ moisture content is 10.6% by wt of solids, a reduction of 14% compared to the laboratory result. The in situ samples were taken from the bottom of the fill slab at the front and at the back of the stope.

The results from table 3 yield an average cement content value of 6.42% by wt of solids which corresponds to sand cement ratio of 16:1. The laboratory results yielded a value of 4.6% cement by wt of solids or a 21:1 ratio. Uniaxial compressive strength test of approximately 2.76 MPa was achieved. A 10:1 hydraulic backfill contains 9.1 wt% cement and has a uniaxial compressive strength in the range of 1 MPa (INCO past testing); an effective cement reduction of 36% was realized. The strengths obtained with the high density alluvial paste fill are much higher than those obtained with hydraulic backfill. Therefore in a mining situation where only normal hydraulic backfill strength is required, the cement content of high density fill could be reduced even further.

With a confinement pressure of 0 MPa at 28 days, eight test cylinders yielded compressive strength values of 1.63 MPa while the 15.24 cm diameter core by 38.1 cm high generated a value of 3 MPa. This value is 84% higher than the laboratory results. The 7.62 cm cores with strengths of 1.62 MPa and 1.59 MPa fell within the range of the test cylinders.

After 28 days with no confinement pressure the average of 8 samples yielded a Young's modulus value of 1.12 GPa while the 15.24 cm diameter core had a response of 0.52 GPa which is 54% lower than the laboratory results.

Each Young's modulus cell was made of: 1 pressure transducer, 1 AD590J temperature sensor, and 1 LVDT cell to which pressure was applied with a manual hydraulic pump. The Young's modulus cells were manually monitored, but from the beginning of the monitoring program the temperature sensors and the pressure transducers were damaged and inoperable. Only the LVTD cells worked properly. The LVTD cells displacement response to mining where plotted against the manually induced pressure. The Young's modulus of the fill was calculated by using the uniformly pressurized penny shaped crack theory. The in situ Young's modulus cells yielded an average value of 1.77 GPa which is 58% more than the 28 days average value and 3.43 times more than the 15.24 cm diameter core. One of the problems encountered with the cells is that they were not properly calibrated prior to their installation and consequently, did not perform satisfactorily. Some further research and development is necessary before this instrument can be used reliably to measure in situ Young's modulus.

The average in situ wet specific gravity was 2.17 down 1.4% from the laboratory results of 2.20. The average dry specific gravity was 1.95 down 0.65% from the laboratory results of 1.96.

4 CONCLUSION

The principal objectives of this project were to evaluate in situ properties and in situ behaviour of a high density fill under load.

To obtain pertinent information concerning the in situ properties of the fill mass, cylinders were cast both at the pour point and on surface and subsequently tested in the laboratory. Furthermore, in situ samples were taken and tested after the stope was mined out. The comparison between the cast samples results and the in situ samples show that the uniaxial compressive strength of the in situ samples is 86% higher than the lab. test samples. This implies that lab test results are conservative compared to in situ fill. The results for the test cylinders moisture content and unit weight agree within + 15% and - 2% respectively with the in situ samples. This is conclusive as far as comparison between in situ fill and laboratory cylinders are concerned. The Young's modulus results from the surface cast test cylinders were within 54% of the in situ results from the sampling program and within 58% of the Young's modulus monitoring program. It is very difficult from these results to pin point the true Young's modulus value of the fill since:

1) the Young's modulus cells were poorly calibrated;

2) the in situ sampling program was not a total success.

Nevertheless, review of the laboratory results from both the cylinders and the in situ samples shows that the Young's modulus of the fill falls between 0.34 and 0.67 GPa.

The total earth pressure cells were laid out in a pattern that would permit analysis of the arching behaviour of the fill. It was found from the in situ stress results that the fill mass did not behave to the expected elastic beam theory with arching. The in situ pressures registered in the fill were too high to agree with the elastic beam theory. Therefore, it was believed that by mining underneath the fill, closure of the side walls was induced which produced increased lateral pressures in the fill mass.

The chemical analysis of the in situ samples revealed that the sand to cement ratio was about 15:1 which is a 35% cement reduction compared to 10:1 hydraulic backfill at 70% density. But, the strength achieved was 160% higher than the hydraulic backfill which suggests that more tests with a lower cement content are warranted.

While mining was progressing it was observed that the fill was of high quality throughout the complete slice. The fill was competent and it performed very well by providing a very straight and solid back for timberless undercut and fill

application. There was no need to install any other type of support within the stope to maintain the fill in place while mining. This reflects a direct productivity increase.

This was anticipated based on the laboratory tests performed on the high density fill. The high average uniaxial and triaxial compressive strengths obtained after 28 days (about 1.38 MPa unconfined and 2.76 MPa confined) indicates the good quality of the fill. This was confirmed by the in situ sample results which had an average uniaxial compressive strength of 2.95 MPa.

The major problems encountered with the in situ monitoring system was the delay in delivery of equipment. This hindered the possibility of verifying the manufacturer instrumentation's calibration before installation in the stope. Therefore, the problems were worked out of the instrumentation while the experiment was in progress and everything was finally put into place by mid-October.

Overall the project demonstrated the possibility and the advantages of using high density fill in a timberless undercut and fill application. It clearly brought a better understanding of the behaviour of high density fill under load and a better knowledge of high density fill properties. The results will permit, with further research, better stope design and relate the cement content of the fill according to the strength requirement to improve safety. This would be achieved by numerically modeling the fill response to excavation made in its vicinity.

5 REFERENCES

1- Landriault, D. and Goard, B.: Research Into High Density Backfill Placement Methods by the Ontario Division of INCO Limited, CIM Bulletin, Volume 80, no. 897, January 1987, pp. 46-50.

2- Brady, B.H.G. and Brown, E.T.: Rock Mechanics for Underground Mining. George allen & Unwin, London, 1985, 527 pages.

3- Hidgon, A. et al.: Mechanics of Materials. John Wiley & Sons, New York, 1985, Fourth Edition, 744 pages.

*Innovations in Mining Backfill Technology, Hassani et al. (eds), © 1989 Balkema, Rotterdam. ISBN 90 6191 985 1*

# Laboratory testing on high density tailing backfill mixes

W.G.Hunt
*Trow Ontario Ltd, Sudbury, Ontario, Canada*

ABSTRACT: Testing procedures and sample preparation techniques are reported for 75 mm by 150 mm backfill specimens. Experimental methods are outlined for uniaxial and undrained triaxial compression tests performed on high density cemented backfill samples. The same methods have been used for a variety of backfills cast in both laboratory and field environments including: classified and unclassified tailings at both high and normal pulp densities, alluvial sand backfills, and backfills using different concrete admixtures. Recommendations with respect to industry wide standardization are made in an attempt to provide consistency and uniformity in test results provided by researchers.

## 1.0 INTRODUCTION

Test results for backfill properties have been reported from many sources including: Weaver and Luka (1970), Mitchell and Smith (1979), Landriault and Goard (1985). Testing procedures and specimen preparations vary with researchers who extract guidelines from ASTM, CSA standards and/or ISRM suggested methods.

The need for careful experimental investigation is critical for engineers solving practical problems. The methods of testing require accuracy but need not be elaborate. The methods that have been used successfully by Trow Ontario Ltd. on recent projects are presented in this paper. It is the intention of the author to present test methods and procedures from projects involving cemented tailings/sand backfills at high pulp densities. The methods outlined are both economic and routine in nature.

## 2.0 COMPRESSION TESTING EQUIPMENT

Selection of a suitable compression testing frame was important for establishing load rate and range.

Hydraulic backfill materials have, for sometime, been tested on standard concrete type compression machines by others, Swan (1985), Davis (1988). Fine grained backfill cylinders of the 150 mm by 300 mm size have been tested successfully; particularly, for higher binder ratios at 28 day periods using these concrete testing machines. However, these relatively stiff concrete testing compression machines are difficult to control at low loads, and specifically at low load rates. This results in weak backfill mixes being prone to considerable scatter in test results using the stiffer concrete test machines.

Trow's experimentation with several types of compression machines has shown that the use of low load capacity soil type compression machines are best suited for testing hydraulic backfill materials for both low and high pulp densities. A standard soils compression machine is now used to conduct both uniaxial and triaxial compression tests on high density backfills.

The compression machine used is a 100 kN Wykeham Farrance with a load rate set at 1.0 mm/min. A variety of load rates can be selected with this machine, and proofing rings in the 200 kg to 5000 kg load range are available. Confining pressures of up to 1000 kPa are available from the laboratory compressor and pneumatic/water control system.

To accommodate the 75 mm x 150 mm specimens, proper fitting lower and upper 75 mm platens were machined from high strength aluminum. A "basic" 150 mm diameter triaxial cell used during both unconfined and confined tests ensuring correct axial alignment of the test specimens at all times.

## 3.0 SPECIMEN PREPARATION

Test specimens are cast both in the labor-

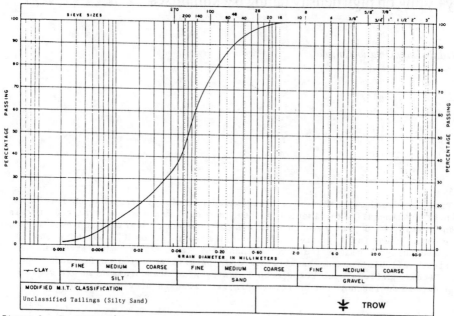

MECHANICAL ANALYSIS

Figure 1.  Typical Grading Distribution For  Suggested Test Methods

atory setting, and at underground locations
in the vicinity of the backfill pour.  Backfill material tested by the method described
herein consisted of a particle size distribution with 100% passing 2.0 mm grain diameter (i.e., #10 sieve size), shown in Figure 1.

Backfill material sent to the laboratory
is sealed in large metal containers and
kept submerged in water prior to mixing.
Mix designs of the various fill: binder
ratios are prepared in general accordance
with CSA A23.2-2C. The relative specific
gravity of the backfill is measured, and
appropriate moisture content corrections
are carried out before batching.

The backfill mixes are poured into
plastic moulds which are constructed from
75 mm diameter PVC schedule V880 322 pipe.
The pipe is cut in 235 mm lengths and is
also cut along its length using a table saw.
This parallel cut allows for removal of the
specimen after the material is initially cured (i.e., usually 1 to 3 days after casting).

The pipe is bound together using two
100 mm diameter hose clamps.  A glass
plate is tapped securely to the bottom
(optional 75 mm plexiglass plate has been
used with small (i.e., 4.5 mm) holes drilled through it to allow for added drainage in some cases).  The casting mould is
shown in Figure 2a.

Backfill is poured into the casting
moulds, angled to avoid air entrapment.
The moulds are filled to the top, vibra-

ted by hand, and covered with plastic bags.
Sets of three are casted for each day of
testing per mix design.  Generally, 3, 7,
14 and 28 day strengths are required. Triaxial tests are also conducted on groups
of three.

When the specimens have cured sufficiently
to handle (i.e., determined by resistance
to light indentation of the thumb) they are
stripped from the moulds.  Typically, for th
20:1 ratio and better, this can usually be
done after 24 hours.

The fresh specimens are carefully placed into a trimming mould as shown in Figure
2b.  The trimming mould is machined from
aluminum and is exactly 75 mm by 150 mm.
The mould fits together by aligning three
pin holes set on the sides of the mould.
A regular hacksaw and laboratory knife is
used to trim the ends of the specimens.

The prepared samples are bagged and carefully placed in a constant temperature
and humidity (i.e., 100%) room to cure until testing.  Curing temperature was selected at 22°C.


4.0  TEST PROCEDURES

Cured specimens are removed from their
sealed plastic bags, weighed, and the
corresponding bulk unit weights are calculated.  After testing to failure, the
moisture contents are recorded from a

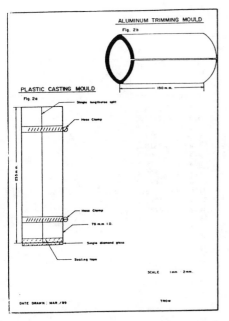

Figure 2a, 2b.  Casting & Trimming Mould

representative central portion of the specimen thereby allowing for porosity to be calculated using the data collected.

Test procedures follow Imperial College triaxial test methods, Bishop (1957) and ASTM D2850-87.  Confining pressures of 170 kPa and 340 kPa were typically used together with the unconfined test result to establish the Mohr envelope and calculate $\phi$ and c values.  The statistical coefficient of regression is computed to evaluate data.

Uniaxial compressive strength results have been very consistent using the triaxial confinement chambers and machined platens with samples prepared in the trimming mould.  Alignment of the proofing ring over the specimen is improved significantly when all uniaxial tests are conducted within the confinement chamber.

Triaxial tests are conducted using the same set-up, with the rubber membranes mounted over the specimens and the platens are sealed with O-rings.  The backfill specimens are partially saturated in nature, with the degree of saturation less than 100%.  Hence, consolidation occurs when the confining pressure is applied and also during axial shear.  The Mohr envelopes for these undrained triaxial tests are therefore were developed at the different confining pressures.

5.0  REPRESENTATIVE TEST RESULTS

Test results obtained from the procedures outlined above have been encouraging.  Table 1 presents typical unconfined compressive strength tests obtained from a recent project.

The 75 mm x 150 mm specimens, cast in the manner described, provide samples with consistent unit weights and the resulting compressive strengths are quite uniform.

Triaxial test data was also very consistent as demonstrated in Figures 3 and 4.  Test specimens cast from the same material as used for the uniaxial tests yielded results typical of those shown for the 32:1 mixes of unclassified tailings to slag/cement binder, respectively.  Confining pressures of 170 kPa and 340 kPa were used.  The coefficient of regression was 0.95 or greater, and c and $\phi$ values calculated from the Mohr envelopes are reasonable.

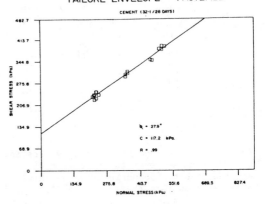

Figure 3

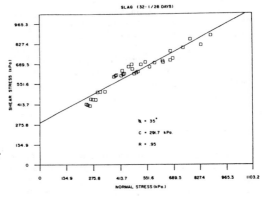

Figure 4.

277

Table 1. Representative Uniaxial Compressive Strengths - Tailings/Binder Ratios at 78% Pulp Densities

| Cylinder # | Binder Type | Ratio | Age Days | Bulk Wt. kg/cm³ | Moisture % | Porosity % | Strength kPa |
|---|---|---|---|---|---|---|---|
| 179 | #10 Cement | 8:1 | 7 | 2.223 | 18.9 | 35.9 | 2598.0 |
| 180 | #10 Cement | 8:1 | 7 | 2.237 | 18.9 | 35.9 | 2467.2 |
| 181 | #10 Cement | 8:1 | 7 | 2.210 | 18.9 | 35.9 | 2634.8 |
| 137 | #10 Cement | 8:1 | 28 | 2.233 | n/a | n/a | 3797.6 |
| 138 | #10 Cement | 8:1 | 28 | 2.202 | n/a | n/a | 3818.6 |
| 139 | #10 Cement | 8:1 | 28 | 2.233 | n/a | n/a | 3907.7 |
| 62 | #10 Cement | 16:1 | 7 | 2.062 | n/a | n/a | 819.8 |
| 63 | #10 Cement | 16:1 | 7 | 2.145 | n/a | n/a | 809.4 |
| 64 | #10 Cement | 16:1 | 7 | 2.086 | n/a | n/a | 824.9 |
| 65 | #10 Cement | 16:1 | 28 | 2.218 | 20.0 | 37.0 | 1288.6 |
| 66 | #10 Cement | 16:1 | 28 | 2.218 | 20.0 | 36.9 | 1215.2 |
| 67 | #10 Cement | 16:1 | 28 | 2.218 | 20.0 | 36.9 | 1288.6 |
| 101 | #10 Cement | 32:1 | 7 | 2.135 | 21.6 | 38.8 | 320.6 |
| 102 | #10 Cement | 32:1 | 7 | 2.146 | 21.6 | 38.8 | 342.7 |
| 103 | #10 Cement | 32:1 | 7 | 2.151 | 21.6 | 38.8 | 452.3 |
| 140 | #10 Cement | 32:1 | 28 | 2.133 | 22.3 | 39.5 | 378.5 |
| 141 | #10 Cement | 32:1 | 28 | 2.130 | 22.3 | 39.5 | 364.8 |
| 142 | #10 Cement | 32:1 | 28 | 2.157 | 22.3 | 39.5 | 410.9 |
| 44 | Slag | 8:1 | 7 | 2.078 | n/a | n/a | 1675.6 |
| 45 | Slag | 8:1 | 7 | 2.075 | n/a | n/a | 1644.7 |
| 46 | Slag | 8:1 | 7 | 2.110 | n/a | n/a | 1582.8 |
| 47 | Slag | 8:1 | 28 | 2.215 | 21.1 | 33.6 | 3624.8 |
| 48 | Slag | 8:1 | 28 | 2.197 | 21.2 | 33.5 | 3488.6 |
| 49 | Slag | 8:1 | 28 | 2.216 | 21.1 | 33.6 | 3588.1 |
| 74 | Slag | 16:1 | 7 | 2.170 | n/a | n/a | 1031.2 |
| 75 | Slag | 16:1 | 7 | 2.218 | n/a | n/a | 1314.7 |
| 76 | Slag | 16:1 | 7 | 2.223 | n/a | n/a | 1294.1 |
| 77 | Slag | 16:1 | 28 | 2.153 | 20.8 | 37.7 | 2310.0 |
| 78 | Slag | 16:1 | 28 | 2.227 | 20.7 | 37.7 | 2414.8 |
| 79 | Slag | 16:1 | 28 | 2.240 | 20.7 | 37.7 | 2414.8 |
| 84 | Slag | 32:1 | 7 | 2.139 | 22.1 | 39.3 | 432.2 |
| 85 | Slag | 32:1 | 7 | 2.140 | 22.2 | 39.2 | 426.9 |
| 170 | Slag | 32:1 | 28 | 2.212 | 22.3 | 39.5 | 1267.7 |
| 171 | Slag | 32:1 | 28 | 2.209 | 22.3 | 39.5 | 1338.6 |
| 172 | Slag | 32:1 | 28 | 2.213 | 22.3 | 39.4 | 1252.1 |

## 6.0 CONCLUSIONS

Test results derived from the above testing standards have proven to be acceptable and repeatable. Comparisons of uniaxial compressive results gathered during different projects, involving similar materials, are far less erratic using the triaxial confinement chamber and soil testing load frame than the results obtained from concrete type compression tests.

Triaxial testing on the same size specimens (i.e., 75 mm x 150 mm) using the same equipment as used for uniaxial tests has led to excellent results. The test equipment and procedures outlined in this paper should be adopted for standardized testing of fine grained, cemented backfills of normal and high densities.

## ACKNOWLEDGMENTS

The author wishes to acknowledge the support and assistance offered through both Falconbridge Limited, Sudbury and Inco Limited, Sudbury in development of the test methods suggested in this paper.

## REFERENCES

Weaver, W.S. and Luka, R. 1970. Laboratory studies of cement-stabilized mine tailing CIM Transactions, Vol 73: 204-217.

Mitchell, R.J. and Smith, J.D. Mine Backfill design and testing. CIM Bulletin: 82-89.

Landriault, D. and Goard, B. 1985. Research into high density backfill placement methods by the Ontario Division of Inco Limited. Ninth District meeting CIM, Thompson, Manitoba, Canada.

Bishop, A.W. and Henkel, D.J. 1957. The measurement of soil properties in the triaxial test. London: Edward Arnold Publishers Ltd.

Davis, R.T.H. 1988. Notes on procedures for hydraulic fill testing. Foseco Inc. Communications, Guelph, Ontario.

Swan, G. 1985. Notes on Denison Mine pillar backfill confinement tests. Div. Report: 85-78 (TR), CANMET, Elliot Lake, Ontario, Canada.

*Innovations in Mining Backfill Technology, Hassani et al. (eds), © 1989 Balkema, Rotterdam. ISBN 90 6191 985 1*

# Some factors relating to the stability of consolidated rockfill at Kidd Creek

T.R.Yu
*Falconbridge Ltd, Kidd Creek Division, Ontario, Canada*

ABSTRACT: In recent years, Kidd Creek has been placing consolidated rockfill at a rate of about 2 million tonnes into approximately 30 mined stopes annually. The backfill in mined openings plays an important role in providing ground stability for maximum extraction of ore. The backfill materials used at Kidd Creek consist of a graded aggregate and a blended binder of Portland cement and flyash. Sand has been occasionally mixed in the fill to serve as a void filler. Several factors affecting the static and dynamic strength of fill products have been evaluated and incorporated in fill specifications for individual stopes during fill operations. Blasting vibration effects on the fill stability by adjacent pillar recovery have been investigated in detail. This paper presents the findings of those studies, together with field test results considered to be useful in designing stable fill exposures.

## 1 INTRODUCTION

The Kidd Creek minesite is located 27 km north of the city centre of Timmins, Ontario. Currently the operation produces 4.3 million tonnes of base metal sulphide ore annually. The mining method for both No.1 and No.2 mines is mainly sub-level blasthole stoping with delayed backfilling.

In No.1 mine, stopes are normally 18 m wide, 22 m to 65 m long, with heights varying from 90 m to 135 m. Vertical rib pillars between stopes are typically 24 m wide and sill pillars are 30 m thick. Pillar recovery is implemented after the consolidated rockfill placed in the mined stopes has cured for a period of at least three months.

No.2 mine is located below No.1 mine. Stopes are typically 15 m wide, 30 m long and 60 m high. The mined stope is immediately filled with consolidated rockfill. After a three week curing period, the adjacent stopes are subsequently mined by retreating toward the footwall, and expanding along the strike on each side of an initial stope.

To fill the mined openings, the mines require 2.4 million tonnes of backfill annually. Approximately 80 % of the fill placed has been consolidated materials, with the remaining 20 % being unconsolidated rockfill or sandfill.

The consolidated rockfill(CRF) was set in 1976 for the strength requirements of 7 MPa which was necessary for the fill to stand over an exposed face 120 m high or more, by 70 m long. To support the gravity loading alone, the fill required a compressive strength of 2.8 MPa. A safety factor of 2.5 was applied to allow for additional blast loading and the reduction of fill strength due to inadequate mixing. It was found that the desired fill product could be achieved by mixing 20 parts of graded aggregate with 1 part of Portland cement (PC) by weight as a binding agent.

In 1982, Kidd Creek began to use ground blast furnace slag as a substitute for a portion of the Portland cement (Yu & Counter 1983). In 1984, the blast furnace slag was replaced by type "C" flyash (FA) for further cost savings (Yu & Counter 1988).

Since May 1976, the mines have placed 15 million tonnes of consolidated rockfill in over 150 mined stopes. Consolidated rockfill has been observed in over 170 stope exposures after the pillars adjacent to the filled stopes were blasted. During this period, various operational procedures have evolved and fill methodologies adopted to produce more stable and cost effective fill. The experience and studies encountered from fill operations have generated informative data which are invaluable in dealing with fill stability problems.

## 2 FACTORS RELATED TO THE FILL SYSTEM FOR QUALITY PRODUCTS

Backfill is used underground for overall ground support, to maintain the stope walls as the broken ore is withdrawn, to confine fractured pillars for extraction at a later date, and to prevent dilution of broken ore whenever adjacent pillars are mined (Yu 1987). For fill to be capable of performing those functions, a quality fill product is necessary. The following are some of the observed factors affecting quality products in connection with the fill system at Kidd Creek.

### 2.1 Aggregate

The open pit phase of the mine produced approximately 55 million tonnes of waste rock mostly consisting of rhyolite and andesite. This waste rock is the primary source of fill material.

Originally aggregate was crushed and graded to -15 cm to +1 cm for coarse, and -1 cm for fines at a proportion of 3 to 1 by weight. Secondary and tertiary crushing of the aggregate in a closed circuit screening loop was used to obtain designed gradation, and to attain the maximum bulk density or minimum porosity of the aggregate (Swan 1985).

During the winter season it was observed that poor coating and delayed initial curing were caused by frozen aggregate. To alleviate this situation, calcium chloride was mixed in the aggregate, at a rate of 2 % cement by weight, to lower the freezing point of the aggregate by $12^{\circ}C$, and to provide additional heat for curing.

As mining progressed deeper, excessive fines created by the attrition of particles passing through longer raises were found in the aggregate, and additional binder was required to coat the extra fines. Hence, the secondary and tertiary crushing and the closed circuit screening have not been incorporated in the crushing system for a few years.

The attrition of aggregate in a fill raise was measured, and the results expressed as the attrition ratio($D_{50}$ at surface/ $D_{50}$ at h) vs. travel distance are plotted in Fig. 1 (Bronkhorst 1986), and also fitted to Eq. 1.

$$D_{50} \text{ at surface } / D_{50} \text{ at } h = 1 + h/1100 \quad (1)$$

where $D_{50}$ = aggregate size corresponding to 50 % passing

h = aggregate vertical travel distance (m)

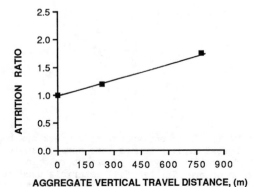

AGGREGATE VERTICAL TRAVEL DISTANCE, (m)

Fig. 1 Attrition rate of aggregate in fill raise

By using the graph or the formula, the relative size of aggregate in terms of $D_{50}$ at any depth can be found. For example, the attrition ratio at h = 900 m can be found from Fig. 1 or Eq. 1 to be 1.8. If the $D_{50}$ of aggregate = 8 cm at surface, the $D_{50}$ of degraded size after passing through a 900 m raise is 4.5 cm.

A slight change of moisture content in the aggregate may affect the fill quality significantly. To prepare a normal batch of CRF, the amount of mixing water requires only 3.9 % by weight of aggregate. An increase of 1 % moisture content in aggregate results in an excess of 25 % mixing water. This excessive water washes off the cement coating of solids, thereby weaking the cementation. Close observations of fill piles during stope filling, therefore, are essential to the adjustment of the pulp density of cement slurry for a proper mixing of fill materials.

### 2.2 Segregation

Segregation of consolidated rockfill when filling a stope is unavoidable. However, segregation can be minimized if fill operations are well planned and closely monitored.

From a series of tests and observations underground, it was found that a zone of fine aggregate tends to occur near the impact area, consuming most of cement paste and leaving a low cement content rockfill at the perimeter of the fill cone.

The segregation phenomena become more pronounced when stopes are filled by conveyor due to the impact velocity caused by the speed of the belt and a free fall. When a stope is filled by mobile vehicles, only the largest particles have the momentum to travel to the stope wall. The rest of the material fill the stope by progressive slumping resulting in a more uniform product.

Excessive water in the slurry or aggregate will wash the cement paste and the cement coating on the aggregate off as noted before, and flush it toward the lowest zone. This causes dilution of cement content in other areas except the lowest zone resulting in more heterogeneous fill. The control and prevention of such segregation would make a strong fill near the stope perimeter, which is desirable during pillar recovery.

## 2.3 Backfill raise

The fill raise constitutes another important factor to be considered for producing a stable fill. The raise should be strategically located and oriented so that there is a uniform distribution of fill material.

After a comprehensive study on the number of raises to be required, including an evaluation of five raises in a single stope, it was concluded that the use of two raises in an opening 18 m wide by 45 m long would be satisfactory without detriment to the segregation.

The orientation of a raise determines the location of the fill cone in the opening to be filled. When the material is in the fill raise, it attains a specific falling velocity which governs the trajectory of the fill into the stope. The trajectory of the material in the stope can be predicted using the motion of projectile. The material in the raise, however, encounters some frictional resistance, deviating from a freely falling body. It has been observed that in determing the initial velocity leaving the raise, a constant $K = 0.42$ should be incorporated in the calculation of the falling velocity in the raise.

The equations are illustrated, together with an example in Appendix A to compute the horizontal trajectory of the fill in a stope.

## 2.4 Mixing

The key to produce a competent consolidated fill is to coat all of the aggregate with the supplied amount of cement slurry in a very short time duration. Prior to starting the fill operations in 1976, an intensive study on this subject was made, including full scale testing of a vibratory mixing conveyor, the use of a slusher for mixing, a baffled mixing culvert, and a drum mixer.

In No.1 mine, currently the aggregate is mixed with slurry in a 1.2 m diameter by 2 m long steel culvert, before being passed through a 0.7 m diameter raise.

The culvert is equipped with three baffles set at 55 degrees to the axis of the culvert. Cement slurry is sprayed on the aggregate as it enters the culvert. The tumbling action of the aggregate as it passes over baffles ensures that all aggregates receive a thorough coating of slurry. A chute replaces the culvert if an opening is directly accessible from a fill level, or the worn culvert cannot be reached with a machine to replace it. The chute is equipped with four baffles set at an angle to divert the aggregate for a proper mixing with sprayed cement slurry.

In No.2 mine, aggregate is passed to fill stations below the 2600 level. At the backfill station, 11.8 tonne trucks are loaded from a feeder. The cementing agent is sprayed on the aggregate and hauled to the desired stope. The fill is poured from a drift access directly into the stope, and coating of the aggregate by the slurry takes place during dumping and rolling of the material in the stope. Conveyors are also used to transport aggregate to some stopes.

If the material is not coated with slurry during the mixing process, it may never be properly coated since:

a) It is impossible to control the flow of free slurry.

b) The slurry does not flow uniformly over the entire backfill cone.

c) The percolation rate of slurry is variable throughout the backfill due to differential settling.

d) Slurry which is not actually used to coat the aggregate acts as a void filler. Since there is insufficient cement to fill all voids, some portion of the backfill may remain unconsolidated.

For remedial measures to the above, an extra batch of cement slurry approximately 10 $m^3$ is usually sprayed onto the stope fill to provide additional cementation if the aggregate becomes too coase or too fine.

The fill operator must closely watch and respond quickly to changes in aggregate coarseness for better quality control.

In the earlier backfilling stages in No. 2 mine, aggregate was mixed in a 5 m$^3$ redi-mix concrete mixer for about 1.5 minutes. A mixing station was equipped with two drums. Each station could prepare 3000 tonnes of well coated fill per day. The mixer processing was regarded to be one of the best methods to provide a well coated rockfill, but was discontinued later as the demand of a greater quantity of fill was required.

## 2.5 Impact damage

This is another important yet easily over-looked factor governing the fill stability. The impact of aggregate falling from the fill raise onto those already placed consolidated rockfill, which has cured beyond its final set time, causes permanently broken bonds.

To reduce the extent of impact damage, a retarding admixture was added to the cement slurry. The admixture was introduced at a ratio of 90 cc retarder per 45 kg Portland cement to delay the final set time so that a minimum 60 cm thick buffer layer of consolidated fill was created. This allowed the shock of dumped fill to be absorbed by the plastic state surface layer preventing impact damage to the cured fill beneath. Another advantage of adding retarding agent to the cement slurry was found to significantly reduce the rate of cement scale build up in the slurry pipelines.

The retarding admixture, however, has not been applied since the use of a blended Portland cement-slag or flyash binder was introduced, as the blended binder exhibits a much slower curing rate compared to the binder of Portland cement alone.

## 2.6 Water quality

The same criteria of mixing water required for the preparation of concrete should be met in order to produce a good quality of consolidated rockfill.

In 1987, a few large scale fill failures were experienced in stopes filled during late 1985 and early 1987. Upon close inspection of the failed backfill, it was observed that the slurry coating of the aggregate had not cured properly. During this period, recycled underground waste water had been supplemented to prepare the cement slurry. Normally, the slurry mixing water consisted of either potable water supplied from a nearby lake, or surface recycled water from the mine's surface treatment pond.

After a comprehensive investigation into several possible causes, including the quality of binders used and various contributing factors mentioned in this paper, it was believed that the quality of the water used was the most feasible explanation for the low strength of the fill.

Test results indicated that the cylinders containing 60 % FA/40 % PC binder, cast with underground recycled water, had only 50 % of the compressive strength compared to those cast with potable or surface treatment pond water. The test results also revealed that the samples cast with underground recycled water were much less cohesive and bled more, especially in the 60 % FA/40 % PC binder mix. This suggested that the hydration of Portland cement and flyash was hindered by the subgrade underground recycled water, resulting in that water intended for hydration being released as bleed water (Henning 1987).

Contaminants found in the underground recycled water which might have affected the fill strength included:

a) high concentration of dissolved solids,

b) build-ups of oil and grease, and

c) cumulation of water treatment chemicals added to water.

The surface treatment pond mix water having a high PH value (10 to 11) due to lime addition was proven to be acceptable as slurry mix water. To improve further the quality of mix water, the following efforts had been made:

1) to remove oils and grease,

2) to co-ordinate the use of water treatment chemicals more carefully,

3) to increase water cycling time by dredging the treatment ponds, and

4) to improve the blending of surface treatment pond mix water with potable water.

## 2.7 Quantity control of fill materials

It is well known that to obtain an acceptable fill product, the quality of binders and aggregate used has to be maintained. It is also equally important that the quantity of binders and aggregate used in filling operations should be held according to the designed criteria. A deviation of the designed quantity can easily occur without being noticed. Such case had happened at Kidd Creek when the accuracy of the weightometers for the aggregate conveyors and/ or the weighing scale of the binder hopper were slightly off. Because the amount of binders designed for consolidated rockfill employed only an average of 5 % by weight of aggregate, any overdosed aggregate or insufficient amount of binders caused shortage of cement slurry to coat each solid particle, resulting in weak fill structure.

On the contrary, an overdosed binder or lower proportion of aggregate may produce a high strength fill but would increase the unit cost of operations.

# 3 STATIC AND DYNAMIC STRENGTH OF THE CRF

When a pillar adjacent to a filled stope is blasted the fill must be able to sustain not only the gravitational loading of the overlying fill material, but also the dynamic effects applied during blasting. It is, therefore, necessary to consider both the static and dynamic fill strength, if the stability of the fill exposure is to be maintained.

## 3.1 Static strength of CRF

The most important parameter governing the stability of gravity loaded fill is the shear strength of the material, which is related to its cohesion and friction angle. The CRF has a relatively consistent friction angle of $33^\circ$.

The cohesion, however, can vary significantly with the parameters of matrix and binders used. From a laboratory direct shear test, the apparent cohesion of the CRF samples at 5 % Portland cement content was determined to be approximately 1 MPa.

In practice, the compressive strength parameter has commonly been quoted as the static strength, because it is relatively easily measured in the laboratory. The compressive strength of 15 cm diameter CRF cylinders at 28 day cure has shown a relationship to the cement content by:

$$Q_u = 1.5 \ e^{0.25C} \quad \text{for } 5 < C < 25 \qquad (2)$$

Where   $Q_u$ = uniaxial compressive strength in MPa
        $C$ = Portland cement content by weight % of minus 4 cm aggregate

A partial substitute of ground blast furnace slag, or type C flyash for Portland cement as a binder mix for cost savings reduces the early compressive strength of the fill products. The allowable curing time of fill material placed in a stope, before blasting of adjacent pillar, should be considered in designing a proper binder mix for the desired strength.

In No.1 mine, wherever a minimum three-

month curing period is available, up to 60 % of Portland cement has been replaced by an equivalent weight of type C flyash with satisfactory results. Due to a shorter curing period available, only 33 % replacement of Portland cement by flyash is allowable in No.2 mine.

Consolidated rockfill exhibits pronounced heterogeneity. Evaluation of large fill samples and in-situ fill strength is necessary to supplement measured results of small samples. Test results indicated that the strength ratio between 30 cm diameter cylinders and 15 cm diameter cylinders was about 66 %.

A great deal of difficulty was experienced in obtaining a representative data from placed fill samples. As shown in Fig. 2, the fill forms alternating layers of coarse and fine aggregate with a repose angle of $36^\circ$. The fill strength varies from 11 MPa in the fine aggregate core of the fill cone area to 1.3 MPa in a segregated coarse aggregate layer.

## 3.2 Dynamic strength of CRF

The parameters governing blast damage to consolidated fill are not fully understood yet. It is postulated that the blasting vibration resistance may be closely related to the dynamic tensile strength of the fill.

CRF, being similar to weak concrete, can be strong in compression, yet weak in tension especially in the outer zone of the fill pile or adjacent to stope walls. The weak zone should be reinforced if the degree of blast damage is to be minimized.

A proper proportion of sand in the consolidated rockfill was found to enhance the strength characteristics.

Test results indicated that addition of sand at the rate of 5 weight % of aggregate increased the compressive strength by 40 %. However, the further addition of sand, up to 30 weight % of aggregate, created "saturated" conditions resulting in compressive strength decreases of up to 66 %.

Feasible explanations of above phenomena are as follows. The 5 % sand with the aggregate thickened the layer of slurry coating on aggregate, and increased contact areas of matrixes, thereby enhancing the overall strength. However, under "saturated" conditions, the overdosed sand robbed most of cement paste leaving little cement slurry to coat aggregate.

Impact tests were conducted on fill material with a large Schmidt hammer, acting as a constant source of dynamic loading. The results indicated that the most impact resistant fill occurred to the aggregate

Fig. 2  Segregation and layering of placed consolidated rockfill

fully surrounded by a sand-cement matrix.

An esker sand, which is coarser than cycloned tailings, yet finer than alluvial sand, is ready available near the minesite. Having 90 - 95 % by weight passing a No.18 mesh screen, and an average percolation rate of 15 cm/hr, this sand when mixed with cement slurry, flows without setting to fill the voids in segregated areas of aggregate.

## 4 BLASTING EFFECTS ON CRF

The understanding of transmitted shock waves from a medium into fill materials is important to design a stable fill structure. When shock waves travel through a medium, depending on factors including the angle of incident waves, the interface conditions, and the characteristic impedances of the media, they may be reflected, refracted, diffracted, scattered, and absorbed. The dynamic loading of shock waves impinging on fill will suddenly add on the static loading. A competent consolidated rockfill, therefore, should be capable of sustaining the combined loading.

### 4.1 Transmission through two media

For simplicity, a normal incidence of seismic waves at a solid interface is to be considered. At a discontinuity, some of plane acoustic waves will be reflected back along the original path in medium 1, and a portion of waves will be transmitted through the boundary into medium 2. The relations for the amplitudes of the incident, reflected, and refracted waves at the boundary can be expressed as follows (Seto 1970):

1. The reflection coefficient for particle displacement, $A_r/A_i$, and for particle velocity, $V_r/V_i$, respectively

$$= (Z_1 - Z_2)/(Z_1 + Z_2) \qquad (3)$$

The transmission coefficient for particle displacement, $A_t/A_i$, and for particle velocity, $V_t/V_i$, respectively

$$= 2Z_1/(Z_1 + Z_2) \qquad (4)$$

2. The reflection coefficient for force,

$$P_r/P_i = (Z_2 - Z_1)/(Z_2 + Z_1) \qquad (5)$$

The transmission coefficient for force,

$$P_t/P_i = 2Z_2/(Z_2 + Z_1) \qquad (6)$$

3. The acoustic power reflection coefficient,

$$W_r/W_i = (A_r/A_i)^2 \qquad (7)$$

The acoustic power transmission coefficient,

$$W_t/W_i = (Z_2/Z_1)(A_t/A_i)^2 \qquad (8)$$

Where $Z_1$, acoustic impedance of medium 1

$= D_1$ (density in medium 1) x $C_1$ (p-wave velocity in medium 1)

$Z_2$, acoustic impedance of medium 2

$= D_2$ (density in medium 2) x $C_2$ (p-wave velocity in medium 2)

It is to be noted from Eq. 4 that the transmitted particle velocity can be higher than the incident particle velocity in a medium of a lower acoustic impedance; yet the transmitted stress can be lower than the incident radial stress as expressed by Eq. 6.

It can be seen from Eq. 5 that a compression wave is reflected as tension at a free end, or at a boundary opposing a lower acoustic impedance medium 2. Consequently, a tensile stress with the magnitude higher than the tensile strength of the medium can cause scabbing failures at the interface. Based on Eqs. 4 and 6, the transmitted force can be expressed in term of the product of the acoustic impedance and the peak particle velocity.

## 4.2 Blasting vibration monitoring

The heterogeneous nature of consolidated rockfill affects not only the static properties but also the dynamic characteristics. At the interface between a rock wall and consolidated rockfill, the boundary condition is generally so complicated that the transmitted amplitudes from a stope blast may deviate significantly from those calculated using above equations. Field measurements, therefore, are necessary to obtain the actual magnitudes for practical applications.

The first seismic measurements were carried out in a backfill test drift excavated into a stope fill from the footwall. Seismic disturbances were generated by using a sledge hammer. Two velocity sensors detected the vibration level which was recorded in a portable storage oscilloscope. The measured results were very inconsistent from one place to another due to fill heterogeneity. The partitioning of seismic waves at the interface due to mismatch of the acoustic impedance, in terms of the transmission coefficient, was in the range between 25 % and 73 %. The absorption of seismic intensity in the fill during propagation was 0.57 db/m in the compact fill, and 0.95 db/m in the segregated aggregate

fill. The seismic frequency at measurements was approximately 500 Hz.

A second series of seismic measurements were conducted at a deeper level of the mine. The measurements were to determine the transmission coefficients of blasting vibration and energy in consolidated rockfill. The amount of explosive used in the blasts ranged from 12 kg to 84 kg per delay for a total of 2150 kg. Monitoring was conducted using two sets of triaxial velocity sensors and a seven channel F.M. instrumentation recorder. Analysis of the data was carried out with a signal analyser. The results indicated that the transmission coefficients for particle velocity and energy between rock and fill averaged 37 % and 56 % respectively.

Both field measurements revealed that the transmission coefficient for particle velocity from rock to fill was always less than unity, which seems to be contrary to the one calculated from Eq. 4. The discrepancy of the results could be attributed to the over simplified boundary conditions in Eq.4 calculation where the two media might not be continuous evenly.

It was observed in a fill drift that a competent consolidated rockfill had subjected to a scabbing failure at an estimated particle velocity of 30 cm/sec. This vibration level may be considered as a maximum allowable particle velocity in planning a safe production blasting against fill. The orientation of blast holes drilled in a pillar adjacent to a fill structure has been found to relate to the extent of fill damage. Stope fill failures were seen less severe for blast holes drilled in parallel to the fill than those drilled toward it in the past.

## 5 CONCLUSIONS

At Kidd Creek, backfill palys an important role in extracting ore safely and economically. Team work has been emphasized to make the fill operation successful. However, there are still numerous implicit factors which may affect the stability of fill exposures.

Generally, the cause of instability can be divided into two types. One is due to the inferior fill caused by system operations, and the other is attributed to the excess of external loading from blasting of adjacent stopes. Some factors relating to the instability of placed fill were identified at Kidd Creek, and proper remedy measures have already been taken, as described in the paper.

It is believed that familiarization of

those factors would be helpful in improving fill quality not only for a similar type of consolidated rockfill, but also for every kind of fill products.

ACKNOWLEDGMENTS

The author wishes to thank the management of Kidd Creek Mines for the permission to publish this paper. Descriptions in the paper have been based on observations and studies conducted by several people involved in fill operations at Kidd Creek. The present fill study group consists of J.G.Henning and M.Scripnick. A fill study project contracted by DSS, Ottawa under the contract No. 09SQ-23440-6-9011 has produced some interesting data which are included in the paper. Mr. A.Annor of CANMET is the scientific authority of the DSS project with which P.Farsangi of Kidd Creek has been involved.

REFERENCES

Bronkhorst, D.L. 1986. Attrition of backfill aggregate in No.2 mine. Kidd Creek Mines internal memo.
Henning, J.G. 1988. Effect of water quality on backfill strength. Kidd Creek Mines internal memo.
Seto, W.W. 1970. Acoustics. McGraw-Hill.
Swan, G. 1985. A new approach to cemented backfill design. CIM Bulletin, Vol.78, No.884, pp.53-58.
Yu, T.R. 1987. Ground support with consolidated rockfill. CIM Special vol.35, pp.85-91.
Yu, T.R. and Counter, D.B. 1983. Backfill practice and technology at Kidd Creek Mines. CIM Bulletin, Vol.76, No.856, pp.56-65.
Yu, T.R. and Counter, D.B. 1988. Use of flyash in backfill at Kidd Creek Mines. CIM Bulletin, Vol.81, No.909, pp.44-50.

APPENDIX A

For a given dip angle and length of fill raise above a stope to be filled, the horizontal trajectory ($X_h$) of the fill material can be found from the following: (See Fig. 3)

$$X_h = V_h \times T \qquad (9)$$

$$V_h = V_v/\tan D = (2g \times L \sin D \times K)^{0.5}/ \tan D \qquad (10)$$

where

$V_h$ and $V_v$ are horizontal and vertical velocity components respectively when discharging fill into stope, m/sec
L = length of the fill raise, m
D = dip angle of the fill raise, degree
K = 0.42 (an empirical constant)
g = grvitational acceleration, 9.8 m/sec$^2$
T = time taken for the fill material to fall from the end of the raise to reach the top of a fill cone, sec

$$= [-V_v \pm (V_v^2 + 196\ H)^{0.5}] / 9.8 \qquad (11)$$

where
H = height of the free fall of aggregate in stope, m

Example;

For a fill raise dipping at 60° with an inclined length of 35 m, the horizontal velocity at the end of the raise from Eq.10 = 9.1 m/sec. If the free falling height is 60 m in the stope, the falling time from Eq.11 = 2.69 sec. Using Eq.9, the horizontal trajectory can be founded to be 24.4 m.

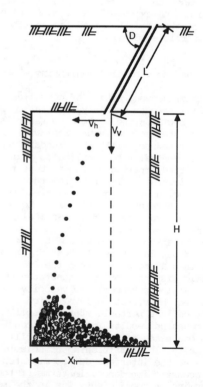

Fig.3 Horizontal trajectory of fill in a stope

# The use of backfill for improved environmental control in South African gold mines

M.K.Matthews
*Chamber of Mines of South Africa, Research Organization, Auckland Park, RSA*

ABSTRACT

The backfilling of worked out areas in the stopes of South African gold mines can effect improvements in environmental conditions. By sealing off large exposed areas of hot rock surface and eliminating leakages of ventilation air, backfill can be used to achieve substantial reductions in stope heat loads and improvements in air utilization efficiency. The effects of backfilling are significant on a mine-wide scale and can lead to savings in ventilation and refrigeration costs. These and other environmental considerations should be taken into account in an integrated design approach to the use of backfill in deep mines.

## 1 INTRODUCTION

The high level of interest in the use of backfill that is currently being shown within the South African gold mining industry has resulted in the recent introduction of backfilling systems in a number of different mining applications. However, it is in the ultra-deep mines, characterized by narrow stoping widths and high virgin rock temperatures, that backfilling has its greatest potential, providing enhanced regional support and facilitating an increased percentage extraction of payable reef. In addition to the rock mechanics benefits it offers, backfill can be used to improve environmental control. As mining depths and heat loads increase, environmental control considerations become an increasingly important factor in the integrated design and economic evaluation of a backfill system, and under certain conditions could constitute the major motivation for the use of fill.

The principal environmental benefits achievable through the use of backfill are the reduction in stope heat load and the improvement in the utilization efficiency of ventilation air that result when fill is used to seal off the often extensive areas of hot rock surface in the mined-out part of a stope. This it does to a degree that cannot be matched by traditional ventilation control methods. There are further, secondary benefits and some potential environmental problems associated with backfilling, and whilst the problems are of a comparatively minor nature, an awareness of their existence is nonetheless necessary.

The various aspects of environmental control that are affected by backfilling have been investigated at the Environmental Engineering Laboratory of the Research Organization of the Chamber of Mines of South Africa (Matthews, 1986), and the purpose of this paper is to give an overview from the findings of that work. In so doing, it is intended to demonstrate the typical magnitudes and relative importance of the various effects of backfilling on the underground environment and to highlight the significance of the improvements and cost savings that can result.

## 2 PHYSICAL PROPERTIES OF BACKFILL MATERIALS

Currently, the most commonly used back-fill materials are prepared from metallurgical plant tailings and crushed waste rock, used either individually or as an aggregate. These backfills have a solid fraction with a range of particle sizes up to a maximum of about 1 mm, and are transported hydraulically. After placement in the stopes, the excess water drains from the fill to leave a porous material composed of quartzite, water, and air. However, given the small particle size and the complex network of particles and voids that is set up, the material is impervious to airflow at the low differential pressures typically occurring in stopes. Hence, backfill may be used to provide a near-perfect ventilation seal.

To understand the effects of backfilling on the pattern of heat flow in the rock surrounding the mine excavations, a knowledge of the thermal characteristics of the backfill is required. The fill material has a thermal conductivity that depends upon:

1. The mineralogy, size and geometrical arrangement of the solid particles.
2. The porosity of the bulk material.
3. The residual water-to-solids mass ratio.

A series of laboratory measurements examining the thermal conductivity of backfill at typical 'in situ' conditions was undertaken using a modified divided bar apparatus (Sass, Lachenbruch and Munroe, 1971). Results are shown in Figure 1, and from these it may be concluded that newly placed backfill, with a conductivity of around one third that of ordinary rock, will have some value as an insulator.

The specific heat capacity of a composite material such as backfill may be calculated from a mixed mean value, based on the specific heats of the constituent substances present, weighted according to their respective mass fractions.

Since the mass of air in the backfill material is usually negligible, the composite specific heat may be estimated from:

$$C_p = 4,18 \left(\frac{w}{w + 1}\right) + 0,84 \left(\frac{1}{w + 1}\right) \qquad (1)$$

where
w is the water-to-solids mass ratio
4,18 kJ/kg K is the specific heat of water,
0,84 kJ/kg K is the specific heat of a typical quartzite.

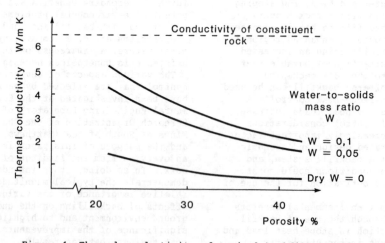

Figure 1. Thermal conductivity of typical backfill material

288

## 3 EFFECT OF BACKFILL ON MINE HEAT LOAD

Heat flow from rock to the ventilation air is the major component of heat load in the stoping areas of deep mines. Part of a typical stope is depicted in Figure 2, and it may be noted that as much as 50 per cent of the total stope heat load can be attributed to heat pick-up by air inadvertently ventilating the worked out areas. This occurs as a result of the problems associated with the sealing of these areas by traditional means. By virtue of its being impervious to airflow, backfill can be used to effectively seal off the rock surfaces in worked out areas and eliminate this unnecessary heat load.

The normal pattern of backfilling is a series of 'ribs' of fill parallel to the working face, extending for the full width of the stope panel. If insufficient material is available for complete backfilling, the same effect may be achieved by closing off 'paddocks' using ribs of fill, although compatibility with rock mechanics requirements must be ensured in such a situation. So, the final heat load reduction depends upon the extent of the worked-out area that is effectively sealed off. Where complete sealing is possible, the distance of the most recently placed fill from the working face is an important parameter. Keeping this fill-to-face distance small minimizes the exposed rock surface area, and the rock in this newly mined region is also the hottest, so prompt backfill placement will give the maximum heat load reduction.

The use of backfill, whilst sealing off hangingwall and footwall rock, creates a conduction path to the new vertical surfaces bordering the gullies and face zone. From these surfaces, heat is transferred to the ventilation air by convection. The effect will be more pronounced in the older parts of the stope where closure has stressed the fill and increased its conductivity. The newly placed fill near the face will have a higher porosity and act to a certain extent as an insulator. These considerations aside, the narrow reef stoping geometry is such that this additional heat flow into the gullies is small in comparison with the overall reduction achieved.

The heat load components for a typical stope panel, both filled and unfilled, were evaluated using a simple proven heat flow model (von Glehn & Bluhm, 1986). The chosen stope layout and the results of this exercise are shown in Figures 2 and 3. The predicted reduction in stope heat load due to backfilling was 48 per cent. The effect of varying fill-to-face distance was also investigated for this typical case, and the results shown in Figure 4 illustrate the importance of keeping the distance small so as to maximize the heat load reduction.

The significance of the heat load reduction resulting from the mine-wide use of backfill can be gauged by comparing the total stope heat load with the magnitudes of other heat sources occurring throughout the mine. This comparison depends upon the design and layout of the workings, and will vary considerably between individual mines. However, such an analysis may be carried out simply using the HEATFLOW computer program (von Glehn et al, 1987), and widely differing situations have been examined by a 'case-study' approach. Predicted reductions in total mine heat loads resulting from complete backfilling throughout the mine have been in the range 10 to 22 per cent.

In the case of a new mine at the planning stage, it will be possible to realize the benefit of backfilling entirely as a reduced cooling requirement, leading to a saving in both capital and running costs for refrigeration and air-cooling plant. Where backfilling is undertaken on an older mine, the design of the installed cooling system may not be sufficiently flexible to accommodate a partial reduction of refrigeration capacity, or alternatively the existing thermal environment may be considered as below the ideal standard. In these cases, part or all of the reduced cooling requirement may be 'traded off' to give lower temperatures in the stopes, with the associated indirect benefits of improved production (Smith, 1984).

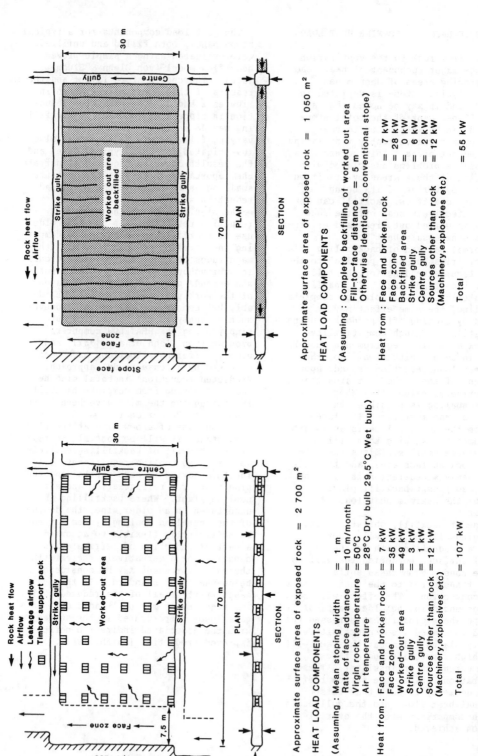

Approximate surface area of exposed rock = 1 050 m²

HEAT LOAD COMPONENTS

(Assuming : Complete backfilling of worked out area
Fill-to-face distance = 5 m
Otherwise identical to conventional stope)

Heat from : Face and broken rock = 7 kW
Face zone = 28 kW
Backfilled area = 0 kW
Strike gully = 6 kW
Centre gully = 2 kW
Sources other than rock = 12 kW
(Machinery, explosives etc)

Total = 55 kW

Figure 3. Backfilled stope panel

Approximate surface area of exposed rock = 2 700 m²

HEAT LOAD COMPONENTS

(Assuming : Mean stoping width = 1 m
Rate of face advance = 10 m/month
Virgin rock temperature = 50°C
Air temperature = 28°C Dry bulb 29,5°C Wet bulb)

Heat from : Face and broken rock = 7 kW
Face zone = 35 kW
Worked-out area = 49 kW
Strike gully = 3 kW
Centre gully = 1 kW
Sources other than rock = 12 kW
(Machinery, explosives etc)

Total = 107 kW

Figure 2. Conventional stope panel with typical dimensions and heat loads

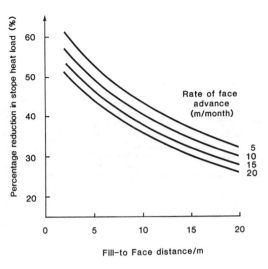

Figure 4. Dependence of stope heat load reduction on fill-to-face distance

## 4 EFFECT OF BACKFILL ON MINE VENTILATION

By preventing leakage through the worked-out areas of stopes, backfilling improves the utilization efficiency of ventilation air, but at the same time raises the resistance of the stoping section to airflow. The actual increase in the utilization efficiency, which is defined as the percentage of the total stope air quantity that reaches the working face, is very much dependent on the quality of sealing that would normally have been achieved by conventional methods. It has been estimated that, under favourable conditions using backfill, as much as 82 per cent of the total air quantity could be made to reach the face, whereas measurements in conventional stopes have shown utilization values as low as 25 per cent. The only paths permitting the passage of air through a backfilled stope, such as that shown in Figure 3, are the open area at the face and the centre gully. The quantity of air at the face will depend upon the resistance of the face zone to airflow, which in turn is governed by the fill-to-face distance. Hence, when fill-to-face distances are small, high air utilization can only be maintained if the centre gully resistance is high. Regulation of airflow in the centre gully is traditionally carried out by means of regularly spaced frames supporting strips of heavy brattice

material. These brattices are prone to damage and their maintenance is generally neglected. In a backfilled stope, the centre gully brattices have a direct bearing on the quantity of air reaching the face and it is essential that their effectiveness be maintained. The dependence of air utilization on both fill-to-face distance and brattice quality for a typical stope are shown in Figure 5.

Despite the improvement in air utilization, the increased resistance of a backfilled stope could cause a situation whereby the stope air quantity would be reduced. With a reduced airflow at the face, the rate of temperature rise of the air will be higher, and the point may be reached at which re-cooling of the air in the stope becomes necessary. If this is unavoidable, the use of an auxiliary airway parallel to the face, formed by leaving a gap between two specific ribs of backfill, has been suggested as providing a convenient cooler location (Gundersen, 1988).

Irrespective of air quantity, the air velocity at the face will always be greater (usually by a factor of 2 or more) than in a conventional stope, and this gives the air an increased cooling power. These findings have been confirmed by the underground measurements made to date (Viljoen, 1986), (Matthews, McCreadie and March, 1987).

Examined in a mine-wide context, the increased resistance of the stoping sections will generally have little effect on the operation of existing fans, since the stopes usually account for only a small proportion, generally less than 7 per cent, of the total mine resistance. It is nevertheless recommended that resistance changes be estimated for individual situations to confirm that the effect is in fact small, and that changes in the operating points and duties of the various fans do not result in a flow instability or possible stall.

In the case of a new mine designed to use backfill, the planned quantity of air to be circulated may be reduced to take advantage of the improved air utilization. Amongst the examples studied, a maximum feasible reduction of 35 per cent was found, leading to lower costs for the purchase and running of fans and the possibility of using smaller airway sizes in an optimization of costs. In the case of an existing mine, reduction of the total air

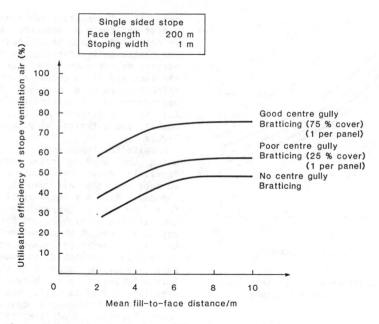

Figure 5 Dependency of air utilization on fill-to-face distance

quantity circulated must be approached with caution, since the ventilation system will also serve areas other than stopes. Hence, without an extensive undertaking to re-distribute airflows throughout the ventilation circuit, certain areas, including stopes that remain unfilled, are likely to suffer from an insufficiency of air. It was concluded from the case-studies that airflow reduction will generally not be practicable for an existing mine.

## 5 SECONDARY BENEFITS OF BACKFILLING

Whilst the principal environmental benefits of backfilling in terms of heat load reduction and air utilization improvement have already been discussed, there are further benefits that are worth noting. These are as follows:
- Reduction of fire risk:
  The greatly reduced usage of support timber, apart from being a cost saving in itself, means that there is much less combustible material present in a back-filled stope. Furthermore, there can be no accumulation of combustible rubbish as often occurs in the back areas of conventional stopes, and the backfill itself will act as a barrier against the spread of a fire.

- Shortening of re-entry period:
  After blasting of the stope face, a certain 're-entry period', during which time workers may not return to the stope, must elapse to allow the levels of pollutants generated by the explosion to fall below legally stipulated threshold values. To a first approximation, the required length of the re-entry period has been shown to be proportional to the volume of the workings (Alexander, Unsted and Benecke, 1987). The effect of backfilling is to dramatically reduce this volume - a reduction of around 85 per cent would be achieved by complete filling of a worked-out area as shown in Figure 3. Hence, a significant decrease in re-entry period can be expected for a backfilled stope, with obvious benefits to production.

## 6 POTENTIAL ENVIRONMENTAL PROBLEMS ASSOCIATED WITH BACKFILLING

To date, no major environmental problems have arisen as a result of backfilling at the various mines where pilot plants have been in operation. There are, however, several points that may warrant attention as backfill is used on a larger scale and at greater depths.

Firstly, the temperature of the fill as it arrives in the stopes should be considered. When the material is piped from surface in slurry form, as is usually the case, the slurry will experience a temperature rise due to the loss of potential energy (Joule-Thompson Effect). A simple energy balance gives:

$$\Delta t = \frac{gz}{c_p} \qquad (2)$$

where $\Delta t$ is the temperature rise of the slurry in °C

$g$   is the acceleration due to gravity in m/s$^2$

$z$   is the change in elevation in km

$c_p$   is the specific heat capacity of the slurry in kJ/kg K

For a typical slurry of specific gravity 1,65, the temperature rise would be 4,6 °C per 1 000 metres depth, were there no heat exchange with the ventilation air. Fill delivery temperatures of 30 °C have already been recorded at depths of around 2 200 m (Matthews, McCreadie & March, 1987) and whilst the additional heat load will not be large, it may be expected that conditioning of the fill to provide an acceptable delivery temperature could be necessary in ultra-deep mines. Cooling the slurry on surface has been shown to be ineffi-cient postionally, and underground con-ditioning using a heat exchanger operat-ing with chilled mine service water offers a possible solution.

Secondly, concern has been voiced regarding the possibility of high levels of airborne dust resulting from the use of backfill. Data from routine stope surveys, however, have not revealed an increase in respirable dust levels in backfilled stopes at the majority of mines. During the earlier backfilling trials, high dust levels could be associated with events such as the collapse of retaining enclosures, or other system problems resulting in large quantities of unconstrained fill material being discharged in the stope. Now that these initial difficulties have been overcome, the only potential source of extra dust will be the relatively small quantity of fine particles that escape from the retaining enclosure with the excess water drainage. There is a tendency for this material to collect in the strike gullies as a thick mud, and whilst in this form it will not give rise to dust, preventive measures may on occasion be required to avoid an excess-

ive mud accumulation. Nevertheless, it should be remembered that one effect of backfilling, as discussed earlier, is to increase air velocities at the stope face. If velocities are allowed to rise too high (greater than 2,5 m/s) then there will be an increased likelihood of dust pick-up, involving not only a fines from the of backfill, but all the parti-culate quartzite generated during the various stoping activities.

A third potential problem area concerns mine water. When backfill is placed hydraulically, a substantial amount of extra water is introduced into the mine and, apart from the cost of this extra water, further financial pen-alties may be incurred in its removal. The capability of drainage and pumping systems should be examined, as some upgrading may be required on an existing mine. The possibility of water conta-mination by backfill is another area for concern, particularly in tailings-based fills with regard to the carry-over of cyanide from the metallurgical plant leaching process. Although monitoring to date has shown that backfill treat-ments in use are adequate in maintaining an acceptably low level of cyanide, it is nevertheless suggested that a regular water sampling programme be adopted.

## 7 CONCLUSIONS

It is evident that backfill can be used to significantly improve the environ-mental conditions in stopes and the extent to which they may be controlled.

The major benefits that can be realis-ed are a reduction in stope heat load and an improvement in the utilization efficiency of ventilation air. These can have significant effects on a mine-wide scale, leading to a reduced overall refrigeration requirement (approximately 10 to 20 per cent less) and the possibl-ity of circulating less air (up to 35 per cent less) in the case of a new mine. Considerable savings in both capital and running costs can result, although the exact magnitudes of these savings will be strongly dependent on the specific mine design. For an exist-ing mine, it will generally not be feas-ible to reduce the air quantity circula-ted and the environmental benefits of backfilling will not necessarily trans-late directly to a capital saving. There will, however, be a great improve-ment in stope conditions, and associated with this will come indirect benefits in terms of increased production and safety.

Further advantages offered by backfilling are a reduced fire risk and the possibility of shortening the re-entry period after blasting, again giving the potential for improvements in production and safety.

Whilst no major environmental problems have been experienced with the use of backfill to date, cautionary notes must be sounded with regard to the possibility of increased dust levels if air velocities rise too high, the need to carefully monitor mine water quality to establish that unacceptable contamination is not occurring, and the possibility of high backfill delivery temperatures at greater depths leading to a need to condition the fill.

Finally, it must be stressed that in view of its effectiveness as a means of reducing stope head load at source, backfilling should always be considered as part of an integrated strategy to approach the heat problem when mining at great depth.

ACKNOWLEDGEMENT

The findings discussed in this paper resulted from work carried out as part of the research programme of the Research Organization of the Chamber of Mines of South Africa.

REFERENCES

Alexander, N.A., Unsted, A.D. & Benecke, K.C. 1987. Controlled recirculation : its effect on blast contaminant decay. Journal of the Mine Ventilation Society of South Africa, Vol. 40, No. 7.

Gundersen, R.E. 1988. Hydro-power : extracting the coolth. Equipment alternatives in underground mining. SAIMM Conference, Johannesburg.

Matthews, M.K., 1986. The implications of backfilling on environmental control. SAIMM Symposium on backfilling in gold mines. Johannesburg.

Matthews, M.K., McCreadie, H.N. & March, T.C.W. 1987. The measurement of heat flow in a backfilled stope. Journal of the Mine Ventilation Society of South Africa, Vol. 40, No. 11.

Sass, J.J., Lachenbruch, H. & Munroe, R.J. 1971. Thermal conductivity of rocks from measurements on fragments, and its application to heat flow determination. Journal of Geophysical Research, Vol. 76, No. 14: 3391-3401.

Smith, O. 1984. Effects of a cooler underground environment on safety, labour and labour productivity. Third International Mine Ventilation Congress, Harrogate. (Institute of Mining and Metallurgy.)

Viljoen, P.L.J. 1986. The application and benefits of underground slime fill. Mine Ventilation Society of South Africa Conference, Mine Vent. '86, Johannesburg.

Von Glehn, F.H. & Bluhm, S.J. 1986. The flow of heat from rock in an advancing stope. GOLD 100, SAIMM, Proceedings of the International Conference on Gold, Vol. 1, Gold Mining Technology, Johannesburg.

Von Glehn, F.H., Wernick, B.J., Chorosz, G. & Bluhm, S.J. 1987. ENVIRON : A computer program for the simulation of cooling and ventilation systems on South African gold mines. APCOM 87. Proceedings of the Twentieth International Symposium on the Application of Computers and Mathematics in the Mineral Industries, Vol. 1, Mining, SAIMM, Johannesburg.

# 6 Soft rock mining

*Innovations in Mining Backfill Technology, Hassani et al. (eds), © 1989 Balkema, Rotterdam. ISBN 90 6191 985 1*

# Preparation of an early-strength fill material for roof support in longwall mining of coal

E.G.Thomas
*University of New South Wales, Sydney, Australia*

R.D.Lama
*Kembla Coal and Coke Pty Ltd, Wollongong, Australia*

K.Wiryanto
*University of New South Wales, Sydney, Australia*

The need for improved longwall mining performance at Kembla Coal and Coke has resulted in a programme of development of improved mining technology, including investigation of a fill system to provide early roof support during extraction of coal.

The paper is presented in two parts. The first describes the proposed mining system and the scope it offers for improvement over existing mining practices.

The second covers a laboratory test program to investigate a fill material capable of providing roof support within say one hour of placement. The fill system studied was based upon a mixture of minus 10 mm coal washplant reject, Type A Portland cement and calcium chloride (as a set and cure accelerator).

## 1 INTRODUCTION

Developments in longwall coal mining technology have reached the stage where individual modern longwalls are routinely producing around 10000 tonne of coal per day in medium thickness seams (2.5 to 3 m) and from face lengths of 200 m. Further improvements applying current technology ensure system capacities of 15000 to 20000 tonne per day with a one-directional cutting operation and 25000 to 40000 tonne per day for a two-directional cutting operation. Drawing board designs are looking at shearers with shearing speeds of 16 m/min which, coupled with compact design and favourable ground conditions, could produce up to 50000 tonne per day, as detailed by Lama (1988).

Underground coal mining today using longwall technology is in a position to compete with surface mining, provided the following conditions are met.
1. Roadways are developed and maintained in good condition.
2. Face environment (gas and dust) is effectively controlled.
3. Equipment is efficient and properly maintained to ensure high system reliability.
   Considering each of these requirements, new conceptual models have been proposed by Lama (1988) to ensure roadway maintenance and improved ventilation.
   Also, developments in gas drainage (Lama, 1980 and Hanson et al, 1987), dust control using water sprays and water infusion (Hewitt and Lama, 1987) and water assisted cutting, dust extractor drums and air curtains (Hewitt and Lama, 1987), along with complete automation of the face operation, will permit targets of 30000 to 40000 tonne per day from a single longwall.
   Further, developments in condition monitoring, fault detection, system analysis, machine maintenance, scheduling systems and training should combine to allow system availabilities of better than 0.8 to be a reality.
   Fast advancing longwall faces combined with high capacity shield supports provide effective ground control at the face. High rates of advance ensure less bed separation and greater continuity of the roof at the face. Higher rates of advance do however increase the front abutment stress at the face and bring the peak closer to the face, though this should not produce problems since the increased peak stress effect is more than compensated by the time dependent strength properties of the rock both at the face and the gate road ends.

## 2 DEVELOPMENT REQUIREMENTS FOR HIGH PRODUCTION LONGWALL SYSTEMS

High production longwall systems usually involve developing gate roadways in advance and working the longwall in retreat. Advancing longwalls present serious problems at the face ends and the technology to develop the face ends and maintain roadways behind the face in the goaf has not matched the production levels achieved on the longwalls.

Use of multiple shearers at the face ends, heading machines and pump packs behind the face to maintain gate roadways in advancing longwalls, have organisational and technical constraints as well as physical constraints and greatly reduce overall production levels from longwall districts.

The usual system of gate roadways for longwall mining is a two-heading development. In certain mines with difficult ground conditions and high gas, three-heading development is practiced. Certain countries (e.g. USA) require four-heading development by law.

Gate roadways are driven with conventional continuous miners with shuttle cars loading onto belt conveyors. Development rates may range from 90 m/shift under exceptionally good conditions, through an average of 20 m/shift in good conditions to as low as 5 m/shift (15 m/day) under adverse mining conditions at depth and with high horizontal stresses.

Development of flexible conveyors, such as the Joy flexible conveyor train, the Klöckner-Becorit mobile conveyor, the Consol continuous haulage system and the DMS angle conveyor, has promised to increase development rates but under adverse conditions these are still far short of acceptable rates for the production targets of future longwalls.

A high production longwall retreats at a very high rate. A longwall face of 200 m length producing 30000 tonne per day will retreat at a rate of almost 42 m/day and a longwall face 250 m long producing 40000 tonne per day retreats at 50 m/day. To meet the requirements of fast retreating longwalls, headings must, for a face length of 250 m and daily production of 30000 tonne, be advanced at 100 m/day for a two-heading development and 125 m/day for a three-heading development. Even higher rates will be required before the full potential of the longwall system is achieved.

Experience both in Australia and overseas has been that inadequate development rates have on occasion forced longwall faces to stop or required additional continuous miner units.

Typical productivity of a continuous miner development panel under adverse mining conditions is 10 to 20 tonne per manshift. That of a longwall face is 200 to 300 tonne per manshift. Though efforts are made to maintain a low ratio of longwall shifts to continuous miner shifts (say between 1 and 2), in deep mines this ratio may exceed 4, resulting in a 4 to 5 fold drop in overall productivity.

## 3 WIDE HEADING DEVELOPMENT SYSTEMS

One possible solution to achieve a high rate of development is to apply longwall technology, which is far more advanced and much simpler than continuous miner technology. The argument is simple, namely, "if a 250 m long face can be advanced at a rate of 35 m/day to produce 30000 tonne per day, then why not advance a shorter face (of say 15 to 30 m length) at an equivalent rate and split this using pack walls into two or three headings, as depicted in Fig. 1?" Such systems also show that they are capable of improving the advance/development ratio by from 50 to 173% over conventional systems.

The technology of wide heading development is well established. Wide heading development (shortwall mining) has been practiced in a number of mines in Europe using longwall technology with ESA-60, ESA-150 and Dosco ISM-200 machines and longwall face lengths of 15 to 60 m. For example, at Stolzenbach lignite mine in West Germany, an overall productivity ranging from 30 to 65 tonne per manshift was achieved, depending upon panel length, which ranged from 180 to 500 m.

Some conceptual designs using longwall technology for development of say a 15 to 18 m heading are given in Figs. 2 and 3, clearly indicating the role of the pump packs in providing roof support.

## 4 TECHNICAL FEASIBILITY OF WIDE HEADING SYSTEMS

Geomechanics studies conducted on wide heading development using FEM have shown that a heading width of up to 20 m is feasible, even with a comparatively modest pack strength.

FEM analysis using elasto-plastic

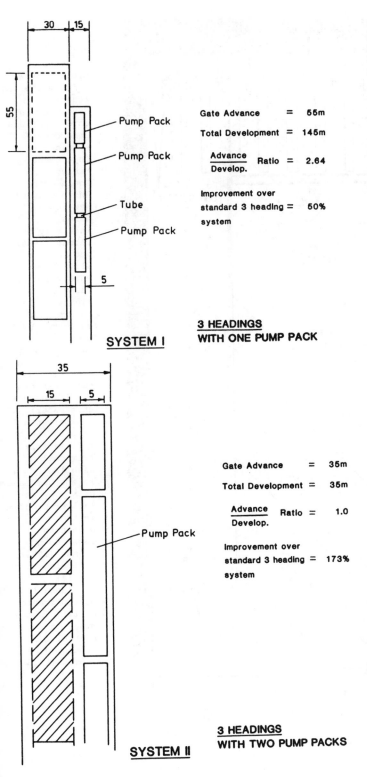

**SYSTEM I**

Gate Advance = 55m

Total Development = 145m

$\dfrac{\text{Advance}}{\text{Develop.}}$ Ratio = 2.64

Improvement over standard 3 heading = 50% system

**3 HEADINGS**
**WITH ONE PUMP PACK**

**SYSTEM II**

Gate Advance = 35m

Total Development = 35m

$\dfrac{\text{Advance}}{\text{Develop.}}$ Ratio = 1.0

Improvement over standard 3 heading = 173% system

**3 HEADINGS**
**WITH TWO PUMP PACKS**

Fig. 1 Future gate road development systems.

modelling showed that a heading with a total width of 20 m using a 10 m wide pack undergoes a mid-point deformation of 85 to 95 mm, depending upon pack properties, as demonstrated in Fig. 4. Field measurements at West Cliff colliery have shown that convergence of roof and floor of up to 100 mm can be tolerated before any visible damage to the roof. Therefore, even a low quality pump pack can be expected to limit convergence to acceptable levels.

In practice, the pump pack material must develop early strength to

1. allow re-location of the slip forms into which it is pumped, as the mining face advances, and

2. minimise roof movement before the roof is supported mechanically.

It is envisaged that in practice the pump pack material will be prepared remote from the face by mixing coal washplant reject lightly crushed to minus 10 mm, Portland cement, cement accelerator and water, and pumped to the face for placement behind the slip forms. The material must be placed at the lowest possible moisture content acceptable to allow pumping. A system akin to concrete pumping is envisaged, with very little if any drainage of water from the placed pack, possibly requiring viscosity modifiers to allow pumping at acceptable head losses. Pump pack placement is however beyond the scope of this paper.

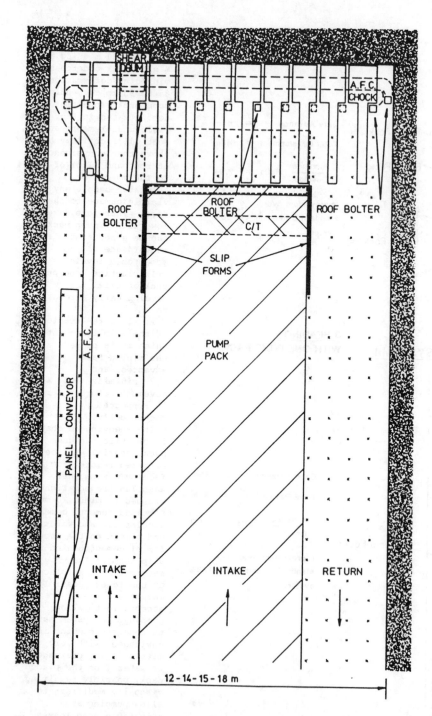

Fig.2 Shortwall entry driveage system with pump packing, using longwall technology.

The test program on West Cliff washplant reject described in this paper was conducted within the Department of Mining Engineering, University of New South Wales, initially by Thomas (1986) and later by Wiryanto supervised by Thomas. (Testwork is continuing, conducted by Munir supervised by Thomas, but no results from such testwork are included in this paper).

Experience in preparation of fill material from coal washplant reject had been gained by an earlier program in the Department, reported by Thomas (1985), following on an initial consideration of application of fill in coal mines in Australia presented by Thomas (1976).

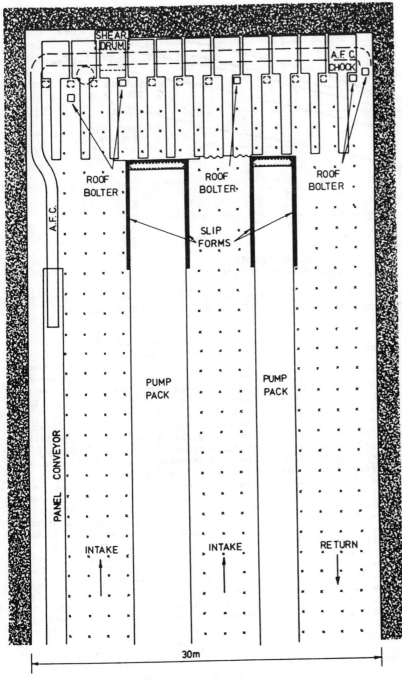

1. Combined coarse and fine reject from West Cliff colliery washplant. Reject was lightly crushed to minus 10 mm, considered to be maximum pumping particle size. All care was taken to ensure a representative sample and to minimise sample deterioration before testing.

2. Type A (ordinary) Portland cement from Blue Circle Southern, Berrima, Australia, initial and final set being respectively 75 and 140 min for the Thomas project and 105 and 180 min for the Wiryanto/Thomas project. Cement addition ranged 5 to 25 mass per cent of mass of (dry) reject plus cement.

3. Darex (brand) calcium chloride, as a 30 mass per cent aqueous solution. This was the only cement accelerator used though obviously scope exists for assessment of alternate techniques for set and curing acceleration. Calcium chloride addition (dry) ranged 3 to 10 mass per cent of Portland cement, well above typical addition levels in concrete practice.

4. Sydney mains water.

Fig. 3 Shortwall entry driveage system with pump packing, driving more than one heading at a time using longwall technology.

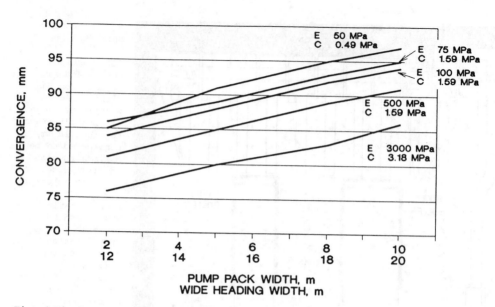

Fig. 4 Final convergence of mid-point of a wide heading as a function of width of pump pack. (C is compressive strength, E is Young's modulus of pump pack material).

## 7 TEST PROGRAM-PROCEDURES AND MEASUREMENTS

1 Batch preparation. Reject at known (natural) moisture content was mixed with Portland cement in a concrete mixer. Calcium chloride in aqueous solution was added, followed by additional mix water carefully controlled to produce a mix consistency reasonably reproducible from mix to mix. The consistency adopted was that of a high-slump concrete, considered suitable for pumping in practice. (In any cemented aggregate mine fill practice, mix consistency, in both laboratory and operation, is a critical parameter.) In all, 40 batches and 260 specimens were prepared, at a specimen replication of two or three.

2. Test specimen preparation. After mixing, batch material was scooped from the mixer into split, plastic curing cylinders 100 mm long and 100 mm diameter. Each specimen was vibrated to remove air pockets and the top surface was scraped level.

3. Test specimen curing. Each specimen was sealed and cured for from 30 min to 28 days, nominally at 20°C.

4. Compressive strength testing. After curing, each specimen was removed from its mould and loaded in uniaxial compression in a 50-tonne capacity, servo-controlled, constant-rate-of-strain Schenck stiff testing machine. The following data were recorded during compression testing.

4.1 Load-deformation curve, extended well into the post-failure range since it is considered that pump packs may well operate in this range in practice. The load-deformation curve allowed calculation of strength (maximum load over original cross-section), pre-failure modulus (tangent), and post-failure modulus (maximum load to 50% maximum load secant). A typical load-deformation curve is included as Fig. 5.

4.2 Specimen lateral deformation, using two different methods, the first, directly using a dial gauge, the second, via circumference increase measured using a Helipot (brand) continuous linear potentiometer. These readings, together with the load-deformation curve, allowed calculation of Poisson's ratio values.

4.3 Specimen sonic velocity, by pulsing a signal horizontally across a specimen diameter at mid-height. Sonic velocity was measured both during curing and compressive strength testing.

4.4 A close-up photograph was taken of each specimen at several significant instants during each test.

5. Each specimen was weighed moist before test and oven-dry after test to allow calculation of density and moisture content.

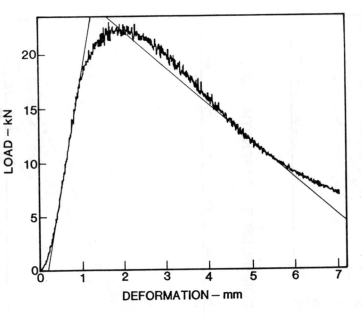

Fig. 5 Typical load-deformation curve, 10 mass % cement, 14
days curing, showing method of calculation of Young's
modulus and post-failure modulus.

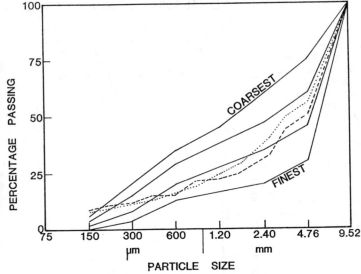

Fig. 6 Particle size analyses of test reject materials
compared with curves recommended for 10 mm concrete
aggregate after McIntosh and Erntroy (1955). (......
Thomas project, ----- Wiryanto/Thomas project).

## 8  TEST PROGRAM-RESULTS AND DISCUSSION

1. Crushed reject particle size
analysis. Particle size analyses for the
Thomas project and the
Wiryanto/Thomas project
are given in Fig. 6,
superimposed upon gradings
for 10 mm concrete
aggregate recommended by
McIntosh and Erntroy
(1955). The reject curves
are seen to sit
comfortably with the
McIntosh and Erntroy
curves, with the possible
exception of an excess of
fines, which could be
accepted in practice as an
aid to pumping or remedied
by lighter crushing of
coarse reject before
pumping.

2. Batch moisture
content at pouring
averaged around 14.6 mass
% on a total mass basis,
ranging from 12.06 to
17.62 mass %, this range
incorporating both
experimental error and
true batch variability
effects.

3. Compressive strength.
Data obtained were far too
voluminous to present in a
paper such as this. Four
plots have been selected.

3.1 From Wiryanto/Thomas
project, at 30 min curing
time and varying calcium
chloride, Fig. 7.

3.2 From Wiryanto/Thomas
project, at 48 h curing
time and varying calcium
chloride, Fig. 8.

3.3 From Thomas project,
at 1 and 2 h curing at 3
mass % calcium chloride,
Fig. 9.

3.4 From Wiryanto/Thomas
project, data superimposed
on a plot prepared from
data tabled by Clark and
Newson (1985) as repre-
senting criteria for pump
pack strengths, Fig. 10.1
(to 24 h curing) and Fig.
10.2 (to 3 h curing). From
Fig. 10.1 it can be con-
cluded that for curing
times of say 9 h and beyond
the Clark and Newson
criteria can be met with 10
mass % cement addition but
Fig. 10.2 indicates 20 mass
% cement is required to
meet these criteria in the

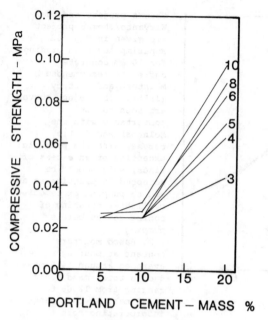

Fig. 7 Relationship between unconfined compressive strength and cement content at 30 min curing and calcium chloride ranging 3 to 10 mass %, as indicated up right hand side of figure.

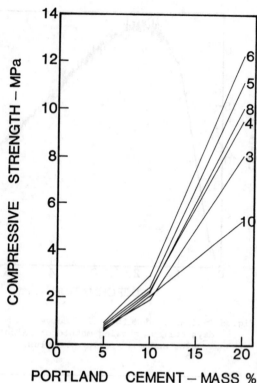

Fig. 8 Relationship between unconfined compressive strength and cement content at 48 h curing and calcium chloride ranging 3 to 10 mass %, as indicated up right hand side of figure.

short-term curing range. Such a cement addition is not tolerable economically and obviously another solution must be sought.

The load-deformation curve in Fig. 5 shows considerable post-failure strength, of significant potential importance in practice. This is especially so when it is considered that test specimen height to diameter ratio of one equates to the maximum ratio of seam thickness (2 m) to pump pack width (2m, Fig. 4) likely to be encountered in practice. It is obvious that a 4 m wide pump pack in a 2 m thick seam will be considerably stronger per unit area than a 2 m wide pump pack in the same seam thickness. Further, load-deformation curves demonstrate considerable residual strengths at specimen deformations equivalent to well in excess of the 100 mm roof to floor convergence mentioned earlier as being still acceptable for roof integrity at West Cliff Colliery. The observation that the cemented coal reject packs can (and in practice probably will) operate in the post-failure state is considered to be one of the significant findings of the current study. The fact that curing is proceeding as loading occurs is also significant.

Figs. 7 and 8 show that, although

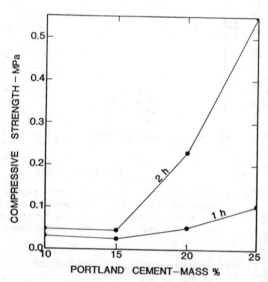

Fig. 9 Relationship between unconfined compressive strength and cement content at 3 mass % calcium chloride

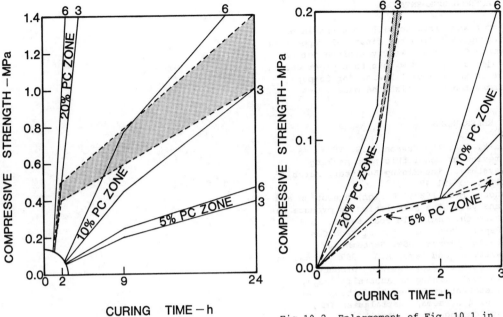

Fig 10.1  Strength test data (at 3 and 6 mass % calcium chloride) compared with packing strength criteria (shaded region) proposed by Clark and Newson (1985).

Fig 10.2  Enlargement of Fig. 10.1 in the short curing time range. (Clark and Newson criteria region is again shaded).

calcium chloride is obviously beneficial as a set and curing accelerator at low curing times and high cement addition levels, its benefits are not so obvious at lower cement addition levels and it could be deleterious at extended curing times. (Project specimen replication becomes inadequate when explanations of detailed parameter interactions are attempted).

These observations are largely in accord with Neville (1973) who has written that

1. the normal process of cement hydration is not changed with calcium chloride, and

2. calcium chloride is generally more effective in increasing the early strength of rich mixes with a low water/cement ratio than of lean ones.

4. Specimen lateral deformation at the time loading was ceased ranged from a low of about 4 mm diameter increase to a high of about 20 mm diameter increase, increasing with decreasing cement content. Failure was mainly by shear, with weaker specimens exhibiting barrelling as well. At the extreme specimen deformation, both longitudinal and lateral, when loading was ceased, massive gaps had developed within the specimens.

5. Specimen longitudinal deformation at the time loading was ceased ranged from a

low of about 2 mm to a high of about 20 mm, increasing with decreasing cement content.

6. Specimen sonic velocity measured during compression testing could be related quite closely with the load-deformation curve. Measured sonic velocity was typically constant up to maximum load and fell as load reduced in the post-failure zone. As failure progressed, a situation was ultimately reached where massive gaps in the specimen prevented the passage of the sonic pulse across the specimen from transmitter to receiver. This would typically occur in the range 60 to 75% of maximum load. Despite such massive gaps, the specimen would continue to support a significant proportion of the maximum load.

Specimen sonic velocity was also measured during curing for the longer curing times and was found to increase slightly as curing proceeded. Maximum sonic velocity measured was 2500 m/sec.

305

## 9 ACKNOWLEDGEMENTS

Thomas and Wiryanto wish to acknowledge with appreciation the financial support of Kembla Coal and Coke Pty Limited for the projects reported within this paper and all three authors likewise for Company permission to publish the paper.

## 10 REFERENCES

Clark, C.A. and Newson, S.R. 1985. A review of monolithic pump packing systems. The Mining Engineer, March 1985, p 491-495.

Hanson, M.E. 1987. Design, execution and analysis of a stimulation to produce gas from thin multiple coal seams. SPE Paper 16860, 62nd Annual Tech Conf, Dallas, Texas, USA, September.

Hewitt, A. and Lama, R.D. 1987. Research on dust control for high production longwall mining in Australia. Symp. 21st Century Higher Production Coal Mining Systems, Australasian IMM, Wollongong, Aust., April.

Lama, R.D. 1988. Developments in face technology and design of underground coal mines. Int. Symp. on Modern Mining Technology, Taian, Shanding, PRC, October.

McIntosh, J.D. and Erntroy, H.C. 1955. The workability of concrete mixes with 3/8 in aggregates. Research Report No. 2, Cement and Concrete Association, London.

Neville, A.M. 1973. Properties of Concrete, London, Pitman.

Thomas, E.G. 1976. Thick seam underground coal mining by a slice and fill method. Symp. Thick Seam Mining by Underground Methods, Australasian IMM, Rockhamption, Aus., September.

Thomas, E.G. 1985. Preparation of mine fill material from coal washplant reject. Mining Science and Technology, 3 (1985) 1-9, Elsevier, Amsterdam.

Thomas, E.G. 1986. Laboratory strength testing of West Cliff washplant reject. Report to Kembla Coal and Coke Pty. Limited, January.

*Innovations in Mining Backfill Technology, Hassani et al. (eds), © 1989 Balkema, Rotterdam. ISBN 90 6191 985 1*

# The flow characteristics through hydraulic filling materials produced from coal washery rejects

G.Şenyur
*Hacettepe University, Ankara, Turkey*

ABSTRACT: This paper is concerned with the water flow characteristics through backfill materials of coal washery rejects with respect to systematic variation in their particle-size distribution. A special permeability test system has been developed. The followings have been inspected: the relationship between pressure gradient (dP/dL) and flow velocity (v), the true permeabilities ($K_0$) and the relationship between Fanning friction factor ($f_k$) and Reynolds number ($R_k$) by using the square root of permeability as characteristic length. An equation which expresses the true permeability ($K_0$) in terms of particle-size distribution parameters has been developed. It has been observed that, the non-linear or transition flow, in which laminar flow dominates, exits through materials which have true permeabilities ($K_0$) greater than $K_0 = 1 \times 10^{-9}$ m$^2$ and complete laminar flow governed by Darcy's law exits through materials which have true permeabilities ($K_0$) smaller than $K_0 = 1 \times 10^{-9}$ m$^2$.

## 1 INTRODUCTION

The water flow through hydraulic backfill in mine stopes have been evaluated as complete laminar flow as described by Darcy's law. However, the mining industry uses filling materials of different origin and composition. The essential factor controlling the drainage properties is the particle-size distribution of material. Even if a single type of material is used variations in granulometry display extensive variation of water flow. A set of experiments was programmed with coal washery rejects which are the most available material in collieries. The samples were taken from coal washery rejects of the Zonguldak colliery/Turkey, and test samples of different granulumetry were prepared by succesive operations of sieving resulting in six size batches mixed in different proportions. The study is supposed to give an idea about the permeability and flow rate depending on the granulumetry and to inspect the water flow from the view point of theory. Ward (1964) reported that the use of the square root of permeability as characteristics length showed a possible friction factor and Reynolds number relationship common to different types of porous media. This relationship at the hydraulic gradient of approximately one, which is the general

case in mining operations, will display the flow regime i.e., laminar flow, transient flow.

## 2 THEORY

Muskat (in Ward 1964) suggested that turbulent flow in porous media could be represented by

$$\frac{dP}{dL} = av + bv^2 \qquad (1)$$

where dP/dL stands for the pressure drop per unit length, v represents the macroscopic velocity, and a and b are constants of the fluid and the porous medium.

Dimensionally, the equation for both laminar and turbulent flow in porous media should be

$$\frac{dP}{dL} = \phi \ (v, \ K, \ \rho, \ \mu) \qquad (2)$$

where $\phi$ is an unknown function, K symbolizes the permeability of the porous medium, $\rho$ denotes the mass density of the fluid, and $\mu$ is equal to the absolute viscosity of the fluid. Designating dimensional relationships by square brackets (Şenyur 1986), then

$$\left| \frac{dP}{dL} \right| = \left| v^a \, K^b \, \rho^c \, \mu^d \right| \qquad (3)$$

Introducing the fundamental units of mass, M, length L, and time, T of the various parameters into equation 3 yields

$$\left| ML^{-2} \, T^{-2} \right| = \left| M^{c+d} \, L^{a+2b-3c-d} \, T^{-a-d} \right| \qquad (4)$$

and solving for b, c and d in terms of a yields.

$$\frac{dP}{dL} = \phi \left( v^a \, K^{\frac{a-3}{2}} \, \rho^{a-1} \, \mu^{2-a} \right) \qquad (5)$$

$$\frac{dP}{dL} = \sum_{a=1}^{a=2} c_a \, v^a \, K^{\frac{a-3}{2}} \, \rho^{a-1} \, \mu^{2-a}$$

$$= c_1 \frac{\mu v}{K} + c_2 \frac{\rho}{K^{1/2}} \, v^2 \qquad (6)$$

in which $c_a$, $c_1$ and $c_2$ are dimensionless constants of proportionolity.

Darcy's law governs laminar flow in porous media and is (Hubbert 1940),

$$\frac{dP}{dL} = \frac{\mu v}{K} \qquad (7)$$

At low velocities, equation 6 should be equal to equation 7, in that the second term ($c_2 \, \rho \, v^2 / K^{1/2}$) becomes numerically insignificant compared with the first term ($c_1 \, \mu \, v/K$) and therefore the value of $c_1$ is one. For the sake of simplicity, let $c_2 = c$, and equation 6 may be written in a simpler form, as follows:

$$\frac{dP}{dL} = \frac{\mu \, v}{K} + \frac{c \, \rho \, v^2}{K^{1/2}} \qquad (8)$$

where c is a dimensionless constant. At high velocities, equation 8 can be approximated by,

$$\frac{dP}{dL} = \frac{c \, \rho \, v^2}{K^{1/2}} \qquad (9)$$

Some investigators e.g., Arbhabhirama and Antonid (1973), Harleman et al. (1963) have used the square root of permeability as characteristic length in dimensionless Reynolds number which defines the flow regime in porous media. The permeability Reynolds number is,

$$R_k = \frac{v \, K^{1/2} \, \rho}{\mu} \qquad (10$$

Similar to the resistance to flow in pipe Fanning friction factor is given to define the friction loss in porous media (De Wiest 1969). The dimensionless Fanning friction factor, $f_k$, is,

$$f_k = \frac{dP}{dL} \cdot \frac{K}{v^2 \rho} \qquad (11$$

Therefore, equation 7 (Darcy's law) can be written in the form

$$f_k = \frac{1}{R_k} \qquad (12$$

and equation 8 can be written as

$$f_k = \frac{1}{R_k} + c \qquad (13$$

and equation 9 can be written as

$$f_k = c \qquad (14$$

According to the above discussions, the plot of $f_k$ values versus $R_k$ values obtained in experiments, will display a relationship expressed by equation 12 in laminar flow and a relationship expressed by equation 13 in transition flow and a relationship expressed by equation 14 in turbulent flow.

## 3 TRUE PERMEABILITIES

Four runs at four succesive increasing head were made on each sample. The permeabilitie K, were calculated by means of equation 18, which is the modification of equation 7. The permeability of porous media, i.e., fill sample, was determined graphically from a plot of 1/K (as determined by equation 18) versus $v\rho/\mu$. This is a straight line plot, and the reciprocal of the value of 1/K, in which the straight line intersects the 1/K axis at $v\rho/\mu = 0$, is the true permeability $K_o$ of the porous media. This straight-line relationship can be expressed as

$$\frac{1}{K} = \frac{1}{K_o} + E \frac{v\rho}{\mu} \qquad (15)$$

or

$$\frac{dP}{dL} \cdot \frac{1}{\mu v} = \frac{1}{K_o} + E \frac{v\rho}{\mu} \qquad (16)$$

and

$$\frac{dP}{dL} = \frac{1}{K} \mu v + E \rho v^2 \qquad (17)$$

Equation 17 is identical to equation 8 when $E = C/K_o^{1/2}$. In laminar regime E will be zero (c = 0), therefore, $v\rho/\mu - 1/K$ line will be parallel to the abscissa axis. (1/K = 1/K_o)

## 4 ENGINEERING PROPERTIES OF THE MATERIALS

Refuse from washing plants is roughly a mixture of siltstone and sandstone grains. Siltstone is dominant in the mixture. The shape of the grains was longish to flat

with different degrees of roundness. The length: breadth ratios were in the order of magnitude 1.3 to 1.7 and breadth: thickness ratios were in the order of magnitude 1.3 to 4.0. The specific gravity of grains was 2.67 on average and water absorption was 2 percent on average. As stated before, the test samples differed to their particle - size distributions. The parameters used to define the granulumetry are the maximum particle-size ($d_{max}$), minimum particle-size ($d_{min}$), effective size ($d_{10}$) (which is the ten percent passing size in granulumetric distribution), coefficient of degree $C_c$ ($d_{30}^2/d_{60} \cdot d_{10}$) and coefficient of uniformity $C_u$ ($d_{60}/d_{10}$) where $d_{60}$ and $d_{30}$ are sixty percent and thirty percent passing sizes respectively (Craig 1976). The bulk porosity, n, which is defined as the ratio of total void volume among the particles to total volume of bulk sample was also determined for each test sample. The samples are enumerated and the corresponding properties of each sample are given (in Table 1).

Table 1. Properties of test samples

| Sample No | Maximum particle size, $d_{max}$ (mm) | 80 % passing size, $d_{80}$ (mm) | Coefficient of uniformity $C_u$ | Coefficient of degree $C_c$ | Effective size, $d_{10}$ (mm) | Minimum particle size, $d_{min}$ (mm) | Porosity n |
|---|---|---|---|---|---|---|---|
| 1 | 13.2 | 6.52 | 3.04 | 0.71 | 0.197 | 0.15 | 0.427 |
| 2 | 19.0 | 9.5 | 17.7 | 0.226 | 0.22 | 0.15 | 0.378 |
| 3 | 19.0 | 11.05 | 14.25 | 1.23 | 0.36 | 0.15 | 0.374 |
| 4 | 19.0 | 10.44 | 13.06 | 2.37 | 0.5 | 0.15 | 0.43 |
| 5 | 19.0 | 8.25 | 3.9 | 0.77 | 0.96 | 0.5 | 0.489 |
| 6 | 19.0 | 9.09 | 5.02 | 0.757 | 0.80 | 0.5 | 0.439 |
| 7 | 19.0 | 15.76 | 12.33 | 0.9 | 0.97 | 0.5 | 0.429 |
| 8 | 13.2 | 8.59 | 7.2 | 1.033 | 0.87 | 0.5 | 0.451 |
| 9 | 19.0 | 14.25 | 6.0 | 1.9 | 1.34 | 0.5 | 0.46 |
| 10 | 19.0 | 12.13 | 8.68 | 1.3 | 1.13 | 0.5 | 0.448 |
| 11 | 19.0 | 10.0 | 4.53 | 1.11 | 1.52 | 0.5 | 0.462 |
| 12 | 19.0 | 12.23 | 4.38 | 1.26 | 2.35 | 0.5 | 0.476 |
| 13 | 19.0 | 12.33 | 3.43 | 1.38 | 3.23 | 0.5 | 0.485 |
| 14 | 19.0 | 15.46 | 3.0 | 1.09 | 4.08 | 3.36 | 0.483 |
| 15 | 19.0 | 12.2 | 3.27 | 1.47 | 3.2 | 0.5 | 0.484 |
| 16 | 19.0 | 15.43 | 5.05 | 0.9 | 2.26 | 0.5 | 0.462 |
| 17 | 19.0 | 16.0 | 2.9 | 0.93 | 4.33 | 3.36 | 0.494 |
| 18 | 19.0 | 12.23 | 2.04 | 1.0 | 6.1 | 3.36 | 0.498 |
| 19 | 19.0 | 16.1 | 1.73 | 0.9 | 7.63 | 6.7 | 0.513 |

Permeability for each batch was measured using the constant-head method. The essential elements are that full saturation is obtained, a steady flow is reached with a constant head of water and the fluid temperature and porosity are recorded. The apparatus used and succesive heads, H, are shown in Figure 1. The diameter, D, of the cylindrical test column was designed according to the maximum particle size, $d_{max}$, of the samples ($D > 10\ d_{max}$) (Şenyur 1985). The length of the sample columns L, was between 0.35-0.40 m. The experimental arrengament and procedure were carried out in parallel with the recommendations of Wayment and Nicholson (1964). The percolation rates v (m sec$^{-1}$) at four water heads ($H_1 = 0.4$ m, $H_2 = 0.6$ m, $H_3 = 0.8$ m, and $H_4 = 1.0$ m.) were measured. The head was raised to the next level by closing the lower open drain. The relationship between pressure gradients, dP/dL (Newton/m$^2$/m), and percolation rates, v (m sec$^{-1}$), were inspected. The permeabilities, K, at four succesive heads, H, were also found and the true permeabilities $K_o$ of the porous media i.e., test samples were found by using the graphical method explained before. The permeabilities K were found according to the following equation

$$K = \frac{Q \cdot L \cdot \mu}{H \cdot A \cdot \rho^2 \cdot g} \qquad (18)$$

where, K is the permeability (m$^2$), Q is the mass rate of flow of fluid through a porous medium (kg sec$^{-1}$), L is the length of the porous medium in the direction of flow (m) H is the water head (m), A is the cross-sectional area of the porous medium, normal to the direction of flow (m$^2$), $\mu$ is the dynamic viscosity of fluid flowing (Pa. sec), $\rho$ is the density of fluid flowing (kg m$^{-3}$) and g is the gravitational acceleration (m sec$^{-2}$).

## 6 RESULTS AND DISCUSSION

The relationships between 1/K and $v\rho/\mu$ where K is the permeability and v is the percolation rate, $\rho$ and $\mu$ are the mass density and dynamic viscosity of water are shown in Figures 2 and 3. The permeability K, and percolation rate, v values have been obtained at four succesive heads. The numbers in the Figures refer to the test samples in Table 1. Reffering to equations 15 and 17, Figure 2 displays that the flow through the samples numbered 1 to 10 is completely laminar (E = 0).

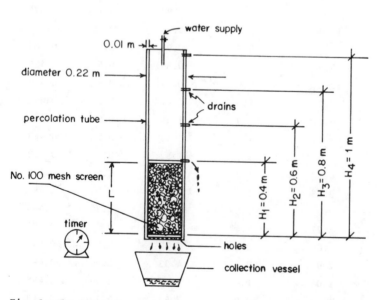

Fig. 1 The laboratory set-up

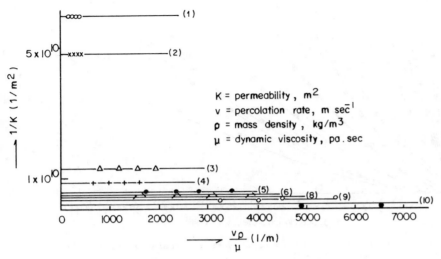

Fig. 2 Graphical determination of the permeability (the numbers refer to the test samples)

Figure 3 shows that the lines demostrating the relationships for samples 11 to 19 are sloped which means that the coefficient E in equation 15 is getting values. In other words, the relationship given by equation 17 is completely applicable. Therefore, the water flow through samples 11 to 19 is not wholly laminar, but, transient which is described as laminar flow mixed with turbulent flow. Figure 4 illustrates the relationship between pressure gradient (dP/dL) and percolation rate, v. The flow through the test samples 1 to 10 was being governed by Darcy's law. However, the parabolic relationships for the other test samples illustrate deviations from Darcy's law.

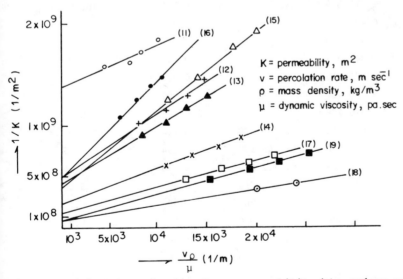

Fig. 3 Graphical determination of true permeability (the numbers refer to the test samples)

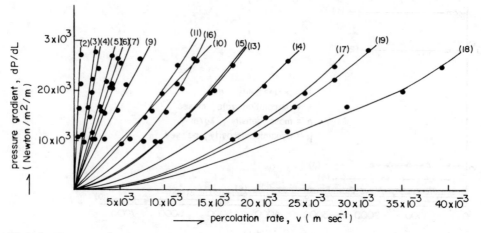

Fig. 4 The graphics of pressure gradient, dP/dL versus percolation rate, v.
(the numbers refer to the test samples)

Table 2. Permeability test results

| Sample No | True permeability $K_o$ ($m^2$) | Pressure gradient, dP/dL (Newton/$m^2$/m)-percolation rate, v(m/sec) function $dP/dL = r\ v + w\ v^2$ r | w | Hydraulic gradient dH/dL $\simeq$ 1 dH: hydraulic head dL: sample length Reynolds number $R_k$ | Fanning friction factor $f_k$ |
|---|---|---|---|---|---|
| 1 | $1.6\ \times 10^{-11}$ | 81 180 017 | – | $4 \times 10^{-4}$ | 2554.0 |
| 2 | $1.97 \times 10^{-11}$ | 59 090 910 | – | $6.44 \times 10^{-4}$ | 1601.0 |
| 3 | $7.64 \times 10^{-11}$ | 14 928 298 | – | 0.007 | 143.0 |
| 4 | $1.17 \times 10^{-10}$ | 11 418 309 | – | $9.7 \times 10^{-3}$ | 106.0 |
| 5 | $1.76 \times 10^{-10}$ | 6 488 476 | – | 0.023 | 43.0 |
| 6 | $2.18 \times 10^{-10}$ | 5 636 596 | – | 0.02453 | 41.0 |
| 7 | $2.24 \times 10^{-10}$ | 5 347 347 | – | 0.045 | 23.0 |
| 8 | $2.51 \times 10^{-10}$ | 5 174 415 | – | 0.024 | 41.2 |
| 9 | $3.4\ \times 10^{-10}$ | 3 620 457 | – | 0.061 | 16.1 |
| 10 | $6.37 \times 10^{-10}$ | 1 976 778 | – | 0.12 | 8.3 |
| 11 | $7.17 \times 10^{-10}$ | 1 633 651 | 42 618 339 | 0.114 | 15.5 |
| 12 | $1.92 \times 10^{-9}$ | 577 200 | 61 745 193 | 0.354 | 5.57 |
| 13 | $2.28 \times 10^{-9}$ | 499 089 | 56 423 520 | 0.398 | 5.44 |
| 14 | $4.29 \times 10^{-9}$ | 303 392 | 35 248 716 | 0.69 | 3.69 |
| 15 | $2.5\ \times 10^{-9}$ | 456 000 | 59 940 000 | 0.39 | 6.17 |
| 16 | $2.0\ \times 10^{-9}$ | 632 500 | 99 970 000 | 0.265 | 8.2 |
| 17 | $6.67 \times 10^{-9}$ | 190 500 | 25 000 000 | 1.06 | 3.0 |
| 18 | $1.25 \times 10^{-8}$ | 96 000 | 12 987 000 | 2.26 | 1.91 |
| 19 | $1.33 \times 10^{-8}$ | 10 252 | 41 970 009 | 1.78 | 3.55 |

Table 2. lists the results of tests. The obtained true permeabilities, $K_o$, the parameters of the flow equations of the graphs in Figure 4, the dimensionless Reynolds numbers, $R_k$, and Fanning friction factors, $f_k$, at a hydraulic gradient of one are given for each test sample. Table 2 and Figure 4 show that a non-linear transient flow regime slowly begins with sample 11, while the true permeability $K_o$ is approaching the value of $K = 1 \times 10^{-9}$ m$^2$. This leads to the generalization that whereas the flow regime through materials having true permeability values below $K_o = 1 \times 10^{-9}$ m$^2$ is laminar and governed by Darcy's law, the flow regime through those having true permeability values greater than $K_o = 1 \times 10^{-9}$ m$^2$ is non-linear.

The true permeability values, $K_o$, have been correlated with the granulumetric parameters given in Table 1. The following relationship has been obtained (coeff. of corr., r = 0.97)

$$K_o = 3.348 \times 10^{-11} (C_c)^{-0.01} (n)^{-2.83} (d_{10})^{2.0}$$

$$(19)$$

The parameters $C_c$, n and $d_{10}$ have been explained before.

The plot of the friction factory, $f_k$, versus the permeability Reynolds number, $R_k$, is seen in Figure 5. The values of $f_k$ and $R_k$ have been calculated for the hydraulic gradient of one, wich is the general appearance in mines. It is obvious from Figure 5 and equation 8 that there is no sharp division between laminar, transition and turbulent flow in porous media; a smooth transition is expected. In Figure 5 the points deviating from the laminar flow regime are located at the initial stage of transition. Therefore, these points represent a flow regime where the flow is laminar in most parts of the porous medium, but there are some few parts where the flow is turbulent.

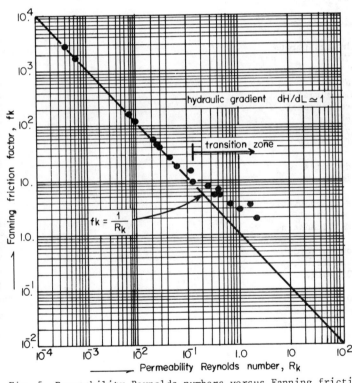

Fig. 5 Permeability Reynolds numbers versus Fanning friction factors

# 7 CONCLUSIONS

The square root of permeability can well represent the characteristic length of the pore-spaces.

For washery rejects, the permeability value $K_o$ (m$^2$) is correlated with the particle-size distribution parameters by the following relationship:

$$K_o = 3.348 \times 10^{-11} (C_c)^{-0.01} (n)^{-2.83} (d_{10})^{2.0}$$

where, $C_c$ is the coefficient of degree, $d_{10}$ (mm) is the effective size and n is the bulk porosity.

The flow through materials having permeability values greater than $K_o = 1 \times 10^{-9}$ m$^2$ deviates from a complete laminar regime with small turbulance.

## REFERENCES

Arbhabhırama, A. and Antonid, A. 1973. Friction factor and Reynolds number in porous media flow. Journal of the Hydraulics Division, Proceedings of the ASCE, Vol. 93, 9784: 901-911

Craig, E.F. 1976. Soil mechanics. Mc Graw-Hill, New York.

De Wiest, R.J.M. 1969. Flow through porous media. Academic Press, New York.

Harleman, D.R.F., Mehlhorn, P.F. and Rumer, R.R. 1963. Dispersion-permeability correlation in porous media. Journal of the Hydraulic Division, Proceedings of the ASCE, Vol. 89, 3459: 67-84.

Hubbert, K.G. 1940. The theory of ground-water motion. The Journal of Geology, Volume XLVIII, 5: 901-911.

Şenyur, G. 1986. Statics and dynamics. Lecture notes. Hacettepe University, Ankara, Turkey.

Şenyur, G. 1985. The behaviour of pneumatic filling materials in one-dimensional compression. Ph.D. thesis, METU, Ankara, Turkey.

Ward, J.C. 1964. Turbulent flow in porous media. Journal of Hydraulic Division, Proceedings of ASCE, Vol. 90, 4019: 11-12.

Wayment, W.R. and Nicholson, D.E. 1964. A proposed modified percolation rate test for use in physical property testing of mine backfill. U.S.B.M. Report of Investigations 6552.

*Innovations in Mining Backfill Technology, Hassani et al. (eds), © 1989 Balkema, Rotterdam. ISBN 90 6191 985 1*

# Backfilling at IMC Canada K-2 potash mine

L.M.Kaskiw & R.M.Morgan
*IMC Canada, Esterhazy, Saskatchewan, Canada*

D.C.Ruse
*Cavern Engineering, Regina, Saskatchewan, Canada*

ABSTRACT: To stabilize inflow areas at the K-2 minesite, it was decided to backfill "troubled" areas with tailings salt. Since the placement of tailings in potash mines at depths of 3200 feet is a unique process, various handling and placement techniques were attempted to achieve a system where salt tailings could be placed underground in the areas of need in the most cost effective and efficient manner. The backfill well was spudded November 28, 1986 and the main casing was cemented on February 9, 1987. Drilling was suspended from December 19, 1986 to January 11, 1987 as the drilling rig was required elsewhere to control an increase in the water inflow. Tailings backfill was first placed underground in March, 1987 and is continuing with no apparent problems in either handling or placement techniques.

## 1 INTRODUCTION

The effect of flowing brine along room pillars resulted in the dissolution of the pillars and a further weakening of the pillars. The weakened pillars allowed greater mine closures resulting in the redistribution of stress that further allowed the development of fractures in the Dawson Bay and Prairie Evaporites. The purpose of the tailings backfill was to provide support to inflow areas retarding closure and subsequent crack development of competent rock providing support.

The major concern for the drilling program was to drill the hole as close to vertical as practical to decrease the wear of any injection string carrying the tailings. Also, consideration had to be given to ensure that all water bearing formations were isolated before mining to the well could commence.

## 2 K-2 BRINE INFLOW PROBLEM

Brine was discovered flowing into the K-2 mine on December 29, 1985. Initially the inflow volume was not determined and it took 10 months to find the inflow point. Extensive remining had to be done in very unstable ground to reach the inflow location (Figure 1). The brine flowing from the Dawson Bay formation into the

mine was saturated with NaCl, but had only low values of K and Mg. This caused large amounts of sylvite and carnallite to be dissolved in the mining horizon (Table 1).

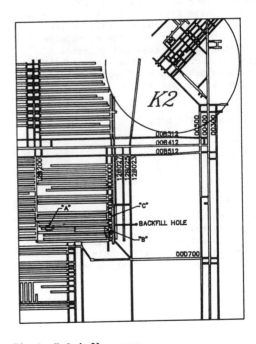

Fig.1 K-2 inflow area

Table 1. Inflow brine chemical analysis.

| | Dawson Bay Brine | Saturated Mine Brine |
|---|---|---|
| K | 0.49 | 5.57 |
| Na | 9.78 | 7.21 |
| Ca | .20 | .21 |
| Mg | 0.076 | .36 |
| % Cl | 15.96 | 17.46 |

Together with the remining, it created open areas far in excess of normal mining rooms, which in turn created very unstable ground conditions. Our microseismic system picked up numerous seismic events in the general inflow area.

Closure measurements as well as visual observations indicated heavy ground movement. It was therefore decided quite early to try backfilling some of the mined and washed out areas to provide a reasonable amount of ground support again.

Attempts to stop the brine inflow by injecting a concentrated $CaCl_2$ solution into the Dawson Bay formation in October, 1986 showed up initially as very effective and brine flow was reduced from an estimated 8,000 gallons per minute to about 1,000 gallons per minute within 10 weeks. Then on December 18, 1986, another high flow of brine into the mine occurred about 2,200 ft. east of the initial inflow point. Again, calcium chloride was injected and caused a gradual decrease in flow. At present, we estimate our inflow at approximately 1,400 gallons per minute.

# 3 K-2 BACKFILL WELL

## 3.1 Location

### 3.1.1 Surface

The backfill well is located approximately one mile south of the K-2 minesite (Figure 2). The surface location of the backfill hole was dictated by the water inflow area.

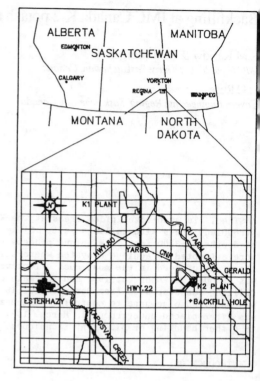

Fig.2  Surface location of backfill well

### 3.1.2 Underground

The backfill well was located 100 feet into the pillar as an extension to 08B052. The location underground was chosen so that backfilling could easily be accomplished underground in 10B052, the water inflow area (Figure 1). A 100 foot pillar was chosen to surround the backfill hole as this would provide a suitable margin of safety.

The first step was to tie in the surface surveying system with the underground system to make sure the final location underground could be determined accurately.

## 3.2 Geology

### 3.2.1 General

The IMC Canada K-2 minesite lies in the northwestern flank of the Williston Basin. Regional dips are general to the southwest towards the centre of the Williston Basin.

The general stratigraphic sequence is shown below and also in Figure 3.

Table 2. General stratigraphic sequence.

| Formation | Subsea Elevation | Log Depth Below K.B. |
|---|---|---|
| Second White Specks | 265.65 M | 245.0 |
| Blairmore | 116.85 M | 393.8 |
| Lodgepole | 56.65 M | 454.0 |
| Bakken | - 17.25 M | 527.9 |
| Three Forks | - 27.35 M | 538.0 |
| Nisku | - 77.05 M | 587.7 |
| Duperow | - 106.50 M | 616.7 |
| Souris River | - 278.35 M | 789.0 |
| First Red Bed | - 376.75 M | 887.4 |
| Dawson Bay | - 388.85 M | 899.5 |
| Second Red Bed | - 432.55 M | 943.2 |
| Prairie Evaporites | - 440.35 M | 951.0 |

K.B. = 510.65 M  Ground Level = 506.45 M

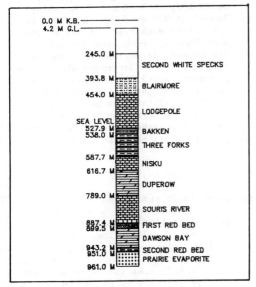

Fig.3   General stratigraphic sequence

### 3.2.2 Dawson Bay Formation

The most extensive published study on the Dawson Bay Formation which includes the IMC Canada property was compiled by Lane (1959).  Overlain by the First Red Beds of the Souris River Formation, and underlain by the potash-bearing Prairie Evaporite Formation, the Dawson Bay Formation is composed mainly of highly fractured limestones and dolomites.  The Dawson Bay Formation has been grouped into six distinct members deposited in Middle Devonian time as described by Lane.  A generalized stratigraphic sequence is shown in Figure 4.  At the

backfill well location, the Dawson Bay is 43.7 M thick.

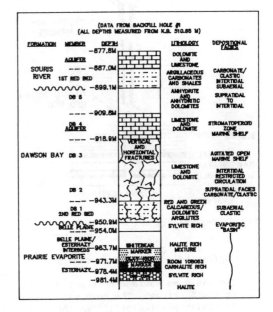

Fig.4   Dawson Bay geology

## 4 DRILLING METHOD

### 4.1 Type of equipment

The drilling rig was one of the small standard double rigs, Simmons Drilling #27, used to drill observation wells, grout wells and disposal wells.  This rig has 450 Kw of power which is more than adequate.  To drill the larger than normal surface (444.5 mm) and intermediate (311 mm) holes additional 229 and 178 mm drill collars were rented and an extra mud pump was rigged up to get the necessary annular velocities. Permission to drill with only a bag type annular blow out preventer rather than the bag type plus mechanical pipe and blind rams was granted by Saskatchewan Energy and Mines.  A full B.O.P. stack for these hole sizes would not fit under the rig substructure.

### 4.2 Well design

The well design was controlled by the need to have the Blairmore formation isolated from the Dawson Bay formation and both isolated from the mine workings by cementing casing into place (Figure 5).

317

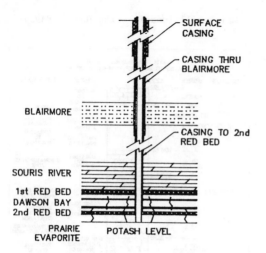

BLAIRMORE

SOURIS RIVER

1st RED BED
DAWSON BAY
2nd RED BED

PRAIRIE
EVAPORITE        POTASH LEVEL

Fig.5  Well design

a)  Surface hole and surface casing

The 444.5 mm surface hole was drilled
to 144.5 metres.  The hole was
initially piloted at 222 mm then 311
mm and then reamed out to attempt to
stay as straight as possible.
Conductor pipe had been set before
moving the rig in.  The 339.7 mm,
71.5 Kg/m H.S., S.T. & C. casing with
a guide shoe and float collar was run
to 144.5 metres and cemented with 20
tonnes Class G cement with 3% $CaCl_2$
and good cement returns were obtained
at surface.

b)  Intermediate hole and intermediate
casing

The intermediate hole was piloted
with 222 mm bits to 605 metres and
then reamed out to 311 metres to keep
the hole as straight as possible, the
concern being that any serious
deviation or change in deviation
would result in excessive wear on the
injection casing when delivering
backfill to the mine.  Mud motor runs
were made as required to correct any
tendency for the well to wander.
With the added pumping capacity, the
hole cleaned well.  The 244.5 mm,
53.62 Kg/m, J-379 casing with float
shoe and float collar was run to
601.0 m K.B. and cemented with 30
tonnes Class G cement mixed with
fresh water.  The intermediate casing
is just into the Duperow formation
which effectively isolates the
Blairmore formation from the mine.

c)  Main hole and long string casing

The main hole is 222 mm from
intermediate casing to 981.8 metres.
For this as well as intermediate
hole, clear flocculated saturated
salt water was used as drilling mud.
This system works very well for
drilling in the Esterhazy area.

The 177.8 mm, 34.26 Kg/m, S.T. & C.,
J-379 casing with a float shoe, float
collar and twelve centralizers was
landed at 981.6 metres and cemented
with 25.5 tonnes Class G cement with
34% NaCl by weight of mix water to
assure salt saturation.

d)  Injection casing

The injection casing is 139.7 mm,
34.23 Kg/m, L.T. & C., N-80 casing
landed at 980 metres with a full bore
L.H. style packer set at 940 metres.
When landing the casing and setting
the packer, directions were received
by phone from underground where mine
personnel could see the end of the
casing.  The annulus between the
177.8 mm and 139.7 mm casing is
filled with brine and kept at a
moderate pressure to monitor for any
injection casing leaks.

4.3 Directional control

a)  Directional tendencies

The well tended to head off in a
northwesterly direction, even on
surface hole.  When drilling, the bit
would tend to turn into the formation
so this indicates the formations are
dipping slightly to the southeast at
the backfill hole location.  By
drilling a 222 mm pilot hole, reaming
to 311 mm and then to 444.5 mm the
maximum inclination was .35° from
vertical at 111 metres and .3° at
140 metres.  All other surveys were
between .08 and .19°.

b)  Use of mud motors

There are two basic styles of mud
motor.  The first was the Turbo Drill
which is just like a turbine.  The
second is a positive displacement
type of motor very similar to a Moyno
pump.  With both systems the drill
string is not rotated, instead the
drilling fluid powers the motor which
turns the bit.  Typically, a 1/2 to

3° bent sub is run above the motor with a monel drill collar above this. There are magnets put in the monel drill collar whose position relative to the "kick" sub are known. A survey instrument is run into the monel collar that reads the magnet's position relative to the angle and bearing of the hole. This permits the mud motor to be oriented and the hole deviated or corrected to the required direction and angle. At Esterhazy, the positive displacement motor was used.

On the intermediate pilot hole the survey at 300 metres was .41° N 54.76 W. The mud motor was run at 330 metres and the hole turned back to .57° S 28.01° E by 406 metres. Normal drilling was then done to 511 metres where the hole had turned to the west with the survey at 499 metres reading .43° S 57.35 W. The mud motor was then rerun and by the survey at 574 metres was .43° N 54.02 E. The well was then drilled with normal rotary drilling to 605 metres. The 222 mm hole was then reamed out to 311 mm.

In the main hole, no corrections were made with the mud motor. At 935 metres the hole was .65 metres N and 1.88 metres E of the surface location.

Cores were cut from 900 metres to 945.2 metres (Dawson Bay) at which time the drilling was interrupted and the well used for injection of $CaCl_2$ brine to assist in controlling an increase in the water inflow.

## 4.4 Testing and completion method

### 4.4.1 Objectives

The first objective was to reduce the permeability in the well bore through the Dawson Bay formation so that had the hole encountered a fracture in the Prairie Evaporite, the resulting flow into the mine would be zero.

The second objective was to end up with a cased hole, open into the mine that was isolated from the Dawson Bay formation and to carry out various tests to confirm isolation before opening the well to the mine.

### 4.4.2 Methods

a) Open hole logging and coring

The standard set of logs were run: Compensated Density Neutron Gamma Ray Caliper; Compensated Sonic Gamma Ray Caliper; and a Dual Induction Laterlog. The logs indicated a normal section for the area. An abnormally high porous streak (.5 metre) is present four metres above the top of the Prairie Evaporite but this feature can be seen in the nearby disposal wells where the Dawson Bay is tight.

Initially coring was done from 900 to 947.6 metres but unfortunately recovery was only 70%. The well was then taken over for $CaCl_2$ injection from December 21, 1986 until January 11, 1987. Coring resumed February 8, 1987 and four cores from 949 to 982 metres were cut. Coring was plagued with problems with junk (bits of steel and pot metal) in the hole from drillable tools used previously. Recovery was 93.1%. Nothing unusual was seen in the recovered core.

b) Sealing and testing of Dawson Bay Formation

The plan was to run and then drill out cement plugs to destroy any permeability in and near the bore hole through the Dawson Bay. The success of the plugs was to be measured by running drill stem tests over the plugged zones. Seven cement plugs totalling 32 tonnes Class G cement both fresh and salt saturated were run. As well two treatments with acrylamide followed by cement and one treatment with just acrylamide was run. A drillable open hole packer was used for these treatments to isolate the zone to be "squeezed."

To check the effectiveness of the squeezes, eleven drill stem tests were run. Although there was some variation in test results, the final test of the Dawson Bay showed no significant reduction in the ability of the total Dawson Bay formation to deliver water into the bore hole.

c) Cementing of 177.8 mm long string casing

The 177.8 mm, 34.26 Kg/m, J-379

319

casing was run with a float shoe and a float collar to 981.6 metres. The casing was cemented with 25.5 tonnes salt saturated Class G cement. Good fluid returns were obtained while cementing and displacing with the fluid in the annulus dropping away when pumping stopped which is not untypical for the wells in the area.

d) Testing of cemented casing

The cement, float collar and float shoe were drilled out with a service rig and the hole circulated clean to 984 metres K.B. with saturated brine. The well was pressured to 12,680 KPa bottom hole pressure (1,840 psi) and the pressure held.

A cement bond log was run which showed excellent bond from the casing shoe to 902.5 metres. Above this there were a few holidays but generally good bond exists up to the cement top at 547 metres.

The final test was to swab the hole as dry as was practically possible and 17.5 m³ of 18.9 m³ brine was removed from the casing. The idea was that if the seal around the casing was poor, fluid would leak into the casing. This did not occur.

## 5 MINING TO BACKFILL WELL

After completion of the backfill well, and the swabbing of the hole, mining was initiated to join the mine workings to the backfill hole. Before mining was initiated, an attempt was made to locate the well by Seismic Tomography. Geophones were set up along the east wall of 10B023 at the intersection of 08B052. A seismic blast was then set off in the backfill well from surface and the seismic travel times were measured by the geophones and from the travel times an accurate calculation could be made as to the distance from the geophones to the well.

Mining was initiated with the Alpine miner cutting one pass 16 feet wide towards the backfill hole. It was estimated by the mine engineering department and confirmed by tomography that 100 feet required excavation. A minimum width was desired so that a precautionary bulkhead could be constructed in case the cement job at the backfill well broke down, causing an inflow in the mine.

When breaking into the location of the backfill well, the operator was covered with brine, as expected, as the well could not be swabbed clean. The well was now ready for installation of the injection string and underground personnel began to install the underground placement system.

## 6 INJECTION STRING

139.7 mm, 34.23 Kg/m, L.T. & C., N-80 casing was run as an injection string after the mine connection was made and landed at 980 metres. There is an L.H. style full bore packer at 940 metres. This compression packer isolates the 177.8 x 139.7 mm annulus and also provides protection in that it would prevent the casing from falling down the hole should it ever part up hole. The full bore is necessary as erosion would be excessive in a restricted bore packer.

The annulus between the injection string and the cemented string was filled with brine and pressured up moderately. By monitoring the annulus, should even a small leak develop in the injection string it can be detected.

Since the well has been in operation two internal casing inspection logs have been run in the injection string and there was no evidence of any significant wear.

## 7 PRECAUTIONARY BULKHEAD

### 7.1 Design

In case of any problems developing with the backfill well, and the overlying formation becoming connected to the mine workings, a precautionary bulkhead was proposed. This bulkhead was located in the backfill access drift, to isolate the backfill well from the mine workings (Figure 1).

After the mining of the backfill access was completed, a 15 foot tapered notch was cut fifteen feet west of the backfill pipe. This notch was excavated in order to place a 10 foot long concrete plug in the drift. The tapered notch was similar to a cork. As the pressure builds up behind the bulkhead, the concrete plug tightens into the notch.

Prior to placement of the concrete, the area was sprayed with epoxy to protect the salt against dissolution. The bulkhead was formed with 2'6"x4'8"x10' steel opening to allow access to the backfill pipe for inspection. The steel opening is complete with a hinged steel door so closing of the bulkhead, if required, can be done. Five other steel lines were also installed in the bulkhead for various reasons. There are three 10" pipes to drain brine from behind the bulkhead to relieve pressure, a 16" line for the backfill line to run through and a 4" line which carries the instrumentation cables (Figure 6).

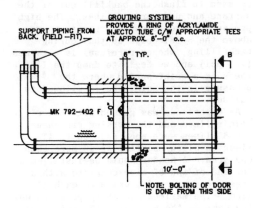

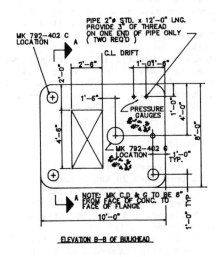

Fig.6  Precautionary bulkhead design

After the concrete was poured and allowed to cure, grouting was done between the rock and concrete interface and between any steel/concrete interfaces.

## 7.2 Instrumentation

The main instrumentation is a brine level

alarm that is tied into dispatch for an early warning system that indicates water behind the bulkhead. If this alarm goes off, the door to the bulkhead is closed and sealed to prevent brine from entering the mine workings.

There are also extensometers, closure meters and piezometers installed to monitor ground movement around the backfill well.

## 8 UNDERGROUND HANDLING AND PLACEMENT

### 8.1 Basic schematic and design

After the completion of the backfill well and the access drift, the installation of the underground handling system was initiated. The well casing was cut off and a 900 # 8" ANSI flange was welded in place in order to install a ceramic lined tee that was designed to withstand the severe abrasiveness that was expected. A ceramic lined reducer from 8" to 6" was installed to adapt to the main underground pipeline. The well casing, tee and reducer, were suspended from the back utilizing spring hangers in three locations. The hangers were designed to allow movement of the pipe as the fluid is moving from the vertical to the horizontal (Figure 7).

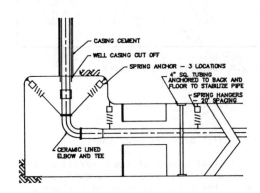

Fig.7  Backfill piping schematic

The main underground pipeline, after leaving the ceramic lined reducer, consists of 6" diameter schedule 160 steel and was designed to allow for a possible pressure in excess of 2000 psi in the event of the well becoming "salted off." A thicker wall pipe was chosen as we were unaware of the abrasive characteristics of the tailings as it fell down the backfill well. The first

321

95 feet of the horizontal pipe runs between 4x4 steel guideposts, alternating bolting between the floor and back, to remove any horizontal movement in the pipe.

After the expansion loop, the pipe is suspended from the back with fabricated pipe brackets. The distance between the support brackets varies from a minimum of 47 feet to a maximum of 95 feet and is a function of the back contour. The supports are rockbolted to the back and supported with four cables to take any movement out of the bracket. Each bracket comes complete with a double adjustment to compensate for any changes in elevation. A chain is also used to support the pipe at 10 foot spacings between the support brackets (Figure 8).

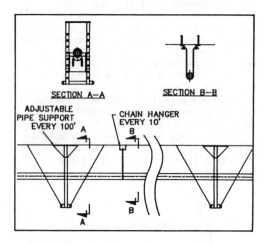

Fig.8 Backfill pipe support

## 8.2 Pipelines

### 8.2.1 High pressure pipelines

The initial design of the backfill system allowed for line pressure of 2000 psi. During the first placement of tailings, the pressures were very low at 0 to 25 psi at the outlet and pressures at the well bore were moderate at 300 to 600 psi. With working pressures substantially less than expected, the switch was made to Sclair pipe for placement.

### 8.2.2 Low pressure pipelines

Sclair pipe was an attractive alternative because of the cost differential and installation considerations. Adaptations were made to the existing schedule 160 steel line to increase to 8". At each adaptation, a 900 psi ball valve was installed and from the valve a connection was made to change over from steel to Sclair. A pressure gauge was also installed downstream of the ball valve so the pressure on the Sclair line was always known. The high pressure valve is used during the dumping process. Whenever we backfill, an area must always be used to flush the backfill out of the surface and underground lines. The high pressure valve allows a pressure range above the Sclair pipe so that during backfilling, whenever refusal is obtained (150 psi) and we begin to dump the line, there is a brief period where the pressure in the main line is allowed to buildup.

As experience was gained with the 8" Sclair lines and the working pressures, the 8" lines were soon reduced to 4" lines to allow backfilling in more than one area. A header would be installed at the end of the 8" line complete with a pressure gauge and a valve to each 4" line. The maximum operating pressure was the same as the 8" line - 150 psi. The 4" lines are also more convenient to install and are able to place them in areas where, otherwise, could not be accessible.

Through our backfilling experience, we have learned that at low density (20%) one 4" backfill line can handle the flow but at high density (30%) we have to use two 4" lines to handle the flow.

## 8.3 Controls

A basic requirement for backfilling is to have all personnel in the backfilling process in communication with each other. There are four people involved in the backfill process; two on surface and two underground.

The personnel underground are located at the backfilling area and at the base of the well. The starting and stopping of the underground placement system is done via direct line with the Central Control on surface. The man in Central Control is also in communication with the backfill operator on surface who ensures there are no problems with the surface line.

During the backfilling process, we generally operated on visual inspections until the bulkhead is closed up and then operation was based on line pressure. There are pressure gauges and a telephone

322

wherever a man is situated underground.

## 9 DEPOSITION METHOD

### 9.1 Bulkheads

In backfilling an area, bulkheads are
required to allow the brine to drain and
also to deposit the tailings where
required.  Early bulkheads were
constructed with 6"x6" rough timbers,
supported to the back with rockbolts and
cable, and 3"x8" rough planks, with a
3/4" gap between the planks.  Each
bulkhead was covered with burlap on the
downstream side (Figure 9).  The burlap
and the gaps in the planks were to allow
the brine to bleed out of the bulkhead.
All backfill lines came back through the
bulkheads and the bulkhead would be built
to a certain height before backfilling.
Once the height was reached, we would
build up the bulkhead as required to
complete the job.

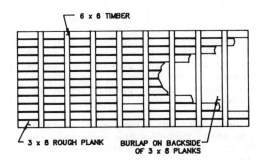

Fig.9  Initial bulkhead design

Early backfilling proved that the brine
would not decant fast enough and the
burlap was becoming plugged with salt.
The brine would then develop a head and
would come out around the bulkhead at an
increased velocity causing erosion of the
tailings recently placed.  These erosion
channels would then cause problems
throughout the backfilling process.
Modifications were then made to combat
this problem.  A 3' notch was constructed
in the middle of the bulkhead to drain
the brine and not allow it to build any
head.  We would continue to backfill
until the return brine contained salt and
we would then drop another plank into the
notch.  The notch proved very successful
and all future bulkheads were constructed
in this fashion.  The 3/4" gap between
the planks in the main bulkhead was kept

as it allowed less planks to be used in
construction and a minor amount of brine
was allowed to decant (Figure 10).

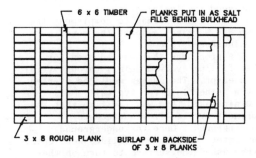

Fig.10  Final bulkhead design

### 9.2 Rehabilitation for backfilling

In order to backfill 10B400 and 10B500,
access to the areas must first be
obtained.  This was done using the Alpine
miner to dress up the area for safe
access and installing backfill and grout
pipes.  Most areas in 10B panel all tend
to collapse in the same fashion.  They
develop an arch and failure occurs along
the arch.  In many areas of
rehabilitation, the arch has already
collapsed and we proceed along the top of
the fallen muck and dress up any minor
fracturing that is still occurring.  The
amount of cutting is kept to a minimum as
we only are attempting to get in to
install our backfill pipes and then
retreat (Figure 11).

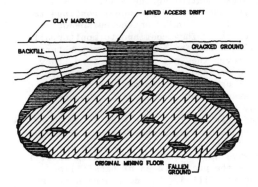

Fig.11  Cross section of typical room

### 9.3 Backfill procedure

The backfill process is generally
controlled by the underground personnel.
They will let Central Control know what
is required and the surface personnel

control the delivery system. In the early days of backfilling, the density of the tailings slurry was typically 20%. Modifications in the surface system which included the installation of an additional 4x3 Warman to complement the two 8x6 Warmans, allowed the density to be increased to 30%.

The procedure that was adopted for backfilling evolved mainly through trial and error. The system would be started by a brine flush for typically 30 minutes. This was the time that the brine travelled from surface to underground. Once the underground operators were sure all lines were open, they asked the surface to turn on the tailings at a specific density. The backfilling would then continue for one hour and then a brine flush would be put into effect for fifteen minutes and this process would continue in a cyclical fashion. The brine flush had two goals to achieve, namely keep all pipelines open and also we found by experience that areas could be backfilled for longer durations (increased tonnage) by the flushing process. The underground personnel would then keep notes on what transpired during their shift. They would monitor time, pressure, density and comment on percent of area filled. The surface personnel would submit a report indicating hours of backfilling, flushing, densities and a backfill tonnage total. Throughout the entire backfilling process, underground personnel are in communication with surface personnel via a direct line to Central Control in the mill.

The underground personnel continue to backfill and build up the bulkheads as required. They continue to monitor all bleed lines for evidence of any returns. Whenever they get a brine return, they have an idea the area is beginning to fill up and begin to monitor their pressures more frequently. The operator may then decrease the backfilling time and increase the flush time in order to keep an area available for backfilling.

In backfilling, the operators must always have another area to dump into, in case the present area of backfilling becomes filled. The dump area must be known to accept backfill before backfilling commences and the operator ensures this by running a brine flush.

Once an area pressures up and the area is designated to be filled, the operator dumps his line to the designated area and flushes the entire system until clean brine is coming out the end of the pipe. The flush generally lasts between 30

minutes to an hour. The next area is now ready to be backfilled and the procedure begins all over again.

## 10 HANDLING EXCESS BRINE

### 10.1 Sumps

In most cases, when using tailings backfill, a sump is located near the bulkhead to allow the removal of the excess brine. These sumps are generally already located and are used to pump inflow brine to surface. A series of sandbag dykes are used to help the salt settle out, but in many cases, the pumps become "salted in" and the scooptram must muck out the sump.

### 10.2 Pumps

In most cases, the Grindex pump is used underground to transfer both inflow and backfill brine to the National pumping station near the shaft. If the sump, that we are backfilling near, handles inflow brine, then additional two Grindex's are installed in the sump to handle the backfill brine of 1000 USGPM. Each Grindex pump is rated at 500 USGPM. In order to get the water from 10B panel, a 8x6 Warman booster pump is employed halfway to the shaft on the main 16" line.

The Grindex pumps have an 8" discharge line which runs from the sumps and tie into the main 16" line which goes to the shaft.

Backfill brine and inflow brine are pumped to the National pump station. At the high pressure pump station, the brine goes through a screen which separates out material that is 1/16" or larger. These screens are set up in parallel so that one can be cleaned while the other one takes the flow. After the brine clears the screens, it goes into one of three 5000 gallon surge tanks. The surge tank also acts as a settling tank to remove any of the fine salt.

### 10.3 High pressure pumps

A 6x4 Warman pump is used from the level controlled surge tank to boost the suction pressure to the high pressure pumps. A pressure of 30 psi is required to open the suction valves on the high pressure pumps.

The high pressure pump system consists

of four banks of six J-275 National pumps
and one bank of four J-375 pumps. The
pumps are positive displacement piston
type, with spring loaded suction and
discharge valves on each piston (Figure
12).

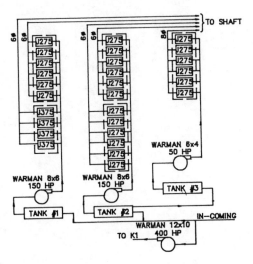

Fig.12  Pumping station schematic

The J-275 pumps are a five piston pump
with a 275 hp electric motor that will
discharge 160 USGPM at a pressure of 2000
psi.  The J-375 pumps are a three piston
pump with a 375 hp electric motor that
will discharge 230 USGPM at a pressure of
2000 psi.  Each pump discharge into a
common header with a check valve to
isolate each pump.

10.4 Shaft pipelines

Each of the high pressure pump banks has
its own independent line up the shaft.
Each bank ties into 4" schedule 160 steel
line that is connected with flanges in
the mine and victaulic clamps in the
shaft.  The pipe is clamped on brackets
every 20' in the shaft.  The shaft lines
discharge into a manifold on surface and
transferred via 16" Sclair line to the
tails pond.

325

*Innovations in Mining Backfill Technology, Hassani et al. (eds), © 1989 Balkema, Rotterdam. ISBN 90 6191 985 1*

# The use of relaxation tests to predict the compaction behaviour of halite backfill

C.J.Fordham, M.B.Dusseault & L.Rothenburg
*University of Waterloo, Waterloo, Ontario, Canada*

D.Mraz
*Mraz Project Consultants Ltd, Saskatoon, Saskatchewan, Canada*

## ABSTRACT

The waste material from the processing of potash, granular halite, is currently being used as a backfill in potash mines. As stopes close, halite backfill will eventually compact to a porosity where it will begin to support load and reduce the closure rate of the stope.

A series of relaxation tests were performed to develop a model describing the compaction behavior of halite backfill. The stress dependency of the compaction data showed grain boundary sliding to be a dominant mechanism of deformation during the tests. At low stresses and porosities, however, it is likely that pressure solution dominates the compaction behavior of halite backfill.

## 1.0 INTRODUCTION

Granular halite, the waste product from potash milling operations, is currently being used as a backfill material in two New Brunswick potash mines and several potash mines in Saskatchewan. Backfilling operations serve three purposes:

   a:  an environmentally acceptable disposal of waste,

   b:  a working platform cut and fill mining and a roof support for second pass mining, and,

   c:  a method of reducing surface subsidence.

Regardless of the reasons for backfilling, backfill will eventually compact to a degree where it will begin to support load. Understanding the compaction behavior will aid in the design of mining and backfilling programs. During early stages of compaction, halite backfill has very little strength, hence much of the observed compaction is due to normal stress effects. In the laboratory, hydrostatic compression tests, either constant stress or relaxation, can be used to develop a model describing this type of behavior.

Compaction rates in halite backfill are a primarily a function of stress, loading rate, and porosity. Other parameters such as grain size, moisture content, and insoluble impurities content can greatly affect compaction rates in granular halite (Spiers et al., 1989; Johnson et al., 1984; Pufahl and Yoshida, 1982), however, in potash mine tailings these parameters do not vary as they are fixed by the physical properties of the ore and processing procedures.

In order to examine the compaction behavior of halite backfill, a series of multi-stage, hydrostatic relaxation tests were performed. During each test, the porosity and stress were monitored with time. The results were used to develop a semi-empirical model of backfill compaction behavior.

## 2.0 TEST PROCEDURES

### 2.1 Halite Backfill

The halite backfill used for this study was obtained from a Saskatchewan potash mine. The backfill contained particles from about 5 mm to less than 0.1 mm in diameter (Figure 1). This reflects the generally large grain size of the potash ore in Saskatchewan. In contrast, granular halite tailings from potash mines in New Brunswick are dominantly between 0.125 and 2.0 mm in diameter, a reflection of the much finer grained ore (Figure 1) (CANMET, 1987).

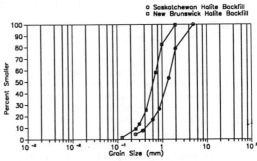

Figure 1: Grain size distribution of halite backfills from Saskatchewan and New Brunswick.

The backfill is almost pure NaCl and has a specific gravity of 2.16. Although insoluble materials such as clay, anhydrite, and quartz have been noted in Saskatchewan potash tailings (Johnson et al., 1984; Pufahl and Yoshida, 1982), none were noted in the test material.

## 2.2 Methodology

A relaxation test is performed by rapidly applying a load to a test specimen in a pressure vessel, then sealing the pressure vessel, and monitoring the pressure drop and specimen compaction with time. The relaxation tests reported here were performed in a hydrostatic consolidation cell designed for fluid/rock interaction studies (Figure 2). The cell applies hydrostatic stresses (hydraulically) to a test specimen while fluid is either pumped into or allowed to flow out of a test specimen.

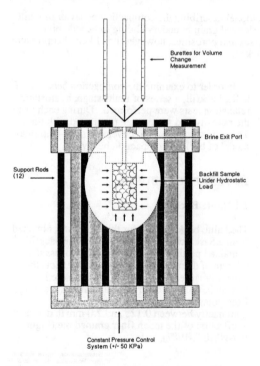

Figure 2: Hydrostatic consolidation cell used to perform relaxation tests.

To perform a test, a cylindrical specimen was prepared by filling a polyurethane membrane of known volume with a known mass of halite backfill and attaching it to the lid of the consolidation cell (Figure 2). The backfill was then saturated under vacuum with a brine formed by dissolving backfill in distilled water. Once saturation was complete, flow lines were attached to the cell to monitor the outflow of brine from the specimen. Outflow volumes were monitored by a series of burettes (Figure 2). Pressure was monitored with a pressure transducer output to a strip chart recorder.

Initial stresses were applied to the test specimens with a large volume, pressure control system. Stresses were applied in about 30-45 seconds. Once the desired stress had been reached, the cell was isolated from the pressure control system and the stress was allowed to drop as the backfill compacted. Brine outflow and cell pressure were monitored with time. Once the volumetric strain rate had dropped to about $10^{-8}$ sec$^{-1}$, the initial pressure was re-applied and a new relaxation stage began. Three tests were performed; two three-stage tests at 2 and 5 MPa, and a four stage test at 2 MPa. The lowest stress measured at the end of a relaxation stage was about 0.25 MPa.

Porosity was calculated by;

$$n = \frac{(G_s \cdot \gamma_w - \gamma_b)}{G_s \cdot \gamma_w} \tag{1}$$

where:  $n$ = porosity
$G_s$ = bulk specific gravity
$\gamma_b$ = backfill density
$\gamma_w$ = density of water

Density was calculated by;

$$\gamma_b = \frac{M_s}{[V_0 - V_{out}(t)]} \tag{2}$$

where:  $\gamma_b$ = backfill density
$M_s$ = Mass of salt
$V_0$ = Initial Volume
$V_{out}$ = Volume of outflow
$t$ = time

Temperatures were not controlled during this study. Laboratory temperatures average 20° C and fluctuate between about 18° and 23° C.

## 3.0 RESULTS

### 3.1 Pressure and Porosity Changes with Time

As the backfill compacted, the pressure in the confining fluid dropped (Figure 3, 4, 5). Compaction rates and associated pressure drops were rapid during the first 24 hours after pressure was applied but levelled off as the backfill porosity dropped. Each successive test stage showed lower initial (24 hour) porosity and pressure drops. The relationship between the applied stress and the porosity of the backfill after 24 hours (an arbitrary measuring point) is shown in Figure 6.

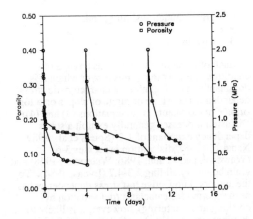

Figure 3: Porosity and pressure changes with time for 2 MPa, three stage relaxation test.

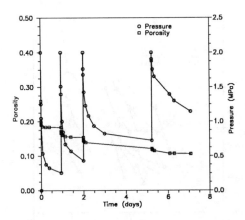

Figure 4: Porosity and pressure changes with time for 2 MPa, four stage relaxation test.

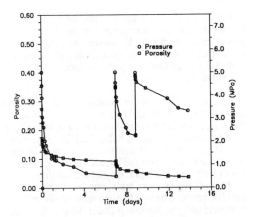

Figure 5: Porosity and pressure changes with time for 5 MPa, three stage relaxation test.

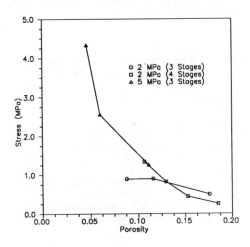

Figure 6: Backfill stress vs. porosity after 24 hours of relaxation.

### 3.2 Compaction Rates

In order to examine backfill compaction rates (i.e. volumetric strain rates), the data was plotted on a log $\sigma$/log $\dot{\epsilon}$ diagram (Figure 7). This diagram shows that a given strain rate can be achieved by different combinations of stress and porosity. Each point on the relaxation curves represents a lower backfill porosity than the point previous to it (higher stress, strain rate) on the curve.

### 4.0 DISCUSSION

### 4.1 Creep of Salt

The mechanical behavior of intact salt (i.e. pillar, floor, and roof material) is time-dependent and it deforms at a constant rate (creeps) under a given stress after a transient creep stage is completed. In general, the applied shear stress ($\sigma_1 - \sigma_3$) governs the rate at which salt creeps although normal stress effects have been noted (Dusseault et al., 1987). Creep of salt is also sensitive to temperature variations.

329

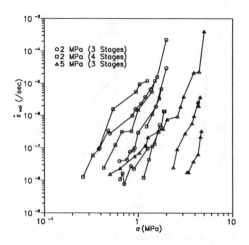

Figure 7: Log stress vs. log volumetric strain rate for all relaxation cycles. Each lower point on the curves represents the backfill at a lower porosity, strain rate, and stress.

The single mechanism, steady state creep of salt can be described by an equation of the form;

$$\dot{\varepsilon} = A \cdot \sigma^m \cdot e^{(-Q/RT)} \qquad (3)$$

where:
$\dot{\varepsilon}$ = strain rate
A = derived constant
$\sigma$ = effective stress
m = slope of curve on log $\dot{\varepsilon}$/log $\sigma$ plot
Q = activation energy of a given mechanism
R = universal gas constant
T = temperature in degrees Kelvin

Under isothermal conditions, equation (3) can be written;

$$\dot{\varepsilon} = B \cdot \sigma^m \qquad (4)$$

where:
B = derived constant which accounts for temperature phenomenon under given isothermal conditions

This equation describes a the isothermal creep behavior of salt due to a single creep mechanism. The mechanisms described and suggested for creep in salt to date include;

a: dislocation phenomenon,

b: diffusion processes including vacancy diffusion and pressure solution,

c: grain boundary sliding, and,

d: cataclasis.

Each of these mechanisms has a range of conditions (temperature/stress) over which it is dominant and each exhibits a characteristic stress dependency (i.e. slope or range of slopes on a log $\sigma$/log $\dot{\varepsilon}$ or exponent 'm' in equations (3) and (4)). Diffusion processes generally exhibit a stress dependency exponent of 1-2 (Spiers et al., 1989; Nabarro, 1948), dislocation processes 3-6 (Wawersik and Zeuch, 1986; Weertman, 1968) and grain boundary sliding 2.3-4.7 (Evans, 1984). To the authors' knowledge, no stress dependency has been reported for cataclasis (deformation by crushing and fracturing), however, it is likely to dominate creep only at high stresses (> 20 MPa) (Fordham et al., 1988) and exhibit a relatively high stress dependency exponent ($\approx$ 5-10).

### 4.2 Creep of halite backfill: compaction creep

Halite backfill is made up of many small pieces of intact salt. As such, it is subject to the same creep processes as intact salt. However, the conditions to which halite backfill is subjected once placed in a mined-out stope limit the possible mechanisms of deformation.

Temperatures in most Canadian potash mines are between 18° and 30° C. Temperature variations of this magnitude have been shown to have little effect on the compaction rate of halite backfill (Spiers et al., 1989). Self-weight stresses are likely to be the only stresses acting on the backfill for long times after placement. At a porosity of 0.25, halite backfill gravity stresses are about 16 kPa/m of backfill depth. The mechanism most likely to dominate compaction creep under these conditions is pressure solution (Spiers et al., 1989).

Pressure solution occurs by the diffusion of dissolved ions through a liquid film from areas of high stress on a grain (grain contacts) to areas of low stress (pores) (Spiers et al., 1989; Raj, 1982). Potash mine tailings generally have a placement moisture content (in the form of brine) of about 7-8% which reduces to about 2-3% in a short period of time. This moisture provides the liquid film through which dissolved ions diffuse.

In order to accommodate shape changes in the grains, grain boundary sliding often accompanies pressure solution (Burton, 1977; Rutter, 1976). Much of the previous data on grain boundary sliding comes from metals which have essentially no porosity. It seems likely that grain boundary sliding is a significant compaction process in a porous material such as halite backfill where the grains can move into pores.

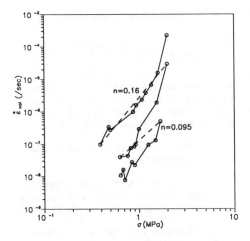

Figure 8: Log stress vs. log volumetric strain rate for 2 MPa, three stage test showing constant porosity lines between stages.

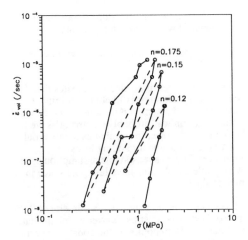

Figure 9: Log stress vs. log volumetric strain rate for 2 MPa, four stage test showing constant porosity lines between stages.

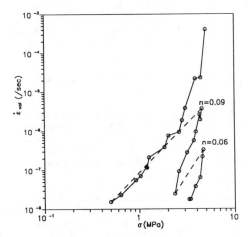

Figure 10: Log stress vs. log volumetric strain rate for 5 MPa, three stage test showing constant porosity lines between stages.

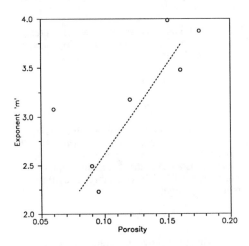

Figure 11: Stress dependency exponent 'm' vs. porosity.

### 4.3 Compaction creep behavior of halite backfill

In order to examine the stress dependency of compacting halite backfill, the data must be examined at constant structure (i.e. porosity) (Spiers et al., 1989). Assuming a relatively instantaneous stress change, constant structure is maintained between the low stress and strain rate at the end of one stage and the higher stress and strain rate of the successive stage. The slope between these two points is indicative of the process governing compaction creep at that porosity.

The final stress/strain rate measurement of each relaxation stage was joined with the first stress/strain rate measurement of the successive stage (Figures 8, 9, 10). It is understood that true constant structure is not maintained between successive relaxation stages. However, because porosity changes were less than 0.01 between any two stages, it is considered a reasonable assumption.

Although scattered, when plotted against porosity, the slopes of the constant porosity lines showed (Figure 11);

a: a general decrease in slope (from about 4 to about 2) with decreasing porosity

b: an increase in slope for the lowest porosity, 0.06.

The range of measured slopes indicates that grain boundary sliding was likely the dominant mechanism of compaction during the tests, although pressure solution was probably occurring as well. The decrease in slopes shows that as the material compacts, grain boundary sliding loses its dominance. This is a reasonable assumption as at low porosities, the grains have less pore space in which to slide.

The increase in slope noted at a porosity of 0.06 is primarily due to the stresses over which the data was obtained. Both points on the constant structure curve were obtained at stresses > 2.0 MPa. Spiers et al. (1989) note similar behavior in pure, fine-grained, granular halite. This may also explain the slight increase in the slope for the 0.09 porosity line as one of the two points on this line was obtained at a stress of 4.8 MPa.

The general decrease in slope can be expressed by an equation of the form (the 0.06 slope has not been included in this regression) (Figure 11);

$$m = 18.75 \cdot n + 0.74 \qquad (5)$$

where: $n$ = porosity

Compaction creep behavior of halite backfill can be described by an equation of the form of (4). The exponent m is a function of porosity as shown by (5). The constant A is the strain rate at a given porosity at $\sigma = 1.0$. The constant A is by the empirical equation;

$$A = 2.5 \cdot 10^{-2} \cdot n^{6.03} \qquad (6)$$

Combining the above, the compaction creep behavior of the halite backfill tested during this study can be described by an equation of the form;

$$\dot{\varepsilon} = 2.5 \cdot 10^{-2} \cdot n^{6.03} \cdot \sigma^{[18.75 \cdot n + 0.74]} \qquad (7)$$

This empirical law is considered valid over the range of $n \approx 0.2$ to 0.05 and $\sigma \approx 0.2$ to 2 MPa for the tested material. At higher stresses, it is likely that the stress exponent increases as evidenced by the n=0.06 porosity line. At stresses lower than 0.2 MPa, where grain boundary sliding is less likely to dominate creep, stress exponents will decrease to about 1.

## 4.4 Support capabilities of halite backfill

*In situ*, stopes close at a relatively constant rate (steady state creep) (CANMET, 1987; Dusseault et al., 1987). As the backfill densifies, it will begin to slow the room closure rate by accepting load from the walls, pillar, and floor. The rate at which the backfill accepts load will govern the effectiveness of the waste halite as a support material. One of two types of behavior would be preferred;

a: rapid initial transfer of load leading to quick, full roof support, or,

b: little support and therefore little impedance to room closure until the backfill reaches a certain porosity, followed by rapid support.

Any other behavior will slow the room closure without the backfill being able to adequately support the roof during the economic life of a mine.

Stress development in the backfill under constant strain rate was calculated by rearranging (7) in terms of stress under the following conditions;

a: above $\sigma$=0.2 MPa, the stress dependence of the backfill follows (5) with the limits $1 \leq m \leq 4$,

b: below $\sigma$=0.2 MPa, the stress dependence of the backfill is one.

A schematic diagram of this behavior is shown in Figure 12. Although it appears stress exponents increase at higher stresses, an increase in exponent to 4 for all porosities above $\sigma$ = 2.0 MPa had little effect on the behavior of the model so it was abandoned for simplicity.

The support capabilities of the halite backfill tested during this study under conditions of constant strain rate are shown in Figure 13. Also shown in Figure 13 are the processes assumed to govern backfill compaction. At low stresses and porosities, pressure solution dominates creep. At higher porosities, grain boundary sliding and pressure solution are both operative, and at high stresses and porosities, grain boundary sliding alone may be the dominant mechanism of creep compaction. Under the conditions found within most Canadian potash mines, pressure solution will be the dominant mechanism of compaction.

The type of behavior shown in Figure 13 suggests that the halite backfill tested during this study will support little stress until it reaches a porosity of about .05 under the closure rates found in most Canadian potash mines ($\dot{\varepsilon} \approx 10^{-12}$ to $10^{-10}$ sec$^{-1}$). It will gain strength rapidly and begin to support the closure of the room only at very low porosities.

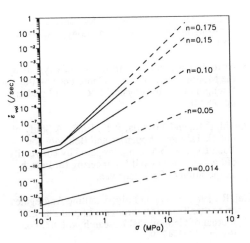

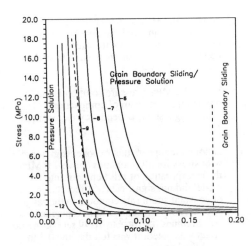

Figure 12: Schematic diagram of model used to describe compaction of halite backfill. Note, lines are dashed above 2 MPa as slopes are likely to increase in this region, however, have been kept at constant slopes for simplicity.

Figure 13: Stress vs. porosity at constant strain rate for halite backfill. The lines were produced by rearranging equation (7). Numbers next to curves represent the logarithm of the strain rate. The curves are valid for the tested backfill. Curves for backfill of different grain size can be determined by (10).

## 4.5 Extension to other halite backfills

The main difference between halite backfills from different Canadian potash mines is the grain sizes which they contain (Figure 1). Diffusion mechanisms such as pressure solution are extremely sensitive to fluctuations in grain size (Spiers et al., 1989; Turcotte and Schubert, 1982; Shor et al., 1981). Although all halite backfills will qualitatively behave in a similar manner, they will exhibit a wide variation in compaction rate at a given stress and porosity.

The relationship between grain size and compaction rate in granular halite is of the form (Spiers et al., 1989; Raj, 1982);

$$\dot{\varepsilon} \propto \frac{1}{d^3} \qquad (8)$$

where:  d = grain size

A scaling constant can be derived for other backfills assuming a mean grain size of 0.0014 m for the tested backfill and a constant relationship for stress dependence and porosity (i.e. equation (5) is constant for all halite backfills). The scaling constant can be determined by calculating the compaction rate at σ = 1.0 (ignoring the porosity term in equation (6));

$$C = 2.5 \cdot 10^{-2} \cdot (0.0014)^3 \qquad (9)$$

where:  C = scaling constant
= $6.86 \cdot 10^{-11}$ (units are ignored for simplicity)

The general equation describing compaction in halite backfill is;

$$\dot{\varepsilon} = \frac{C}{d^3} \cdot n^{6.03} \cdot \sigma^{[18.75 \cdot n + 0.74]} \qquad (10)$$

## 6.0 CONCLUSIONS

The stress dependency of creep data, can be used to identify the mechanism(s) involved in deformation. The relaxation data showed that at constant structure, the stress dependency exponent of halite backfill decreases from about 4 to about 2 with porosity decreases from about 0.175 to 0.09. At low stresses and porosities, the exponent is expected to be about 1.

Halite backfill will compact under low stresses (≈ 16 kPa/m) and temperatures (18° to 30° C) in Canadian potash mines. Under theses conditions, pressure solution will be the dominant mechanism of deformation. At higher stresses and porosities, grain boundary sliding becomes an increasingly more important compaction mechanism.

The model developed here adequately describes the compaction behavior of halite backfill based on the following assumptions;

333

a: between 0.2 MPa and 2.0 MPa grain boundary sliding and pressure solution dominate compaction with stress dependency exponent increasing from 1 at n≈.014 to 4 at n≈.175,

b: below 0.2 MPa, pressure solution is the dominant mechanism of compaction exhibiting a stress dependency exponent of 1,

c: although it appears that exponents increase with stress for all porosities above 2.0 MPa, the incorporation of this behavior into the model did not significantly affect the shape of constant strain rate contours in the range of conditions expected within Canadian potash mines ($\dot{\varepsilon} \approx 10^{-12}$ to $10^{-10}$ sec$^{-1}$). As no effect was noted, it was felt that the 0.2 to 2.0 MPa behavior can be extended to 20 MPa.

Halite backfill will not support significant load until it has reached a low porosity. The porosity at which the backfill will begin to support load and reduce closure rates in a mining stope is strain rate and grain size dependent.

REFERENCES

Burton, B., 1977. Diffusional Creep of Polycrystalline Materials. Trans Tech Publications, Aedermannsdorf, Switzerland. 119 p.

CANMET, 1987. Use of Backfill in New Brunswick Potash Mines. Prepared by Denison Potacan Potash Co., Mraz Project Consultants Ltd., and University of Waterloo, for the Canada Centre for Mineral and Energy Technology, Dept. of Energy Mines and Resources. Five Volumes.

Dusseault, M.B., Mraz, D., and Rothenburg, L., 1987. The design of openings in saltrock using a multiple mechanism viscoplastic law. Proceedings of the 28th U.S. Rock Mechanics Symposium, Tucson, Arizona. 10 pp.

Evans, H.E., 1984. Mechanisms of Creep Fracture. Elsevier Applied Science Publishers, New York. 319 p.

Fordham, C.J., Dusseault, M.B., and Mraz, D., 1988. Strength Development in Halite Backfill. Proceedings of the 41st Canadian Geotechnical Conference, Waterloo, Ontario. pp. 194-200.

Johnson, R.F., Lamb, K.N., Hart, R.T., Reid, K.W., 1984. Investigation of the Physical and Chemical Characteristics of Potash Tailings in Saskatchewan. Prepared by potash Corporation of Saskatchewan, Saskatoon, for Canada Centre for Mineral and Energy Technology, Project No. 350204.

Nabarro, F.R.N., 1948. Deformation of Crystals by the Motion of Single Ions. Conference on the Strength of Solids, Physical Society of London. pp. 75-90.

Pufahl, D.E., and Yoshida, R.T., 1982. A Study of Potash Tailings: Properties and Behavior for Total Extraction Mining. Prepared by University of Saskatchewan Civil Engineering for Potash Corporation of Saskatchewan.

Raj, R., 1982. Creep in Polycrystalline Aggregates by Matter Transport through a Liquid Phase. Journal of Geophysical Research, vol. 87, no. B6, pp. 4731-4739.

Rutter, E.H., 1976. The Kinetics of Rock Deformation by Pressure Solution. Philosophical Transactions of the Royal Society of London A, vol 283, pp. 203-219.

Shor, A.J., Baes, C.F.Jr., and Canonico, C.M., 1981. Consolidation and Permeability of Salt in Brine. Prepared by Oak Ridge National Laboratory, Oak Ridge, Tenn. for U.S. Dept. of Energy, Contract No. W-7405-eng-26.

Spiers, C.J., Peach, C.J., Brzesowsky, R.H., Schutjens, P.M.T.M., Liezenberg, J.L., and Zwart, H.J., 1989. Long Term Rheological and Transport Properties of Dry and Wet Salt Rocks. In press, Nuclear Science and Technology EUR Series, vol. 11848, Office for Official Publications of the European Communities, Luxembourg.

Turcotte, D.L., and Schubert, G., 1982. Geodynamics. John Wiley and sons, New York. 450 p.

Wawersik, W.R. and Zeuch, D.H., 1986. Modelling and Mechanistic Interpretation of Creep of Rock Salt Below 200° C. Tectonophysics, vol. 121, pp. 125-152.

Weertman, J., 1968. Dislocation Climb Theory of Steady-State Creep. Transactions of the ASME, vol. 61, pp. 681-694.

*Innovations in Mining Backfill Technology, Hassani et al. (eds), © 1989 Balkema, Rotterdam. ISBN 90 6191 985 1*

# Utilization of salt tailings backfill at Denison-Potacan Potash Company

G.Herget
*Canada Centre for Mineral and Energy Technology (CANMET), Ottawa, Ontario, Canada*

S.R.Munroe
*Denison-Potacan Potash Company, Sussex, New Brunswick, Canada*

ABSTRACT: Backfilling of mined out sections of the potash orebody with waste salt might offer the key to increased extraction ratios. Laboratory characterization, numerical modelling and in situ testing have however shown that present backfill will not provide the required roof support within a five year time frame. To recover barrier pillars by secondary extraction, stopes on either side filled with dry backfill probably require topping off by slurry or hydraulic backfill to compensate for shrinkage and provide moisture for improved consolidation and recrystallization. At this point it appears that dewatered backfill topped off with hydraulic or reslurried waste salt and compacted by convergence from aggressive primary mining, would provide a good target to guide further studies for the identification of backfill support.

## INTRODUCTION

The Denison-Potacan Potash Company (DPPC) mine site is located southwest of Sussex, New Brunswick as shown in Figure 1. The Company's Clover Hill deposit was first identified by surface exploratory drilling in 1973 by the New Brunswick Department of Natural Resources. An extensive underground exploration program led to a decision in 1983 to develop the property into a producing mine capable of producing 1.3 million tonnes of product per year. Potash production commenced in July, 1985.

DPPC uses a room and pillar method with long barrier pillars to mine the inclined potash orebody. Both continuous mining and drill and blast techniques are used. An important component of mine operations at Denison-Potacan is the placement of tailings in the mined out underground workings. The placement of tailings backfill underground reduces environmental impact from potash processing and potentially allows increased recovery of potash resources.

The neighbouring Potash Company of America mine at Sussex currently uses tailings backfill as an integral part of their cut and fill mining operation in the steeply dipping potash ore. Saskatchewan potash mines do not use backfill as part of normal operating

practices in their tabular orebody and large waste salt tailings piles are stored on surface. In order for Denison-Potacan to assess the support potential of tailings backfill, a number of technical questions were required to be answered:
- strength and deformation characteristics of tailings backfill material and determination of achievable in situ densities,
- strength and deformation characteristics of intact saltrock,
- development of constitutive laws to characterize saltrock and backfill behaviour,
- numerical modelling of a number of excavation layouts,
- in situ stress and deformation monitoring of mining excavations with backfill.

Funding for the research work was obtained from Denison Potacan Potash Company and under the Canada/New Brunswick Mineral Development Agreement. MRAZ Project Consultants of Saskatoon, Saskatchewan provided technical expertise and carried out the numerical modelling. Laboratory testing was conducted at the University of Waterloo's salt testing laboratory (WATSALT). Dr. G. Herget monitored progress for the research funded under

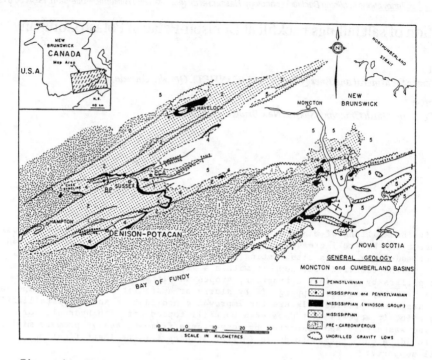

Figure 1 - Geology of Moncton subbasin in New Brunswick.

the Canada/New Brunswick Mineral Development Agreement.

GEOLOGY

The Denison-Potacan potash deposit is part of the Cassidy Lake Evaporite Formation of the Windsor Group. The deposit is confined to the Marchbank Syncline which is the most southerly major structural feature of the Moncton Subbasin (Figure 1). The evaporite interval has been subdivided into four members: Basal Halite, Middle Halite, Potash, and Upper Halite.

The Potash Member consists of the Lower Gradational Bed (LGB) and the Sylvinite Bed. The LGB is comprised of medium to fine grained brown to orange halite with sylvite clusters. The Sylvinite Bed is composed of fine-grained sylvite and halite with minor disseminated interstitial clays and only sporadic traces of carnallite. The Sylvinite Bed is subdivided into five generally distinct zones based on mineralization, percent insolubles and colour. These have been designated from oldest to youngest as the A, B, C, D and E Zones with the B and C Zones comprising the main ore bed (Figure 2). The geology of the Denison-Potacan deposit has been

previously documented in detail by Waugh and Urquhart(1).

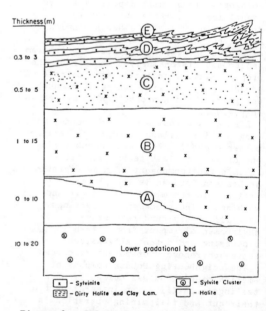

Figure 2 - Schematic stratigraphy of potash member in Cassidy Lake Formation.

# PHYSICAL CHARACTERISTICS OF TAILINGS BACKFILL

The waste salt tailings from the surface process plant consist of a coarse and fine tails section. The coarse and fine dewatered tailings, at approximately 8-10% moisture content by weight, are combined and then transported for disposal as backfill. Prior to February, 1987 process plant tailings were reslurried and pumped to the surface tailings pond for temporary disposal. Since February, 1987 dewatered tailings have been disposed of underground in mined out workings. The temporary disposal of tailings on surface was required to allow for the tailings disposal system to be installed and for sufficient disposal capacity to be created underground.

Coarse and fine tailings components were sampled as well as the combined stream tailings. Precautions were taken to ensure that the process circuit was operating at steady state conditions, that the plant feed was representative, and that the moisture content of the tailings was typical. The samples were sealed in epoxy coated steel drums for shipment to the University of Waterloo where the testing was conducted.

## Grain Size Analysis

Grain size analysis of the tailings was performed by wet sieving in the presence of acetone. The results showed that the tailings are uniformly graded with 95% of the particles between 2 mm and 0.125 mm in size. In comparison with other mine tailings the Denison-Potacan material is coarser and possesses negligible slimes (< 200 mesh or 0.075 mm).

## Mineralogy

The mineralogy of the tailings was studied by microscope and by x-ray diffraction. Under the microscope most of the tailings showed up as orange, orange-yellow, yellow and clear halite. Approximately 1% of the tailings below 1 mm in diameter was sylvite. X-ray difraction confirmed that the only minerals which appeared on traces were halite and sylvite. In contrast to typical Saskatoon area potash tailings no insoluble components were noted. The presence of insolubles was observed, however in small amounts in the surface tailings pond.

## Specific Gravity

The specific gravity of the tailings material was calculated as 2.14 $g/cm^3$ based on the mineralogical composition.

## LABORATORY AND FIELD DENSITY TESTING OF TAILINGS BACKFILL

The prime objective of the testing was to evaluate various placement methods and to provide density values for the planning of underground backfilling procedures. The benefits from investigating the placement density were to maximize the disposal of process plant tailings and to minimize the porosity of the placed backfill.

## Laboratory Density Placement Testing

Three placement density testing methods were investigated at the University of Waterloo. The standard Proctor test was used to simulate mechanical spreading with mobile equipment. Drop tests at various heights were used to simulate flinger placement, and vibration tests were used to simulate compaction with vibro-compaction equipment. A summary of the results obtained at 8% moisture are presented below:

| Compaction Method (Tests) | Dry Density $(g/cm^3)$ | Compaction Variable |
|---|---|---|
| Proctor | 1.36 | --- |
| Drop | 1.07 | 10.84 m/s vel. |
| Vibration | 1.38 | 7.36 kPa load |

Consolidation load tests were also performed to determine the time related response of backfill consolidation. The results revealed that the consolidation of the tailings backfill occurs rapidly after initial placement, and that further consolidation is relatively slow once the mechanism of intergranular friction takes over.

A very significant point was observed that moisture is necessary to aid in the consolidation process as shown in Figure 3, where halite backfill is consolidated under 10 MPa with 0% and 5% moisture. The 5% moisture backfill achieves a 30% higher density after 30 days. Figure 4 shows the strain rates

337

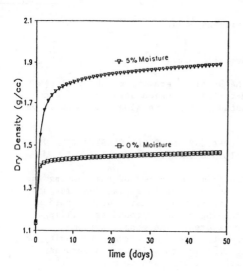

Figure 3 - Consolidation testing of DPPC halite backfill at 10 MPa and moisture contents of 0% and 5%.

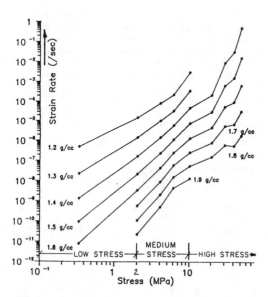

Figure 4 - Behaviour of backfill at different densities for different normal loads.

for different densities (1.2 to 1.3 g/cm$^3$) under 0.35 to 50 MPa compaction stresses.

Surface Placement Density Trials

Field placement density testing was conducted on surface at Denison-Potacan

to obtain preliminary information on the expected placement densities using actual full scale equipment. Three compaction methods were evaluated:

1.  Slurry sedimentation - Dewatered tailings were reslurried and allowed to settle due to gravity on the surface tailings pond. This process was considered analogous to the placing of tailings slurry underground and therefore the material of the surface storage pond was sampled after a few days and tested.

2.  Mechanical spreading with vibro-compaction by a portable vibratory roller.

3.  Mechanical spreading with a ST8A scooptram with compaction from tire pressures in the surface backfill storage building.

Density values were determined on the surface tailings storage pond and in the surface backfill storage building by using the ASTM Sand Cone Method of In Situ Field Density Determination. The following results were obtained:

| Compaction Method | Average Dry Density (g/cm$^3$) |
|---|---|
| Slurry Sedimentation | 1.330 |
| Scooptram | 1.379 |
| Vibro-compaction | 1.611 |

It should be noted that the sand cone method of density determination samples only a limited depth of the material, and an alternative density determination method is required to sample a full cross section of the material. A moisture/density probe has been obtained for this purpose.

Underground Placement Density Trials

Underground placement density testing was conducted after backfill placement commenced in February 1987. Three compaction methods were evaluated:

1.  Mechanical spreading with ST8 scooptrams placing backfill material on lifts with direct compaction from tire pressures.

2.  Large lift dumping with ST8 scooptrams.

338

3. Pneumatic stowing with a Triton 50 pneumatic stowing machine.

Vibro-compaction equipment was not evaluated. The following results were obtained with the ASTM sand cone method:

| Compaction Method | Average Dry Density (g/cm$^3$) |
|---|---|
| Scooptram (mechanical spreading) | 1.558 |
| Scooptram (large lift dumping) | 1.225 |
| Pneumatic | 1.380 |

## INTACT ROCK TESTING

The collection and laboratory testing of intact rock samples was required to determine the physical characteristics of the roof, floor and ore materials. These results established mechanical and rheological properties as input into numerical models for:

- creep
- brittle yield
- elastic modulus

of the Upper Halite, the Sylvinite Bed, and the Lower Gradational Bed.

The laboratory testing program at the University of Waterloo required that 100 mm diameter core samples be obtained. A Craelius T6S 131 mm x 1500 mm core barrel and a Corborit tungsten carbide bit were obtained for

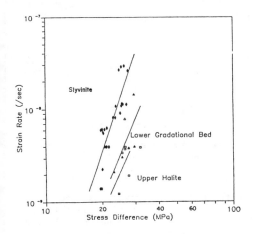

Figure 5 - Laboratory results on creep rate variation with stress difference for three evaporites.

use with a Model 200 Superdrill coring drill rig. The Superdrill is used at Denison-Potacan for exploratory core drilling and uses brine to flush drill cuttings.

The coring program was conducted in July, 1986 with representative roof (Upper Halite), floor (Lower Gradational Bed), and ore (Sylvinite Bed) samples obtained from Panel 10 Back Entry. An attempt was made to obtain all core samples perpendicular to bedding planes. The samples were shipped only for testing if their length to width ratio was greater than two and if no cracks or fractures were visible in the sample. The following results were obtained:

| Rock Type | UCS[1] | T[2] | E[3] |
|---|---|---|---|
| Sylvinite Bed | 26.3 | 1.7 | 5.46 |
| Upper Halite | 29.2 | 2.0 | 10.21 |
| Lower Gradational Bed | 28.4 | 1.5 | 6.91 |

1) Unconfined compressive strength (MPa)
2) Tensile strength (MPa)
3) Elastic modulus (GPa)

The elastic modulus is usually measured directly on the loading curve, however this procedure is inadequate for determining the short term response of salt rocks. An unload/load modulus is generally used for modelling the elastic behaviour of salt rock.

Creep tests established steady state strain rates at various principal stress differences. The results are shown graphically in Figure 5.

The deformation behaviour of potash was described by a complex material model composed of a classical Mohr-Coulomb rupture criterion, and the description of deformation with time based on viscosity and plasticity. The material behaviour model as developed by Dusseault, Mraz and Rothenburg is detailed in the final project report(2) and by a paper given by Mraz et al. in this symposium. The rupture criterion is given below:

$$\sigma_1 = \frac{\sigma_2}{2} \left[ (1-MR) + \sqrt{(1+MR)^2 + 4M\frac{\sigma_3}{\sigma_c}} \right]$$

$$R = \frac{\sigma_T}{\sigma_c}$$

Where:

$\sigma_1$ ... maximum principal stress
$\sigma_3$ ... minimum principal stress

$\sigma_c$ ... unconfined compressive strength
$\sigma_T$ ... tensile strength
M ... empirical constant

The following values gave the best fit equation to the laboratory data:

| | R | M |
|---|---|---|
| Sylvinite Bed | 0.0629 | 53.8 |
| Upper Halite | 0.0684 | 500.8 |
| Lower Gradational Bed | 0.0521 | 34.2 |

Time dependent deformation from in situ measurements by Denison-Potacan were fitted by regression to the power law:

$$\varepsilon_{ss} = \varepsilon_K \left( \frac{\Delta\sigma}{2K} \right)^n$$

Where:

$\varepsilon_{ss}$ ... steady state strain rate
$\varepsilon_K$ ... steady state strain rate at Prandtl Limit
$\Delta\sigma$ ... principal stress difference
K ... Prandtl Limit

The measured in situ strain rates were two to three orders of magnitude lower than those obtained in the laboratory.

The magnitude of "n" above the stress level of "2K" was determined by the power law fit and the magnitudes of "$\varepsilon_K$" were derived from extensometer readings in the Test Room and the Test Stope. The magnitude of exponent "m" below the stress level of "2K" was estimated. The properties used in numerical modelling varied as follows:

| | 2K (MPa) | $\varepsilon_K$ ($s^{-1}$) | n $\Delta\sigma>2K$ | m $\Delta\sigma<2K$ |
|---|---|---|---|---|
| Sylvinite Bed | 17.6 | $3\times10^{-11}$ | 6.0 | 1.0 |
| Upper Halite | 20.0 | $3\times10^{-11}$ | 4.5 | 1.0 |
| Lower Gradational Bed | 18.8 | $3\times10^{-11}$ | 4.8 | 1.0 |

NUMERICAL MODELLING

The numerical models VISCOT and FLAC were developed/modified and applied to the mining geometries of Denison-Potacan. A good agreement was obtained from a comparison of numerical modelling output and in situ deformation measurements by extensometers and convergence meters. At a depth of 850 m the overburden load had been measured by overcoring in the anhydrite as 20 MPa with horizontal stresses of similar magnitude.

BACKFILL/EXCAVATION INTERACTION

A mined out room (stope 11200), approximately 42 m wide and 18 m high, was instrumented with pressure cells and extensometers in the roof and one wall prior to backfilling. Earth pressure cells were placed in various locations of the room after backfilling commenced. Some instruments in the floor had to be abandoned due to the presence of pooled brine from an adjacent backfilling site. The room containing the instrumentation was filled by scooptrams in several large lifts with approximately 75,000 tonnes of dewatered (8-10% moisture) backfill.

Several interesting observations were made. Figure 6 shows the measured vertical (0.17 MPa) and lateral (0.12 MPa) stresses shortly after filling at floor level in the center. For the 15 m backfill column this translates into an average density of 1.16 $g/cm^3$ which is close to the density values obtained during in situ density testing for the large lift dumping by scooptrams. The reduction of pressures after backfilling is possibly due to the drainage of brine out of the backfill.

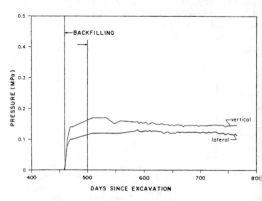

Figure 6 – Backfill pressure cell readings in stope 11200.

Site inspection showed that the backfill shrinks with time and pulls away from the side walls and the roof in the order of centimeters over a few weeks. Shrinkage away from the roof was sufficient to allow an earth pressure cell to be placed on top of the backfill

75 cm from the roof. The backfill therefore loads the floor and only the lower part of the walls immediately. The wall extensometer showed virtually no change in convergence rates after the room had been backfilled.

Figure 7 shows a potentially disturbing reaction of the roof after backfilling. Up to a depth of 5 m, roof strata convergence accelerates. Both higher moisture content in the mine air during backfilling and possibly increased temperatures from diesel equipment operating close to the roof accelerated roof deformation.

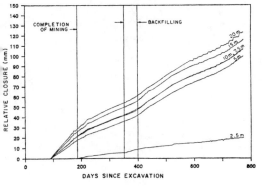

Figure 7 - Roof extensometer readings in stope 11200.

Preliminary numerical modelling has shown that overall convergence rates are significantly reduced in tightly backfilled stopes (Denison-Potacan 1987). At this point it appears very unlikely that the present backfill (high porosity, shrinkage cracks) will provide any significant roof support within a 15 year time span, given the present mining geometry.

D. Fulda et al. (1988) addressed the question of support capacity from backfill in East German potash mines. Testing of hydraulic waste salt backfill which had been in place for periods of 15 to 60 years showed strength values in excess of those of the in situ evaporites at Denison-Potacan. The strength of the 15 year old backfill was the highest because of compaction due to accelerated excavation convergence during a retreat pillar recovery phase.

Fulda et al. suggest that wall and floor confinement can improve pillar strength by at best 40 to 70% when using dewatered backfill. Hydraulic backfill however can assume the function of in situ pillars under favourable conditions

within a five year time frame. They advise that dewatered waste salt backfill should be topped off with hydraulic or reslurried salt tailings to compensate for shrinkage.

CONCLUSIONS

The research work on potash mining with dewatered salt backfill at Denison-Potacan has provided new insights to reduce the mining and mine planning risks associated with the aim to increase in situ recovery based on the support function of backfill.

1) A constitutive law to model the strength and the deformation characteristics of in situ evaporites has been developed to allow numerical analysis of planning options for the potash orebody.

2) The present backfill will have limited roof support capability prior to gap closure and significant elimination of porosity. The time required to achieve this backfill compaction from room convergence is unacceptably long. Gap closure time could probably be greatly reduced by topping off existing backfilled stopes with hydraulic or reslurried backfill.

3) Laboratory testing of dewatered backfill has shown that increased moisture content will improve compaction and recrystallisation and thus achieve higher strength.

4) Additional studies are required to achieve an optimum of resource recovery and return on investment:

4.1) Quantify the interaction of various salt backfills (different moisture contents, additives) and mine excavations.

4.2) Compare the benefits from more aggressive primary mining (higher extraction ratio, increased room convergence, improved backfill compaction) with the potential of increased costs to maintain access and haulage facilities.

ACKNOWLEDGEMENTS

The authors appreciate the excellent cooperation provided by all participants

during this project and thank
Denison-Potacan management for the
permission to publish this paper.

REFERENCES

1.  Waugh, D.C.E. and Urquhart, B.R.
    "The Geology of Denison-Potacan's
    New Brunswick Potash Deposit", Sixth
    International Symposium on Salt -
    Volume 1, Toronto, p. 85-98, 1985.

2.  Denison-Potacan Potash Company. Use
    of Backfill in New Brunswick Potash
    Mines - Volumes 1-5. DSS File No:
    14SQ.23440-5-9031, DSS Contract
    Serial No: OSQ85-00208, CANMET
    Project No: 1.4.09.05, (May, 1987).

3.  Denison-Potacan Potash Company.
    Progress report Phase II April-
    September, 1988.

4.  Fulda, D., Jäger, G., Kutscha, T.
    "Is placed salt waste backfill
    capable to assume the geomechanical
    function of barrier pillars in
    potash mines?" (German) Neue
    Bergbautechnik, Jg 18, vol. 7,
    261-264, 1988.

*Innovations in Mining Backfill Technology, Hassani et al. (eds), © 1989 Balkema, Rotterdam. ISBN 90 6191 985 1*

# Field monitoring of salt tailings used as backfill in cut-and-fill potash mining

R.J.Beddoes, B.V.Roulston & J.Streisel
*Potash Company of America, New Brunswick, Canada*

ABSTRACT:

Potash Company of America in Sussex, New Brunswick, operates a cut-and-fill stoping system, unique in potash mining, in which all mill tailings are returned directly to the mine for permanent storage. Roadheader-type miners cut the stope back to a height of 6.7 m and salt tailings is stowed behind them to within 2.4 m of the back using conveyors belts and scoop trams. Tailings, which provide the only means of stope support, are compacted by the wheel load of the scoop.
    A potash stope has been instrumented with earth pressure cells and extensometers. Monitoring has revealed that the fill is subjected to only small pressures and that arching between the walls results in lower than anticipated consolidation. Shrinkage caused by drainage of pore brine contributes to the low support pressure.
    A simple constitutive model for tailings is proposed and its application in finite element analysis of the instrumented stope is illustrated.

## 1 INTRODUCTION

In the Sussex area of southeastern New Brunswick, potash is found in three local synclines within the Upper Mississippian age Moncton sub-basin. Potash Company of America mines salt and potash at the western end of a pronounced north-easterly trending salt pillow. Salt is known to occur over a strike length of at least 25 km in this anticlinal feature, which displays the early stages of diapirism. Due to halokinesis, the salt body is in places over 1000 m thick, coming within 100 m of the surface, and contains complex asymmetric overfolding (Roulston and Waugh, 1984). Internal flowage has led to extreme foliation of the ore, boudinaging and rapid changes in dip as illustrated in Fig. 1.

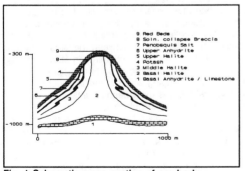

9 Red Beds
8 Soln. collapse Breccia
7 Penobsquis Salt
6 Upper Anhydrite
5 Upper Halite
4 Potash
3 Middle Halite
2 Basal Halite
1 Basal Anhydrite / Limestone

Fig. 1 Schematic cross section of ore body

To meet the challenge of the complex ore body, PCA has developed a flexible cut-and-fill mining method using roadheader miners to cut stopes upwards in lifts of 4 m from sill cuts at 700 m and 560 m below surface. Extraction over the stoping interval is continuous along the strike and approaches 100%. Salt tailings resulting from milling of the ore, mixed with fines from the crushing of rock salt, provides both a working surface for the mining machines as they progress upwards in the stope, and the only means of stope support. It is conveyed to the stopes by belts and then spread using scoop trams. The tire load of the scoops is the only method of compaction employed, although vibrating rollers have been tried without any clear increase in final density. Backfill behaviour is thus a crucial aspect of long term mine development.
    The behaviour of salt tailings has received attention in several countries due to its importance in the proposed disposal of nuclear wastes in salt deposits. Numerous laboratory studies of the consolidation and strength development of fine salt under a variety of load and moisture conditions have been performed. Several studies (e.g. Pfeifle et al (1984), Hesterman (1979)) have shown that the degree of consolidation achieved during loading is dependent upon the maximum stress and time under load. The presence of moisture is important in facilitating consolidation, possibly because pressure solution is a contributing mechanism. However, the results of Miller (1983) suggest that the actual moisture content may be of secondary importance. Strength development is strongly influenced by time under load; a sample

which is quasi-statically loaded to 21 MPa possessed no measurable strength (Holcomb and Hannum, 1982), whereas one consolidated for 3 weeks under the same load developed an unconfined strength of 11 Mpa (Beddoes, 1984). The importance of secondary crystal growth, during the period of creep, in the development of strength was shown qualitatively in electron micrographs by Beddoes (ibid.). Based upon laboratory testing, relationships between creep rate, stress, time and porosity (Korthaus, 1984) and between bulk modulus and stress (Holcomb and Hannum) have been proposed. They each have limitations in describing in situ behaviour where load controlled conditions do not prevail.

Studies of the in situ characteristics of salt tailings are limited and provide little guidance for design purposes. Kappei and Gessler (1984) have reported on the deformation and permeability of 60 year old fill. Although it contained other industrial wastes which have unknown influence, a cohesion of 1 MPa had developed and the angle of internal friction was 60°. Distinct consolidation was noted during creep tests of recovered core despite the length of time over which the fill had been subjected to rock loads from closure.

The application of salt backfill for strata support in potash mining has become a topic of concern in Saskatchewan, where longwall methods have been considered, and in New Brunswick, where high extraction mining is practised. However, the studies so far published (e.g. Denison-Potacan, 1987) have been of insufficient scope to yield useful information on the long term support which the fill might provide.

## 2 IN SITU INVESTIGATIONS

As part of a continuing rock mechanics programme, excavations in the potash, hanging wall and footwall salts and Basal Halite have been monitored. Results of these studies, along with those from laboratory tests, have been used to derive relevant mechanical properties for each of the rock types. In order to check the validity of using these properties for the prediction of large scale mine stability, and to assess the contribution to stability made by the backfill, a potash stope was extensively instrumented during its development.

Tailings delivered to the stope corresponds in gradation to a poorly graded, fine sand with a coefficient of uniformity ($D_{60}/D_{10}$) of 4.8 (Fig. 2). The moisture content on arrival in the stope is 6-8%. This has proven to be the optimum value - if it is any wetter the tailings are dilatant and become quick under wheel loads, if drier, a lower degree of compaction can be achieved. The free brine drains readily from the tailings and, as suggested by laboratory tests, a moisture content of around 2% is rapidly reached. Figure 3 shows the distribution of moisture content in tailings which has been in place from 6 to 10 months.

The compacted density of the fill depends significantly on its location in the cross section of the stope. Near the footwall and in the centre, the scoop trams traverse the tailings many times during the

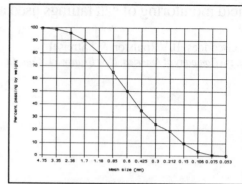

Fig. 2  Particle size gradation for salt tailings

placement operation. On the hanging wall side, where headroom is often limited by the low dip of the ore, fill is pushed against the wall by the ejector plate on the

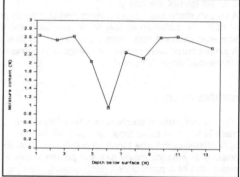

Fig. 3 Moisture content profile in compacted tailings

scoop, but not driven upon and thus has a much lower density. This is illustrated by Figure 4 which depicts the results of sand cone tests taken approximately 12 hours after placement. After several weeks, during which time there is much vehicular traffic along the centreline of the stope, it is common for surface densities in that region to be up to 10% higher than those near the footwall. The reduced density adjacent to the hanging wall represents a significant problem. Drying shrinkage and settlement combine to create a void between the fill and the wall, which varies from a few centimetres in stopes dipping at 45° to a metre or more in parts of stopes dipping at 25°. Thus hanging wall support is generally not available where it is most needed until some fracture and collapse has taken place.

Buried instrumentation to measure wall deformations and fill pressures was installed in the 1900-2 stope on lifts 4, 5 and 10 at section 140+00. This location was chosen because it provided the opportunity to study a stope throughout its development and provided reasonable egress for the instrumentation cables.

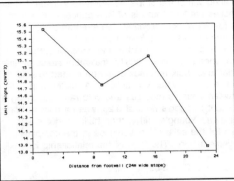

Fig. 4 Backfill densities on a cross section of the potash stope

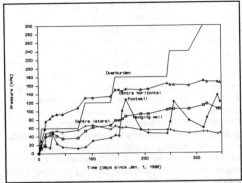

Fig. 6 Earthpressure cell readings on lift 4, 1900-2 stope.

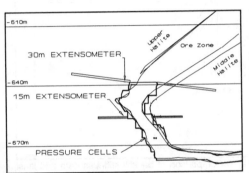

Fig. 5 Cross section of the ore body at instrumentation site

Figure 5 is a cross section of this location, showing the instruments installed. It may be noted that the ore body is overturned at this section, so the footwall will in some respects be acting as the hanging wall would in the normal orientation.

Hydraulic earth pressure cells, equipped with transducers were packed in silica sand and buried in the fill. One was placed against each wall, one horizontally at the centre of the stope and one vertically perpendicular to the centreline. A similar arrangement was placed on both lift 5 and lift 10. The cells have a diameter of 230 mm and a maximum rated pressure of 6.9 MPa. Their orientations were chosen to permit measurement of the support pressure exerted by the fill on the stope walls, and the vertical and lateral pressures within the fill. As can been seen in Figure 6, the vertical pressure of tailings has reached an almost stable value after 4 lifts of fill have been placed over the cell location, suggesting that arching between the walls is responsible for supporting the major portion of the incremental fill load. The calculated tailings overburden pressure is included in the Figure for comparison. Lateral

pressure is also constant at 36% of the vertical pressure; assuming this to be the at-rest earth pressure, an angle of internal friction of $39°$ can be deduced. Pressure on the overturned hanging wall continues to increase with increasing depth of fill, possibly indicating movement of the wall into the fill, but also reflecting the geometry of the stope. The footwall has experienced a series of transient pressure increases which have occurred as the mining machine cut out a new lift vertically above. These are believed to result from slumping of fill into voids along the overturned footwall as a result of vibration and increased pressure on the fill surface. The depth is too great for the imposed static surface load alone to cause such transients. Further evidence for fill failing in a similar manner to an earth slope has been noted at other locations where tension cracks have appeared over great lengths, some 2 m from the footwall.

Grouted multi-point extensometers, equipped with potentiometers for remote reading, were installed as shown in Figure 5. Anchors are at depths of 3, 6, 9,

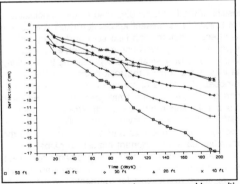

Fig. 7 Hanging wall deformations measured by multi-point extensometer

12, and 15 m on lift 5 and at 6, 12, 18, 24 and 30 m on lift 10. Total stope closure is also monitored remotely on lift 10, and has been measured consistently by tape extensometer on the open lift during mining. Extensometers in both walls indicate that deformations on the 5th lift extend beyond the 15 m depth of the deepest anchor. Also, contrary to evidence found elsewhere in the mine, the hanging wall is expanding more rapidly than the footwall in response to mining. This was thought to be an indication of bed separation in the hanging wall, where clay partings and interbedded anhydrite are common. Mining induced hanging wall failure nearby has provided additional evidence of the weakness of bedding planes and shear features in the Upper Halite. However, numerical modelling has provided evidence that this may be simply a geometric effect and that extensometer readings are consistent with the known material properties.

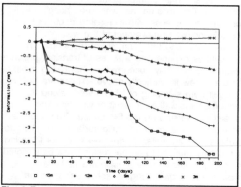

Fig. 8 Footwall deformations measured by multi-point extensometer

## 3 BACKFILL PROPERTIES

Results of the in situ studies lead to the conclusion that backfill in its present application does little to alter the stress or deformation field surrounding an active stope. Its main ground control function is to limit failure to relatively small bed separations, thus limiting the loss of support for the rock mass. For the purposes of analysis of a stope during its relatively short operational life, the small pressure exerted by the fill is small compared to the in situ stress of about 14MPa at this depth. However, the tailings fill a majority of the mined void and over the long term their properties have an important impact on global mine behaviour. Hence for stability assessments and mine planning in more general terms, it is important to have a constitutive relationship for the backfill which can be effectively implemented in a numerical scheme.

Of primary importance is the compressibility, or bulk modulus, of the backfill. Since it is completely

confined within the stope, its ability to withstand shear stresses (its "strength") is of little concern. Study of the results of consolidation and strength tests performed on tailings from a Saskatchewan potash mine (Beddoes, 1984) led to the conclusion that, independent of initial density, maximum stress or time under load, bulk modulus could be related to void ratio (e = Vol. voids/Vol. solids). A relationship between Young's modulus and void ratio, assuming a constant Poisson's ratio of 0.28, was formulated from this data. Young's, rather than bulk, modulus was used for the sake of convenience in the computer implementation. The form of the relationship is:

$$E = 34/e^{1.7} \quad MPa$$

Figure 9 compares this relationship with the laboratory data. Although all other material models employed in

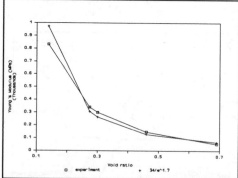

Fig. 9 Comparison of backfill stiffness data and least squares fit

the finite element program (GEOROC, Fossum et al, 1985) are visco-elastic, omitting the time variable from the backfill model is felt to be justified because the rate of deformation of the fill will always be controlled by the rate of rock deformation. The relationship is readily implemented in a finite element program because the volume of every element is calculated routinely, thus void ratio at any time can be found if its value at the time of placement is known. In order to reduce the number of times at which the stiffness matric must be recalculated as the fill becomes stiffer, a user-specified ratio between the newly calculated and previous stiffness value must be exceeded. This ratio is normally 1.1. In addition, the fill stiffness is checked before being updated to ensure that it has not become greater than the host salt.

Figure 10 illustrates the typical effect of backfill, described by the proposed relationship, upon the rate of closure of a long cavity with a 15 m X 25 m cross-section, typical of the salt stoping area of the PCA mine. Closure of a stope in the highly plastic Basal Halite is effectively halted within 5 years provided the space is filled tightly.

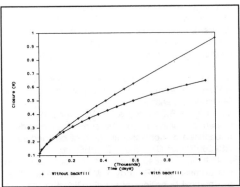

Fig. 10 Closure of a rectangular cavity with and without backfill

## 4 MODELLING OF THE INSTRUMENTED STOPE

A finite element model of the instrumented stope was created for the purpose of validating the behavioural models for the various evaporite units against field data. Each of these units was described by a power law of the form:

$$s = A \, \sigma^m t^n \quad \text{where s is strain}$$

A, m and n are constants, $\sigma$ is the deviatoric stress. m was found to decrease to a value of about 1 with decreasing stress, which is in agreement with theories regarding a change in creep mechanism from dislocation climb to diffusion (Rischbieter, 1984; Dawson, 1979). The power law for each material was

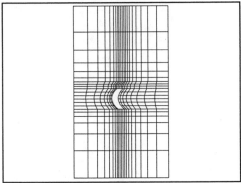

Fig. 11 Finite element mesh used for simulations of instrumented stope

derived from a combination of laboratory tests and instrumentation of small test rooms. Similar laws were used by Denison-Potacan (1987). The mesh used is illustrated in Figure 10, in which the backfill elements within the stope have been omitted for clarity.

A series of analyses with this model were performed to demonstrate the effects of varying geometry,

material properties and the absence of backfill. A significant observation was that a greater proportion of the total closure was derived from the footwall although the expansion within the first 15 m was much greater in the hanging wall. This supports the field observations presented earlier. The analyses also predicted a net contraction in the first 6 m of the footwall; this behaviour in the footwall extensometer set at 3 m had previously been assumed to be erroneous. Figure 12 illustrates the predicted footwall deformations and shows that as time progresses an increasing depth of footwall is contracting. This is caused by tension in the dip direction and would be eliminated if tensile cracks developed.

The model predicts a rate of stress increase in the fill which is somewhat greater than that measured to date. After 5 years the fill pressure would be 4 MPa and it would increase very little during the following 10 years. At 5 years the rate of stope closure would be 50% of its value if backfill had not been used.

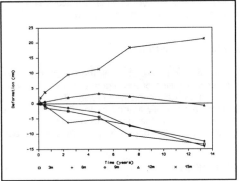

Fig. 12 Predicted deformations in the footwall

Extrapolation of the current measured pressures results in pressures lower than those predicted. This is probably due to the tailings being assigned too high a stiffness value. Tests on core from the PCA backfill will be conducted in the future to define the stiffness function more accurately.

## 5 CONCLUSIONS

Salt tailings have been shown in practice to provide passive support to the stopes, preventing caving from regressing into the walls. Without tailings support hanging wall caving would limit the maximum stope size considerably. However, arching between the stope walls appears to develop very rapidly, reducing the anticipated increase in fill pressure with depth. Consolidation and stiffening of the fill is thus caused primarily by visco-elastic closure of the stope and pressures at the wall-fill interface increase slowly. Over the anticipated 5 year life of a stope, fill pressures will probably not exceed 10-15% of the in situ stress and total closures will be reduced by less than 30%. In the long term, fill is likely to be effective in reducing total closures by approximately 70%, improving the global stability of the mine and reducing surface subsidence. Direct measurements of deformations and pressures in the fill have proven to be invaluable in gaining an understanding of the contribution of fill to stope behaviour.

The behavioural model for backfill developed from laboratory tests provides reasonable agreement with measured fill behaviour and its simplicity is believed to be an asset for practical applications in which funds for sampling, laboratory testing programmes, and numerical analysis are often limited. The model will be refined in the future when core samples of undisturbed tailings can be collected.

Predictive modelling has been a useful tool in the interpretation of in situ measurements. Although the numerical model was not able to provide a completely accurate picture of fill-stope interaction a priori, the approach of developing the model from field measurements of excavations in single rock types and applying it to large scale, multi-material stopes has been validated. A site-specific model capable of reliable analysis of more complex situations is a realistic expectation in the near future.

## 6 ACKNOWLEDGEMENTS

The authors would like to thank the Potash Company of America, Division of Rio Algom Ltd. for permission to publish the data contained herein. The work reported was partially funded by the Province of New Brunswick, through the Mineral Development Agreement, and its support is gratefully acknowledged.

## REFERENCES

Beddoes, R.J. 1984. Determination of engineering properties of waste salt for backfilling in underground potash mines. CANMET project no. 310104 Calgary:RE/SPEC Ltd.

Dawson, D.E. 1979. Preliminary deformation-mechanism map for salt (with application to WIPP). Sand-79-0076. Albuquerque:Sandia National Laboratories.

Denison-Potacan Potash Company 1987. Use of backfill in New Brunswick potash mines. Report to CANMET for contract no. OSQ85-00208 Sussex, N.B.:Denison-Potacan.

Fossum, A.F., R.R.Brander, J.D.Chieslar & H.Y.Tammemagi 1985. GEOROC: a numerical modelling package for designing underground openings in potash. Report to CANMET for contract no. OSQ83-00269 Calgary:RE/SPEC Ltd.

Hesterman, L. 1979. A brief study of compaction and water expulsion in compressed potash mine tailings. Report No. E79-8 Saskatoon:Saskatchewan Research Council.

Holcomb, D.J. & D.W.Hannum 1982. Consolidation of crushed salt backfill under conditions appropriate to the WIPP facility. SAND 82-0630 Albuquerque N.M.:Sandia National Laboratories.

Kappei, G. & K.Gessler 1984. In situ tests on the behaviour of backfill materials. In H.R.Hardy &

M.Langer (eds.), The mechanical behaviour of salt II. Clausthal-Zellerfeld:Trans Tech Publ.

Korthaus, E. 1984. Effect of backfill material on cavity closure. In H.R.Hardy & M.Langer (eds.), The mechanical behaviour of salt II. Clausthal-Zellerfeld:Trans Tech Publ.

Miller, H.D.S. 1983. Use of rocksalt as backfill. In R.M.McKercher (ed.), Potash technology. p 341-345 Saskatoon, Canada.

Pfeifle, T.W., P.E.Senseny & K.D.Mellegard 1984. Influence of variables on the consolidation and unconfined compressive strength of crushed salt. RSI-260 Rapid City, S.D.:RE/SPEC Inc.

Reischbieter, F. 1984. Stress distribution and flow in salt domes. In H.R.Hardy & M.Langer (eds.), The mechanical behaviour of salt II. Clausthal-Zellerfeld:Trans Tech Publ.

Roulston, B.V. & D.C.E.Waugh 1985. Stratigraphic comparison of the Mississippian potash deposits in New Brunswick, Canada. In B.C.Schreiber & H.L.Harner (eds.), Sixth international symposium on salt. p 115-130 Alexandria Va:Salt Institute.

# 7 System design

*Innovations in Mining Backfill Technology, Hassani et al. (eds), © 1989 Balkema, Rotterdam. ISBN 90 6191 985 1*

# Backfilling on the base metal mines of the Gold Fields Group

D.A.J.Ross-Watt
*Gold Fields of South Africa, Johannesburg, RSA*

ABSTRACT: The Gold Fields of South Africa (GFSA) Group operates three major mechanised base metal mines. Black Mountain Mineral Development Company (Pty) Limited (BMM), O'okiep Copper Company Limited (OCC) and Tsumeb Corporation Limited (TCL). Backfilling on all three mines is addressed. In particular the changeover from blasthole stoping to cut and fill at BMM, the successful use of modified slag as a cementitious binder at OCC and the progress toward a cemented unclassified tailings backfill at TCL are discussed.

## 1.0 INTRODUCTION

The base metal mines of the Gold Fields of South Africa group annually mine and treat some 5,4 million tons of ore. Metals and minerals contained in the final products are copper, lead, zinc, tin, silver, gold, cadmium, arsenic, antimony and pyrite. In particular three major mechanised base metal mining companies fall within the control of the group: Black Mountain Mineral Development Company (Pty) Limited (BMM), O'okiep Copper Company Limited (OCC) and Tsumeb Corporation Limited (TCL).

## 2.0 BACKFILLING DETAILS

The three mining companies mentioned above have different backfilling requirements based on the types of orebodies, the mining methods, the value of ore and other factors. The materials available for backfilling differ significantly as did the economic considerations applied from time to time in the design of suitable backfills. TCL was a pioneer in cemented backfilling in Southern Africa, OCC designed backfills to suit extremely tight economic constraints and BMM has made significant advances in a short period of time to maximise extraction of a high grade orebody in a most cost effective manner.

### 2.1 Backfilling at Black Mountain

The ore reserves at the Broken Hill deposit at Black Mountain are contained in two orebodies extending over a total strike length of 1 400 metres and plunging from a surface outcrop in the west to a depth of approximately 800 metres in the eastern extremity.

Blasthole open stoping was employed where the dip is steep enough to allow gravity cleaning and cut and fill stoping is used elsewhere.

The backfill placed in blasthole stopes must be of sufficient strength to allow a free-standing wall during pillar extraction. The maximum height of such a wall was in fact an exposure over four levels of 140 metres. Due to the dip of the stope however the maximum vertical height between footwall and hangingwall was less than 50 metres. For this criterion the failure of backfill under its own weight alone was considered.

Initially an unconfined compressive strength (UCS) of 700 kPa at 90 days was recommended. This strength criterion was subsequently reduced to 400 kPa based on extensive large scale model testing underground. It is of significance that the blasthole stoping area comprised the highest grade section of the orebody where complete extraction had to be achieved but where 50 percent of the stopes had been extracted and backfilled prior to the exposure of a backfill wall.

In addition the backfill had to be sufficiently resilient to withstand the effect of pillar stope blasts. The cement ratio required for a UCS of 700 kPa is in fact too high to allow for resilience. This also resulted in fairly severe vertical shrinkage cracks. In the early stages however the risk associated with inadequate strength was considered to be greater than that associated with blast damage or shrinkage effects. The change to a UCS of 400 kPa improved the properties of the fill in this regard.

The backfill in the base of the stope had to be sufficiently strong to allow the crown pillar of the cut-and-fill stope to be mined out completely.

The backfill in all blasthole stopes and in the cut-and-fill stopes must be sufficiently strong to allow for overall stability of the mine as extraction progresses. Due to the width of the orebody it is unlikely that any redistribution of stress will in fact take place via the backfill. Rather, as mining progresses stresses will be distributed to the major dip stability pillars, the unmined portion of the orebody and the host rock on the boundaries. The backfill material will however contain and prevent progressive scaling of the excavations and in this way assist in maintaining the strength of the rock mass immediately adjacent to the excavations. This bulk criterion was the major property required for backfill placed in blasthole pillar stopes.

In cut and fill stoping the backfill is required to provide a readily available working surface as well as sufficient bulk for overall stability. In addition the backfill must be sufficiently strong broken ore. Test work indicated that

with the available loading equipment and expected blast effects a backfill material with a UCS of 800 kPa at 24 hours would be necessary. Below the capping a low strength bulk fill was required on which persons could safely work. A 12 hour strength of 200 kPa was required.

In all backfill the strength and the drainage properties had to be such that retained water was kept to a minimum so that no possibility existed of backfill flowing once a holing was made into a backfilled area. In the case of cemented backfill this is not a problem and accordingly a small proportion of cement was used throughout the blasthole pillars. For cut and fill uncemented bulk backfill is placed. Drainage characteristics are however good and in addition the capping layers have the effect of stabilising the overall backfill mass.

The following materials are available for backfilling. The recovery plant produces mill tailings which consist primarily of very fine magnetite particles with a relative density of 3,4. Relatively coarse dune sand is abundant, but has a low percolation rate and a high residual moisture content. A limited resource of river sand was available. This was very coarse and had high percolation rates and a low residual moisture content.

No suitable materials which could be used to supplement or substitute for Portland cement were readily available.

The concentrator tailings had to constitute the bulk of the backfill. The nature of stoping in the blasthole area was such that large voids would be opened which would subsequently have to be backfilled at a comparatively high rate. As stoping progressed the rate of backfilling became more uniform but in the early stages high backfilling capacity was required and insufficient tailings were available even if unclassified tailings were used. With some degree of classification the balance would be even more unfavourable. At least during the backfilling of the blasthole stopes and pillars a further material source was required from the point of availability alone. The material chose for this purpose was the coarse dune sand.

The viscosity of the backfill is of importance for transportation. Early bench tests showed that dune sand alone had a low viscosity and would tend to settle out in pipe columns. Conversly classified and unclassified tailings and mixtures of tailings with dune sand indicated a rapid increase in viscosity at around 78 percent pulp density (solids by weight). The concentrator tailings had a high specific gravity and these pulp densities thus represented very large volumes of water in the backfill. The water required for transport was excessive for the hydration of the cement and thus good drainage of the backfill was essential. Also it was necessary to place all backfill with as high a pulp density as possible. In practice pulp densities of the order of 75 percent were employed initially.

At the design stage it was considered important that the backfill should be sufficiently permeable to allow percolation of excess water through the backfill to drainage points. The backfill design was such as to maximise percolation. The required percolation rates quoted at the time were around 100mm per hour. It was considered that while reduction in the fines content of the backfill would reduce strength, a good percolation property was essential and any reduction in strength could be made up with appropriate additional cement. In retrospect more fines could possibly have been retained allowing increased strength and lower cement contents but without effecting the overall drainage properties.

In practice most of the water was drained by decanting due to the relatively low rates of rise. Very little percolation occurred through the cemented backfill.

UCS tests were carried out using unclassified tailings, classified tailings and combinations of classified tailings and dune sand with various cement contents. A combination of 50 percent classified tailings and 50 percent dune sand with 1:13 cement ratio of solids by weight produced the required strength of 700 kPa after 14 days for the bulk of the blasthole stope backfill.

Further testing using a 1:14 scale model resulted in the strength criterion being reduced to 400 kPa and the cement content being reduced to a ratio of 1:20. A cement content of 1:13 was used for the backfill placed in the troughs at the base of the blasthole stopes and pillars. A cement content of 1:30 together with the same bulk constituents was planned for the blasthole pillars. In practice as much waste as possible was tipped into the pillar voids and both dune sand and cement content were reduced.

Backfill for cut and fill stopes is required both as a bulk fill and as a working surface which must have rapid setting properties. This backfill will be transported over progressively longer horizontal distances as stoping progresses upwards and down plunge.

The daily backfill rise was expected to be in excess of 1,5 metres thus requiring good percolation rates. This was particularly important where uncemented bulk backfill was used.

Initially the same bulk constituents as for the blasthole stopes were used for cut and fill but with appropriate cement ratios. For a 3 metre stope rise approximately 2,25 metres of bulk backfill was placed followed by a capping of 0,75 metres to act as a working surface. This consisted of 50 percent deslimed tailings and 50 percent dune sand with a 1:30 cement ratio for the bulk backfill and a 1:10 cement ratio for the capping. As for the blasthole stopes most of the drainage was by decantation. This backfill system however contained significant cement and was expensive. Testwork was carried out on a laboratory scale and in addition various full-scale tests were conducted in producing cut-and-fill stopes. Cappings with as low as 1:20 cement ratios were placed on a test basis and in extreme cases very poor working surfaces resulted with significant loss of fill to the ore trammed. The classified tailings dune sand combination did not have sufficient percolation to be used as an uncemented backfill. Uncemented dune sand alone was tested but the fine fraction prevented any percolation. The sand retained excessive water which resulted in an effective quick sand. Work was however now in progress to reduce cement by careful investigation of the bulk materials available for backfill.

The mill tailings, whether cycloned, uncycloned or combined with sand, had such low percolation rates that this material could not be used as an effective uncemented backfill. The percolation rates of dune sand were low, normally less than 10mm/hour. The sand in ancient river beds however, had a much greater percolation rate of around 360mm/hour making this a good material for uncemented bulk backfill. A total reserve of 600 000 tons of river sand was available within some 8km of the backfill plant. The channels of sand were however of shallow depth, limited extent and were mined with some degree of difficulty. While the transport characteristics of the classified tailings dune sand based backfill were good due to a relatively high viscosity, the low viscosity of the river sand required placement of backfill at higher velocities and lower densities. This was also necessary due to the increased horizontal distances involved. A velocity of 3 metres/sec was employed at a relative density of 1,65 which corresponds to low pulp density of 65%. The high percolation rates of the placed river sand made drainage relatively easy. The large quantities of percolation water produced in the stope however placed severe strain on the mines settling and pumping system which required upgrading.

The cost of river sand was relatively high and as this resource became depleted it was necessary to establish the technology required to use dune sand as a bulk uncemented backfill. The dune sand reserves of some 5 000 000 tons occurred in large easily mineable deposits approximately 7 km from the backfill plant. Laboratory and large scale testwork indicated that unclassified dune sand could not be used as an uncemented backfill owing to the low percolation rates and the associated high residual moisture content of the backfill. This high moisture content resulted in high pore pressures and instability with liquifaction occurring readily. It was therefore essential, prior to using dune sand as an uncemented backfill to improve its percolation rate and to reduce the residual moisture content of the placed backfill. The main cause of the poor percolation was the presence of very fine material. The dune sand had 3 percent material less than 75 $\mu$m compared to 1 percent for river sand. Testwork indicated that if the percentage of material less than 75 $\mu$m could be reduced from 3 percent to 1 percent then the percolation rate and residual moisture content would be acceptable. Both cyconing and hindered settling were considered with the latter method proving too costly. Various flocculants and chemicals were tested to study the effect on the residual moisture content. A two-stage hydrocycloning plant was installed with a capacity of 250 tons of backfill per hour at a relative density of 1,75. The material less the 75 $\mu$m is kept below 1 percent and the total reject of sand feed is maintained at below 3 percent. The deslimed dune sand is placed by gravity and generally requires high transport velocities and high water contents due to the low viscosity of this backfill. Wear in the reticulation system was greatest for river sand followed by dune sand and least for cemented tailings. The dune sand requires additional drainage towers due to lower percolation rates. The beaching effect of the dune sand is less than that for river sand and this reduces the amount of cemented capping backfill that would otherwise be used in filling the hollows left by steep beaches. Where this effect is excessive a small amount of 1:30 cement ratio backfill is placed to level off the surface prior to placement of the final 1:10 cement ratio capping.

The initial backfill used for the capping was subsequently changed to uncycloned tailings with a 1:10 cement ratio. This is suitable as percolation in the thin capping is not critical. Ore loss in cappings is relatively low at less than 0,5 percent with cleaning by bulldozer at the end of each lift. Resulting loss of the capping results in a dilution of ore by backfill of some 2,5 percent by weight of the total ore mined.

A crown pillar is formed between the lowest cut of the stope and the stope advancing upwards from below. Due to the high ore grades it is essential that close to total extraction of the crown pillars is achieved. This is partially a function of the strength of the backfill placed in the lowest cut of the upper stope. Several systems have been employed incorporating strong cemented backfill with up to a 1:6 cement ratio, 19mm grouted cables, and 8 gauge diamond wire mesh. These have been partially successful with some difficulty still being encountered in extracting crown pillars. One of the problems has been separation where backfill has been placed over relatively extensive areas in thin layers. A new development is backfilling in paddocks so as to create full lift layers in excess of 2 metres with each backfill placement. The second important aspect however is the mining of the crown pillar from below. Techniques have not yet been perfected. For wide stopes no amount of backfill and reinforcement strength will be adequate for large unsupported areas and an effective method of cutting and tight filling before extraction of secondary pillars must be developed.

## 2.2 Backfilling at O'okiep Copper Company

Early backfilling at the various mining operations of the OCC was for the purpose of general stability. Deslimed tailings material produced at the Nababeep, O'okiep and Carolusberg mills was employed. Backfilling at the Hoits mine became necessary around 1980 when the large voids created by stoping started to result in severe overall stability problems on the mine. A cemented backfill was used for the purpose of filling existing voids and for the cut and fill method adopted for further mining. Initially backfill consisted of a cemented mix of Carolusberg mill tailings and dune sand from the Hoits mine area in a ratio of 3 to 1 by weight. This was reticulated at a pulp density of 75 percent. Subsequently a cemented backfill using dune sand only as a bulk material was used. This backfill was reticulated at a pulp density of 80 percent. A 1:17 cement ratio resulted in an UCS of 650 kPa.

Carolusberg is the largest individual mine at OCC. A vertical shaft to a depth of 862 metres was sunk in 1961 to service the ore reserves now known as the Carolusberg Shallow area. Subsequently the discovery of the Carolusberg Deep orebody resulted in the sinking of a vertical shaft to a depth of 1690 metres together with an appropriate sub-shaft and spiral decline. The orebody contained some 17 million tons of ore grading at around 2 percent copper. The individual orebodies constituting the Carolusberg Deep deposit have strikes up to 150 metres, a width of 120 metres and a height of up to 650 metres. The ore deposit bottoms out at 1540 metres below surface.

The Carolusberg ore body is characterised by a horizontal maximum principal field stress of the order of twice the vertical stress and which reached over 92 MPa at the bottom of the Carolusberg Deep orebody. The orebodies are in addition heavily jointed. This has necessitated well designed stoping methods and a high level of rock mechanics control. The latter being achieved by a continuously updated three dimensional computer model of the mine.

In the Carolusberg Shallow orebody primary stopes were mined by sub-level open stoping methods followed by mass blasting of pillars. With increasing depth major rock mechanics problems were experienced. Knowledge that conditions would worsen in the Carolusberg Deep orebody led to the testing and acceptance of the vertical crater retreat (VCR) mining method for this orebody. The stope back is advanced upwards and only sufficient broken rock is extracted to allow a void for blasting. During blasting of the stope the walls are supported by the broken rock and stoping crews work safely in the drill drives remote from the void. Once extraction is complete the broken rock is removed as rapidly as possible and the stope is filled with cemented hydraulic backfill. The orebody is mined in a checkerboard fashion with typical stope dimensions of 20 metres to 30 metres in cross-section and heights up to 150 metres. The backfill must be sufficiently strong to withstand mining of an adjacent stope and as mining increases to be free-standing for heights of up to 150 metres. Testwork and experience indicated that an appropriate strength criterion for the backfill was a UCS of 850 kPa after 90 days.

Experience had been gained with cemented hydraulic backfill at Hoits and initially a decision was made to use Carolusberg tailings and Hoits dune sand as bulk materials. Testwork showed that a backfill consisting of a ratio of 3,5:1 Carolusberg tailings to Hoits dune sand by weight, a 1:15 cement ratio and reticulated at a pulp density 75% would provide the required strength criterion. The Carolusberg Deep orebody project however involved significant capital expenditure and was set against a background of depressed copper prices. In order to achieve adequate returns mining working costs would have to be kept to a minimum. Even with a possible reduction of cement content as experience was gained the high cost of using Portland cement as a binder for the backfill was not appropriate.

The copper reverbatory slag produced at the OCC smelter was a readily available low cost possible substitute for Portland cement. Copper reverbatory slags are however less suitable for this purpose than are iron blast furnace slags and significant investigation and testwork was required to define the optimum final slag product to be made available for backfilling.

The common objective of cements whether conventional Portland cement, blast furnace or reverbatory furnace cement is the formation of stable, insoluble cementitious hydrated calcium silicates. The essential common factor in these materials is CaO, $SiO_2$ and $H_2O$ or C-S-H. Other cementitious compounds are $Al_2O_3$, $SO_4$ and $Fe_2O_3$ variations. The optimum slag will contain all of the above cementitious compounds in the correct proportions and where deficiencies exist supplements must be made. Abundant iron and manganese such as are present in reverbatory slags have a detrimental effect on the cementitious properties of a slag. This feature together with lower contents of the essential elements CaO, $Al_2O_3$ and $SO_4$ are the reasons for reverbatory slags being less suitable than blast furnace slags. In particular additives are required for the OCC reverbatory furnace slag.

In addition to the ability of the slag to supply the fundamental cementitious elements the following physical characteristics of the slag are significant. To ensure maximum reactivity of a slag all its phases should be in a glass or amorphous state. Quenching of the O'okiep slag produces 70 percent to 90 percent glass and thus it is suitable with respect to this property. The grain size, or specific surface area of the slag expressed in blaine has a considerable influence on its reactivity. The higher the specific area the higher the reactivity. A fineness of 3 000 blaine is achieved in the O'okiep slag milling process. The presence of $AlO_2$ in the slag reduces the silica bond strength of the glass and thus increases reactivity. The heat treatment of the slag has an influence with higher temperatures prior to cooling resulting in higher reactivity. The hardness of the slag has an effect on subsequent grinding costs. In this respect wet grinding is less energy intensive than dry grinding but has the disadvantage that hydration may start prematurely.

Various additives with the purpose of affecting the start of the chemical reaction, supplying additional elements participating in the formation of cementitious materials or controlling the speed of reaction were investigated.

External features such as water quantity and quality, temperature and humidity of the environment, ratio of slag to tailings, lime addition to slag and the handling properties of the backfill produced were considered.

In all some twenty variables were investigated to test the suitability of the local reverbatory slag and to produced an optimum cementitious material capable of generating the backfill strengths required.

Laboratory and large scale testwork showed that while the locally produced slag is low in calcium, aluminium and sulphate and high in silica a suitable cementitious material can be produced by the addition of lime and gypsum. The addition of a 6 to 7 percent by weight of this modified slag to the backfill was shown to produce the required strength at

90 days thus totally subsituting for a similar percentage of Portland cement.

A description of the backfill production for the Carolusberg Deep orebody follows. The various quantities quoted refer to a backfill production of 90 000 tons of solids per month.

A granulated reverbatory furnace slag screened to minus 3 mm is trucked from the O'okiep smelter on a backhaul against the concentrates. The composition of the 5 000 tons per month of slag consumed is as follows:

| % | % | % | % | % |
|---|---|---|---|---|
| Copper | Ash | $SiO_2$ | FeO | $Al_2O_3$ |
| 0,6 | 0,8 | 35,9 | 41,0 | 4,0 |
| % | % | % | % | |
| CaO | MgO | Pb | Zn | |
| 9,0 | 4,8 | 0,08 | 3,9 | |

Gypsum is trucked approximately 300 km and is added at a ratio of 1:5 to slag, thus 1 000 tons per month. These constituents are milled together through two ball mill stages to produce at 3 000 blaine product. The secondary mill discharge is treated by flotation to recover 50 percent of the contained copper prior to thickening to a pulp density of 60 percent at a relative density of 1,7. This material is stored in a holding tank at the backfill plant. Carolusberg current mill tailings are progressively thickened to a pulp density of 71 percent at a relative density of 1,86. Dry tailings from the Carolusberg slimes dam have replaced Hoits dune sand to make up the total backfill material requirements. Lime and the slag/gypsum constituent are mixed with the bulk materials to produce backfill at a rate of 172 tons of solids per hour at a pulp density of 75,5 percent and a relative density of 2,0. The solid constituents of the backfill material are as follows:

Wet Tailings 120 tons per hour (69,75%)
Dry Tailings 40 tons per hour (23,%)
Slag and Gypsum 11 tons per hour ( 6,5%)
Lime 1,2 tons per hour ( 0,75%)

Reticulation to the stopes is relatively easy with short horizontal distances in relation to the available vertical head.

## 2.3  Backfilling at Tsumeb Corporation

Mining at the Tsumeb pipelike orebody started in 1906 and by 1940 had progressed to 22 level some 650 metres below surface. The bulk of this mining was in the narrower section of the orebody with extraction by means of a full longitudinal cut across the orebody. Backfilling in these areas was by means of quarried waste rock. The orebody widened from 18 level and a full longitudinal cut was no longer possible. Transverse stopes with intermediate pillars were employed.

Initially the stopes were filled with quarried rockfill with an associated loss of ore. Pillars were extracted using square set timbering and uncemented hydraulic backfill. This type of backfill was first employed in 1967 and consisted of classified tailings with red soil to make up bulk requirements. As this backfill was uncemented, percolation was essential hence the removal of a significant proportion of the fines. This backfill was used in both primary stopes and in square set mining of pillars prior to being replaced by cemented hydraulic backfill. The uncemented backfill was not in fact satisfactory with fluid fill running away on occasion when holings were effected into backfilled cavities and with significant material loss into the general infrastructure of the mine causing very poor working conditions. Loss of ore into the backfill was a problem and these factors together with increasing mechanisation made a changeover to cemented backfill essential.

Cemented backfill production was started in 1974. This backfill was initially employed in pillar extraction improving support conditions substantially and allowing for significant reductions in timber support. Subsequently cemented hydraulic backfill became standard for all cut and fill stoping.

Initially the bulk materials employed were the same as for the uncemented backfill with percolation still being regarded as important. The proportions shown below illustrate the large amount of material which had to be handled to achieve the required filling of voids.

1 ton of ore produced required 0,64 tons of backfill.
1 ton of backfill placed required:
0,85 tons of red soil hauled
0,52 tons of red soil after slimes and oversize removed
0,42 tons classified mill tailings
0,06 tons cement on average

In fact of every ton of tailings produced only 0,35 tons were utilised in backfilling. For every ton of backfill employed 0,07 tons of development waste were packed.

The bulk backfill in the cut and fill stopes utilised a cheaper 1:20 cement ratio. A high strength layer of 1:8 cement ratio was placed in a capping of between 0,3 metres and 0,5 metres thick. The UCS of the backfill containing classified tailings and red soil was as follows:

1:8 cement ratio - 930 kPa at 7 days
1:20 cement ratio - 110 kPa at 7 days

This backfill was however expensive due to the cost of handling large quantities of red soil and due to the high cement content needed to achieve the required strength. Substantial testwork was carried out in order to investigate the suitability of various materials as a substitute for the red soil and to try to achieve the required strength with a reduced cement content. It was also recognised that percolation was not an essential characteristic for cemented backfill and that a higher strength could be obtained by retaining fines and achieving a higher density. Most of the water is in fact drained by decantation. Tests were carried out on the currently used materials, ground dolomite from the old rockfill quarry, as well as the tailings underflows from the primary and from the secondary cyclones and unclassified tailings.

The most suitable material was the unclassified tailings which showed the following average strengths.

| Cement Ratio | UCS kPa at 7 days |
|---|---|
| 1:8 | 2 900 |
| 1:12 | 830 |
| 1:15 | 650 |
| 1:20 | 500 |
| 1:25 | 240 |
| 1:30 | 240 |
| 1:35 | 120 |

In general strength increased with fines content of the tailings as is evident below:

| | Percent Minus 45μm | UCS kPa at 7 days |
|---|---|---|
| Secondary underflow | 3,8 | 765 |
| Primary underflow | 15,0 | 800 |
| Unclassified | +45,0 | 2 900 |

This was however not true for the ground dolomite which was unclassified and had up to 33 percent minus 45 μm. The material did not lose any water, did not set at all, and after 7 days was still too wet to test. The explanation is that the ultra-fines were still present in the ground dolomite whereas in the unclassified tailings this fraction had been removed with the flotation process.

Strength also increased with increasing pulp densities as is shown in the following table for a 1:12 cement ratio.

| Pulp density before cement addition % | UCS kPa at 7 days |
|---|---|
| 82 | 1 690 |
| 80 | 1 190 |
| 78 | 1 000 |
| 76 | 920 |
| 75 | 830 |
| 74 | 760 |
| 72 | 690 |
| 70 | 565 |

Pulp densities for transport purposes should however not exceed 78 percent for strong mixes and 80 percent for weaker mixes. In practice it is difficult to consistantly achieve pulp densities of over 75 percent.

With the new backfill design the equivalent proportions to those listed above are as follows:

1 ton of backfill required:
0,96 tons dewatered tailings
0,04 tons cement
Of every ton of tailings 0,98 tons should be utilised for backfilling.
This change in fact had the effect of decreasing backfill costs by over 30 percent.

The current backfill plant is designed to produce 0,7 tons of backfill for each ton of ore produced. In order to achieve this, direct and continuous delivery of total mill tailings is necessary and sufficient storage capacity must be provided. Five storage silos are filled until overflow occurs. Flocculant is added and overflow is monitored to ensure that the relative density does not exceed 1,05. The cycle for a 400 m$^3$ silo is approximately 4 hours filling, 6 hours settling and 2 hours transfer to a holding tank. Tails are pumped at a relative density of 1,95 to a holding tank and then to a mixing tank where cement is added. Backfill with a final relative density of 1,97 is passed down the mine via boreholes. The horizontal transfer is generally relatively short in relation to the vertical head available, being restricted to less than 150 metres.

Two types of backfill are currently used with specifications as follows:

| Cement Ratio | UCS kPa 24 Hours | UCS kPa 7 Days |
|---|---|---|
| 1:12 | 83 | 830 |
| 1:25 | 35 | 240 |

The higher strength backfill provides a surface which is available for production after 48 hours.

Two aspects are given very high priority at Tsumeb. One is the maximum recovery of fines in the tailings and the second is placing the fill at the highest possible pulp density. With respect to the second aspect the viscosity can be readily measured by a simple practical cone test.

When mining at Tsumeb extended to below 1 000 metres depth an en-echelon pattern of transverse cut and fill stopes replaced the primary and pillar cut and fill system previously used. Stresses had increased significantly and the blocky ground conditions necessitated the change in the stoping system as well as substantial cable bolt roof support. The en-echelon system consists of a set of immediately adjacent transverse cut and fill stopes each 10 metres wide and 3 metres high. The pattern is led by the central stope with successive stopes lagging by one lift. This forms an arched shape which tends to be self supporting from an overall point of view. The central stope will have solid walls on both sides while all other stopes will have a solid wall on the outer side and a backfill wall on the inner side. The backfill must therefore be able to form a stable 2,5 metre high exposed wall as well as providing bulk fill and a suitable working surface for mechanised equipment. In addition to the backfill support, comprehensive grouted cable support is essential. This is in the form of 7 strand 28mm cable bolts 18 metres long and at 2,8 metre spacing and grouted into the hangingwall. As the stope advances upwards the cable is progressively cut off and when less than 5 metres of cable support remains a new set of 18 metre long cables are installed.

The mining of the crown pillar immediately below a mined out area is a significant problem. Ore grades are relatively high and total extraction is aimed at. Two aspects are most important in achieving high extraction rates. Firstly the preparation on the sill level of the stope above and secondly the mining methods used from below in extracting the crown pillar. On the sill level the footwall is completely cleaned and a grid pattern of 28 mm cables at 3 metre centres is laid on the footwall. This is covered by weldmesh reinforcing and a 1:12 cement ratio, high strength backfill is placed for the entire 2,5 metre lift. In mining from below the pillar is extracted using 3,5 metre wide transverse cuts, starting from the centre and working outwards across the orebody. The success of the mining of these pillars has been due to the ability to tightfill each cut with high strength backfill prior to mining of the adjacent cut.

## 3. CONCLUSIONS

The intention of this paper was not necessarily that of a comprehensive description of backfilling on the GFSA base metal mines but rather to highlight novel and innovative aspects. The foregoing discussion does however illustrate certain important backfilling fundamentals some of which are listed below:

a) The need to define carefully exactly what is required from the backfill, and what is not required.

b) The need to have a good understanding of the properties of all materials available for backfill.

c) The relatively low importance of permeability in cemented backfill.

d) The relatively high importance of permeability in uncemented backfill.

e) The increase in strength associated with denser unclassified material, with an associated loss of permeability.

f) The role of ultra fine material in total loss of permeability and entrapment of water.

g) The large water quantities associated with high specific gravity materials.

h) Transport problems associated with highly viscous backfill such as dense cemented tailings.

i) Transport problems associated with low viscosity backfill such as uncemented sand.

j) The increase in strength associated with placement at high bulk density.

k) The ability to modify slag to obtain a suitable cementitious binder.

l) The relative ease of handling a cemented backfill compared to uncemented backfill. The problems of spillage of uncemented backfill and the associated poor working conditions.

m) The importance of mining methods including tight filling in extracting crown pillars.

ACKNOWLEDGEMENT

The author would like to thank Gold Fields of South Africa for permission to publish this paper and to acknowledge the contribution of various published documents and internal mine memoranda which represent substantial technical work done over an extended period.

REFERENCES

De Jongh, C.L. 1980. Design parameters used and backfill materials selected for a new base metal mine in the Republic of South Africa. Application of Rock Mechanics to Cut and Mill Mining.

Smith J.D., de Jongh C.L., Mitchell R.J. 1983. Large scale model tests to determine backfill strength requirements for pillar recovery at the Black Mountain Mine. International Symposium on Mining with Backfill.

Ross-Watt, D.A.J. 1983. Initial experience in the extraction of blasthole pillars between backfilled blasthole stopes. Proceedings of the International Symposium on Mining with Backfill.

Kinver, P.J. 1985. Cut and fill mining of base metals at Black Mountain Mine. J. S.A.I.M.M. Vol. 85 No. 2.

Brownrigg, J.F. 1985. Uncemented hydraulic backfill at Black Mountain Mineral Development Company (Pty) Limited. A.M.M.S.A.

Kinver, P.J. 1988. A review of backfilling at Black Mountain Mine. Backfill in South African Mines. S.A.I.M.M., A.M.M.S.A.

Brownrigg, J.F. 1988. Consolidated backfill at O'okiep Copper Company. Backfill in South African Mines. S.A.I.M.M., A.M.M.S.A.

# Fill operating practices at Isa Mine – 1983-1988

J.D.McKinstry & P.M.Laukkanen
*Mount Isa Mines Ltd, Mount Isa, Australia*

ABSTRACT: Since 1983 a number of changes to fill operating practices have occurred at Mount Isa Mines. The introduction of heavy medium reject aggregate required changes to the system used to reticulate fill underground. To further optimise filling operations sophisticated monitoring systems have been introduced. This paper describes the various innovations.

## 1 INTRODUCTION

Mount Isa Mines Limited produces 5.5 million tonnes/year of copper ore and 4.5 million tonnes/year of silver-lead-zinc (known as lead) ore from open stope and cut-and-fill operations at the Isa Mine. Open stoping methods, which aim for 100% recovery of the orebody, require the filling of primary stopes to enable adjacent pillars to be recovered. Isa Mine's requirement for large vertical exposures and rapid re-entry to pillars requires a high performance fill with sufficient free standing ability and strength to provide the regional support for continued safe and productive open stoping. Mechanised cut-and-fill operations in narrow orebodies require rapid and free draining fill to maintain high production through the mining cycles. Fill practises up to 1983 are described in Neindorf, 1983.

## 2 FILL TYPES

### 2.1 Hydraulic fill

Hydraulic fill (HF) is produced from de-slimed and de-watered mill tailings at a specified density of 69% solids by weight, supplying two lines at 200-240 tonnes per hour. HF is used mainly in cut-and-fill operations.

Cemented hydraulic fill (CHF) is produced at the same plant. The cement addition is a slurry comprising 3% by weight Portland Cement and 6% by weight pozzolan (ground Copper Reverberatory Furnace Slag). The monitoring of the Portland Cement addition is essential to both the economics of the filling operation, and to the strength of the fill. Monitoring is carried out by weighers linked into the process control system.

Mechanical star feeders provide feed to one of two mixing tanks. In the mixing tanks the Portland Cement is added to the pozzolanic slag slurry and then pumped across to the mixing cones to be added to hydraulic fill. CHF is used in conjunction with either rock or aggregate fill.

The type and usage of various materials for stope filling for year 1987/88 are illustrated in Table 1.

Wet fill densities and flow rates are constantly monitored by nuclear density gauges and magnetic flow meters linked to the process control system. Information is available on a historical, or an instantaneous basis, at Mine Control.

Sand utilisation has improved considerably from 75% in 1983/84

Table 1. Fill usage by type - 1987/88

| Fill Type | Dry Tonnes (x1000) | % to Cu Stopes |
|---|---|---|
| Hyd. Fill (HF) | 1292 | 14 |
| Cem. Hyd. Fill (CHF) | 1242 | 85 |
| Hyd. Agg. Fill (HAF) | 14 | -- |
| Cem. Agg. Fill (CAF) | 1246 | -- |
| Rockfill (used with CHF) | 1120 | 100 |
| Agg. Fill (AF) | 842 | 97 |
| Mullock Fill | 139 | 2 |
| TOTAL | 5895 | 54 |

Table 2. Change in waste product usage.

| | 83/84 | 87/88 |
|---|---|---|
| Deslimed tailings | | |
| available | 3027 | 4065 |
| used | 2286 | 3390 |
| utilisation | 75% | 83% |
| HMS Rejects | | |
| produced | 1340 | 1499 |
| used | 405 | 1354 |
| utilisation | 30% | 90% |
| Rejects Application | | |
| dry placed | 214 | 1120 |
| hyd. placed | 191 | 234 |

Figure 1. Aggregate fill mixing station.

to the current 85% due to increasing demand. This figure is expected to drop slightly, as a result of a finer grind in the lead circuit. To provide a free draining fill, sand is classified, with less than 9% passing 10 microns.

2.2 Aggregate

The commissioning of a heavy medium reject plant in late 1982 required a review of operating practices in underground filling at the Isa Mine. The rejection of approximately 1.2 Mt of coarse aggregate resulted in a considerable reduction in the amount of available sand. The loss of sand was partially compensated for by the addition of the <20mm aggregate back into the fill system.

From the first full year of operation of the plant, 1983/84, the pattern of concentrator waste products usage has changed considerably, as illustrated by Table 2. Far greater use of dry aggregate has made a significant difference to overall utilization.

The aggregate is conveyed from the Heavy Medium Plant stockpile to the aggregate mixing tower (Figure 1). In this tower the aggregate is either added to the wet fill via launders, or placed onto a surface conveyor to be transferred to one of two surface fill passes for placement underground.

The rate of addition of aggregate to the wet fill is regulated by weightometers on variable speed conveyors. This is currently fixed at 25 Wt%. Pumping trials in 1977 indicated that 30 Wt% addition could be achieved, however, early problems with blockages has led to the more conservative rate. Aggregate is not added until 10 minutes after the running of hydraulic fill has commenced, and is shutdown 10 minutes before the fill has finished. This minimises the possibility of blockages due to the presence of aggregate.

Aggregate is used in both cemented and uncemented forms in the lead orebodies. The lead stopes are steeply dipping, regular shaped and well suited to hydraulically placed fill. Cemented aggregate fill (CAF) is used in primary stopes where fill will be exposed by adjacent stopes, and hydraulic aggregate fill (HAF) is used in pillars for regional stability. Excessive horizontal runs prevent the running of wet fill containing aggregate to the 1100 Orebody.

3 PROCESS CONTROL SYSTEM

The process controller used for the monitoring of fill operations is a Toshiba TOSDIC 264 which has been applied to various mining operations (Welbourn & Budd,1988). The processor is located on the surface at Mine Control which is manned 24 hours a day (Figure 2). The processor is also used to monitor surface fans, mine dewatering and high voltage electrical reticulation.

The wet fill station provides information on mass flow, density and cement addition. Data is recorded at least once every three seconds and is continuously analysed giving moving averages, frequency and cumulative frequency displays. Graphical output is available for all these parameters (Figure 3). Although the Mining Division has no operational input to the fill station, alarms activated by the process controller can be relayed back to the fill station operators. It allows ready checking of the wet fill specifications, by mine personnel, to ensure a consistent, high quality fill.

Research is in progress to provide on-line detection of the presence of cement in the CHF slurry stream. In 1983, over a period of several months, low cement percentages were added to the CHF delivered underground. This was due to a fault in the

cement weighing system. The consequences were large volumes of weak fill in the stopes being filled at the time. Pillar recovery operations adjacent to these stopes have been affected by fill dilution caused by the collapse of weak fill. The objective is to monitor the CHF such that the presence of Portland Cement can be verified by conductivity or pH measurements.

Laboratory tests indicate that significant differences in pH can

Figure 2. Toshiba TOSDIC 264 mine process control computer.

be detected for various cement contents. Laboratory tests on conductivity showed a linear relationship with cement content, but the low correlation factor indicated this method would not be precise enough for measurement purposes. Conductivity instruments have proved to be a rugged, simple, low cost, cement/no cement detector.

4 FILL APPLICATIONS

4.1 Open stope filling

Four types of dry fill placement are used in open stope filling at

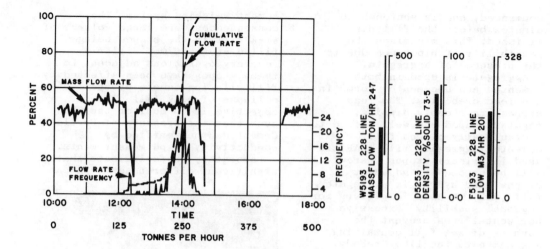

Figure 3. Typical graphical display available on TOSDIC computer.

Isa Mine; either singly, or in conjunction with hydraulic fills.

- Aggregate
- Development waste
- Cemented rockfill (CRF)
- Differentially Placed Aggregate with CHF.

All stoping areas at Isa are filled to provide regional stability. Aggregate and development waste fill are used only in remnant voids where fill will not be exposed. Aggregate is used where possible in the 1100 orebody, since the material is cheap, readily available and easily handled via the fill pass and fill conveyor system. By adding hydraulic fill, the natural angle of repose (37 degrees) is reduced to approximately 5 degrees. This allows tight filling. Development muck is used where possible, but waste in choke-fed fill passes tends to hang-up. In the upper level lead stopes, waste fill is tipped through raise-bored holes from surface whenever possible. For details of lead open stoping refer to Goddard,1981.

For the cemented rockfill (CRF), used in the 1100 Orebody, graded rock from the nearby Kennedy Siltstone Open Cut (KSOC) is mixed with CHF. Placement methods have remained virtually unchanged since

the practice was introduced in the mid-1970s.

Rockfill is conveyed dry to a fill pass at the top of each stope, and the CHF is run simultaneously through the pass, to mix during the fall. The rock serves several purposes:

- improves strength of fill
- hydraulic fill alone is not sufficient to satisfy filling requirements
- provides faster fill rates
- reduces the amount of water introduced underground.

CRF is used in primary stopes and in secondary pillar stopes where fill exposures are planned. For details on open stopes and pillar recovery refer to Alexander and Fabjanczyk,1981.

The quality of the rock plays an important role in fill strength. KSOC provides a high quality rock that resists deterioration in the transfer from surface to underground. The rock is graded to between 25mm and 300mm ensuring optimum strength from the cement addition. A higher percentage of fine material requires a higher cement content to achieve the target strength of 1 MPa. Rock larger than 300mm causes handling problems on conveyors underground.

Monitoring of the surface fill passes is conducted using

accelerometers attached to rock bolts (Figure 4). These are placed on various levels and detect the movement of the rock, or aggregate, in the fill pass. An alarm is given, warning of a blockage, if movement is not detected during a preset time. This alarm is given to the operators underground to prevent pulling the pass empty. The information is also transmitted to the surface by telemetry.

In order to determine optimum rock to CHF ratios, extensive use is being made of a computer program, developed by Mount Isa Mines Limited, called FILPAK (Cowling, 1989). Having input the co-ordinates of the stope at various levels through the height of the stope, the program enables a profile of the rock cone to be determined for various pass configurations. To provide maximum strength, rock is not allowed to build up on walls to be exposed. By inputting the walls that rock is not allowed to build up on, the program gives optimum filling ratios at various levels in the stope.

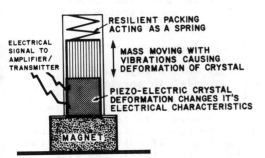

Figure 4. Accelerometer mounted on rock bolt.

FILPAK enables mine planning engineers to trial various pass configurations to get the best possible ratios. On completion of extraction, true stope dimensions are obtained by survey pick-ups. The final dimensions are input to FILPAK enabling the operating section to determine the ratios to be run in the stope. Regular visual inspections ensure filling proceeds as planned.

Fill rates are determined by the CHF flow rate and the filling ratio for the stope. Vibrating feeders on the fill pass chutes allow control of the rock onto the conveyors. The main fill horizon, 15 Level, utilises two main trunk conveyors, on the footwall and hanging wall (Figure 5), which are rated at 1000 tph. Rock is tripped either directly off the trunk conveyors, or onto cross-conveyors, into fill passes.

In 1985 a trial was carried out using CHF addition to dry placed aggregate fill referred to as In-situ CAF. The ratio of CHF to aggregate was 1:1. On completion of curing, a development heading was mined into the fill to determine the strength of the resulting fill mass. The results, although disappointing, proved valuable.

Fill exposed comprised beds of un-consolidated aggregate, where slumping of the cone had washed the aggregate out to the walls. CHF was unable to penetrate the aggregate effectively. More recently efforts have switched to cementing single walls, ensuring a 5m beach of CHF against the wall to be exposed. This has been called Differentially Placed CAF to distinguish the two methods. The aggregate is placed via 200mm diameter ITHH holes designed to spread evenly around the stope, or through a 1.2m x 1.2m longhole winze. Small diameter holes are prone to blockages and restrict the filling rate.

## 4.2 Cut-and-Fill

Large scale cut and fill mining operations commenced at Mount Isa in 1964. A number of cut and fill mining methods have evolved through the years culminating in the introduction of mechanised cut and fill in the lower areas of the mine. Filling is carried out using HF reticulated through the orebodies via 150 mm diameter galvanised pipe. The cyclic nature of cut and fill mining requires priority to be given on one of the two lines available from surface. The filling and drainage of cut and fill stopes is described in more detail by Bourke and Wright, 1988.

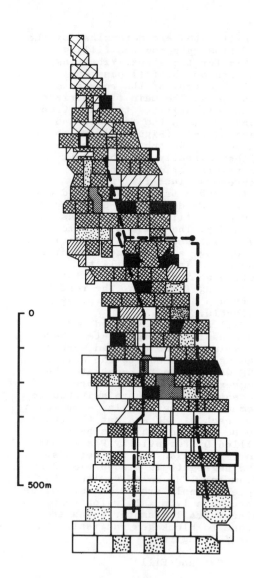

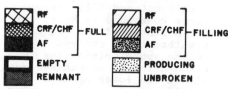

Figure 5. Plan of 1100 orebody showing location of rockfill conveyors.

5 RETICULATION

5.1 Pipes

All wet fill is reticulated underground via any two of four surface fill holes. The fill is transferred along the levels through 3m lengths of flanged 150mm nominal bore, lined steel pipe. Since CAF/HAF reticulation commenced, in late 1982, various trials have been undertaken to determine the most cost effective means of handling the highly abrasive aggregate.

Rubber lined pipes have proved to be the most cost effective means of reticulating CAF/HAF. Two types of rubber lining have been trialled. One is a hot cured rubber which is vulcanised directly to the pipe. The bond between the pipe-walls and the rubber is suspect following bubbling of the rubber in certain areas. Bubbles can enlarge to the point where they block the flow, or tear off in large strip causing blockages further down the line.

The second, and preferred type, is a pre-cured rubber, glued to the mild steel pipe. Two grades of rubber have been trialled; one with a higher cut resistance, the other with high resilience. The high cut resistance rubber is now the standard lining for all 3m lengths of pipe. The ability to reline pipes with this rubber has resulted in considerable cost savings.

5.2 Fittings

Fittings have always been the weakest point in any hydraulic fill distribution system. With the introduction of CAF/HAF, various fittings were trialled in an effort to reduce line maintenance and prevent fill spillages. On the basis of those trials the life expectancy of 90 degree bends was as follows;

| | |
|---|---|
| Rubber lined | 70 hours |
| Polyurethane lined | 100 hours |
| Hardened steel lined | 200 hours |
| Hard-faced | 400 hours |

The hard-faced fittings were found to be the most cost effective until the idea of replacing the bend with a drop-box was adopted in late 1985 (Figure 6). The drop box allows the fill to work against itself through turbulence. The turbulence slows the fill and breaks the flow pattern, resulting in reduced line wear and more stable running conditions. Minimum life expectancy of the drop-box is around 4000 hours.

For all other fittings, hard-faced pipe is used. The extent of the fill system requires a high degree of standardisation as far as fittings are concerned. A total of six types of bends and six different length straight pieces are used. Valves are rarely used due to high maintenance, and almost all line changes are done using fittings.

## 5.3 Fill holes

All fill is reticulated underground via vertical fill holes from surface. Various methods have been used to line the first 90m of weathered rock. The standard practice for surface fill holes is to drill with a raise-borer and then concrete line. This is achieved by down-reaming the top 90m at 735mm diameter then re-drilling back through at 300mm.

Figure 6. Dropbox replacement for ninety degree bend.

Camera surveys conducted during maintenance are used to indicate

significant wear in the fill hole. Any badly worn sections are re-concreted and re-drilled. There are currently four such holes, each 220m long.

## 5.4 Fill reticulation monitoring

All underground fill monitoring is by visual inspection. This is labour intensive; two men per shift, three shifts per day, seven days per week. Efforts are being made to use the process control system for fill line monitoring. The four main main objectives of monitoring are:

. detect line blockages
. determine fill destination
. detect fill type
. optimise flushing water .

The vibration in the fill pipes can be used as a means of monitoring. Accelerometers cemented onto small magnets are attached to fill pipes. Due to the simple nature of the devices, instrument technicians need not be involved in the installation or maintenance of the devices. Outputs are transmitted to the process control system, where a mimic panel indicates a flow/no flow diagram (Figure 7). The close correlation between mass flow and pipe vibrations enables detection of burst lines and line blockages. Eventually, software within the processor will be used to generate alarms to indicate abnormalities in the flow.

| CHF/HF – LINE 225 | OT/H; LINE 226 |
| AGG – LINE 1 | IT/H; LINE 2 |
| 254T/H; LINE 227 | OT/H; LINE 228 256T/H |
| IT/H; LINE 3 | IT/H. |

| | SURFACE | | | | | | |
| 4 LEVEL | P65 | 1 | 2 | 3 | 4 | |
| 6 LEVEL | P62 A B | | | | R66 B A | |
| 9 LEVEL | R66 A B | | | | Q67 A B | |
| 10 LEVEL | P58 A B | | | | M70 N70 | |
| 11 LEVEL | | | | | | |

Figure 7. Schematic of fill reticulation monitoring.

To detect whether aggregate is present in the fill or not, the

367

frequency spectrum for the fill is used. Practice has shown the amplitude of the noise generated by fill with aggregate to be much greater than normal flows of fill. Flushing water also provides a different signal. The marked difference in the signal will be used to relay information back to the fill station operators. This will be used to reduce the amount of water used, increase the fill strength and reduce cure time for stopes (Cowling et al,1989).

The turn around time for stopes is particularly important as primary stopes are becoming fewer and pillar extraction is necessary to maintain production rates. Advances in this area have reduced the turn around time for the CRF stopes by up to 30 days (Grice, 1989).

ACKNOWLEDGEMENTS

The permission of Mount Isa Mines Limited to publish this paper is gratefully acknowledged.

REFERENCES

Alexander,E. and Fabjanczyk,M.W. 1981. Extraction design using open stopes for pillar recovery at Mount Isa. Proc. Conf. Caving and Sub-level Stoping. Denver.

Bourke,P. and Wright,J. 1988. The evolution of cut and fill mining at Isa Mine. Proc. Underground Operators Conf. Mount Isa.

Cowling,R., Voboril,A., Isaacs,L.T., Meek,J.L. and Beer,G. 1989. Computer models for improved fill performance. Proc. Symp. Mining with Backfill. Montreal.

Cowling,R., Grice,A.G. and Isaacs,L.T. 1988. Simulation of hydraulic filling of large underground mining excavations. Proc. Conf. Numerical Methods in Geomechanics. Innsbruck.

Goddard,I. 1981. The development of open stoping in lead orebodies at Mount Isa Mines Limited. Proc. Conf. Caving and Sub-level Stoping. Denver.

Grice,A.G. 1989. Fill research at Mount Isa Mines. Proc. Symp. Mining with Backfill. Montreal.

Neindorf,L.B. 1983. Fill operating practices at Mount Isa Mines. Proc. Symp. Mining with Backfill. University of Lulea.

Welbourn,K. and Budd,B.1988. The introduction and use of a mine monitoring system at Isa Mine. Proc. Underground Operators Conf. Mount Isa.

*Innovations in Mining Backfill Technology, Hassani et al. (eds), © 1989 Balkema, Rotterdam. ISBN 90 6191 985 1*

# Total tailings backfill properties and pumping

J.D.Vickery & C.M.K.Boldt
*Spokane Research Center, US Bureau of Mines, Spokane, Wash., USA*

ABSTRACT: The Bureau of Mines conducted a series of tests using dewatered total tailings from three U.S. metal mines to determine the engineering and pumping characteristics for backfilling. The tests were unique because pulp densities of the samples were all above 75-pct solids, creating nonsettling, homogeneous slurries or pastes.

The engineering properties tested included dry density; slump; percent settling after 28 days of curing; tensile strengths after 28, 120, and 180 days of curing; and unconfined compressive strengths after 7, 28, 120, and 180 days of curing. Properties of backfill mixtures made from tailings from the three mines were analyzed using linear and nonlinear statistical methods. Graphical results provided a predictive tool with which to select binding materials and proportions to create a backfill with the desired engineering properties.

Six large-scale pumping tests were conducted using cemented and uncemented paste backfills made from two of the tailings included in the property tests. The pastes were mixed to similar consistencies to compare the friction losses at various flow rates through a loop consisting of pipes of three different diameters.

## 1 INTRODUCTION

In 1964, the Bureau became involved in backfill research to define the properties of hydraulically placed tailings backfill (Nicholson and Wayment, 1964). From 1961 to 1970, researchers tested a multitude of mixes utilizing portland cement and mine tailings (Weaver and Luka, 1970). Because of the available transport methods, tests were limited to backfill slurries with <70-pct solids.

New technological advances in positive displacement pumps and pneumatic stowers may allow stopes to be backfilled with >80-pct solids, total tailings material. Such a capability no longer limits the mix matrix to 70-pct solids or the inclusion of only the sands fraction of mill tailings. The resulting decrease in water improves the material's strength, homogeneity, and curing time, while reducing indirect costs of mine dewatering

and cleanup, thus making lean, cemented total tailings backfill an attractive option.

This report summarizes laboratory work done by the Bureau to define the strength and pumping characteristics of paste backfills using total tailings as aggregate with varying amounts of cement, additives, and water (Boldt et al., 1989). The pumping tests were undertaken to provide information that could be used to overcome some of the problems of engineering a mine-wide paste backfill transport system. There are several different thoughts on the mechanics of vertical transport, but there is a consensus on the importance of understanding this phenomenon if paste is to be used for backfilling in deep mines. The term "total tailings" as used in this manuscript includes the full grain-size gradation of tailings produced by a mill, typically from 0.001 to 0.6 mm in diameter.

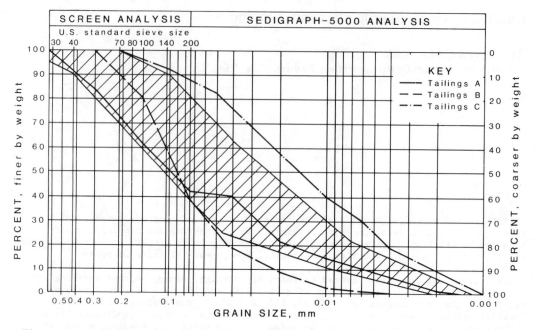

Figure 1. Grain-size distribution of tested tailings. Hatched area represents the general range of U.S. metal mine tailings (adapted from Soderberg and Busch, 1977).

## 2 ENGINEERING PROPERTIES

### 2.1 Mix composition

The mill tailings used as the basic aggregate in this test series came from three underground metal mines. Tailings A is from a deep silver mine in Idaho, tailings B came from a lead-zinc mine in Missouri, and tailings C is from a copper-silver mine in Montana. Tailings A was partially classified during dewatering, while B and C were total mill tailings. The grain-size gradation curves of the tailings are shown in figure 1.

Commercially available Class F fly ash, Category B superplasticizer, pit-run and ground smelter slag, kiln dust, and oil shale retorted waste were incorporated in the test mixes to determine their influence on the physical properties of the backfills.

Various amounts of fly ash were added to the cemented backfills to determine whether the pozzolanic influence would be sufficient to decrease the required amount of cement and still maintain the unconfined compressive strength. The superplasticizer was added to a few mixes

to determine the superplasticizer's ability to decrease the water-to-cement ratio, yet maintain a pumpable consistency. Ten times the manufacturer's dosage for concrete was needed to increase the slump, so no further tests were attempted. Oil shale retorted waste was used as an additive to determine if its cementing properties could be used in the backfill. Previous oil shale research had documented the cementing properties of certain retorted wastes (Marcus et al., 1985).

A commercially available Type I-II portland cement and tap water were used throughout the test series.

### 2.2 Mix handling

Combinations of each oven-dried tailings sample and various additives were mixed for a minimum of 2 min and visually checked for homogeneity. A slump measurement was then taken to quantify the paste consistency (ASTM C 143-78, 1986). Samples were then packed into 3- by 6-in standard, waxed cardboard cylinders to eliminate large voids and cured in a fog-room (ASTM C 192-81, 1986). The slurry density was calculated from the mix proportions, while the 28-day wet density was measured after 28 days of curing. Eight cylinders were made from each test mix. Backfill mixes from tailings B and C were cast into gang molds to determine tensile strength by pull tests. These tensile briquets were also cured in a fogroom.

## 2.3 Laboratory testing

The test matrix included 95 mixes from tailings A, 45 mixes from tailings B, and 47 mixes from tailings C. Laboratory tests included measuring unconfined compressive strengths after 7, 28, 120, and 180 days of moist curing. Each of the strength tests was run on a duplicate cylinder sample and the two strength readings averaged to minimize errors. For each mix from tailings A, eight cylinder samples were tested: two each for 7-, 28-, and 120-day cured, unconfined compressive strengths (ASTM C 39-86, 1986); one for the 120-day cured, confined compressive strength tests; and one for determination of dry density. For

each mix from tailings B and C, eight samples were tested: two each for 7-, 28-, 120-, and 180-day cured, unconfined compressive strengths. In addition, the test results from three briquet specimens were averaged to determine 28- and 120-day cured tensile strengths for tailings B and C (ASTM C 190-82, 1984). The water-to-cement ratio was calculated as a proportion of the weight of the water to the weight of the cement. The slurry density of the mixes was determined by dividing the weight of the solids by the weight of the solids and water. The tensile and unconfined compressive strengths were the average of multiple test samples.

## 2.4 Data analysis

The data showed that the water-to-cement ratio, regardless of additives, could be used to predict the unconfined compressive strength for various curing periods. An exponential function was selected to best fit the data.

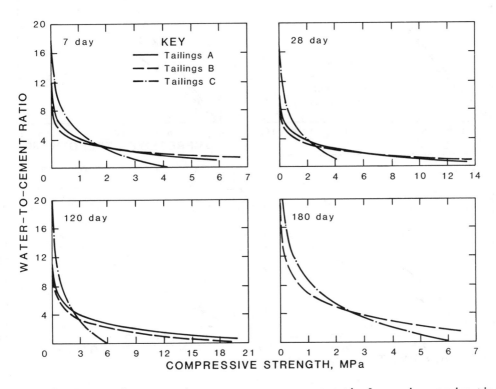

Figure 2. Compressive strength versus water-to-cement ratio for various curing times for total data base.

Plots of the compressive strength versus water-to-cement ratio for each of the tailings types with and without additives are presented in figure 2. The index of determination, I, for 7-day compressive strength of the total data base is 0.796. As the data base became more selective, the goodness-of-fit improved.

A linear function was selected to model the relationship between the 7-day compressive and tensile strengths and those of subsequent days of curing. The goodness-of-fit for the data base ranged from a low of 0.89 to a high of 0.98.

The results (based on least-squares fitting) between 7-day compressive strength and 28-, 120-, and 180-day compressive strengths and between 7-day compressive strength and 28-day tensile strength for the total data base is presented in figure 3. The formulation of these types of relationships can bypass the need for long-term testing.

2.5 Discussion of results

This test series of backfill materials was initiated to determine what engineering properties could be expected from several different types of total tailings. The incorporation of additives was meant to define the extent of increased strength or workability of the resultant mix. In some cases, such as those where oil shale retorted waste was added to the tailings, a full range of mixes was not attempted because the goal was merely to determine whether or not the retorted waste was a detriment to the mix, thereby indicating possible uses of oil shale retorted waste.

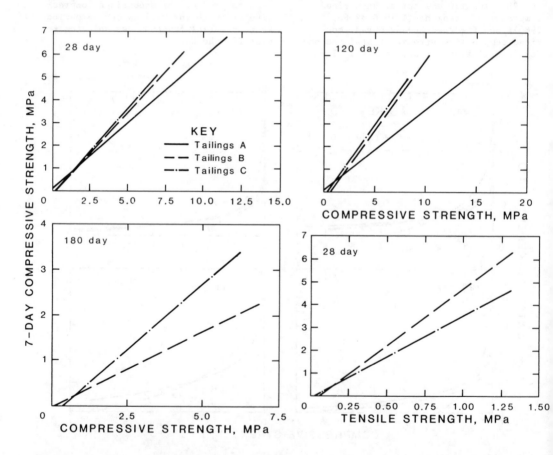

Figure 3. Relation between 7-day compressive strength and other curing strengths for total data base.

The results indicated that the addition of oil shale retorted waste without cement produced compressive strengths on the order of 717 kPa in 28 days. The cementing properties of the retorted waste were greater for the finer, grain-sized particles of tailings C. The addition of fly ash improved the compressive strength of the total tailings aggregate. As the tailings grain-size fraction greater than 200-sieve increased, the influence of the fly ash decreased. The 28-day compressive strength of tailings A was increased by 25 pct, tailings B by 48 pct, and tailings C by 98 pct over the compressive strengths gained by the use of cement alone.

The compressive strength-to-tensile strength ratios for the various days of curing ranged from 4.4 for the total tailings not containing additives to 4.8 for the total tailings data base.

## 3 PUMPING PROPERTIES

### 3.1 Paste description

The term "paste" is used to refer to pumpable slurries having the consistency of a measurable slump. Slurry concentration alone is not a good indicator of consistency. Slump tests were found to be quite helpful in establishing upper and lower limits for paste slurry densities. In the future, laboratory tests will be conducted to quantify the slurry density of the tested pastes and determine the relationship between slurry density and slump. Pastes with low slumps require higher pumping pressures, while those with high slumps have a greater tendency to cause "hammering" at the pump and then settle when left stationary in the line. Uncemented as well as cemented pastes with slumps from 10.8 to 17.8 cm (30.5-cm slump cone) were pumped through the test loop.

Concrete pumping experience shows that proper selection and grading of aggregates is the most important consideration in pumping mix designs (Crepas et al., 1984). Tailings A and C were selected for pump testing because their grain-size distributions closely bracket the outside limits of domestic metal mine tailings as shown in figure 1.

### 3.2 Test loop

An instrumented pipe loop was constructed with horizontal sections of 150-, 125-, and 100-mm diameter steel pipe and a 20-m-high vertical section of 150-mm pipe. The vertical section was bypassed during some tests to reduce pump pressures. The total length of the loop was

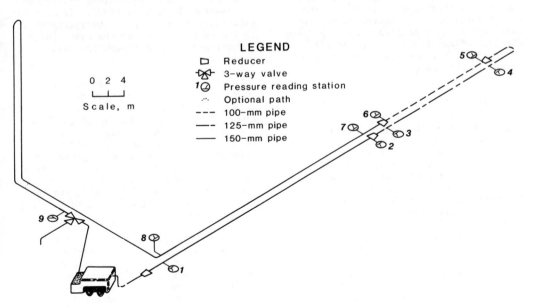

Figure 4. Schematic of pumping loop and pressure monitoring locations.

373

approximately 170 m.  Figure 4 shows the
most common locations of the nine move-
able electronic pressure transmitters
that monitored gauge pressures to establish the hydraulic profile.

Several varieties of positive displacement pumps have been used to transport
slurries similar to those tested here,
but a piston pump used for pumping concrete was selected for these tests.  The
nature of these pumps cause them to produce a pulsed flow, and application of
common dampening methods was found to be
difficult because of the energy requirements.  Test results are assumed to be
representative of similar flow conditions
and not steady-state flow.

## 3.3  Pumping test procedure

Dry tailings and cement were loaded into
a ready mix concrete truck and mixed pri-
or to adding water to create the desired
slump.  Additional water was required at
the pumping site to maintain the target
slump.  The pastes were then pumped into
the pipe loop, which had been previously
wetted with water and flushed by compressed air and sponge ball.  Once the
loop was full and the material began re-
circulating, the pump speed was increased
every few minutes until it achieved maximum output.  A strip chart recorder continuously plotted the gauge pressures at
the nine monitoring locations.  At the
end of testing, all cemented pastes were
cast into a concrete form 1.2-m$^2$ by 2.4-m
high for uniaxial compression testing.
The results of these tests are not presented in this paper.

## 3.4  Data analysis

Flow rates were calculated by counting
the number of strokes occurring during a
given time period and multiplying by the
pump cylinder displacement.  The pump
hopper was kept completely full and the
paste consistency did not appear to be
too thick to prevent 100 pct pumping efficiency.  Maximum pressure values from
transmitter pairs at the beginning and
end of the straight sections were sub-
tracted to determine the differential
pressure for the three pipe sizes at the
various flow rates.  Pressure gradients
were then calculated by dividing the differential pressures by the pipe length
between transmitter pairs.

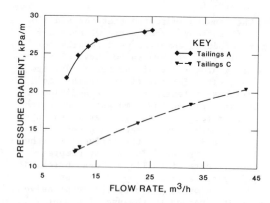

Figure 5.  Pressure gradients of uncemented pastes with 11.4-cm slump.

## 3.5  Discussion of results

Two tests were conducted with uncemented
tailings A and C, each prepared to a
11.4-cm slump.  Even though both pastes
appeared to have the same consistency,
but different slurry concentrations, figure 5 shows the pressure gradients were
substantially greater for tailings A, the
coarser and less uniformly graded of the
two materials.  Pump loop testing of gold
slimes at high slurry concentrations
showed that head loss increased with decreasing $d_{50}$ particle size for slurries
in laminar flow at equivalent slurry concentrations (Sauerman, 1982).  It must be
emphasized that the paste backfills shown
in figure 5 were not at the same slurry
concentrations even though they had
equivalent slumps.

Three pump tests were made with tailings A pastes having similar consistencies (15.2- to 17.8-cm slump) with additions of 0, 4, and 6 pct by weight Type
I-II portland cement.  Figure 6 shows
that pressure gradient increased with decreasing pipe diameter for a given flow
rate.  While the pressure gradient was
expected to decrease with the addition of
cement, an increase that could have been
caused by the slight differences in slump
was observed.

Tailings C with 6-pct cement at a
11.4-cm slump was also pump tested.  Figure 7 suggests that the pressure gradient
of the 125-mm pipe was lower than that in
the 150-mm pipe at the same flow rate.
However, the pressure gradient for the
150-mm pipe was obtained from a pair of
transmitters on a pipe section containing

374

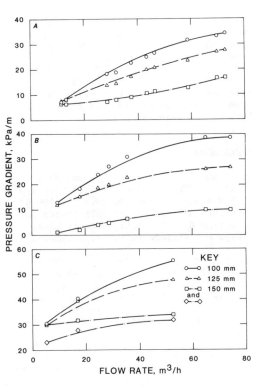

Figure 6. Pressure gradients for pastes made from tailings A. _A_, uncemented with a 16.5-cm slump; _B_, 4-pct cement with 17.8-cm slump; _C_, 6-pct cement with a 15.2-cm slump. Note data from 2-150 mm pipe sections.

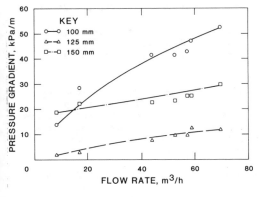

Figure 7. Pressure gradient for tailings C paste with 6-pct cement at 11.4-cm slump.

three 90° elbows and the actual, rather than the equivalent, length was used in the calculation. All of the data will be further analyzed to determine the friction loss at elbows and pipe diameter transitions and to establish equivalent lengths.

## 3.6 Conclusions

All U.S. metal mine tailings that are relatively uniformly graded and have grain-size distributions within the range bordered by those of tailings A and C could potentially be used as paste backfill. However, there may be major differences in pumping characteristics because of the differences in physical properties of the tailings aggregate. The pressure gradients indicated that some of the pastes tested in this series could be pumped nearly 1,000 m horizontally at high flow rates. Pressure gradients for a given flow rate increased with decreasing pipe diameters. Small changes in the slump, or slurry density, of a paste may have a dramatic effect on pressure gradient. Small additions of cement may not have as dramatic an effect on pumping as do increases in water unless the tailings are highly deficient in fines.

## 3.7 Application of results

Presently, the authors are not aware of any documentation of a method to determine how a particular paste backfill will flow through pipes prior to pumping tests. Landriault (1987) used a Bingham plastic rheologic model to interpret the results of high-density backfill pumping tests. Horsely (1982) found that a high-density gold slime slurry exhibited thixotropic properties (time-dependent shear thinning). Preliminary viscosity measurements by a rotational viscometer indicated that a Bingham plastic model may closely define the rheology of a tailings paste. The Rabinowitsch-Mooney relation can be used to develop a rheogram for any fluid based on the pressure gradient obtained from loop testing (Rabinowitsch, 1929; Mooney, 1931). The accuracy of viscometer tests will be checked by this approach. If a Bingham plastic model is appropriate for total tailings paste backfills, the Buckingham equation can be used to predict the pressure gradient for any paste with known rheologic properties in laminar flow through any pipe of given length and diameter. A Moody diagram was developed to predict friction factors of Bingham plastics in laminar and turbulent flow (Hanks and Dadia, 1971). This or other validated flow models will provide a means to evaluate analytically how a backfill with particular rheologic properties will flow through a transport network to the stopes. However, loop tests would still be necessary to complete the final design of most full-scale backfill systems.

## 4 GENERAL DISCUSSION

Design methods based on a thorough understanding of paste flow characteristics are required for systems with long horizontal distances and/or large vertical descents in order to control line pressures and flow rates. Chemical additives are available to reduce friction loss (Horsely, 1982), but high friction loss may not be a detriment during vertical descents. Venting the top of vertical pipelines may control the fall of pastes by trapping a cushion of air below the falling plug that will eventually build up pressure great enough to produce a countercurrent through the plug, thus reducing velocity (Brackebusch, 1988). The pump will only act as a means of feeding paste to the vertical pipe. A vent will also allow any trapped air to escape from the pipeline. Potentially, systems can be designed to produce a given flow rate while maintaining a constant height of slurry in the vertical column by balancing the pressure gradients for the vertical and horizontal pipe lengths (Brackebusch, 1988; Landriault, 1988). Paste transport methods are being used to backfill mines having transport distances up to 2,000 m by using two pumping stations.

The trade-offs between transportability and in situ properties should be carefully weighed in the selection and design of paste backfill systems. Mine layout, physical properties of the tailings, and desired in situ qualities of the backfill will place limitations on the design of paste backfilling systems. Not all sites will have conditions that make paste backfilling economical and desirable. Pipeline sizing and mix proportioning to create specific rheologic properties must be carefully considered to minimize operational costs and problems. The work reported here is intended to lead to the development of analytical evaluation and design methods based on quick, inexpensive laboratory tests to create a paste backfill that meets the requirements of the mining method, yet will also be economic to transport. The ability to transport higher quality backfill may open doors to safer and more efficient bulk mining methods.

## REFERENCES

American Society for Testing and Materials (ASTM). 1984. Tensile Strength of Hydraulic Cement Mortars. C 190-82 in 1984 Annual Book of ASTM Standards: Section 4, Vol. 04.01, Cement; Lime; Gypsum. Philadelphia, PA, pp. 200-206.

American Society for Testing and Materials (ASTM). 1986. Slump of Portland Cement Concrete. C 143-78 in 1986 Annual Book of ASTM Standards: Section 4, Vol. 04.02, Concrete and Mineral Aggregates. Philadelphia, PA, pp. 109-111.

American Society for Testing and Materials (ASTM). 1986. Making and Curing Concrete Test Specimens in the Laboratory. C 192-81 in 1986 Annual Book of ASTM Standards: Section 4, Vol. 04.02, Concrete and Minerals Aggregates. Philadelphia, PA, pp. 142-151.

American Society for Testing and Materials (ASTM). 1986. Compressive Strength of Cylindrical Concrete Specimens. C 39-86 in 1986 Annual Book of ASTM Standards: Section 4, Vol. 04.02, Concrete and Minerals Aggregates. Philadelphia, PA, pp. 24-29.

Boldt, C.M.K., P.C. McWilliams, and L.A. Atkins. 1989. Backfill Properties of Total Tailings. RI in press, U.S. Bureau of Mines.

Brackebusch, F.W. 1988. Personal communications regarding draft paper, Development of a consolidated fill system at the Lucky Friday Mine, by F.W. Brackebusch and S.D. Lautenschlaeger, submitted to Min. Eng.

Buckingham, E. 1921. Proc. of ASTM, 21, 1154.

Crepas, R.A., et al. 1984. Pumping Concrete: Techniques and Applications. Concrete Construction Publications, Inc. Addison, IL.

Hanks, R.W., and B.H. Dadia. 1971. Theoretical Analysis of the Turbulent Flow of Non-Newtonian Slurries in Pipes. AIChE J., 17:554.

Horsely, R.R. 1982. Viscometer and Pipe Loop Tests on Gold Slime Slurry at Very High Concentrations by Weight, with and without Additives. Hydrotransport 8, 8th International Conference of Hydraulic Transport of Solids in Pipes, Johannesburg, South Africa, BHRA Fluid Engineering, Cranfield, Bedford, England. Aug. 25-27. Paper H1.

Landriault, D. 1988. Personal communication.

Landriault, D. 1987. Preparation and Placement of High Density Backfill. Underground Support Systems. Canadian Inst. of Min. and Metall., Special Volume 35.

Marcus, D., D.A. Sangrey, and S.A. Miller. 1985. Effects of Cementation Process on Spent Shale Stabilization. Min. Eng. (NY), 37:1225-1232.

Mooney, M. 1931. Explicit Formulas for Slip and Fluidity. J. Rheol. 2:210.

Nicholson, D.E., and W.R. Wayment. 1964. Properties of Hydraulic Backfills and Preliminary Vibratory Compaction Tests. RI 6477, U.S. Bureau of Mines, 31 pp.

Rabinowitsch, B. 1929. Uber die Viskositat und Elastizitat von Solen. Zeit. Physik. Chem., A145:1.

Sauerman, H.B. 1982. The Influence of Particle Diameters on the Pressure Gradients of Gold Slimes Pumping. Hydrotransport 8, 8th International Conference of Hydraulic Transport of Solids in Pipes, Johannesburg, South Africa, BHRA Fluid Engineering, Cranfield, Bedford, England. Aug. 25-27. Paper E1.

Soderberg, R.L., and R.A. Busch. 1977. Design Guide for Metal and Nonmetal Tailings Disposal. IC 8755, U.S. Bureau of Mines, 136 pp.

Weaver, W.S., and R. Luka. 1970. Laboratory Studies of Cement-Stabilized Mine Tailings. Paper in Research and Mining Applications in Cement Stabilized Backfill (pres. at 72nd Ann. Gen. Mtg. of the Can. Inst. of Min. and Metall., Toronto, Ontario, Apr. 1970). CIM, pp. 988-1001.

377

*Innovations in Mining Backfill Technology, Hassani et al. (eds), © 1989 Balkema, Rotterdam. ISBN 90 6191 985 1*

# The role of hydrocyclone technology in backfilling operations

A.L.Hinde & C.P.Kramers
*Chamber of Mines of South Africa Research Organization, Johannesburg, RSA*

ABSTRACT: Hydraulic backfilling is a relatively new operation in South African gold mines. Most backfill is currently prepared by desliming metallurgical plant tailings using hydrocyclone classifiers. In order to guide the design of desliming plants, investigations were undertaken to establish the effects of classification parameters on the placement, hydraulic transport and mechanical support characteristics of backfill products. Consideration was also given to the effect changes in the classification parameters can have on the overall economics of the backfilling operation. The results of the investigations were used to provide a rational basis for selecting hydrocyclone geometries, operating parameters, and classification circuit flowsheets.

## 1 INTRODUCTION

The use of hydraulically placed backfill as a medium for stabilizing underground excavations to improve safety and enhance the exploitation of ore bodies has only recently gained wide acceptance in the South African gold mining industry (de Jongh and Morris 1988). Currently, some 80 per cent of backfill placed is derived from metallurgical plant tailings that have been deslimed using large diameter hydrocyclones. The advantage of the desliming operation is that it yields a product which can be transported underground and placed hydraulically without the need for pumps, mechanical dewatering devices or cementitious binders. Unfortunately, no clear consensus of opinion has emerged with regard to the optimal design and operation of backfill hydrocyclone plants in relation to backfilling needs. A contributory factor to this uncertainty has undoubtedly been the paucity of reliable data relating hydrocyclone classification parameters to the behaviour and performance of the resultant backfill.

This lack of data has prompted the initiation of a number of comprehensive sampling campaigns of single-stage and multi-stage hydrocyclone plants on mines. Results have shown that the mass of backfill solids produced rarely exceeds 30 per cent of the plant feed. Because the amount of material required to backfill stopes on a mine-wide scale to their practical maximum is between 40 and 50 per cent of the tonnage milled, consideration has been given to finding ways of boosting mass recoveries. One approach is to reduce the classification cut-size by using small diameter hydrocyclones. Whilst there are obvious merits in following this approach, a high mass recovery is not the only criterion to be satisfied when designing classification plants. The

backfill must also be readily transportable, drain rapidly when placed, and have a high resistance to mechanical deformation. Additionally, consideration must be given to economic issues. The principal aim of this paper is to reconcile the pros and cons of changing the classification parameters and to establish some guidelines on the implementation of backfill hydrocyclone technology in South African gold mines.

## 2 CHARACTERIZING HYDROCYCLONE PERFORMANCE

### 2.1 Terminology

The classification performance of a hydrocyclone is best characterized by a plot of the recovery of the feed solids to the underflow, expressed as a function of particle size. Such a plot, called a partition or recovery curve, is shown in Figure 1. Given the partition curve and the feed size distribution, it is possible to establish a mass balance of solids across the hydrocyclone and the size distribution of the underflow and overflow products. The recovery of solids, $R$, of size, $d$, is usually assumed to be influenced by two effects, namely the short-circuiting or entrainment of solids in the water reporting to the underflow, and size-dependent classification arising from the centrifugal and drag forces induced by the incoming feed pulp. This combination of effects can be expressed in terms of an equation of the form:

$$R(d) = R_f + (1 - R_f)R_c(d)$$

where $R_f$, which approximates the recovery of water (Yoshioka and Hotta 1955), is referred to as the bypass parameter. The function, $R_c(d)$, can be expressed mathematically by any one of a number of two-parameter probability functions with a value of zero at zero particle size and asymptotically tending to unity at large particle sizes. The exponential sum proposed by Lynch and Rao (1975) is commonly used:

$$R_c(d) = \frac{\exp(\alpha d/d_{50c}) - 1}{\exp(\alpha d/d_{50c}) + \exp(\alpha) - 2}$$

The parameter $d_{50c}$ is called the cut-size (corrected for water split), and the parameter, $\alpha$, called the sharpness of classification, is a measure of the steepness of the partition curve in the region of the cut-size. Whilst the parameters $d_{50c}$ and $R_f$ are sensitive to geometric and operating variables, the parameter $\alpha$ is relatively insensitive to such variables. The sharpness of cut is essentially a characteristic of the material being classified, irrespective of the hydrocyclone size, inlet and outlet dimensions or operating conditions.

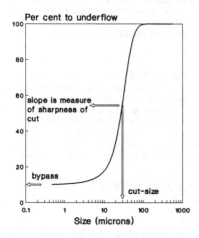

Fig.1 Hydrocyclone partition curve

## 2.2 Modelling

Over the years, many attempts have been made to model the relationship between the partition curve parameters and independent variables such as hydrocyclone geometry, feed density, feed particle size distribution, and feed flow rate. Probably the most popular model is the one developed by Plitt (1976). Whilst the model has enjoyed considerable success, it can yield estimates of underflow pulp densities which are physically unrealistic, especially when the underflow stream is in a "roped" state (Flintoff, Plitt and Turak 1987). This is unfortunate as underflow densities can have a critical effect on backfill placement behaviour. To overcome this problem, a more robust model was developed which constrains estimates of underflow densities to realistic values (Hinde, 1985). This model is based on the assumption that the cut-size of a hydrocyclone

is determined mainly by the solids-handling capacity of the spigot (Fahlstrom, 1963). The robust model predicts the underflow volume fraction of solids, $\phi_u$, in terms of an equation of the form:

$$\phi_u = \phi + \frac{\phi_m - \phi}{1 + f}$$

where $\phi_m$ provides an upper bound on the estimated underflow solids concentration and $f$ is a positive function (the product of power and exponential terms) of the independent variables. It can be seen that the equation constrains the minimum value of the underflow density to that of the feed density, $\phi$. The solids flow rate through the spigot can be calculated readily from the flow split (the ratio of underflow to overflow volumetric flow rates calculated from one of Plitt's equations) and underflow density. A unique value for the cut-size can then be computed from the underflow or spigot solids flow rate, the sharpness of cut, and the feed size distribution.

It should be emphasized that the above models are empirical in structure and contain a number of parameters which must be identified by appropriate test work. The results of sampling campaigns initiated on production and pilot plants are currently being compiled into a data base to facilitate the development of computer models which can be used for predicting mass balances and size distributions for design, control and optimization applications.

## 2.3 The influence of classification parameters on backfill size distributions

It has been established that the size distribution of the product from a hydrocyclone plant is determined by the feed size distribution and the plant partition curve. Whilst the feed size distribution is essentially constant and constrained by metallurgical requirements, much scope exists for effecting changes to the partition curve. This can be done by changing hydrocyclone design geometries, feed flowrates and feed pulp densities, or by using several stages of classification. The influence of changes in the classification parameters on the size distribution of the backfill product can be modelled readily. It turns out that the desliming performance of a hydrocyclone plant (its ability to remove slimes with particle sizes nominally less than 10 microns) is determined mainly by the bypass parameter, and is insensitive to the sharpness of classification. The effect of changes in cut-size on the size distribution is somewhat more subtle. It can be seen in Figure 2 that for a given bypass, a reduction in cut-size leads to marked changes in the shape of the plotted cumulative underflow distributions, and that they can cross each other. It is also important to note that the properties of backfill, especially those associated with drainage and consolidation, are often more appropriately expressed in terms of the distribution of particle surface area. Figure 3 compares the volume and surface area frequency distributions for metallurgical plant tailings before classification. It can be seen that the median size of the area distribution is about one micron, or more than an order of magnitude smaller than that of the volume distribution. It follows that the measurement of size distributions down to one micron, or even less, is highly desirable for proper characterization of the compositional properties of backfill. Such measurements can be done routinely using commercially available particle size analysers employing laser-diffraction techniques.

area. Since the surface area can be calculated from the size distribution of the backfill, it is possible to predict the effects of cut-size and bypass fractions on the coefficient of permeability. The results of such an analysis are shown in Figure 4. Also shown are measured permeability coefficients for samples prepared synthetically by blending full plant tailings with sieved material. It can be seen that measurements closely follow the trend predicted by theory.

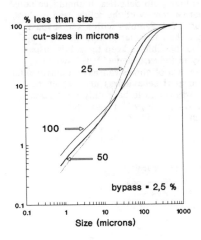

Fig.2  Backfill size distributions

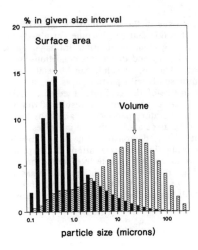

Fig.3  Volume and surface area size distributions

## 3 THE EFFECT OF CLASSIFICATION PARAMETERS ON BACKFILL PROPERTIES

### 3.1 Placement and drainage properties

One of the main reasons for using hydrocyclones in the preparation of backfill is to provide a product which when placed as a slurry can settle, drain and consolidate rapidly prior to taking load. This implies that the material has a high coefficient of permeability and reaches a state of partial saturation during or immediately after placement. Tests have shown that when the fill is partially saturated it is mechanically stable and has high shear strength. If the fill drains slowly and remains in an unstable saturated state after placement, there is a high risk that mechanical shock or inadvertent rupturing of geofabrics used for containment will lead to catastrophic failures and mud-rushes. Such failures obviously erode productivity and constitute an unacceptable safety hazard.

The coefficient of permeability of backfill is mainly a function of specific surface area, defined as the surface area of the constituent particles per unit volume of solids. There are theoretical arguments (Herget and de Korompay, 1978) which indicate that the coefficient of permeability is inversely proportional to the square of the specific surface

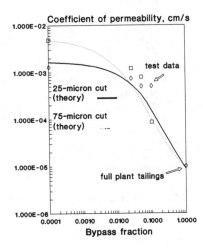

Fig.4  Effect of classification parameters on permeability

381

The data presented in Figure 4 show that for a bypass greater than about 3 per cent, finer cut-sizes result in marginally higher permeabilities. At low bypass fractions, significantly higher drainage can be achieved at large cut-sizes. It can also be seen that the coefficient of permeability is particularly sensitive to the bypass fraction. Sampling campaigns conducted on existing production plants have shown that bypasses as low as about two per cent are being achieved with two or three stages of classification, whereas bypass values closer to ten per cent are currently being achieved with single-stage plants.

An important issue in the placement of backfill is the amount of solids that can become entrained with water arising from decantation drainage. These solids cause slippery and dangerous conditions in working areas. The loss of solids can be related to the fraction of solids with particle sizes smaller than the mesh size of the geofabric used to contain the backfill. Figure 5 presents the results of drainage tests in a laboratory paddock. High solids losses are shown to be associated with fine cut-sizes. These results clearly demonstrate one of the disadvantages of using fine cut sizes, although the problem can be resolved in principle by placing the backfill at high relative densities. Tests have shown that solids losses can be virtually eliminated if the placed backfill has a relative density greater than about 1,85.

best determined by conducting tests in production-scale pipelines, simple viscosity measurements provide an indication of the rheology of pulps when only small samples are available.

Figure 6 shows the results of viscosity tests done using a modified Stormer viscometer (Van der Walt and Fourie, 1957) on synthetically prepared samples. It can be seen that the slurries have yield-pseudoplastic characteristics (Slatter and Lazarus, 1988) and that for a given pulp density and shear rate, a reduction in the bypass fraction can lead to a significant increase in shear stress, especially at high pulp densities. It should be noted that the relative density of the solids used in these tests was about 2,7. Samples with a zero bypass were found to be particularly difficult to handle and it was not possible to keep the solids suspended satisfactorily in the viscometer. It follows that excessive removal of slimes can have a detrimental effect on transport behaviour. Cut-size, on the other hand, was found to have only a marginal effect on slurry rheology (for corrected cut-sizes ranging from 25 to 75 microns).

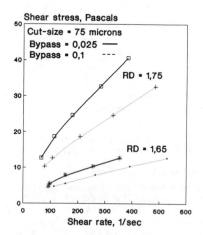

Fig.6  Effect of bypass on rheology

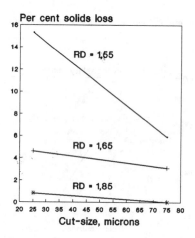

Fig.5  Effect of cut-size on placement solids losses

### 3.2 Transport properties

Backfill pulps are often transported over many kilometres on surface and underground. Thus the transport behaviour of backfill slurries and the pressure drops generated in pipelines are of prime importance in the design and operation of any distribution network. Whilst transport behaviour is

### 3.3 Stress-strain properties

The stress-strain properties of backfills depend primarily on their placed porosities (Briggs 1988). Placed porosities are a function of both placement density and particle size distribution. Low placed porosities are usually associated with high resistance to mechanical deformation. The lowest or minimum placed porosities potentially achievable can be determined in the laboratory using the placement properties test described by Clark (1988). In this test, increments of water are mixed with the backfill solids in a measuring cylinder. The cylinder and its contents are then vibrated and the porosity of the backfill is recorded. A plot of porosity as a function of density can then be used to estimate the minimum porosity and corresponding optimum placement density. The effect of the cut-size and bypass parameters on minimum porosities are presented in Figure 7. It can be seen that the minimum placement porosity increases with a decrease in cut-size, for cut-sizes in the range 75 to 25 microns, and increases with a decrease in the bypass fraction. Under production conditions, placed porosities are significantly higher than the minimum. For current placement relative densities of about 1,65 , placed porosities are roughly 25 per cent higher than the minimum. In order to establish an upper bound on the placed porosities, samples were prepared and mixed in a measuring cylinder at a pulp relative density of 1,65. The solids were allowed to settle and consolidate under gravity in the cylinder without the use of any vibration. The resultant porosities, which were considered to be the maximum likely to be encountered in practice, were found to follow a similar trend to that reported for the minimum porosities with low cut-sizes and low bypass fractions yielding the highest placed porosities.

porosities achieved under production conditions are roughly proportional to the minimum porosity, it turns out that the classification parameters have only a marginal influence on the net stress-strain behaviour in practice. A reduction in the cut-size and bypass fraction tended to give the backfill slightly more resistance to deformation at low stress levels, but the converse was true at high stress levels. At very high stresses, above 100 MPa, the measured strains were found to asymptotically approach values directly related to the placed porosity.

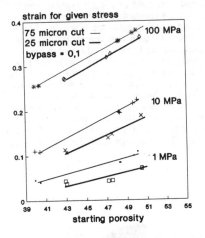

Fig.8 Effect of cut-size on deformation behaviour

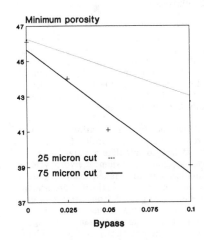

Fig.7 Effect of classification parameters on porosities

Once the minimum and maximum porosities had been established, the samples were subjected to confined uniaxial compression tests to simulate their response to stope closure underground. Plots of strain as a function of initial placed porosity and stress are given in Figures 8 and 9. The porosities plotted in these graphs include the minimum and maximum values described above. Figure 8 shows that for a given bypass and starting porosity, a reduction in cut-size increases the resistance to deformation. Figure 9 shows that for a given cut-size, a reduction in the bypass also increases the resistance to deformation. However, when cognizance is given to the observation that placed

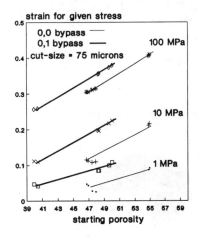

Fig.9 Effect of bypass on deformation behaviour

383

## 4 ECONOMIC CONSIDERATIONS

### 4.1 Capital and working costs

In order to maximize profitabilities, cognizance must clearly be given to economic issues. A theoretical approach was adopted to estimate the capital and working costs of systems based on current backfilling practice for a hypothetical deep-level mine producing 50 000 tons of backfill per month. The flowsheet for backfilling operations on the mine was based on three-stages of classification using large diameter hydrocyclones. Provision was made for the treatment of overflow residues prior to placement on slimes dams, and for the storage, distribution and placement of the backfill.
The estimated first costs or capital costs of construction for the hypothetical system are given in Figure 10, and working costs are given in Figure 11. It can be seen that the capital costs of handling and treating backfill residues are greater than the costs of preparation. The reason for this is that the discarded overflow from the hydrocyclone plant has a density which is too low for slimes dam building directly. Consequently, thickening to increase slurry densities to about 1,3 t/m³ is necessary. It can also be seen that the handling of drainage water is also a significant contributory factor to the first costs. The implication of these findings is

that careful consideration should be given to minimizing the amount of water which must be handled per ton of backfill solids produced. It also follows from Figure 10 that any diseconomy of scale that could arise from a change to small diameter hydrocyclones to boost mass recoveries is unlikely to have a major impact on the overall capital costs.

The cost data given in Figure 11 clearly demonstrate that working costs are dominated by the costs of placement. The bulk of these costs are attributable to the consumables of paddock construction and labour. Whilst it is difficult to establish a quantitative relationship between these costs and the properties of the placed fill, it is not unreasonable to suppose that backfill which settles, drains and consolidates rapidly is likely to have a more favourable impact on costs than one which does not have these characteristics. For example, if the backfill drains and consolidates immediately after placement, scope exists for the development of reusable materials for paddock construction.

It is also pertinent to note that the working costs of hoisting drainage effluents is almost as much as the costs of preparing the backfill, yet again emphasizing the importance of minimizing water requirements in the classification process.

### 4.2 Hydrocyclone circuits and water balance

In the previous section importance was given to minimizing the amount of water needed to produce a given amount of backfill and to maximizing the density of the backfill product. Underflow relative densities achievable on a routine basis typically range from 1,7 to 1,8 for hydrocyclones currently used in South African gold mines.

If the underflow density, classification parameters and feed size distribution are specified, it is a straightforward task to calculate a water balance around a hydrocyclone or system of hydrocyclones. Such a calculation was undertaken, using computer simulation techniques, to ascertain if there are significant benefits to be gained from using two or three stages of classification on the overall water balance over a single-stage plant. The two-stage circuit chosen was the one shown in Figure 12 in which the feed to the secondary is the primary underflow, and the secondary overflow is recycled to the primary feed. In the calculations, the cut-size of each hydrocyclone was changed to maintain an overall target solids recovery of 50 per cent and the bypass parameter was varied to effect changes in the calculated coefficient of permeability of the final backfill product. In each simulation run, the relative density of the underflow pulp of each hydrocyclone was kept at a value of 1,75 . The results of the simulations are given in Figure 13 where the estimated coefficient of permeability of the backfill is plotted as a function of the amount of fresh circuit feed water handled per ton of backfill produced. It can be seen that the benefits of a two-stage circuit are very considerable. By making use of a two-stage circuit it is possible to have a coefficient of permeability as high as

**Costs : Rands (millions)**

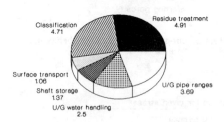

Classification 4.71
Residue treatment 4.91
Surface transport 1.06
Shaft storage 1.37
U/G water handling 2.5
U/G pipe ranges 3.69

Fig.10　System capital costs

**Rands per ton placed**
( depreciation excluded)

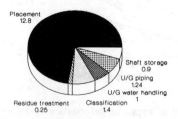

Placement 12.8
Shaft storage 0.9
U/G piping 1.24
U/G water handling 1
Residue treatment 0.25
Classification 1.4

Fig.11　System working costs

1.0 x 10⁻³ cm/s for a water demand of only 5 tons per dry ton of backfill product. On the other hand, for the same water demand, a single-stage circuit will yield a backfill product with far inferior drainage characteristics. Similar simulations were also undertaken for a three-stage plant. However such a plant offered only marginal benefits over a two-stage one.

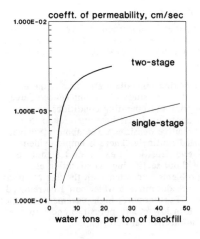

Fig.13  Effect of circuit configuration on water balance

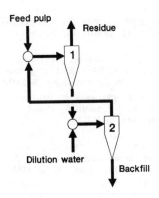

Fig.12  Two-stage flowsheet

## 5 HYDROCYCLONE GEOMETRIES AND OPERATING CONDITIONS

From the tests described above it is evident that a reduction in cut-size has no detrimental effect on the transport behaviour and mechanical support capabilities of backfill. The only significant negative factor of decreasing the cut-size is that the loss of fines with the decantation water can be enhanced, especially if underflow densities are not maximised. It follows that the selection of a cut-size less than that needed to produce the desired recovery of solids is undesirable. A cut-size, corrected for water split, between 30 and 40 microns is probably an acceptable target for gold mine tailings. It is also evident that whilst the bypass has little effect on the mechanical properties of backfill, it has a marked influence on the coefficient of permeability and the placement behaviour of backfill. A bypass of between 2 and 3 per cent appears to be adequate for acceptable drainage behaviour without problems of settlement during storage and transport. It has also been argued that two stages of classification are needed to ensure that water requirements are kept to practical levels. Hydrocyclone geometries and operating conditions were then sought to satisfy the criteria spelt out above.

Test work done by Uys (1988) has shown that the given criteria can be satisfied using hydrocyclones with diameters in the range 127 mm to 165 mm.

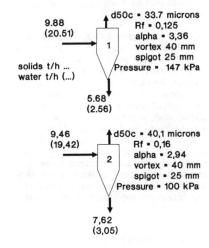

Fig.14  Hydrocyclone specifications

Details of design geometries and operating conditions for commercial 5-inch hydrocyclones are given in Figure 14, together with information on the measured classification efficiencies and mass balance. Unfortunately, tests were done in open loop and so it was not possible to validate the effect of recycling the secondary overflow product back to the primary feed. Even so, it is interesting to note that only 6 tons of water were needed to produce a ton of backfill product. The coefficient of permeability of the secondary underflow product was 1,3 x 10⁻³ cm/sec and the results of confined compression and rheological tests agreed well with predictions based on the results of the tests done on the synthetic samples.

385

## 6 FUTURE DIRECTIONS

Whilst this paper is concerned mainly with established hydrocyclone technology, it is perhaps useful to give an indication of areas deserving attention in the future.

One area meriting immediate attention is the execution of a systematic study of the influence of cyclone geometries and operating conditions on underflow densities. It has been mentioned that underflow relative densities up to about 1,8 have been reported routinely. There is ample evidence that if relative densities in excess of 1,8 could be achieved without sacrificing classification efficiencies, problems associated with the loss of solids entrained with decantation water could be reduced significantly. Furthermore, high placement densities would result in low in situ porosities and improved resistance to mechanical deformation.

Work on the development of strategies for controlling underflow densities and particle size distributions also needs to be pursued actively. Whilst finer cut-sizes can yield higher mass recoveries, the equal feeding of several small hydrocyclones in parallel clusters could well lead to pumping and distribution problems.

There are possible benefits to be gained from the formulation of methods of improving classification efficiencies to obviate the need for secondary hydrocyclones and recycle streams. Improved classification efficiencies have been claimed for a twin vortex hydrocyclone (Anonymous, 1987) which makes use of injected wash water. The nominal flow rate of the wash water is about 30 per cent of the feed pulp. Methods for improving efficiencies in ultrafine sizes by injecting wash water into the spigot of conventional hydrocyclones have also been claimed by a number of manufacturers. However, investigations need to be done to establish whether or not the improved performance can be brought about without increasing the total amount of water handled.

## 7 CONCLUDING COMMENTS

In conclusion, it should be stressed that improvements in the implementation of hydrocyclone technology are likely to be realised only if proper consideration is given to the monitoring of hydrocyclone performance in both pilot-scale and production operations. All too often hydrocyclones are installed with little regard for adequate sampling facilities or the means for accurately measuring circuit feed densities and volume flow rates. It is also important that attention is focussed not only on the size distribution of the final backfill product but also on other streams comprising the plant so that a complete mass balance can be established and partition curves generated for each stage of classification. Only by paying attention to such detail can meaningful action be taken to ensure a consistent product which meets the prime backfilling objectives of improved exploitation of the ore body being mined and better safety.

## ACKNOWLEDGEMENTS

The material embodied in this paper formed part of the research programme of the Chamber of Mines of South Africa. Contributions made by colleagues of the authors are acknowledged most gratefully. Special thanks are due to Mr. C Uys of Gold Fields Laboratories (Pty) Ltd for making available unpublished test data.

## REFERENCES

Anonymous. 1987. Hydrocyclone upgrades classification efficiency. World Mining Equipment, 11:34-35

Briggs, D.J. 1988. The load-bearing properties of fill materials (uncemented fills). Backfill in South African Mines, p. 35-60. Johannesburg, SAIMM.

Clark, I.H. 1988. The properties of hydraulically placed backfill. Backfill in South African Mines, p. 15-33. Johannesburg, SAIMM.

de Jongh, C.L. & Morris A.N. 1988. Considerations in the design and operation of backfill systems for narrow tabular ore bodies. Backfill in South African Mines, p. 355-368. Johannesburg, SAIMM.

Fahlstrom, P.H. 1963 Studies of the hydrocyclone as classifier. Proceedings 6th International Mineral Processing Congress, Cannes.

Flintoff, B.C., Plitt, L.R. & Turak, A.A. 1987. Cyclone modelling: a review of present technology. Canadian Institute of Mining Bulletin, 80:39-70

Herget, G & de Korompay, V. 1978. 12th Canadian Rock Mechanics Symposium, CIM Special Volume 19, p. 117-123

Hinde, A.L. 1985. Classification and concentration of heavy minerals in grinding circuits. 114th AIME Annual Meeting, New York City, paper 85-126

Lynch, A.J. & Rao, T.C. 1975. Modelling and scale-up of hydrocyclone classifiers. Proceedings 11th International Mineral Processing Congress, Cagliari, p. 114-123.

Plitt, L.R. 1976. A mathematical model of the hydrocyclone classifier. Canadian Institute of Mining Bulletin, 69:114-123

Slatter, P.T. & Lazarus, J.H. 1988. The application of viscometry to the hydraulic transport of backfill material. Backfill in South African Mines, p. 263-285. Johannesburg, SAIMM.

Uys, C.J. 1988. Personal communication.

Van der Walt, P.J & Fourie, A.M. 1957. Determination of the viscosity of unstable industrial suspensions with the aid of a Stormer viscosimeter. Journal of the South African Institute of Mining and Metallurgy, 57:709-723

Yoshioka, N & Hotta, Y.K. 1955. Classification with hydrocyclones. Chemical Engineering, Japan 19:632-635.

*Innovations in Mining Backfill Technology, Hassani et al. (eds), © 1989 Balkema, Rotterdam. ISBN 90 6191 985 1*

# Hydraulic transportation of high density backfill

C.P.Kramers, P.M.Russell & I.Billingsley
*Chamber of Mines of South Africa Research Organization, Johannesburg, RSA*

ABSTRACT: The paper briefly describes the backfilling hydraulic transport systems currently in operation in the South African deep level mines and goes on to describe the results of a comprehensive research programme to optimize transportation. Hydraulic transportation at high relative densities is identified as an area where scope exists for improving the efficiency of backfill placement and underground working conditions. It is shown that transportation at relative densities higher than used in current operations is possible and that capital and working costs for systems can be reduced.

## 1 INTRODUCTION

Large quantities of backfill are required in the stopes of the South African gold mines each month to provide regional and/or local support which will reduce the possibility of rockbursts and improve mining conditions. By 1995 it is estimated that 30 gold mines will be placing about two million tons of backfill per month. Currently, depending on various factors, but primarily the stoping depth, various backfill materials with different support qualities are used. These are typically the hydrocyclone classified tailings (CCT), the dewatered full plant tailings (FPT), comminuted development waste (CMW) and centrifuged tailings (CNT).

The extent and range of backfill reticulation into the stopes in South African gold mines is fairly complex in relation to the backfill systems used abroad. Reasons for this are that in South Africa:
(i) the average depth of mining is about 2 000 m below surface with the deepest workings at a depth of 3 800 m. Mining depths of 4 500 m are planned in the future.
(ii) tabular deposits at various angles of dip cover extensive lateral areas, and backfill distribution systems range over horizontal distances of up to four kilometres.

(iii) a longwall mining method is adopted most often and thus backfill needs to be placed close to the face and concurrently as it advances.
(iv) direct access to the stopes via boreholes from surface is often impractical due to the great mining depths,and the logistics of backfill transportation is complicated by having to use existing vertical and subshafts, main haulages, footwall haulages and gullies, where space availability is limited.

These factors pose problems in the design of extensive backfill transportation systems. Mechanical handling methods have been avoided to date and gravity assisted hydraulic transportation to convey the backfill as a slurry into the stopes is preferred.

This paper describes the results of a comprehensive research programme to optimize the hydraulic transportation of backfill in deep mines. The first section of the paper reviews the hydraulic transportation systems currently in operation. This is followed by a review of the results of measured pressure losses and critical deposition velocities in test pipe loops and details of the rheological behaviour of high density backfill slurries. Predictive equations for determining the energy gradient and critical deposition velocities for the backfill slurries over a range of relative densities and in pipelines of various diameters are compared with the measured results.

The relative merits of high density transport of the classified and full plant tailings backfills and of pipes in vertical shafts are discussed.

## 2 TYPES OF BACKFILL IN USE IN SOUTH AFRICA

The cumulative particle size distribution curves for the various types of backfill currently in use in

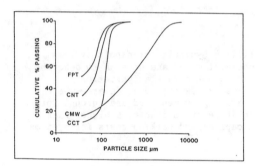

Figure 1. Cumulative particle size distribution curves for various back-fills

South Africa are shown in Figure 1. Gold plant tailings typically have a size distribution where 80 % of the particles, by mass, pass through a 75 micrometre mesh. In preparing fill from gold plant tailings the particle size distribution is altered by a variety of filtering and dewatering processes.

The derivative fills differ significantly in their physical and rheological properties. Centrifuged tailings (CNT) typically contain about 30 % passing 38 micrometres compared to 50 % for the full plant tailings. Hydrocyclone classified tailings contain between 6 and 15 % passing 38 micrometres. Comminuted waste backfill derived from development waste produced underground by two stages of crushing and a rod mill, produces a well graded gravel with a top size of about 10 mm and a fines content of which 10 to 15 % passes 38 micrometres.

## 3 HYDRAULIC TRANSPORT OF BACKFILL IN SOUTH AFRICA

Fourteen mines are currently placing hydrocyclone classified tailings. Back-

fill reticulation in all the systems is based on gravity assisted hydraulic transport in dedicated pipelines or ranges, usually with a 50 mm internal diameter. The selection of this diameter is based on a volumetric flow rate of about 16 m$^3$/h and a slurry relative density of 1,65 (solids concentration by volume $C_v$=38 %). Dedicated pipe ranges are used to serve each level on a mine.

The system pressure is self regulating in each pipe range and is dependent on the hydraulic head, which is built up in the vertical shaft piping, to drive the backfill over horizontal distances from 3 000 m to 4 000 m in length. Flow rate is controlled by maintaining a constant level in a small surge cone at the head of each pipe range. Restriction orifices are installed at the base of the supply header to provide an equal flow rate to the various ranges. The backfill slurry free falls to an interface region dictated by the hydraulic head required for horizontal flow, and is then transported at full flow velocities of between 2 m/s and 3 m/s. To prevent pipeline blockages due to particle settlement, pipelines are flushed with water at the beginning and end of each filling shift. Rupture discs are installed in the piping at the bottom of the vertical section to avoid pipes bursting in the event of a column becoming completely filled due to a blockage between the shaft and the stopes.

The dewatered full plant tailings backfills have a slower rate of water drainage and are placed into the stopes at relative densities of about 1,85 ($C_v$=50 %). Preparation requires dewatering operations by means of filtration or centrifugation either on surface for the shallower mines, or underground for the deep mines. The dewatered product is hydraulically transported to the underground workings by means of gravity assisted flow or from central underground dewatering and storage points by positive displacement pumps. The comminuted waste backfill preparation plant is located underground in a position close to the stopes (±1 000 m max) and positive displacement pumps are used to transport the milled product to the stopes. The system has the advantage that development waste material produced underground need not be hauled to surface.

# 4 THE CHAMBER OF MINES HYDRAULIC TRANS-
## PORT RESEARCH FACILITY

The Chamber of Mines Research Organization (COMRO) has developed a comprehensive hydraulic transport research test
facility at one of the mines. The
facility incorporates pipe loops of
25 mm, 38 mm, 50 mm, 100 mm and 150 mm
nominal bore, with loop lengths of up to
300 m. The pipe loops are equipped with
pressure loss measurement equipment,
accurate to within 3 % for gradients
above 1 kPa/m. Various centrifugal and
positive displacement pumps are
available for use with the various backfill slurries. Facilities for investigating the resuspension of settled
slurries, the wear characteristics of
pipelines, and the placement behaviour
of backfills, are also available. Site
glasses are used to estimate critical
deposition velocities of the slurry in
the pipelines. An ultrasonic doppler
velocimeter, developed in collaboration
with the University of Cape Town, is
used for measuring the bed load velocity
of high concentration slurries.

# 5 PRESSURE LOSSES AND CRITICAL DEPOSI-
## TION VELOCITIES

The effects of the particle size distribution of the full plant tailings (FPT),
the comminuted waste (CMW) and hydrocyclone classified tails (CCT) on slurry
pressure loss are presented in Figures
2, 3, 4, 5, 6, 7 and 8. Measured pressure losses in kPa/m are plotted on the
y-axis and mean slurry flow velocities
on the x-axis. For each material type,
test runs were conducted over a range of
solids concentrations by volume ($C_v$)
in 50 mm, 100 mm and 150 mm diameter
pipes. Each slurry test run was preceded by a test run using water to

establish a reference pipe friction
factor - Reynolds number relationship.
The water pressure loss curves are shown
in each case. Particle size distributions typical of the various backfills
are shown in Figure 1.

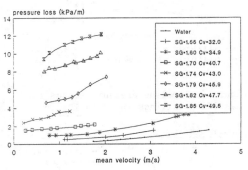

Figure 3. Pipeline pressure losses for
full plant tailings in 100 mm ID pipe

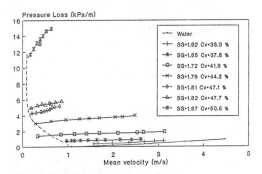

Figure 4. Pipeline pressure losses for
full plant tailings in 155 mm ID pipe

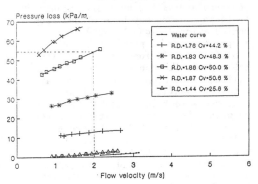

Figure 2. Pipeline pressure losses for
full plant tailings in 50 mm ID pipe

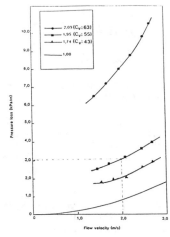

Figure 5. Pipeline pressure losses for
comminuted waste in 50 mm ID pipe

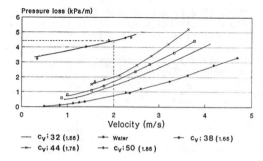

Figure 6. Pipeline pressure losses for classified tailings in 50 mm ID pipe

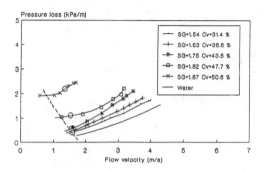

Figure 7. Pipeline pressure losses for classified tailings in 100 mm ID pipe

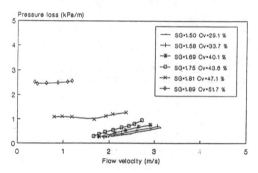

Figure 8. Pipeline pressure losses for classified tailings in 155 mm ID pipe

A range of slurry flow velocities were selected to define the flow curves from the vicinity of the critical deposition velocity to velocities well into practical operating regions.

It is apparent from the curves for full plant tailings and comminuted waste that an increase in the median size has a marked effect in reducing the pressure losses. This effect is illustrated by

considering a slurry velocity of 2 m/s in Figures 2 and 5. For $C_V$=50 %, a pressure loss of 54 kPa/m was measured for the full plant tailings whereas for the comminuted waste this was only 3 kPa/m, even at the higher $C_V$ value of 55,2 %. For the uniform particle size distribution of the hydrocyclone classified tailings (Figure 6), the measured pressure loss was 4,4 kPa/m for similar slurry velocity and concentration conditions.

Similar effects are apparent for the full plant tailings and classified tailings in the 100 mm and 150 mm diameter pipelines (Figures 3, 4, 7 and 8).

The effect of slurry concentration on pressure losses for the various backfills is demonstrated by an increasing divergence of the pressure loss curves from the water pressure loss curve with increasing slurry concentration. The effect is most prevalent for the full plant tailings followed by the comminuted waste and then the hydrocyclone classified tailings.

Critical deposition velocities were calculated using a modified Durand equation (Bain and Bonnington, 1970) in which the relative density of the carrier fluid consists of water and the minus 38 micrometre fraction of the slurry.

A comparison of the calculated critical deposition velocities indicated by circles, and the observed deposition velocities, the locus of the dashed line, are shown in Figure 7. For slurry concentrations by volume of less than 40 %, calculated critical deposition velocities agree fairly well with observed data. At concentrations greater than 40 %, however, large differences are apparent between these values. Predicted critical deposition velocities are higher than those observed and viable hydraulic transport in these regions may be excluded from reticulation designs because of a misplaced fear of pipe blockages.

Similar conclusions were drawn from the testwork conducted on full plant tailings. At high solids concentrations (greater than about 47 %), these products were transported at very low velocities even in the large diameter piping as illustrated in Figure 4.

The pressure loss and critical deposition velocity measurements conducted in production-scale pipelines have provided a better understanding of the transport behaviour of backfills and

390

invaluable information for distribution network design and operation. The results indicate viable hydraulic transport of the materials at solids volume concentrations in excess of 45 %. This offers tremendous benefits for backfill placement and support, and in this context, and in the context of backfill system design, the findings are particularly significant.

## 6 EQUATIONS FOR PREDICTING PIPELINE PRESSURE LOSSES

In relation to the ever increasing depths of mining, wider stoping widths and difficult underground conditions, material selection for backfilling is becoming more and more specialized. Bulk samples of materials are not always available for production-scale testing. Computer programmes to investigate the validity of various models for predicting pipeline pressure losses over a range of particle size distributions and backfill slurry relative densities, have thus been developed.

Four models were selected for making comparisons with the measured data; the well-known Durand model (Bain and Bonnington, 1970), the Wasp calculation method (Wasp, 1977), the pseudo fluid method which assumes that the slurry behaves as a homogeneous Newtonian fluid (Horsley, 1982), and a generalized mechanistic model developed by Lazarus (Lazarus, 1988). To illustrate these models, Figures 9 and 10 show comparisons of the predicted pressure losses and measured values in 100 mm diameter pipelines for hydrocyclone classified backfills.

For a solids concentration of 37 % by volume (Figure 9) good agreement between the measured results and the predictive methods of Lazarus and the pseudo method was obtained. However for a solids concentration of 50 % by volume (Figure 10) no agreement is apparent and all the models considered under-predicted pressure losses. Many computer runs for the different backfill types over a range of concentrations have been conducted. It has been established that the known equations for predicting pipeline pressure losses can only be used for slurry concentrations by volume up to 40 %. As yet no equation based on basic principles, has been developed which will predict pipeline pressure losses for concentrations above 40 % within a reasonable degree of accuracy.

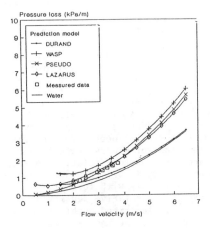

Figure 9. Predicted pressure losses for classified tailings in 100 mm ID pipe slurry relative density 1,63 ($C_V$=37 %)

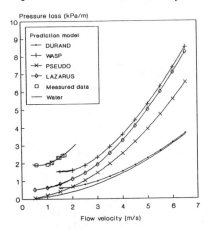

Figure 10. Predicted pressure losses for classified tailings in 100 mm ID pipe slurry relative density 1,87 ($C_V$=50 %)

Empirical correlations have been fitted to the data which can be used for prediction.

The rheology and flow behaviour of high concentration slurries were investigated using a balanced beam tube viscometer (Neill, 1988). The rheology of the lower concentration tailings (less than $C_V$=38 %) was successfully characterized using the yield-pseudoplastic model (Horsley, 1982). At high solids concentrations characterization was not possible and the laminar flow regions of the pseudo-shear diagrams varied with tube diameters. A suitable method for correcting the data has not yet been established.

In Figure 11 measured pressure losses for the full plant tailings and hydro-cyclone classified products are plotted against the relative density of the slurry, at constant velocity, in a 100 mm pipeline. A close to linear increase in pressure loss with increasing relative density occurs for the full plant tailings for slurry velocities of 1 m/s and 2 m/s up to a relative density of about 1,7 ($C_v$=40 %). Pressure losses rise rapidly with further increases in the relative density of the slurry. A similar trend occurs with the classified tails, but the rate of increase in pressure loss is lower.

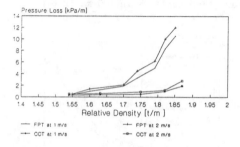

Figure 11. Pressure losses as a function of relative density in a 100 mm pipe

## 7 MERITS OF HIGH DENSITY TRANSPORT OF BACKFILLS

An important issue in the placement of backfill is the amount of solids that can become entrained with the decanted water. These solids cause muddy and dangerous conditions which can adversely affect productivity.

Tests (Clark, 1988a) have shown that solids losses can be virtually eliminated if the placed backfill has a relative density greater than about 1,85 ($C_v$=50 %). It has also been established (Briggs 1988) that the stress-strain properties of backfills depend primarily on their placed porosities. The lower the porosity, the better the support quality. The placed porosities in turn are a function of both the placement density and the particle size distribution (Clark, 1988b). At high relative densities the effects of water drainage from the fill are minimized and filling of the stope with solids is more efficient. Technically, therefore, a

strong incentive exists to hydraulically transport and place the various backfill types at the highest possible relative density.

To maximize profitabilities, consideration must also be given to economic issues. A theoretical approach was adopted to estimate the first or capital costs and working costs of systems based on current backfilling practice (Kramers, 1988). The cost data in Figure 12 demonstrates that the contribution of the transport related costs to the overall first costs is in the order

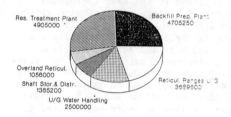

TOTAL FIRST COST 18221050

Figure 12  Overall first costs for a 50 000 t/month classified tailings system

of 50 per cent for a typical 50 000 t/month classified tailings system. By increasing the relative density of the slurry from 1,65 ($C_v$=38 %) which is current practice, to 1,80 ($C_v$=47 %) significant reductions in the costs of piping and storage capacity were estimated due to more efficient utilization, in terms of unit costs per ton of solids moved and water handling. For backfill reticulation in 50 mm diameter piping at a slurry velocity of 2 m/s, 15,4 t/h of solids are moved at a relative density of 1,65, whereas 22,3 t/h (45 % higher) are moved at a relative density of 1,8. A significant reduction of water entering the stopes, from 8,5 m³/h at a relative density of 1,65 to 5,9 m³/h (31 % lower) at a relative density of 1,8, reduces the first costs of underground water handling considerably.

Overall working costs for the system are illustrated in Figure 13. It can be seen here that backfill placement operations are by far the greatest contributor to working costs. From the break down of placement working costs on the basis of consumables and labour, as shown in Figure 14, it is apparent that labour contributes just under 50 per

392

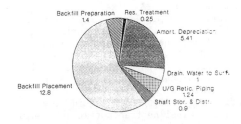

TOTAL WORKING COST R23,04/t [1152000/m ]

Figure 13. Overall working costs for a
50 000 t/month classified tailings
system

cent of the total. At a relative
density of 1,65, two filling shifts per
day are required to meet placement
needs; at a relative density of 1,8
filling can be achieved in one shift.
This represents a significant saving in
working costs.

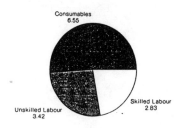

Figure 14. Backfill placement working
cost for a 50 000 t/month classified
tailings system

8 IMPACT OF HIGH DENSITY TRANSPORTATION
  ON RETICULATION DESIGN

Pipeline wear is a serious problem
particularly in the vertical section of
piping in shafts. It has been
identified that this wear is related
mainly to the high slurry velocities in
the free fall region of the piping and
to corrosion-erosion. The short-term
approach to alleviating the wear problem
has been to control slurry velocities by
means of ceramic orifices or smaller
diameter pipelines and giving careful
attention to pipe verticality, pipe
handling, pipe alignment and bore match-
ing. A programme is currently being
conducted to identify the suitability of

specialized materials of pipe construction and selected pipe linings by monitoring in-line wear spools installed at
various mines.

In the longer term, serious consideration should be given to designing
gravity assisted reticulation systems in
'full flow' pipelines, where a combination of increasing the relative density
of the slurry and interposing breaks by
way of intermediary storage points and
energy dissipators will be used to control slurry velocities and maximum pressures. An advantage of high slurry
relative density is that higher pipeline
pressure losses can be counterbalanced
against hydraulic heads thus requiring a
lesser number of break points in the
vertical piping. Controlled slurry
velocities which are at lower values
than present for the equivalent solids
tonnage output, will result in reduced
pipe wear, shaft disruption and hence
cost. The operating pressure of a
classified tailings distribution system
in 50 mm diameter piping, with a
horizontal distance of transport of
3 000 m, is about 6 MPa at a relative
density of 1,65. A hydraulic head of
about 350 m builds up to drive the
system. At a relative density of 1,8
an operating pressure of 11 MPa is
required which is equivalent to a head
of 614 m, and this is within the safe
working pressures of ASTM mild steel
A106 grade B, schedule 80 piping.
Similar arguments apply to full plant
tailings.

9 CONCLUSIONS

The paper has described the results of a
comprehensive research programme where
careful measurements of the pressure
losses and critical deposition
velocities of typical South African
backfills were measured. Giving
particular emphasis to the transport
behaviour of high relative density
backfill slurries, measurements indicate
that viable hydraulic transport of the
backfills at relative densities
considerably higher than currently in
use, can be achieved.

Equations based on the rheology and
particle size distributions of small
samples of material can only be used for
predicting pipeline pressure losses and
critical deposition velocities for backfill slurries with relative densities of
less than 1,7 ($C_v$=40 %). For slurries
with relative densities higher than

this, transport behaviour is best determined by conducting tests in production-scale pipelines.

By transporting and placing backfills at the highest possible relative density, solids losses during placement may be reduced and simultaneously the support quality of the material may be improved. Water losses into the stopes would also be minimized resulting in significant reductions in the first costs of water pumping to surface operations.

Further potential benefits of high relative density transportation are that reticulation pipeline capacities may be maximized, offering greater rates of filling and hence allowing increased mining rates. Working costs would thus be reduced, particularly with regard to the reduced labour required to place the fill.

ACKNOWLEDGEMENTS

The material embodied in this paper formed part of the research programme of the Chamber of Mines of South Africa. Contributions made by colleagues of the authors are acknowledged most gratefully.

REFERENCES

Bain, A.C. and Bonnington, S.T. 1970. The hydraulic transport of solids by pipeline. Perganon Press Ltd. Oxford.

Briggs, D.J. 1988. The load-bearing properites of fill materials (uncemented fills). Backfill in South African Mines, p.35-60. Johannesburg, SAIMM.

Clark, I.H. 1988a. An evaluation of the placement and drainage characteristics of backfills. Unpublished report. Chamber of Mines Research Organization, Johannesburg.

Clark, I.H. 1988b. The properties of hydraulically placed backfill. Backfill in South African Mines, p.15-33. Johannesburg, SAIMM.

Horsley, R.R. 1982. Viscometer and pipe loop tests on gold slime slurry at very high concentrations by weight, with and without additives. BHRA, Hydrotransport 8.

Kramers, C.P. 1988. An evaluation of the theoretical first costs and working costs of a classified tailings backfill system. Unpublished report in preparation. Chamber of Mines Research Organization, Johannesburg.

Lazarus, J.H. 1988 Mixed regime slurries in pipelines - Part 1, mechanistic model. American Society of Civil Engineers, Hydraulics Division.

Neill, R.I.C. 1988. M.Sc. thesis. University of Cape Town. The rheology and flow behaviour of high concentration mineral slurries.

Wasp, E.J. 1979. Solid liquid flow slurry pipeline transportation. Guld Publishing Co. Houston.

*Innovations in Mining Backfill Technology, Hassani et al. (eds), © 1989 Balkema, Rotterdam. ISBN 90 6191 985 1*

# A simulation model for efficient hydraulic fill transport

A.Sellgren
*Department of Water Resources Engineering (WREL), Luleå University of Technology, Luleå, Sweden*

Abstract: The hydraulic characteristics of a filling system have been described in a simulation model. Input data were mainly based on viscometric tests and pilot-scale pipe loop results with two representative Swedish tailings products. Maximal capacities and the corresponding solids concentrations and operating velocities were simulated for simple backfilling systems. The results demonstrated how specific properties of the products, such as the particle size distribution, strongly influenced the maximum possible capacity. The schematic results presented here are to be followed by more detailed investigations.

## INTRODUCTION

Swedish complex sulphide ores (copper, gold, lead, zinc etc) are usually extracted from underground mines in conjunction with cut and fill mining methods. These methods are used in about 15 mines and about two Mtonnes of tailing products per year are filled hydraulically.

The fill product are normally mixed with water in a plant located underground close to the ground surface, see Figure 1.

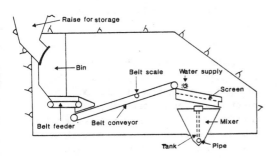

Figure 1 Mixing plant for hydraulic fill. From Krauland et.al (1986)

The mixture is gravitated through pipes or boreholes down and out to the stopes. The vertical lengths varies from 100 to 700 m vertically and 100 - 3000 m horizontally. The pipe diameters are normally in the range of 75 mm to 150 mm. Capacities are typically 75 - 250 tonnes of solids per hour.

## OBJECTIVES AND SCOPE

The overall objective has been to evaluate hydraulic design parameters for representative mine tailings fill products in a computational model and to simulate different operating conditions in hydraulic backfilling systems. The results in this study have mainly been based on pilot-scale experiments with two tailings products in an open loop test facility with a pipe diameter of 0.105 m and a total pipeline length of about 25 m. Detailed results and a description of the facility can be found in Väppling (1989). The fine-grained product has also been investigated rheologically through viscometric measurements, see for example Andreasson

(1989). The two products investigated are shown in Figure 2 together with the normal range of particle size distributions in hydraulic fill systems in Sweden.

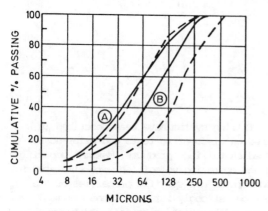

Figure 2. The dashed lines represent range of particle size distributions for tailings products used as hydraulic fill in Sweden. A and B are data for the products investigated here. Solids density = 2850 kg/m$^3$.

The results presented here are preliminary and the simulation model is so far only applied to schematic and simplified examples.

## HYDRAULIC CHARACTERISTICS

### Minimum velocity

The velocity at which the particles settle out at the bottom of a horizontal pipe (the deposition velocity, $U_D$) is generally determined by visual observations of the flow through a transparent pipe section in pilot-plant studies. The relationship most often proposed for determination of $U_D$ goes back to Durand (1953), who related his experimental results to a modified Froude number expressed by a diagram factor, $F_L$.

$$\frac{U_D}{[2gD(s_S-1)]^{0.5}} = F_L \qquad (1)$$

where D is the pipe diameter and
$$s_S = \varrho_S/\varrho_0$$
$\varrho_S$ = density of solids
$\varrho_0$ = density of water

For the systems considered here, particles are nearly uniformly distributed in the pipe section, i.e. the settling rate of the solid component is low under typical flow conditions. Particles of sizes of less than about 0.05 mm take on the viscous fluid properties of the slurry, while larger particles act heterogeneously through (turbulent) flow phenomena. Thus, the contribution of the solid phase to the operating conditions exhibit a combination of hydraulic and rheological effects. These effects are schematically represented in Figure 3, in which Parzonka et.al (1981) related experimental results to the $F_L$ parameter in Eq. (1).

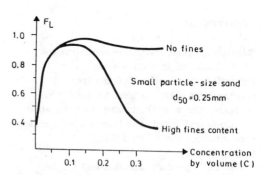

Figure 3. Overall variations of $F_L$ with solids concentration and particle size for slurries of sand and water. Pipe diameters in the range of 0.05 to 0.5 m. C is concentration by volume. After Parzonka et.al (1981)

The reduction in $F_L$ with a large concentration of fine particles (Figure 3) expresses viscous effects on the settling of larger particles. The rheology of the system becomes important and the condition is more related to the transitional velocity $U_T$. In a purely viscous slurry (homogeneous flow) $U_T$ increases with increased values of the solids concentration. Therefore, a region is reached where neither $U_D$ nor $U_T$ nor their relationship with the solids concentration is well defined.

The marked influence of C on $U_D$ in fine-particle systems in Figure 3 is normally not so clearly observable in industrial systems.

It follows from the definition of the $F_L$-factor in Eq. (1) that $U_D$ increases exponentially with the pipe diameter. The exponent is 0.5, which means that the deposition velocity increases by about 40% if the pipe diameter is doubled.

The effects of D on $U_D$ is of great importance, because it dictates the possibility of scaling up results obtained in experimental tests.

Wasp et.al (1977) reviewed published data on deposition velocity for sand slurries, mainly ungraded products. They suggested a modified form of Eq. (1)

$$F_L' = \frac{U_D}{[(2gD(s_s-1)]^{0.5}} \left(\frac{d}{D}\right)^{-0.17} \quad (2)$$

where the empirical constant, $F_L'$, was about 2.5 for concentrations of industrial interest. Eq. (2) relates $U_D$ to the pipe diameter with an exponent value of 0.33 which is in agreement with experimental results for a large varity of industrial slurries. Sellgren et.al (1986) applied Eq.2 to sand and tailings products and related the $F_L'$ factor to a representative particle size, see Figure 4.

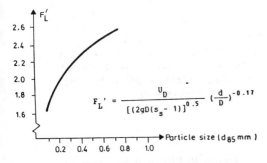

Figure 4. Values of the factor $F_L$ in Eq. (2) based on experimental results for sand and tailings products. The diameter d in Eq. (2) has here been related to $d_{85}$. From Sellgren et.al (1986).

Thus, if a deposition velocity $U_D$ is observed in a pipeline with a diameter of $D_1$ then the following approximate scale-up criterion is suggested:

$$U_D = U_{D_1} \left(\frac{D}{D_1}\right)^{0.33} \quad (3)$$

The criteria for deposit-free operation dictate the lower limit of the operating velocity in a hydraulic backfilling system. From a design standpoint it is advisable to operate at a somewhat higher velocity. If a safety margin of 0.3 m/s is used then

$$U > U_D + 0.3 \quad (4)$$

where $U_D$ is obtained from Eqs. (1) or (3).

With the particle size distributions shown in Figure 2 and filling in normal pipe distribution systems then the critical velocity does not excreed about 2 m/s.

Friction losses

With the laminar flow of a Newtonian fluid in a pipe the Darcy-Weisbach friction factor, f, is a function of Reynolds number only.

$$Re = \frac{U D \rho}{\mu} \quad (5)$$

where U stands for velocity, D for pipe diameter, $\rho$ for density, and $\mu$ for viscosity.

For turbulent flow conditions the absolute roughness, k, of the pipe wall also has to be considered, which leads to the relationship:

$$f = \Phi \left(\frac{k}{D}, Re\right) \quad (6)$$

where $\Phi$ marks "function of" and where k/D is the relative roughness. For full turbulent flow, Eq.(5) is independent of Re, i.e.

$$f = \Phi (k/D) \quad (7)$$

However more generally (Eg.6), it is not clear how the viscosity of the mixture should be defined and applied in the calculation of Reynolds number in Eq.(6). The flow of tailings slurries can be hydraulically described by coupled effects of rheology and sedimentation. In an attempt to obtain a simple correlation tool for practical use, Sellgren (1986) defined a hypothetical viscosity including three empirical coefficients of which two were related to viscous effects.

A preliminary hydraulic analysis of a typical simple fill system indicated that conditions of maximum solids capacity per hour corresponded to an operating velocity considerably larger than the minimum velocity required (Eq. 4). It was also indicated that the flow should be in a turbulent state for the flow conditions and slurries studied here.

Under turbulent pipe flow conditions Newtonian methods of friction loss calculations may apply. One method is to determine the Reynolds number in Eq.(6) with a limiting viscosity obtained as an asymptotic value, $\mu_R$. With concentrated tailings slurries, Sellgren (1986) used the following relationship to calculate $\mu_R$.

$$\mu_R = \mu_0 \left(1 - \frac{C}{\alpha}\right)^{-\beta} \qquad (8)$$

where $\alpha$ and $\beta$ are coefficients determined by pilot-scale test loop data and rheological experiments. Based on data by Andreasson (1989), Sellgren (1986), and Väppling (1989) for the two products studied here, representative viscosities were determined as follows (Figure 5).

CALCULATION MODEL

The following schematic configuration has been used to describe the hydraulic fill system, Figure 6

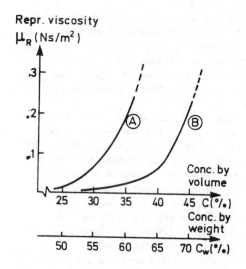

Figure 5. Evaluation of test loop data and rheological tests in terms of $\mu_R$ (Eq.8) for the two tailings products studied here.

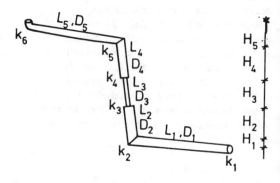

Figure 6. Description of the configuration of the hydraulic fill system in the model. Five different elevations, pipeline diameters and lengths can be chosen.

The system in Figure 6 operates at the flow rate at which the kinetic energy and the energy losses caused by pipe friction, bends etc are balanced by the gravity head, $H_a$, defined by the

398

difference in elevation between the discharge point at the stop and at the free slurry surface in the mixing tank. It follows from Figure 6 that

$$H_a = \Sigma \, H_i \qquad (9)$$

where $H_i$ is the vertical distance of a pipe segment i. If it is assumed that the available head, $H_a$, is completely used then the above mentioned balance can be expressed by:

$$H_a = \Sigma \, \frac{f_i \, L_i \, U_i^2}{2g \, D_i} + \Sigma \, \frac{K_i \, U_i^2}{2g} + \frac{U_1^2}{2g} \qquad (10)$$

where

$f_i$ = Darcy-Weisbach friction factor
$L_i$ = length of pipe segment
$U_i$ = velocity in a pipe segment
$D_i$ = inner diameter of a pipe segment
$K_i$ = loss coefficient in segment connections etc.
$U_1$ = velocity at the discharge end

The operating velocity in a segment, is related to the flow rate of slurry, Q, and the corresponding pipe diameter through the following formula

$$U_i = \frac{Q \cdot 4}{\pi D_i^2} \qquad (11)$$

where the calculated velocity must fullfill Eq. (4).

The capacity or the mass flow rate of solids, $M_s$, is related to the concentration by volume, C, and to Q by the following equation

$$M_s = Q \, C \, \varrho_s \qquad (12)$$

The concentration by weight, $C_w$ is calculated by the following relationship:

$$C_w = \frac{C \, \varrho_s}{\varrho} \qquad (13)$$

Reynolds number (Eq.5) is related to $\mu_R$ in Eq. (8) and expressed by:

$$Re_i = \frac{U_i \, D_i \, \varrho}{\mu} \qquad (14)$$

where $\varrho$ is the slurry density

$$\varrho = \varrho_0 \, (1 + C \, (s_s - 1))$$

The turbulent friction factor in Eq. (6) is then calculated with the Darcy-Weisbach ordinary relationship, known as Colebrook's formula $\qquad (15)$

$$f_i^{-0.5} = -2 \log \left( \frac{k_i}{3.71 \, D_i} + \frac{2.51}{Re_i f_i^{0.5}} \right)$$

where $k_i$ is the absulute roughness in segment i.

Eq. (15) is used for $Re_i > 3000$, no values are calculated in the laminar/turbulent transition region ($Re_i = 1000 - 3000$) because of the unstable flow conditions. In the first version of the model the friction factor for laminar flow ($Re_i < 1000$) is simply related to the ordinary relationship for Newtonian fluids

$$f_i = \frac{64}{Re_i} \qquad (16)$$

The equations discussed here were assembled systematically in a simulation model. Input is given by the geometrical configuration of the system, $D_i$, $H_i$, $K_i$, $L_i$, $k_i$, and to $\alpha$, $\beta$, $\varrho_s$, $\varrho_0$ and $\mu_0$. An interval is then chosen where the solids concentration is allowed to vary. The mass flow rate of solids in tonnes/h is then calculated through iterative procedures for every solids concentration considered.

Output is given in the form of systematic tables where input data are listed together with calculated values of velocity, Reynolds number, friction factor, and absolute pressure in every pipe segment. Examples of results are shown in Figures 7 and 8 for two system configurations.

399

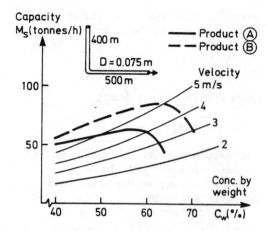

Figure 7. Simulated results of mass flow rate versus solids concentration by weight for the two products studied. Vertical pipeline length = 400 m. Horizontal length = 500 m. Pipe diameter = 0.075 m.

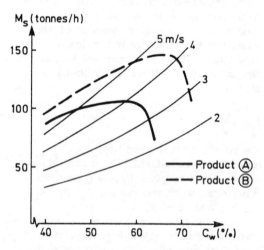

Figure 8. Simulated results of mass flow rate versus solids concentration by weight. Vertical pipeline length = 400 m. Horizontal length = 1000 m. Pipe diameter = 0.1 m.

## CONCLUSIONS

The simulated results for the two tailings products schematically studied here showed that the maximum capacities varied from 60 to 80 tonnes/h in a backfilling system with a pipe diameter of 75 mm and vertical and horizontal lengths of 400 and 500 m respectively (Figure 7). In a system with the same vertical distance and a pipe diameter of 100 mm and a horizontal length of 1000 m the maximum capacities varied from 100 to 145 tonnes/h (Figure 8). With the simple backfilling systems considered here the maximum capacities corresponded to relatively high velocities and turbulent flow conditions.

The results also demonstrated that specific properties of the tailings products, such as particle size distribution etc, have a strong influence on the maximum possible capacity in a backfilling system.

This study will be followed by more detailed investigations of important hydraulic transport parameters which are to be included in the simulation model.

ACKNOWLEDGEMENTS

This work was financially supported by the National Swedish Board for Technical Development, the Swedish Rock Engineering Research Foundation and by the Swedish mining industry through a joint reseach program called Mining Technology For The Year 2000. The author wishes to acknowledge and thank these organizations for their support. Mr Lennart Väppling did the programming work and Mr Mikael Eriksson took part in the computer evaluations. To these and all others involved I am very thankful.

REFERENCES

Andreasson, P., 1989: Rheological Characteristics of Mineral Slurries, Analysis of Measurements in Capillary Tube and Rotational Viscometers, Internal Report, Dept. of Water Res. Luleå Univ. of Tech.

Krauland, N., Nilsson, G. and Magnusson, I., 1986. Backfilling in the Boliden Mines - Why and How? Report Boliden Mineral AB.

Durand, R., 1953. Basic Relationship of
the Transportation of Solids in Pipes-
Experimental Research Proc., Minnesota
Int. Hydraulics Conf., Minneapolis,
Minn., USA.

Parzonka, W., Kenchington, J. M. and
Charles, M. E., 1981. Effects of
Solids Concentration and Particle
Size on the Deposit Velocity, The
Canadian Journal of Chemical
Engineering, Vol 59, June.

Sellgren, A., 1986. Hydraulic Behaviour
of Mine Tailings Products, Paper, The
First World Congress on Particle
Technology, Nuernberg, April, BRD.

Sellgren, A., and Väppling, L., 1986. The
Choice of Solids Concentration in Mine
Tailings Pipelines, Paper, Proc. of The
11th Int. Conference on Slurry Tech.,
USA.

Wasp, W. J., Gandhi, R. L. and Kenny, J.
P., 1977. Solid-Liquid Flow, Slurry
Pipeline Transportaton, Trans. Tech.
Publ.

Väppling, L., 1989. Slurry Transport
of ores and Industrial Minerals By
Centrifugal Pumps. An Experimental
Investigaton In A Horizontal Pipe Loop.
Licenciate Thesis, Dept. of Water Res.
Eng. Luleå Univ. of Tech.

# 8 General topics

Innovations in Mining Backfill Technology, Hassani et al. (eds), © 1989 Balkema, Rotterdam. ISBN 90 6191 985 1

# A study on liquefaction potential of paste backfill

K. Aref
*Klohn Leonoff, Consulting Engineers, Vancouver, BC, Canada*

F.P. Hassani
*Department of Mining and Metallurgical Engineering, McGill University, Montreal, PQ, Canada*

D. Churcher
*Dome Mines, Timmins, Ont., Canada*

ABSTRACT: The backfilling of underground stopes with cemented classified tailings has long been practiced in the mining industry. Lately, considerable research has been undertaken to improve backfill material. The new concept of "paste backfill" was recently introduced as an advanced method for underground applications.

The utilization of paste backfill possesses several economic and mechanical advantages. However, paste backfill contains a high proportion of ultra fine solids which results in a very low permeable backfill material. Hence, the question of liquefaction susceptibility of this material as a result of different loading conditions becomes important.

This investigation was directed primarily towards the behavior of paste backfill when subjected to static and dynamic loading conditions. A testing program was conducted to define physical and mechanical characteristics of this weakly cemented soil . The main objective of this study was to evaluate the liquefaction resistance of paste backfill. The results of comprehensive laboratory triaxial tests and in situ studies for liquefaction evaluation are reported. Liquefaction potential was evaluated based on the principal of "steady state of deformation". These results indicate that paste backfill under specified conditions does not appear to be susceptible to liquefaction.

## 1. Introduction

Tailings are the waste product from mining operations. in normal milling processes, the crude ore is crushed and then ground in mills to a particle size small enough to free the valuable materials from the waste. After removal of the valuable minerals, the waste is either pumped into tailing disposal ponds or classified for use as hydraulic backfill in the mine. Classification results in some portion of fine material being removed, thereby increasing the settling rate and permeability of hydraulic backfill. The lack of true strength restricts the scope for various mining applications. Therefore classified tailings are normally mixed with Portland cement prior to placement in underground stopes.

The high expense of Portland cement as a backfill additive has led to evaluation of other cementing agents and approaches . For example, Thomas reported on the pozzolanic properties of cementing agents such as quenched and finely ground copper reverberatory furnace slags. Other materials such as fly ash and iron blast furnace slags are also known to be pozzolanic, and suitable for incorporation in a backfill mass to augment the cohesion conferred by Portland cement. As an alternative approach, Patchet (1978) and Wayment (1978) described a paste backfill system where centrifuge units were utilized for dewatering unclassified tailings prior to underground placement. These systems have been successfully implemented in several operations around the world.

The Tailspinner system has been introduced as a means of paste backfill production. The pulp density (weight of solid /total weight of mix) of unclassified tailings is raised considerably in this system, prior to cement addition and placement in a stope. This system of backfilling has several mechanical and economical advantages over conventional hydraulic backfill which are discussed in the next section. However , the paste backfill contains a high proportion of ultra fine material normally considered to be destabilizing in hydraulic backfills. This has raised concerns for safe utilization of this material for underground applications. Liquefaction of this material may occur as a result of static or dynamic conditions such as production blasting, rock bursts or earthquakes. Such sources of energy could cause liquefaction and subsequent catastrophic damages.

Limited information on the physical and mechanical behavior of this material is currently available. The main purpose of this paper is to extend the knowledge of "Paste Backfill" behavior when subjected to static and dynamic loading conditions as well as to describe its physical properties.

Fig. 1 Paste backfill in an Underground Stope.

## 2. Paste Backfill for Underground Applications

Application of paste backfill concept has been evaluated in several mines around the world. Recently the dewatering machine trademarked "Tailspinner" was developed as a means of paste backfill production (Wayment, 1978). This system dewaters the full stream tailing materials from approximately 55% water content to approximately 24% water content. Fig. 1 shows a view of the Tailspinner product with approximately 3% cement admixture. In conventional operations tailings are classified prior to placement in the stopes. Classification results in some portion of the minus 200 mesh material being removed, thereby increasing the settling rate and permeability. On the other hand, paste backfilling using full stream tailing increases the volume of the dewatered material for underground operations. The ability to handle entire tailings provides a great potential for utilization of materials that would ordinarily be disposed on the surface.

This densified tailing material is mixed with cement prior to placement in a stope. The lower moisture content of the backfill results in a higher compressive strength than equally consolidated hydraulic backfill and therefore a stronger support system can be generated. In addition to economic advantages, this high strength material can be used where special applications are required .

Probably the biggest problem associated with conventional backfill is "water", which should percolate through the backfill. This condition requires extensive preparation prior to backfilling and bulkhead construction in larger stopes. This part of the mining operation may be quite elaborate and expensive . The paste backfill operation does not require retaining bulkheads. In addition to economic benefits, this is a major advantage where intersecting drifts are no longer accessible for bulkhead construction.

Basically, the performance of paste backfill is greatly influenced by shear strength which is directly controlled by the pore water pressure according to the effective stress law. The sudden build up of pore water pressure as a result of static or dynamic loading conditions could cause paste backfill liquefaction with serious consequences. Therefore, great care must be exercised in design to ensure the stability of paste backfill after placement.

The successful application of this type of backfilling has promoted the rapid expansion of this system for underground hard rock mining operations. Application of this method, however, requires comprehensive evaluation of geotechnical properties and liquefaction potential of this new method of backfilling. Some aspects of this investigation with respect to liquefaction study are reported in this paper.

## 3. Basic Properties of Paste Backfill

Basic properties of unclassified material were determined on a series of representative samples collected randomly. The x-ray diffraction analyses were used to identify the different composition of samples. Results of these analyses show the material consists of predominately quartz and dolomite with minor amounts of Muscovite and Clinochlore concentrated in the fine particles. Pyrite is also present as a minor constituent. The breakdown of mineral content is:

    Quartz...................... 75%
    Dolomite...................18%
    Muscovite and
       Clinochlore...........3% - 4%
    Pyrite..................... 1%

A series of morphological analyses was conducted on representative samples of backfill. The results of these analyses show that differences in the sample preparation did not affect the results. Additionally, variations were not observed in grain distribution with respect to size, shape and angularity. In general, the grains are angler, display moderate sphericity, and are consistent throughout all size ranges. Fig.2 shows a typical morphology of the tailings scanned by the electron microscope.

Fig. 3 shows the gradation of tailings obtained through a series of grain size analyses. The specific gravity of tailings was determined to be 2.81.

### 3.1 Response of Paste Backfill to Load

Comprehensive laboratory and in situ investigations were carried out to ascertain the geotechnical characteristics and liquefaction potential of paste backfill. These investigations are described in following sections.

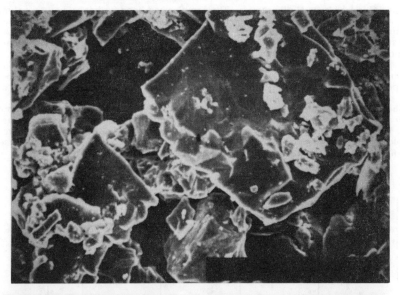

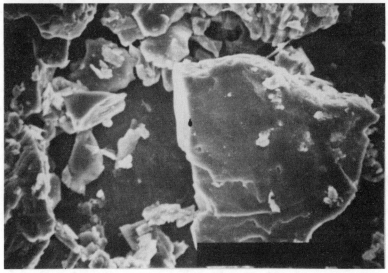

Fig. 2 (x500 & x750) Microphotograph displaying
two grain sizes. Smaller ranging in Size
from 10 to 27 microns and larger from 120
to 135 microns. Both types display
equidimentional to elongate sphericity,
angular and some observable cleavage.

407

An extensive number of triaxial tests at different confining pressures was conducted on both undisturbed and reconstituted samples. These tests were intended to evaluate paste backfill response for different loading conditions. The properties of paste backfill were also evaluated in situ. A series of piezometer friction cone tests was carried out in an underground stope at Dome mines. Geotechnical parameters for paste backfill were measured to various depths. The dynamic properties were investigated through a series of blasts adjacent to the backfilled stope. The resultant energy transfer and peak particle velocity transmitted through the rock and paste backfill was monitored with a series of accelerometers.

The sampler had been further refined to include tungsten carbide cutting edges. A total of 23 meters of undisturbed samples with an 11.43 centimeter diameter was successfully obtained by this method.

The reconstituted samples were cast in cylindrical molds having a length and diameter of 175 and 50 mm respectively. Due to shirinkage of cast specimens, extended lengths of molds were required. This extra length was trimmed according to the standard of each test. The PVC pipes with narrow longitudinal splits on two sides were used for sampling and curing. Clamps were used to keep the split molds together.

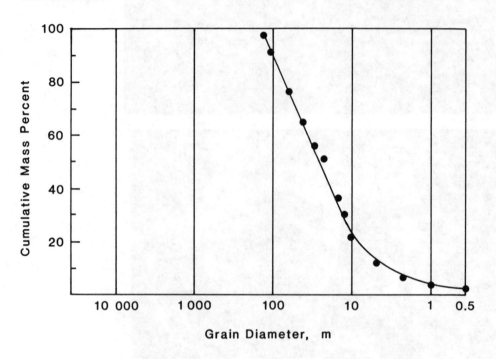

Fig. 3   Grain size distribution for tailings (underflow).

### 3.2 Sample Preparation

Presently there is not a standard method for recovering undisturbed samples of mine backfill. Therefore, several techniques used in soil sampling were modified and implemented for this purpose. Evaluation of retrieved samples show that diamond drilling methods, utilizing air and water flushing techniques do not appear to provide an effective and economical means for paste backfill sampling. Disturbance of retrieved samples was repeatedly observed with this technique. Consequently, the continuous flight auger device along with a modified split barrel sampler was tested.

The paste backfill samples were compacted into a mold in layers with uniform density and water content unless otherwise specified. Because of the many factors that could affect the ultimate strength of the mix, the procedure for specimen preparation was adhered to strictly. Repeated testing of duplicate samples showed that uniform results could be obtained by this sample preparation procedure. Subsequently, the molds were wrapped and stored in the controlled temperature and humidity room for curing. After the curing period, the clamps were released and the sample slid freely from the mold. Then, the sample was trimmed by a rotary blade to a standard length for each test. Capping was not required since parallel and smooth surfaces were cut with a rotary saw.

## 3.3 Triaxial Compression Test

A series of consolidated undrained triaxial tests was conducted with pore water pressure measurements performed at several confining pressures. The following procedures were carefully adopted for each test. Back pressure was used to achieve complete saturation. Saturation was checked by means of conventional B-check method (Bishop and Henkel, 1962). Once full saturation was achieved, the chamber pressure was increased to the particular value specified for isotropic consolidation. The consolidation pressures were chosen according to the typical values obtained from an underground stope at Dome Mines. After completion of saturation and consolidation, all drainage lines were closed. Subsequently, samples were sheared at a constant rate of axial strain, which was sufficiently slow to ensure that the pore water pressure equalized throughout the sample. An average period of seven days was required to complete a triaxial test. During the test, applied axial load, axial deformation and pore water pressure were continuously recorded by a computer data acquisition system.

Typical stress strain curves for paste backfill with cement:tailing ratio of 1:30 are given in Fig. 4a. The data were obtained in triaxial compression tests performed at different confining pressures. The common characteristic of all test results is that the stiffness and peak strength increases with increasing confining pressures. The paste backfill samples show brittle behavior in the unconfined compression condition, and application of low confining pressure is sufficient to generate ductile behavior. This behavior could be traced from the relative contribution of cementation and frictional components. At low confining pressures, the

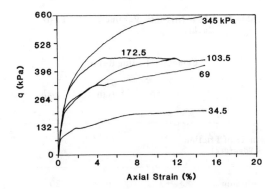

Fig. 4a Typical stress strain curves for C.U.T. tests on samples with 1:30 cement:underflow ratio.

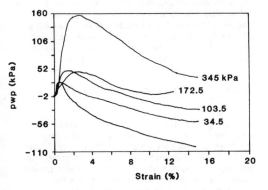

Fig. 4b Typical pore water pressure–strain curves for C.U.T. test on samples with 1:30 cement:underflow ratio.

cementation component is more significant than the frictional component and, paste backfill exhibits a brittle behavior. At high confining pressure, the frictional components becomes more dominant parameters, and backfill behaves as a non cemented frictional material. Similar behavior was also observed for conventional hydraulic backfill by others (Mitchell and Smith, 1979 and Mitchell and Wong, 1982).

Pore water pressure changes are also presented in Fig. 4b. Higher confining pressures initially causes higher pore water pressure raises at small strains. These differences do not appear to be significant at low confining pressure. However confining pressure of 345 kPa generates a high positive pore water pressure. Generally, at greater strain pore pressure drops sharply and becomes negative.

The contribution of cement to backfill resistance was evaluated for three cement: tailing ratios at various confining pressures. The influence of cement content on stress-strain curves at 69 kPa confining pressure is given in Fig. 5. Evaluation of results shows that the peak and residual strength increases with degree of cementation. Additionally, a higher modulus was obtained for samples prepared at higher cement contents. This type of behavior was consistent throughout all of the confining pressures applied to paste backfill samples.

The peak strength values for paste backfill at different cement:tailing ratios are plotted in Fig. 6. The data for tested samples may be fitted by straight lines which have the same slope but different cohesion intercepts. The values of friction angles and cohesion intercepts are given in Table 1.

Examination of peak strength envelopes reveals that paste backfill with different cement:tailing ratios exhibited similar angles of internal friction with different cohesion intercepts. The addition of cement shifts the peak strength values to a higher value at the same angle of friction. However, little differences were obtained in cohesion values for samples prepared at cement:underflow ratio of 1:40 and 1:50.

409

Table 1 - Shear Strength Parameters for Paste Backfill

| | Cement : Tailing Ratios | | |
|---|---|---|---|
| | 1:30 | 1:40 | 1:50 |
| Angle of Friction (degrees) | 31 | 33 | 33 |
| Cohesion (kPa) | 53 | 20 | 20 |

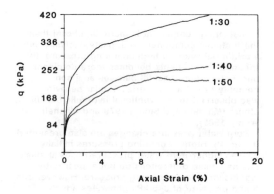

Fig. 5 Typical stress strain curves for C.U.T.tests at 69 kPa confining pressure.

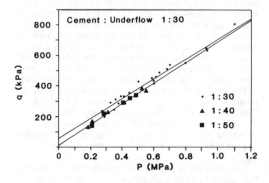

Fig. 6 Failure envelope for paste backfill.

## 4. In Situ Study Results

It has been recognized that knowledge of the characteristics of placed backfill is essential for a realistic underground design. In situ evaluation is probably the only means to establish the mechanical characteristics, environmental and hydraulic properties of placed backfill. Hence, an in situ testing program which consisted of a series of piezometer friction cone test and blasting test, were conducted on paste backfill.

The in situ testing program is the subject of separate publications and only relevant results are discussed in this paper.

The piezometer cone penetration tests conducted on paste backfill at Dome mines were unique in terms of location and underground environment. The objective of the cone testing was to assess variation in strength and evaluate the pore pressure changes in paste backfill. Several geotechnical parameters including relative density, angle of internal friction and dynamic pore water pressure were measured for paste backfill. The average in place void ratio of e = 0.93 was calculated based on these information. The most significant finding from the piezometer friction test data was the fact that no positive pore water pressures were encountered in any of the soundings. All the Piezometer friction soundings indicated paste backfill as unsaturated material with dilative behavior.

A series of long blast holes with different diameters and lengths were drilled parallel to the footwall from the scram area. The resultant energy transfer, peak particle velocity and peak particle acceleration transmitted through the rock and paste backfill were monitored with a series of triaxial accelerometers. The paste backfill exhibits amplitude as well as frequency attenuating properties for small single holes from 5 kg to 55 kg of explosive and distance range studied up to 35 m. The maximum resultant acceleration recorded just inside of paste backfill was 2.3 g. The typical acceleration of 1.0 g was recorded into the backfill. The corresponding frequency was reduced to 155 Hz from 500 Hz.

## 5. Liquefaction Evaluation Procedures

Review of literature pertaining to geotechnical engineering and related areas reveals a large number of publications on the subject of "Liquefaction". This interest is due to the catastrophic damages to human life, environment and property associated with this phenomenon. A comprehensive literature search of all relevant information on liquefaction was conducted by the authors (Dome Mines, 1986). Particular attention was given to those procedures which could be applied to this saturated weakly cemented material known as paste backfill. The study reviewed laboratory and in situ procedures and their possible applications for determining the liquefaction potential of paste backfill.

A review of published papers shows that several

mechanisms have been advanced concerning the failure of saturated cohesionless soils during and after seismic events. The diversity of these mechanisms have been the cause of confusion among engineers. It appears however, that, there are at least two major approaches to evaluate the liquefaction susceptibility of a particular mass:

(I) Liquefaction evaluation procedure based on the concept of "steady state of deformation". This approach is an extension of Casagrande's concept of "critical void ratio". The steady state of deformation was developed by Poulos (1971 and 1981) and further extensively applied for evaluation of liquefaction of soil masses:

"Liquefaction is a phenomenon whereby the shear resistance of a mass of soil decreases when subjected to monotonic, cyclic or dynamic loading at constant volume. The mass undergoes very large unidirectional shear strain- it appears to flow - until the shear stresses are as low or lower than the reduced shear resistance."

(II) Liquefaction evaluation procedure based on evaluation of the cyclic stress or strain conditions likely to be developed in the field by a proposed design earthquake and a comparison of these stresses or stains with those observed to cause liquefaction of representative samples of the deposit in some appropriate laboratory test which provides an adequate simulation of field conditions or which can provide results permitting an assessment of the soil behavior under field conditions. This procedure was first proposed by Seed and Idriss (1967):

"Peak cyclic pore pressure ratio of 100% with limited strain potential or cyclic mobility denotes a condition in which cyclic stress applications cause limited strains to develop either because of the remaining resistance of the soil to deformation or because the soil dilates, the pore pressure drops, and the soil stabilizes under the applied loads."

A logical first step in the analysis of a particular problem such a liquefaction potential of paste backfill is to evaluate whether or not failure by mechanism I (steady state of deformation) is possible. The liquefaction evaluation procedure suggested by Polous et al., (1985) should be followed. Based on this concept, a particular mass sustaining static shear stress greater than the steady state undrained shear strength is susceptible to liquefaction. If paste backfill is potentially unstable then, it is also necessary to consider those possible seismic events that could cause liquefaction.

### 5.1 Liquefaction Evaluation Analyses

In order to judge whether paste backfill will behave satisfactorily when subjected to dynamic loading conditions, the following question must be answered:

"Is the paste backfill susceptible to liquefaction?"

Based on the concept of steady state of deformation (Poulos, Castro and France, 1985), only soils that tend to decrease in volume during shear, i.e., contractive soils, can suffer the necessary loss of shear resistance to result in liquefaction. Soils that tend to increase in volume due to shear, i.e., dilative soils, are not susceptible to liquefaction because their undrained strength is greater than drained strength. Hence, those mass that can be shown to be dilative are not susceptible to liquefaction under the first mechanism (I).

As noted by Poulos et al.,(1985), in situ property of the mass under investigation should be known. The undrained steady state strength, which controls susceptibility to liquefaction, is a very sensitive function of void ratio. Therefore, the in situ void ratio should be measured accurately. At the initial stage of project, no information on void ratio distributions of paste backfill in the stope was available. After a period of approximately three years, backfilling of the stope was completed. It is very difficult to measure accurately the in situ void ratio in an underground stope for paste backfill. Several approaches were considered. Finally, this was successfully accomplished by application of piezometer friction cone which was discussed previously.

According to the suggested method for liquefaction evaluation procedure, (Poulos, et. al., 1985), a series of consolidated undrained triaxial tests was conducted on reconstituted and undisturbed samples at various void ratios. The similar procedure as outlined in the triaxial compression test with large axial strain at different effective consolidation pressures was followed. After a curing period of 28 days, an average period of nine days was required to prepare, conduct and do analysis of each triaxial test.

Typical results of tests on reconstituted samples are presented in Fig. 7. The data of each undrained test was normally plotted in three diagrams: (a) stress - strain; (b) stress path and (c) effective stress. These tests were conducted on saturated samples with void ratios ratio ranging from 0.68 to 0.98 and different cement:tailings ratios. The liquefaction susceptibility of paste backfill was examined for cement:tailing ratios of 1:30, 1:40 and 1:50 . The effective consolidation pressures of 35 , 70, 105, 175 and 350 kPa were selected for this series of experiments.

The common characteristics of samples prepared at 1:30 cement:tailing ratio and void ratio up to 0.96 can be expressed as an initial increase in pore water pressure with increasing shear stress (Fig. 7). Then, as shearing continues the material begins to dilate and the pore water pressure decreases. As shown in Fig. 7, pore pressure decrease continues to approximately 12 percent axial strain at which point negative pore pressure is measured. The steady state is reached at approximately 28 percent strain where the shear stress and the effective minor principal stress are no longer changing as deformation continuous. Dilative behavior with no peak shear stress prior to steady state deformation was observed for all samples prepared at high void ratios. Additionally, piezometer friction cone tests indicated that paste backfill to be unsaturated with dilative behavior. Stronger dilative behavior was observed at lower

411

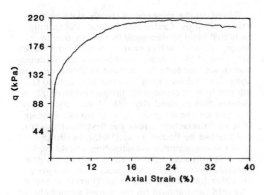

Fig. 7a Stress-strain.

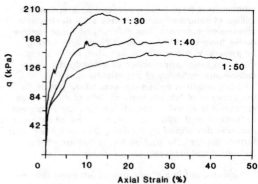

Fig. 8a Stress-strain.

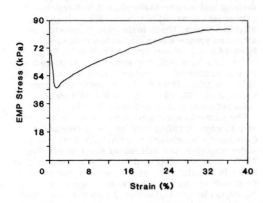

Fig. 7b Stress path.

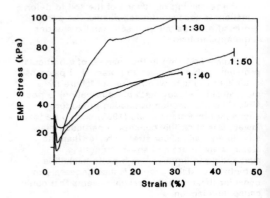

Fig. 8b Stress path.

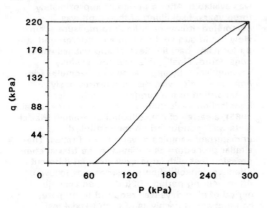

Fig. 7c Effective stress.

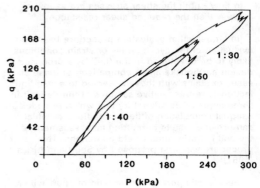

Fig. 8c Effective stress.

Fig. 7  Typical results for paste backfill sample
        prepared with 30% water content under
        69 kPa confining pressure.

Fig. 8  Influence of cement content variations on
        undrained response of paste backfill at
        34.5 kPa confining pressure.

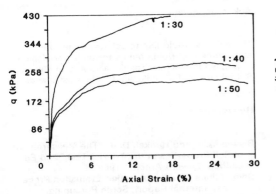

Fig. 9a Stress-strain.

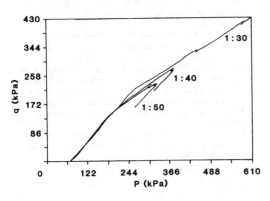

Fig. 9b Stress path.

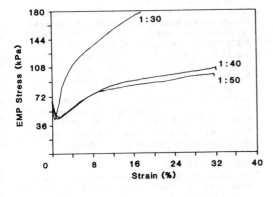

Fig. 9c Effective stress.

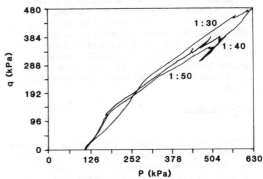

Fig. 10a Stress-strain.

Fig. 10b Stress-path.

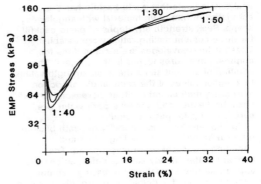

Fig. 10c Effective stress.

Fig. 9  Influence of cement content variations on undrained response of paste backfill at 69 kPa confining pressure.

Fig. 10  Influence of cement content variations on undrained response of paste backfill at 103.5 kPa confining pressure.

413

void ratios and higher cement:underflow ratios. As discussed previously, dilative soils are not susceptible to liquefaction. Hence, paste backfill at 1:30 cement:tailings ratio and void ratio up to 0.96 is not susceptible to liquefaction.

In order to compare the susceptibility of paste backfill to liquefaction at different cement contents, results are plotted in Figs. 8, 9 and 10. This series of plots show the results obtained at 1:30, 1:40 and 1:50 cement:tailing ratios. An examination of these results indicate that all tests start at a specified effective consolidation pressures. The pore water pressure initially builds up at the start of the tests. However, paste backfill samples begin to dilate and therefore cause a drop in pore water pressures. As shearing progresses, the stress path moves towards the failure line envelope with higher peak values for higher cement contents. Similarly, the minimum effective principal stress drops sharply for samples with higher cement content as axial stress increases and then, increases as dilatancy generally sets in. Evaluation of these results demonstrate clearly that cementation increases the liquefaction resistance. However, at the lower effective consolidation pressures and lower cement: tailing ratios paste backfill is weakly dilatant (Fig. 8, 9 and 10). This condition may be encountered shortly after backfill placement in a stope.

## 6. Conclusions

Paste backfilling has been introduced as a superior backfilling system for underground mining application. However, very limited information is available on geotechnical characteristics and liquefaction potential of paste backfill. The results of extensive laboratory studies on the physical and mechanical properties of paste backfill were presented.

The results of particle size and morphological analyses indicate that paste backfill has the characteristics of fine grain material with angular shape. Peak strength envelopes of paste backfill for three cement :tailing ratios were established. These failure envelopes are straight lines for confining pressures tested in this project. The addition of cement shifts the peak strength values to a higher values at the same angle of friction. Increasing the cement content increases the stiffness, brittleness and negative pore water pressure build up for paste backfill.

The liquefaction susceptibility of paste backfill under different cement: tailing ratios and confining pressures were investigated. The concept of the "steady state of deformation" was applied for this evaluation. Increasing cement content and relative density can strongly increase the liquefaction resistance. Dilative behavior was observed in situ and also for all reconstituted samples tested in laboratory. Based on the analyses and tests reported in this paper, liquefaction of paste backfill with 1:30 cement:tailing ratio and void ratio up to 0.96 is unlikely to occur.

## Acknowledgment

The authors would like to acknowledge the management of Dome Mines for their permission to publish this paper. Thanks also due to Messrs., S. Nicholos and K. Shikatani for their assistance.

## References

Bishop, A.W. and Henkel, D. J., "The Measurement of Soil Properties in the Triaxial Test," 2nd Ed., London, Edward Arnold., 1962.

Dome Mines Ltd., "Liquefaction Evaluation Procedures," Internal Report, South Porcupine, Ontario, 1986, pp. 1-99.

Mitchell R.J. and Smith J.D., "Mine Backfill Design and Testing," Canadian Institute of Mining and Metallurgical Bulletin, 72, No.801, 1979, pp.82-88.1

Mitchell R. J. and Wong B. C., "Behavior of Cemented Tailing Sands," Canadian Geotechnical Journal, 19, 1982, pp.289-295.

Patchet, S.J., "Fill Support System for Deep Level Gold Mines - Prototype Installation and Economics Analysis, Proceeding 12 Canadian Rock Mechanics Symposium, Canadian Institute of Mining and Metallurgical, Special Volume 9, 1979.

Poulos, S. J., "The Stress-Strain Curves of Soils," Geotechnical Engineers Inc., Winchester, Mass., Jan., 1971, pp. 1-82.

Poulos, S. J.,"The Steady State of Deformation," Journal of the Geotechnical Engineering Division, ASCE, Vol.107, No. GT5, May, 1981, pp. 553-562.

Poulos, S. J., Castro, G. and France, J.W., "Liquefaction Evaluation Procedure," Journal of Geotechnical Engineering Division, ASCE, Vol.11, No. 6, June 1985, pp. 772-792.

Seed,H. B., "Soil Liquefaction and Cyclic Mobility Evaluation for level Ground During Earthquake," Journal of the Geotechnical Engineering Division, ASCE,Vol. 105,No. GT2, 1979, pp. 201-255.

Seed, H. B., and Lee, K. L., "Liquefaction of Saturated Sands During Cyclic Loading," Journal of the Soil Mechanics and Foundation Division, ASCE, Vol.92, No. SM6, 1966, pp. 105-134.

Thomas, E. G., and Cowling R., "Pozzolanic Behavior of Mount Isa Mines Slag in Cemented Hydraulic Mine Fill at High Slag/Cement Ratios. Mining with Backfill, Proceeding 12th Canadian Rock Mechanics Symposium, Sudbury , Montreal, Canadian Institute of Mining and Metallurgical, 1978, pp. 129-132.

Wayment W. R.,"Backfilling with Tailings - A New Approach," 12 Canadian Rock Mechanics Symposium, Canadian Institute of Mining and Metallurgy, Special Volume 19, 1978, pp. 111-117.

414

*Innovations in Mining Backfill Technology, Hassani et al. (eds), © 1989 Balkema, Rotterdam. ISBN 90 6191 985 1*

# The time effect on flow through mine backfill materials

G.Şenyur
*Hacettepe University, Ankara, Turkey*

ABSTRACT: The time effect on the characteristics of water flow through hydraulic backfill materials has been studied. The samples were prepared from coal washery rejects. A functional decrease by time in the rate of flow as well as in permeability has been found. These relationships have been expressed by exponential functions and the kinetics of the phenomenon has been explained. Statistical analysis has been done for the relationship between the ratio of the decrease in the permeability to its final value and the particle size distribution parameters.

## 1 INTRODUCTION

The rejects were crushed to under 2 cm particles and the product was seperated into six batches of fractions. The test samples were obtained by mixing the fractions in varying proportions. Hence, the problem has been studied according to the particle-size distribution. Constant head permeability tests have been carried out. The observed phenomenon is the change in rate of flow by time. In all the experiments, a functional decrease in the flow rate of water by time under a constant hydraulic gradient of approximately one was recorded.

It has been supposed that the porous backfill media rebuilds its structure under the influence of flowing water. Some of the particles change their positions to improve the resistance of the system to flow. The process consists of clogging or colmatage, i.e., the retention of the particles, and declogging or decolmatage, i.e., the breaking away of previously retained particles.

## 2 DESCRIPTION OF THE ELEMENTARY MECHANISM

Before the beginning of flow some of the particles are in suspension. The retention may occur in different ways as explained by Herzig et al. (1969). The mobilised particles can not penetrate into a pore of a smaller size than its own. The small particle is retained in a sheltered area, a small pocket formed by several grains. The fluid pressure may be considered as the main retention force. Decolmatage may occur if local variations in pressure or flow rate change the flow in the neighbourhood of retained particles or if a mowing particle collides with a retained particle.

### 2.1 The kinetic equation

Consider the porous bed element of area $\Omega$ and depth $\Delta z$ (see Figure 1). During the small time interval $\Delta t$, the retention $\Omega\sigma\Delta z$ increases to $\Omega(\sigma+(\partial\sigma/\partial t)\Delta t)$ $\Delta z$ and a volume $\Omega u_m$ y $\Delta t$ of suspended particles enters the interstices of the porous media, where, $\sigma$ is the retention, i.e., the volume of deposited particles per unit of filter volume, y is the volume fraction of particles in the suspension i.e., the volume of particles per unit of suspension volume and $u_m$ is the approach velocity of suspension.

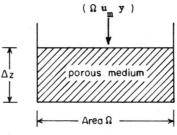

Fig. 1 Porous medium element

The retention probability k of a particle in this element will be defined by

$$k = \frac{\Omega\Delta z \frac{\partial\sigma}{\partial t} \Delta t}{\Omega \, u_m \, y \, \Delta t} = \frac{\Delta z}{u_m y} \, \frac{\partial\sigma}{\partial t} \qquad (1)$$

while the retention probability per unit of depth, $k_c$, is obtained as,

$$k_c = \frac{1}{u_m y} \, \frac{\partial\sigma}{\partial t} \text{ or } (\frac{\partial\sigma}{\partial t})_c = k_c u_m y \qquad (2)$$

and the probability of a decolmatage of a particle, $k_r$, per unit time will be;

$$k_r = - \frac{\Omega \, \Delta z (\frac{\partial\sigma}{\partial t})}{\Omega \, \sigma \, \Delta z} = -\frac{1}{\sigma} (\frac{\partial\sigma}{\partial t})$$

or

$$(\frac{\partial\sigma}{\partial t})_r = -k_r \, \sigma \qquad (3)$$

Under these conditions the retention per unit time will be,

$$\frac{\partial\sigma}{\partial t} = k_c \, u_m \, y - k_r \sigma$$

or

$$\frac{\partial\sigma}{\partial t} + k_r \, \sigma = k_c \, u_m \, y \qquad (4)$$

This equation correlates the retention, $\sigma$ with volume fraction of particles in suspension, y and flow velocity $u_m$. Measurement of volume retention and volume of particles in suspension at any time is experimentally impossible. But the flow velocity i.e., percolation rate can be measured.

## 3 MEASUREMENT OF FILL PERMEABILITY

### 3.1 Definition of permeability and percolation rate

Permeability and percolation rate are related mathematically as described below, each being a measure of the ease of fluid flow through a porous medium.
   Permeability is defined by the equation (Thomas 1978),

$$K = \frac{Q \cdot L \cdot \mu}{H \, A \, \rho^2 \, g} \qquad (5)$$

where, K is the permeability ($m^2$), Q is the mass rate of flow of fluid through

porous medium (kg $sec^{-1}$), L is the length of porous the medium (m), H is the water head (m), A is the cross-sectional area of the porous medium normal to the direction of flow ($m^2$), $\mu$ is the dynamic viscosity of fluid flowing (Pa. sec), $\rho$ is the density of fluid flowing (kg $m^{-3}$) and g is the gravitational acceleration (m $sec^{-2}$).
   Percolation rate is defined by the equation,

$$v = \frac{q}{A} \qquad (6)$$

where, v is the percolation rate (m $sec^{-1}$), q is the rate of flow through porous medium ($m^3 \, sec^{-1}$) and A is the cross-sectional area of porous medium normal to the direction of flow ($m^2$). Percolation rate also applies to the flow of a unique fluid at a unique temperature, with the added restriction that the hydraulic gradient, h/L is taken as unity.

## 4 MATERIALS AND METHODS

### 4.1 Samples

Refuse from washing plants is always a mixture of grains from different rock types. The composition of the washery dirt, which has been used for the tests, was as fallows:
- 16 % (by mass) sandstone,
- 30 % sandy shale,
- 44 % sandfree shale,
- 10 % others and coal
   The shape of the grains was longish to flat with different degrees of roundness. The length/thickness ratios were in the order of 2:1 to 6:1. The length/width ratios were measured between 1:1 and 5:1. The material had the specific gravity of 2.57, and water absorption of 2 %. Six size batches (19.0 mm - 13.2 mm), (13.2 mm - 9.5 mm), (9.5 mm - 6.7 mm), (6.7 mm - 3.37 mm), (3.37 mm - 0.5 mm), (0.5 mm - 0.15 mm) were established from crushed products by sieveing. Test samples having different granulumetric compositions were obtained by mixing the batches in varying amounts. The parameters used to define the particle-size distribution are;
$d_{max}$: maximum particle-size in the sample mm
$d_{10}$ : effective size i.e., ten percent passing size, mm
$C_u$ : coefficient of uniformity, $d_{60}/d_{10}$
$C_c$ : coefficient of degree, $d_{30}^2/d_{60} \cdot d_{10}$
where, $d_{30}$ (mm) and $d_{60}$ (mm) are thirty percent and sixty percent passing sizes

Table 1. Some engineering properties of the samples subjected to percolation rate tests

| Sample No | Maximum particle size, $d_{max}$ (mm) | Coefficient of uniformity $C_u$ | Coefficient of degree $C_c$ | Effective size, $d_{10}$ (mm) | Void volume n |
|---|---|---|---|---|---|
| 1 | 13.2 | 3.4 | 0.71 | 0.197 | 0.4265 |
| 2 | 19.0 | 17.7 | 0.226 | 0.22 | 0.378 |
| 3 | 19.0 | 14.25 | 1.23 | 0.36 | 0.374 |
| 4 | 19.0 | 13.06 | 2.37 | 0.5 | 0.43 |
| 5 | 19.0 | 3.9 | 0.77 | 0.95 | 0.49 |
| 6 | 19.0 | 5.02 | 0.757 | 0.80 | 0.44 |
| 7 | 19.0 | 12.33 | 0.9 | 0.97 | 0.43 |
| 8 | 13.2 | 7.2 | 1.03 | 0.87 | 0.45 |
| 9 | 19.0 | 6.6 | 1.93 | 1.64 | 0.46 |
| 10 | 19.0 | 8.68 | 1.3 | 1.13 | 0.45 |
| 11 | 19.0 | 4.53 | 1.11 | 1.52 | 0.462 |
| 12 | 19.0 | 5.05 | 0.9 | 2.26 | 0.46 |
| 13 | 19.0 | 20.25 | 5.23 | 0.5 | 0.43 |
| 14 | 19.0 | 10.5 | 1.23 | 1.20 | 0.476 |
| 15 | 19.0 | 35.38 | 3.47 | 0.29 | 0.362 |
| 16 | 19.0 | 22.47 | 1.37 | 0.44 | 0.41 |
| 17 | 19.0 | 3.4 | 1.16 | 2.28 | 0.481 |

respectively. The other parameters, void volume, n, which is the ratio of the void volume to the total volume of the sample has been given due to the fact that, in the literature, void volume and void ratio are taken in predictive expression for percolation rate (Bates and Wayment 1967, Jacob 1972). The above engineering properties of the samples subjected to percolation rate tests are given in Table 1.

## 4.2 Constant-head permeameter

The laboratory set-up was designed and tests were performed according to the recommendations of Wayment and Nicholson (1964) and Thomas (1978). Şenyur (1985) introduced a criterion to reduce the wall effect (Thomas 1978) and stated that the diameter of the test cup should be at least ten times the maximum particle-size of the sample. The diameter of the permeameter was designed in line with this recommendation. Figure 2 shows the set-up and describes the symbols used in calculations.

## 4.3 The experiment

The permeability tests were begun with fully saturated samples. The flow measurements were taken at five minutes intervals beginning from the initial flow. The progressive change in flow rate was recorded for each sample.

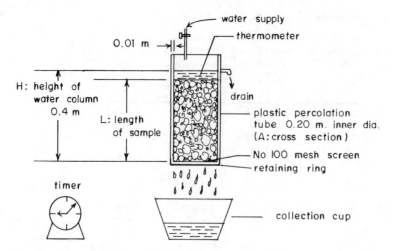

Fig. 2  Laboratory percolation rate test

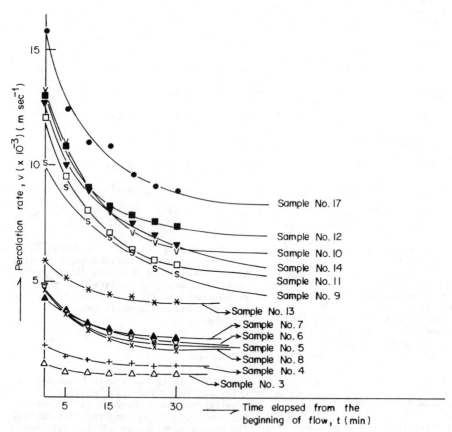

Fig. 3  Flow diagrams

Table 2. Results of permeability tests

| Sample No / H:Water head, L:Sample length, $T_w$:Temp. of water | Time elapsed from the beginning of flow, t, (minutes) | | | | | | |
|---|---|---|---|---|---|---|---|
| | 0 | 5 | 10 | 15 | 20 | 25 | 30 |
| | Percolation rate, v, (m sec$^{-1}$) at time t | | | | | | |
| 1  H/L:1.1  $T_w$:10°C | $0.194 \times 10^{-3}$ | $0.166 \times 10^{-3}$ | $0.151 \times 10^{-3}$ | $0.142 \times 10^{-3}$ | $0.137 \times 10^{-3}$ | $0.134 \times 10^{-3}$ | $0.132 \times 10^{-3}$ |
| 2  H/L:1.1  $T_w$:14°C | $0.256 \times 10^{-3}$ | $0.25 \times 10^{-3}$ | $0.235 \times 10^{-3}$ | $0.225 \times 10^{-3}$ | $0.222 \times 10^{-3}$ | $0.215 \times 10^{-3}$ | $0.21 \times 10^{-3}$ |
| 3  H/L:1.1  $T_w$:15°C | $1.45 \times 10^{-3}$ | $1.16 \times 10^{-3}$ | $1.0 \times 10^{-3}$ | $0.97 \times 10^{-3}$ | $0.92 \times 10^{-3}$ | $0.91 \times 10^{-3}$ | $0.91 \times 10^{-3}$ |
| 4  H/L:1  $T_w$:9°C | $2.26 \times 10^{-3}$ | $1.73 \times 10^{-3}$ | $1.61 \times 10^{-3}$ | $1.5 \times 10^{-3}$ | $1.42 \times 10^{-3}$ | $1.32 \times 10^{-3}$ | $1.28 \times 10^{-3}$ |
| 5  H/L:1.12  $T_w$:15°C | $4.75 \times 10^{-3}$ | $3.51 \times 10^{-3}$ | $2.92 \times 10^{-3}$ | $2.73 \times 10^{-3}$ | $2.57 \times 10^{-3}$ | $2.25 \times 10^{-3}$ | $2.14 \times 10^{-3}$ |
| 6  H/L:1.1  $T_w$:13°C | $4.28 \times 10^{-3}$ | $3.6 \times 10^{-3}$ | $3.1 \times 10^{-3}$ | $2.8 \times 10^{-3}$ | $2.56 \times 10^{-3}$ | $2.4 \times 10^{-3}$ | $2.3 \times 10^{-3}$ |
| 7  H/L:1.0  $T_w$:15°C | $4.65 \times 10^{-3}$ | $3.67 \times 10^{-3}$ | $3.0 \times 10^{-3}$ | $2.9 \times 10^{-3}$ | $2.8 \times 10^{-3}$ | $2.5 \times 10^{-3}$ | $2.5 \times 10^{-3}$ |
| 8  H/L:1  $T_w$:10°C | $4.7 \times 10^{-3}$ | $3.5 \times 10^{-3}$ | $2.8 \times 10^{-3}$ | $2.5 \times 10^{-3}$ | $2.2 \times 10^{-3}$ | $2.1 \times 10^{-3}$ | $2 \times 10^{-3}$ |
| 9  H/L:1.1  $T_w$:12°C | $10.4 \times 10^{-3}$ | $9.0 \times 10^{-3}$ | $7.5 \times 10^{-3}$ | $7.0 \times 10^{-3}$ | $6.3 \times 10^{-3}$ | $5.5 \times 10^{-3}$ | $5.3 \times 10^{-3}$ |
| 10  H/L:1.08  $T_w$:11°C | $13.3 \times 10^{-3}$ | $11 \times 10^{-3}$ | $9.0 \times 10^{-3}$ | $8.0 \times 10^{-3}$ | $7.0 \times 10^{-3}$ | $6.7 \times 10^{-3}$ | $6.5 \times 10^{-3}$ |
| 11  H/L:1  $T_w$:14°C | $12 \times 10^{-3}$ | $9.5 \times 10^{-3}$ | $8.0 \times 10^{-3}$ | $7.0 \times 10^{-3}$ | $6.1 \times 10^{-3}$ | $5.8 \times 10^{-3}$ | $5.6 \times 10^{-3}$ |
| 12  H/L:1.05  $T_w$:11°C | $13 \times 10^{-3}$ | $10 \times 10^{-3}$ | $9.0 \times 10^{-3}$ | $8.1 \times 10^{-3}$ | $7.8 \times 10^{-3}$ | $7.5 \times 10^{-3}$ | $7.3 \times 10^{-3}$ |
| 13  H/L:1  $T_w$:10°C | $5.9 \times 10^{-3}$ | $5.13 \times 10^{-3}$ | $4.6 \times 10^{-3}$ | $4.4 \times 10^{-3}$ | $4.2 \times 10^{-3}$ | $4.0 \times 10^{-3}$ | $4.0 \times 10^{-3}$ |
| 14  H/L:1  $T_w$:13°C | $12.7 \times 10^{-3}$ | $10.9 \times 10^{-3}$ | $9.8 \times 10^{-3}$ | $8.12 \times 10^{-3}$ | $7.5 \times 10^{-3}$ | $6.9 \times 10^{-3}$ | $6.4 \times 10^{-3}$ |
| 15  H/L:1  $T_w$:14°C | $1.93 \times 10^{-3}$ | $1.79 \times 10^{-3}$ | $1.73 \times 10^{-3}$ | $1.50 \times 10^{-3}$ | $1.475 \times 10^{-3}$ | $1.42 \times 10^{-3}$ | $1.38 \times 10^{-3}$ |
| 16  H/L:1  $T_w$:14°C | $1.12 \times 10^{-3}$ | $0.93 \times 10^{-3}$ | $0.845 \times 10^{-3}$ | $0.833 \times 10^{-3}$ | $0.808 \times 10^{-3}$ | $0.795 \times 10^{-3}$ | $0.785 \times 10^{-3}$ |
| 17  H/L:1  $T_w$:14°C | $15.8 \times 10^{-3}$ | $12.4 \times 10^{-3}$ | $11.0 \times 10^{-3}$ | $10.9 \times 10^{-3}$ | $9.52 \times 10^{-3}$ | $9.15 \times 10^{-3}$ | $8.8 \times 10^{-3}$ |

## 5 RESULTS AND DISCUSSION

The test results are given in Table 2. The sample numbers correspond to the samples in Table 1. The hydraulic gradients, H/L, where H is the heigth of the water column and L, is the length of the porous medium, are about one and the temperature, T, of flowing water are given. The flow diagrams showing percolation rates at elapsed times from the beginning of flow are seen in Figure 3. All the flow diagrams display a functional reduction in percolation rate by time. A general regression equation describing the phenomenon has been found; the exponential equation is

$$v = v_L + v_o e^{-mt} \qquad (7)$$

419

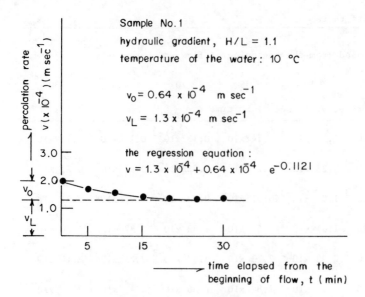

Sample No.1

hydraulic gradient, $H/L = 1.1$

temperature of the water: $10\ °C$

$v_0 = 0.64 \times 10^{-4}\ m\ sec^{-1}$

$v_L = 1.3 \times 10^{-4}\ m\ sec^{-1}$

the regression equation:

$v = 1.3 \times 10^{-4} + 0.64 \times 10^{-4}\ e^{-0.1121}$

time elapsed from the
beginning of flow, t (min)

Fig. 4  The flow diagram of the first sample

where, t is the time elapsed from the beginning of flow (min), v is the percolation rate at time $t(m\ sec^{-1})$, $v_L$ is the approach percolation rate $(m\ sec^{-1})$, $v_0$ is a constant $(m\ sec^{-1})$ and m is a constant. Figure 4 illustrates the flow graph and its equation with the first sample as an example. The results obtained with other samples are evaluated in Table 3.

Now the question is how the theoretical approach for kinetics of flow and the experimental result are to be combined? As discussed earlier, the related parameters, σ, y and $u_m$ in equation 4 are dependant on time or are function of time. The flow velocity i.e., percolation rate has been defined by equation 7. If this expression is inserted into equation 4,

$$\frac{\partial \sigma}{\partial t} + k_r \sigma = k_c (v_L + v_0 e^{-mt})\, y \qquad (8)$$

will be obtained. An assumption for the correlation between the volume fraction of particles in suspension, y, and the time elapsed from the beginning of flow, t can be made. The equation describing the phenomenon for velocity, v (equation 7) implies a similar approach for the volume of particles in suspension, y. Let the following function is assumed for y,

$$y = y_L + y_0 e^{-\beta t} \qquad (9)$$

where $y_L$ is the volume fraction of particles which finally remained in the suspension or escaped from the media and the term $y_0 e^{-\beta t}$ indicates the amount of reduction in time t, $y_0$ and β are constants. If this expression for y is inserted into equation 8

$$\frac{\partial \sigma}{\partial t} + k_r \sigma = k_c (v_L + v_0 e^{-mt})(y_L + y_0 e^{-\beta t}) \qquad (10)$$

will be obtained. The integration of this equation with respect to time t will give,

$$\sigma = \frac{k_c v_L y_L}{k_r} - \frac{k_c y_0 v_L}{\beta - k_r} e^{-\beta t} - \frac{k_c v_0 y_L}{m - k_r} e^{-mt}$$
$$- \frac{k_c v_0 y_0}{m + \beta - k_r} e^{-(m+\beta)t} - c\, e^{-k_r t} \qquad (11)$$

As discussed before, permeability, K and percolation rate, v are related mathematically. Therefore, the functional reduction in percolation rate displays the functional reduction in permeability. Calculations for percolation rates and permeabilities were made together. Equations showing reduction in permeability, for the test samples are given in Table 4.

420

Table 3. Regression equations for percolation rate, obtained from the experimental data

| Samples No | Regression equation* $v = v_L + v_o e^{-mt}$ | | | $r^2$ Coefficient of regression |
|---|---|---|---|---|
| | $v_L$ | $v_o$ | $m$ | |
| 1 | $0.13 \times 10^{-3}$ | $0.064 \times 10^{-3}$ | 0.1121 | 0.99 |
| 2 | $0.19 \times 10^{-3}$ | $0.066 \times 10^{-3}$ | 0.035 | 0.98 |
| 3 | $0.9 \times 10^{-3}$ | $0.55 \times 10^{-3}$ | 0.145 | 0.97 |
| 4 | $1.2 \times 10^{-3}$ | $1.06 \times 10^{-3}$ | 0.084 | 0.99 |
| 5 | $2.0 \times 10^{-3}$ | $2.75 \times 10^{-3}$ | 0.1 | 0.98 |
| 6 | $2.0 \times 10^{-3}$ | $2.28 \times 10^{-3}$ | 0.07 | 0.99 |
| 7 | $2.3 \times 10^{-3}$ | $2.35 \times 10^{-3}$ | 0.09 | 0.81 |
| 8 | $2.0 \times 10^{-3}$ | $2.7 \times 10^{-3}$ | 0.118 | 0.99 |
| 9 | $4.0 \times 10^{-3}$ | $6.4 \times 10^{-3}$ | 0.053 | 0.98 |
| 10 | $6.0 \times 10^{-3}$ | $7.3 \times 10^{-3}$ | 0.09 | 0.99 |
| 11 | $5.0 \times 10^{-3}$ | $7.0 \times 10^{-3}$ | 0.083 | 0.99 |
| 12 | $6.8 \times 10^{-3}$ | $6.2 \times 10^{-3}$ | 0.082 | 0.99 |
| 13 | $3.92 \times 10^{-3}$ | $2.0 \times 10^{-3}$ | 0.1 | 0.97 |
| 14 | $5.0 \times 10^{-3}$ | $7.7 \times 10^{-3}$ | 0.056 | 0.98 |
| 15 | $1.25 \times 10^{-3}$ | $0.68 \times 10^{-3}$ | 0.0555 | 0.95 |
| 16 | $0.78 \times 10^{-3}$ | $0.408 \times 10^{-3}$ | 0.1432 | 0.97 |
| 17 | $8 \times 10^{-3}$ | $7.786 \times 10^{-3}$ | 0.0763 | 0.98 |

*t: the time elapsed from the beginning of flow, min;
v: percolation rate at time t, m sec$^{-1}$; $v_L$: approach percolation rate, m sec$^{-1}$; $v_o$: constant, m sec$^{-1}$;
m: constant

As is observed from Table 4, the functional relationship between permeability, K and time, t is expressed by an exponential equation;

$$K = K_L + K_o e^{-mt} \qquad (12)$$

where, t is the time elapsed from the beginning of flow (min), K is the permeability at time, t, (m$^2$) and $K_L$ is the approach permeability (m$^2$), $K_o$ is a constant (m$^2$) and m is a constant.

The relative reduction, Y in permeability is expressed as the ratio of constant, $K_o$ in equation 12 to the approach permeability, $K_L$. It has been considered that the relative reduction, Y in permeability K is dependent on the particle-size distribution of the material. It is obvious that the shape and the size of the particles, as well as the kind of material, effect the relative reduction. But, the samples were prepared from the same material i.e., washery rejects, meaning that the above parameters were not changing. The particle-size distribution was systematically changed. Then, relative reductions, Y have been correlated with the particle-size distribution parameters of the samples given in Table 1, by regression analysis by

421

Table 4. The regression equations for permeability obtained from the experimental data

| Sample No | Regression equation[*] $K = K_L + K_o e^{-mt}$ | | | $r^2$ coefficient of regression | Y $(K_o/K_L)$ |
|---|---|---|---|---|---|
| | $K_L$ | $K_o$ | m | | |
| 1 | $1.59 \times 10^{-11}$ | $7.83 \times 10^{-12}$ | 0.1121 | 0.99 | 0.492 |
| 2 | $2.0 \times 10^{-11}$ | $7.06 \times 10^{-12}$ | 0.035 | 0.98 | 0.353 |
| 3 | $7.65 \times 10^{-11}$ | $4.68 \times 10^{-11}$ | 0.145 | 0.97 | 0.61 |
| 4 | $1.17 \times 10^{-10}$ | $1.03 \times 10^{-10}$ | 0.084 | 0.99 | 0.88 |
| 5 | $1.76 \times 10^{-10}$ | $2.42 \times 10^{-10}$ | 0.1 | 0.98 | 1.375 |
| 6 | $2.18 \times 10^{-10}$ | $2.48 \times 10^{-10}$ | 0.07 | 0.99 | 1.14 |
| 7 | $2.24 \times 10^{-10}$ | $2.28 \times 10^{-10}$ | 0.09 | 0.81 | 1.02 |
| 8 | $2.52 \times 10^{-10}$ | $3.4 \times 10^{-10}$ | 0.118 | 0.99 | 1.35 |
| 9 | $3.4 \times 10^{-10}$ | $5.44 \times 10^{-10}$ | 0.053 | 0.98 | 1.60 |
| 10 | $6.37 \times 10^{-10}$ | $7.74 \times 10^{-10}$ | 0.09 | 0.90 | 1.22 |
| 11 | $7.17 \times 10^{-10}$ | $1.00 \times 10^{-9}$ | 0.083 | 0.99 | 1.4 |
| 12 | $8.31 \times 10^{-10}$ | $7.58 \times 10^{-10}$ | 0.082 | 0.99 | 0.91 |
| 13 | $5.13 \times 10^{-10}$ | $2.62 \times 10^{-10}$ | 0.1 | 0.97 | 0.51 |
| 14 | $5.97 \times 10^{-10}$ | $9.2 \times 10^{-10}$ | 0.056 | 0.98 | 1.54 |
| 15 | $1.31 \times 10^{-10}$ | $7.14 \times 10^{-11}$ | 0.0555 | 0.95 | 0.545 |
| 16 | $8.35 \times 10^{-11}$ | $4.36 \times 10^{-11}$ | 0.143 | 0.97 | 0.522 |
| 17 | $9.44 \times 10^{-10}$ | $9.19 \times 10^{-10}$ | 0.0763 | 0.98 | 0.973 |

[*]t: the time elapsed from the beginning of flow, min; K: permeability at time t, $m^2$; $K_L$: approach permeability, $m^2$; $K_o$: a constant, $m^2$; m: constant

residuals (Ingels 1980). The following expression has been obtained with the data of 17 samples (coefficient of regression, $r^2 = 0.92$)

$$Ln\ Y = -1.23 + 2.30\ d_{10} - 0.809\ d_{10}^2$$
$$- 0.0123\ \frac{1}{c_c^2} + 1.17\ \frac{1}{c_u^2} \qquad (13)$$

where, Y is the relative reduction in permeability, $K_o/K_L$ (see equation 11), $d_{10}$ is the effective size or ten percent passing size (mm), $C_c$ is the coefficient of degree and $C_u$ is the coefficient of uniformity. It can be seen that the reduction is chiefly controlled by the effective size, $d_{10}$ of the granulumetric

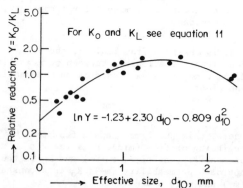

Fig. 5 The relationship between the effective size, $d_{10}$ and the relative reduction, Y in permeability

distribution (see Figure 5).

# 6 CONCLUSIONS

It has been observed that the water flow through backfill prepared from coal washery rejects decreases to a certain quantity with time. This effect should be taken into consideration when designing beckfill stopes.

The decrease in both percolation rate and permeability has been expressed by exponential functions;

$$v = v_L + v_o e^{-mt}$$

$$K = K_L + K_o e^{-mt}$$

where, t is the time elapsed from the beginning of flow (min), v is the percolation rate at time t (m $sec^{-1}$), $v_L$ is the approach percolation rate (m $sec^{-1}$), $v_o$ is a constant (m $sec^{-1}$), K is the permeability at time t ($m^2$), $K_L$ is the approach permeability ($m^2$), $K_o$ is constant ($m^2$) and m is constant.

The relative reduction, which is defined as $Y = K_o/K_L$ where $K_o$ and $K_L$ are as given above, in permeability is chiefly controlled by the effective size $d_{10}$ of the granulumetric distribution.

## REFERENCES

Bates, C.R. and Wayment, W.R. 1967. Laboratory studies of factors influencing waterflow in mine backfill. U.S.B.M., Report of Investigation 7034.

Herzig, J.P., Lecrerc, D.M. and Le Graff, P. 1969. Flow of suspensions through porous media. Flow through porous media, American Chemical Society Publications, Washington D.C.: 128-157.

Ingels, R.M. 1980. How to use the computer to analyse test data. Chemical Engineering August 11, 145-156.

Jacob, B. 1972. Dynamics of fluids in porous media, American Elsevier Pub.Co.Inc., New York.

Şenyur, G. 1985. The behaviour of pneumatic filling materials in one-dimensional compression. Ph.D. thesis, M.E.T.U., Ankara, Turkey.

Thomas, E.G. 1978. Fill permeability and its significance in mine fill practice. Mining with Backfill 12 th Canadian Rock Mechanics Symposium, CIM special volume 19, Ontario: 139-145.

Wayment, W.R. and Nicholson, D.E. 1964. A proposed modified percolation-rate test for use in physical property testing of mine backfill. U.S.B.M., Report of Invest. 6562.

*Innovations in Mining Backfill Technology, Hassani et al. (eds), © 1989 Balkema, Rotterdam. ISBN 90 6191 985 1*

# The basics of preparation of deslimed mill tailing hydraulic fill

E.G.Thomas & P.N.Holtham
*University of New South Wales, Sydney, Australia*

ABSTRACT: Worldwide, much mine fill is prepared by desliming mill tailing and transporting and placing it underground as a pulp of around 70 mass % solids, using centrifugal pumps where necessary to aid gravity. Indeed, such practice is so widespread that
it may be regarded as conventional fill practice.

Where mill grind is relatively coarse, fill preparation usually presents few problems.
However, where mill grind is relatively fine, say 30 or more mass % minus 10µm, problems
are perceived and current practive is not optimum, so scope exists for improvement.

This paper is presented in two parts. The first lists and discusses the various criteria
to be considered in preparation of deslimed mill tailing hydraulic fill, including pulp
density of deslimed mill tailing product, permeability performance with fill cementing
agents and percentage recovery of fill from tailing. The fine mill grind situation is
stressed throughout.

The second part of the paper reviews the various procedures and plant available for fill
preparation from tailing. Accent is naturally on hydrocyclones and cyclone configurations and circuitry. Other less conventional approaches are also considered.

## 1 INTRODUCTION

What may (by virtue of widespread utilisation) be regarded as conventional mine
fill practice, entails the use of hydrocyclones to deslime and dewater mill
tailing. The basic criteria for this
process may be listed as follows.

1. The extent of removal of fine material (fines or slimes) from the tailing
must be adequate to render the resultant
fill (sands) inherently safe with regard
to possible liquefaction after placement
and to allow it to drain sufficiently
rapidly after placement to allow the mining operation to proceed.

2. The pulp density of the resultant fill
must be the optimum for fill transportation and placement. This optimum varies
from fill to fill but may be taken as
somewhere close to 70 mass % solids.

Two further criteria may also apply, in
addition to the criteria already listed.

1. Percentage recovery of fill from
tailing may under certain circumstances
need to be maximised, to ensure an adequate supply of fill. (This requirement
may appear to conflict with the first
criterion listed above and the purpose of
this paper is largely to attempt to resolve
this seeming conflict.)

2. If the fill is to be cemented, its
particle grading may need to be optimised
with regard to cement usage, a slightly
finer fill usually performing better in
this respect. (If cement is used, liquefaction potential and drainage rate reduce
in significance.)

Taken overall, if mill grind is not
excessively fine and percentage removal of
ore as concentrate is not excessively high,
all four criteria introduced above can be
met, without too many problems. The scenario of opposites also applies, to which
this paper is addressed.

## 2 FILL PLACEMENT PULP DENSITY

This is not the main topic of this paper,

425

so it is introduced and dispensed with quickly.

For any filling operation (tailing properties, transportation and placement system characteristics and mining method details) there exists an optimum fill placement pulp density. It is the contention of author Thomas that most fill systems around the world are run at pulp densities below the optimum. For fill placement by centrifugal pump, pulp consistency should be that of a thick, creamy soup, not of a thin, watery soup. The consequences of operating at too low a pulp density have been described by Thomas, Nantel and Notley (1979).

There is a natural tendency for a fill plant operator to add water to a fill line in an attempt to prevent the occurrence of line blockages. Such practice may in fact encourage blockages by allowing faster settling of coarse particles in a lean fill pulp. Where excessive water is added, blockage occurrences are reduced by line velocity increase, compounding pipe wear problems. The importance of maximising fill placement density, hence minimising line velocity, is particularly evident in hydraulic placement of fill consisting of a blend of deslimed mill tailing and rock aggregate (say minus 20 mm), as reported by Thomas (1977, 1978a).

3 BASIC PROBLEM TO BE ADDRESSED

The basic problem to be addressed in this paper is how to recover an adequate, possibly maximum, proportion of fill from tailing, to cater for those operations where grind is fine and/or a significant proportion of ore is removed as con-centrate.

Evidence has been presented by Thomas (1978b) that it is the very fine particles (say minus 10 $\mu$m) that contribute most to control of fill permeability and by Thomas (1983) that such particles also contribute most to control of fill performance with Portland cement additions. Further experimental evidence along these lines is presented later in this paper. (Selection of 10 $\mu$m as the magic division between coarse particles and fine particles is somewhat arbitrary, resulting largely from the fact that C5 on the Warman cyclosizer, frequently used for sub-sieve particle size analysis usually falls at around 10 $\mu$m, unless feed water is very warm (20$^\circ$C plus) and sample solids relative density is very high (3 plus). The actual division point is probably as low as 5 $\mu$m, where some soil classifications divide fine silt and clay. (This arbitrariness will no doubt be resolved with time.)

Fig. 1 may be used to discuss classification of particles into coarse and fine, such as in hydrocycloning of tailing to produce coarse fill and fine overflow.

Fig. 1 is referred to as a partition curve. A perfect partition curve separates feed into coarse and fine at a particular particle size, with no coarse material in the fine product and vice versa. In practice, a perfect partition is highly unlikely, if not impossible. Real partition curves approaching the perfect curve produce what are regarded as sharp splits, and vice versa. The further the real partition curve departs from the perfect, the more fine material will report to cyclone underflow, and vice versa.

It is the contention of author Thomas that, the closer the real partition curve for fill preparation from tailing approaches the perfect curve, the finer can the particle size of the split be made. For a perfect partition curve, the particle size of the split can be as low as 5 to 10 $\mu$m (refer to discussion above), without excessive reduction of resultant permeability of placed fill. Reduction of the particle size of the split of course provides potential for increased percentage recovery of fill from tailing.

Therefore, in operations where mill grind is fine and/or proportion of ore removed as concentrate is high, best fill economics may be to ensure that the technique used to classify tailing into fill and fines produces a sharp split, at a particle size below what would be the case if the split were not sharp.

As an aside, a metallurgist faced with the dilemma of such a task may be able to backtrack and review his decision that such a fine ore grind was in fact necessary, given the facts that it probably increases metallurgical recovery only minutely, at great grinding power expense and to the utter detriment of downstream fill preparation and availability. In this respect, metallurgists should be made more accountable. They should be required to justify their blind pursuit of recovery at all costs.

4 WHAT CONSTITUTES A "FINE" GRIND?

In 25 years of worldwide travels, author Thomas has visited numerous operations with "one of the finest grinds in the world". The following classification is offered in an attempt to modify such claims and to standardise tailing description. (Note the restriction of the suggested classification to the minus 10 $\mu$m fraction of the tailing. This results from the importance of fines to fill behaviour, together with the significance of the cyclosizer C5 to particle size analyses.)

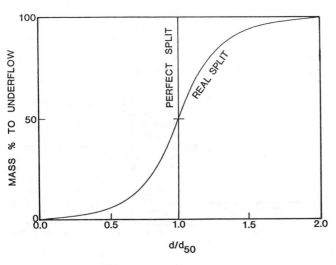

Fig 1 Classifier partition curve.

| Minus 10 µm fraction - mass % | Tailing description |
|---|---|
| 0-10 | Very coarse |
| 10-20 | Coarse |
| 20-30 | Medium |
| 30-40 | Fine |
| 40-50 | Very fine |
| 50+ | Extremely fine |

"Very coarse" tailings occur only infrequently. "Very fine" tailings are almost exclusive to complex base metal sulphides and more recently fine gold ores. "Extremely fine" tailings are probably confined to soft, weathered ores, such as surface gold in Western Australia.

"Very fine" tailings are probably the most challenging from a fill recovery viewpoint, especially since most complex base metal sulphide operations produce multiple metal concentrates and sometimes a pyrite concentrate as well. Another problem with "very fine" tailings is that their 40-50 mass % minus 10 µm interacts jarringly with the range 40-50 mass % of ore required as fill for complete stope filling, considering that dry density of placed fill is usually about half that of in-situ ore and a minor volume of stoping void is usually never filled.

Fill preparation from tailing in a cold climate is more challenging than in a hot climate, because of the dependence of water viscosity on temperature.

## 5 FURTHER EVIDENCE THAT FILL FINES CONTROL FILL PERMEABILITY

As mentioned above, substantial experimental data have earlier been reported by author Thomas that fill fines control fill permeability and that fills may be made finer without sacrifice of permeability if tailing fines (minus say 10 µm material) can be effectively removed. Results from two further laboratory studies are now summarised to reinforce this observation even further.

The first study, conducted and reported by Chen (1988), supervised by author Thomas, was on a base metal sulphide tailing from a mine in New South Wales, Australia. Results in Fig. 2 show that, at low percentages of minus C5 material, permeability is very sensitive to minus C5 percentage. In this range minus C5 percentage needs to be reduced only slightly to produce a significant increase in permeability.

The second study, conducted and reported by Carlile (1988), supervised by author Thomas, was on blends of deslimed mill tailing from a gold mine in Western Australia, and natural dune sand from the mine locality. The de-slimed mill tailing had 7.33 mass % minus 10.4 µm and the natural sand 0.90 mass % minus 10.7 µm. Results in Fig. 3 again show how sensitive permeability is to percentage of fines, at low fines content. For each study, fill permeability was measured in a simple percolation tube.

## 6 METALLURGICAL PROCESSES AND PLANT FOR FILL PREPARATION FROM TAILING

The first part of this paper has demonstrated that a sharp classification in desliming of tailing can allow higher recovery of fill from tailing, without sacrifice of fill permeability. Scope has been indicated for successful deslimed mill tailing fill practice, based upon tailings which may hitherto have been regarded as too fine.

The second part of the paper now reviews metallurgical processes and plant for fill preparation from tailing, emphasizing the theme of sharp splitting of fine tailing.

## 7 FILL PREPARATION FROM MILL TAILING

### 7.1 The Hydrocyclone

The most widely used device for the preparation of mine backfill by the classification of mine tailing continues to be the hydrocyclone. The mechanical features and principles of operation of the cyclone applied to typical mineral processing operations (including fill preparation) are

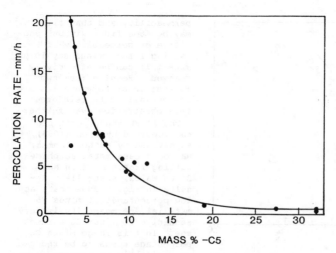

Fig 2  Percolation rate plotted against mass % minus cyclosizer cone number 5.

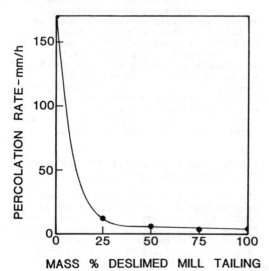

Fig 3  Percolation rate plotted against mass % of deslimed mill tailing (fine material), the balance of each sample tested being natural dune sand (coarse material).

outlined by Tarr (1984), and discussed more comprehensively by Svarovsky (1984), the latter being the first new specialist cyclone textbook to have appeared for a number of years.

As general purpose classifying devices in the minerals industry cyclones have a number of advantages. They are mechanically simple, are physically small compared with mechanical classifiers, and their capital and operating costs are low. Unfortunately, cyclones also have several disadvantages, the principal one being

costs are low. Unfortunately, cyclones also have several disadvantages, the principal one being that they are a relatively inefficient in terms of sharpness of separation, in particular allowing fine slimes to be misplaced into the coarse underflow stream. This problem is made more acute when the feed solids vary in density as well as size; the cyclone, in common with all hydraulic classifiers, separates particles on the basis of both size and density. In addition, once installed, cyclones can be operationally inflexible, the performance being very dependent on feed dilution and feed flow rate.

Cyclones are available in diameters ranging from smaller than 20 mm to greater than 500 mm, with cut sizes from 2 to 200 μm and typical pulp throughputs in the range 2 to 1000 metres$^3$/hour respectively. It will be seen that for the purposes of fill preparation, where high throughput desliming with a sharp separation at say 5 to 10 μm is required, the properties of the cyclone result in contradictory requirements. This leads to the necessity of using large numbers of very small cyclones operating in parallel from a common feed manifold, with the attendant problem of ensuring that each cyclone in the cluster receives the same feed in terms of pulp density and size distribution. Methods of manifolding multiple cyclones are covered by Svarovsky (1984, p79).

## 7.2 Less conventional methods of fill preparation

Although cyclones are widely used for fill preparation, alternative methods of classifying mill tailing may be worth considering, such as the various forms of mechanical classifier. Although they have been largely replaced in their former area of dominance (closed circuit grinding operations) by cyclones, mechanical classifiers such as the rake and spiral classifier retain some advantages in a desliming application, particularly for the very coarse and coarse tailing as defined earlier. The advantages are good solids recovery to underflow with few misplaced fines leading to higher classification efficiency, and lower operating costs compared with a multiple cyclone/feed pump arrangement, especially where very abrasive material is being classified. The principal disadvantages are the higher capital cost, the larger floor area requirement, and the inability to achieve the very fine cutpoint of the small diameter cyclone needed to deslime medium to extremely fine tailing. Feed to mechanical classifiers is normally around 30 mass % solids, and the deslimed underflow can be up to 80 mass % solids.

There are also other alternative forms of classification which may be of interest, and these are discussed in older texts, such as Prior (1965). For example, conventional thickeners may be operated as classifiers by allowing fine slimes to overflow, instead of the clear water normally required, in fact this may often happen accidently. The large volumetric capacity provided by thickeners allows particles to settle under reasonably free conditions, with the very slow settling fines being displaced into the overflow. It is difficult however, to find data on either cutpoint or sharpness of separation, but a thickening type fill preparation system has been operated by at least one Australian mine.

Further alternatives worthy of consideration, perhaps in conjunction with other more conventional processes, may be listed as follows.

1. Sieve bend, used elsewhere for dewatering and desliming, although the normal minimum operating particle size of 50 $\mu$m may be too coarse for application to fine tailings.

2. Filtration, perhaps applied to a portion of the deslimed products from say hydrocyclones, to effect a (slight) pulp density increase.

3. Flocculation, which has been used in fill practice in the past, but is not considered suitable in this application, for many reasons.

## 7.3 Assessment of efficiency

The objective of any fill plant is to recover the maximum amount of deslimed material with the desired size distribution from a feed (mill tailing) having a generally wider size distribution, and which, as discussed earlier, often contains undesirable quantities of very fine material. In order to assess the performance of the plant, three main criteria need to be known, these are: the recovery of feed to the fill product, the nominal size at which the cyclone is operating, and the sharpness of the separation. The first may readily be measured with appropriate instrumentation (density gauges and flowmeters) on the feed and underflow, while the second and third criteria are conventionally obtained by sampling the classifier products and constructing the familiar S-shaped efficiency or partition probability curve (Fig 1) which describes the probability of a particle appearing in the product as a function of particle size. Complete details of the sampling, size analysis and construction of the efficiency curve will not be given as they are readily available elsewhere, Svarovsky (1984), AIChE (1980).

As mentioned earlier, cyclones are relatively inefficient and consequently most efficiency curves do not pass through the origin but have an intercept, representing the finite probability of very fine particles appearing with the coarse underflow product. This phenomenon is generally attributed to the fines simply following the fluid and being unaffected by the inertial forces operating in the cyclone. In order to provide an adequate comparison of the performance of different cyclones, the reduced efficiency curve is often used, in which the short circuiting of fines is allowed for by scaling the raw efficiency data by the fraction of the feed water, $R_f$, reporting to the underflow product, allowing the reduced efficiency to be calculated as follows:

$$E_c(x) = (E(x) - R_f)/(1-R_f)$$

where $E_c(x)$ is the reduced efficiency and $E(x)$ the raw efficiency as a function of particle size x. Fig 1 shows a typical efficiency curve, together with the derived parameters $d_{50}$ (the cutpoint) and $d_{75}$ and $d_{25}$, often used as the ratio $d_{75}/d_{25}$ to describe the sharpness of separation.

## 7.4 Cyclone configurations for the preparation of fill

There are several different cyclone configurations which can be used, from the simple single stage, to more complex multi-stage arrangements and these are discussed briefly below. The application of some of these configurations to a particular fill problem in the South African gold industry is covered by Stradling (1988).

### 7.4.1.Single stage cyclone

The simplest means of classifying mill tailing (and the cheapest) is the single stage cyclone. From a feed pulp density of 20 to 30 mass % solids an underflow of around 70 mass % solids will be achieved. The poor sharpness of separation of this configuration results in quite large quantities of fine material appearing in the fill product, and lower recovery of fill due to feed short circuiting into the overflow. The sharpness of separation will be improved by using a number of small cyclones in parallel, rather than fewer large diameter units.

### 7.4.2 Series connection on the underflow

This arrangement is shown in Fig 4, and in mineral processing terms can be considered as a roughing stage, followed by a clean-

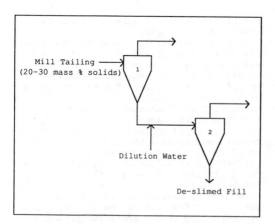

Fig 4   Series connection on the underflow.

ing stage to remove misplaced fines. This results in a sharper separation as well as improved recovery when compared with the single stage configuration. The underflow from the first stage would normally have too high a solids concentration (around 70 mass %) to allow satisfactory operation of the second stage, and thus dilution water is added as shown. The overall efficiency, Eo, of this system is simply given by the product of the individual stage efficiencies:

$$E_o(x) = E_1.E_2$$

Improved performance can be achieved by recycling the second stage overflow back to the first stage feed as shown in Fig 5, in which case the overall efficiency becomes (Svarovsky, 1984):

$$E_o(x) = E_1.E_2/(1-E_1+E_1.E_2)$$

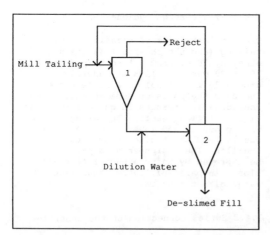

Fig 5   Series connection on the underflow with recycle of the second cyclone over-flow.

## 7.4.3 Series connection on both outlets

This more complex arrangement (Fig 6) is analogous to a rougher, scavenger, cleaner operation in mineral processing and provides a means of further sharpening the separation and improving overall fill recovery. Once again, dilution of the first stage underflow will be necessary to provide optimum operation of the following cyclone.

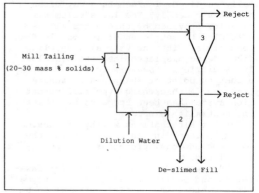

Fig 6   Series connection on both outlets.

## 8 MODELLING AND SIMULATION

The process of designing any new plant today is greatly assisted by the ability to simulate the performance of the plant prior to construction using mathematical models of each piece of equipment. Unfortunately good fundamental models of cyclone classifiers operating at fairly high solids concentrations are not yet available, although their development is an active research topic in a number of centres and progress is being made (Pericleous & Rhodes, 1986). The alternative is the empirical model, of which a number have been developed, the most widely used being those of Plitt (1976) and Lynch (1977). These models typically provide equations for estimating cyclone operating data such as the cutpoint (generally the reduced cutpoint $d_{50c}$), capacity, water split, and pressure drop in terms of fixed design parameters such as inlet, outlet and vortex finder dimensions, and a number of empirically determined constants.

Stratton-Crawley and Agar (1979) used the empirically developed Plitt equations to aid in the optimisation of a three stage (roughing, scavenging, cleaning) fill preparation plant, the objective being to increase recovery of tailing to fill while maintaining the existing cut size. The model allowed the variation of feed pulp density, operating pressure and cyclone apex diameter to be studied, and demonstrated the need to 'tune' the empir-

ical parameters of the model prior to using
it for predictive purposes.

REFERENCES

American Institute of Chemical Engineers
(AIChE). 1980. Particle size classifiers
- a guide to performance evaluation,
AIChE, New York
Carlile, M.G. 1988. An investigation of
underground mining methods for a shal-
low-dipping reef deposit and subsequent
detailed mining production study,
geotechnical analysis and fill development for post pillar stoping. BE Thesis, University of New South Wales.
Chen, J.Z. 1988. Study of a small diameter
hydrocyclone in mine fill preparation.
PhD student progress report, University
of New South Wales.
Lynch, A.J. 1977. Mineral Crushing and
Grinding Circuits, Elsevier, Amsterdam
Pericleous, K.A., Rhodes, N. 1986. The hydrocyclone classifier - a numerical approach, Int. J. Min. Process., 17,
pp23-43
Plitt, L.R. 1976. A mathematical model of
the hydrocyclone classifier, CIM Bul-
letin, 69, pp114-123
Prior, E.J., 1965. Mineral Processing, (3rd
Edition) Elsevier, Amsterdam
Stradling, A.W. 1988. Backfill in South
Africa: developments to classification
systems for plant residues, Minerals En-
gng. 1, No 1, pp31-40
Stratton-Crawley, R., Agar, G. 1979. Mul-
tistage hydrocyclone circuit optimisa-
tion by computer simulation, Proc. 11th
Annual Meeting of Canadian Mineral Pro-
cessing, Ottawa
Svarovsky, L. 1984.Hydrocyclones, Technomic
Publishing Co, London
Tarr.D.T. 1984. Hydrocyclones, in Weiss,
N.L (Ed.), SME Mineral Processing Hand-
book, Society of Mining Engineers of
AIME, New York
Thomas, E.G., Nantel, J.H. and Notley, K.R.
1979. Fill Technology in Underground
Metalliferous Mines. Kingston, Canada:
International Academic Services.
Thomas, E.G. 1977. H.M.S. Float Fill De-
velopment. Stage 1 - Pumping. Mount
Isa Mines Limited Technical Report RES
MIN 48.
Thomas, E.G. 1978a. Cemented Aggregate
Fill - Underground Placement. Report
prepared by Unisearch Ltd for Mount Isa
Mines Limited.
Thomas, E.G. 1978b. Fill permeability and
its significance in mine fill practice.
In Mining with Backfill. 12th Can. Rock
Mech. Symp., Montreal, CIMM.
Thomas, E.G. 1983. Characteristics of ce-
mented deslimed mill tailing fill pre-
pared from finely ground tailing. In
Mining with Backfill, Proc. Int. Symp.
Lulea. Rotterdam, Balkema.

*Innovations in Mining Backfill Technology, Hassani et al. (eds), © 1989 Balkema, Rotterdam. ISBN 90 6191 985 1*

# The convergence resistance of mine backfills

P.H.Oliver & D.Landriault
*Inco Limited, Copper Cliff, Ontario, Canada*

ABSTRACT: The questions of whether or not fill resistance to convergence can become a part of the fill specification process where new fill types are under investigation and whether the simple plasticity and strain-softening numerical models currently available are adequate for modelling fill behaviour are investigated. It is concluded that there is insufficient data available to answer these questions. The laboratory and field work required for a second assessment of these questions is outlined.

## INTRODUCTION

The fills used in underground mining operations have evolved from the early, loose dumped rock and alluvial fills to today's improved convergence resistant, hydraulically transported cemented classified mill tailings and alluvial gravels. The dumped rock fills of today have been either cemented or consolidated with cemented hydraulic fill to increase deformation resistance.

With each change in fill materials, the resistance to closure has improved. However, the rationale for the changes in fill materials and delivery systems has not been justified on the basis of improved support but rather to reduce material handling costs. The fact that each new fill material has provided better support has been a plus which has had, however, no real impact on the financial justifications for the changes. Hydraulic fills significantly reduced fill transportation and distribution costs. The cost of the cement used to consolidate hydraulic fills was more than offset by the savings in timber for fill floors and gob fences.

Rock fill continues to be used in the absence of readily available or suitable mill tailings and alluvial gravels, or where it is necessary to keep the stope opening full while the broken ore is extracted. Again, the justification for rock fill, consolidated or otherwise, is based on factors other than the deformation resistance of the material.

We are entering a new era in fills. Cemented hydraulic fills, have become a production bottleneck in the new total extraction blasthole mining systems. New fill types and transportation methods are being investigated to speed up filling cycles by eliminating fill barricades and water decant systems and by increasing average fill delivery rates. Both cemented paste type fills and low water, low cement content, dumped fills can be used with zero drainage and minimal fill barricades and are therefore candidates to replace present fills and fill systems.

Both the paste type and dumped fills can be designed for specific strengths to satisfy free standing height requirements. But, for a given strength, the two fill types will not be equal in resisting opening closures. Paste type fills, with their high fines contents, a prerequisite for Bingham Flow in pipelines, will have low porosities and be resistant to opening convergence should the initial cement bonds break. The low water and cement content dumped fills, on the other hand, will have porosities dependant to a large extent on the sizing of the fill. Well graded fills will have low porosities and be resistant to convergence, whereas poorly graded fills will have high porosities and a low resistance to convergence following cement bond failure.

The first question is, should we continue as before and ignore deformation resistance in the fill selection process or should we upgrade the process by specifying acceptable pre and post failure fill property ranges? This leads us to the second question of how do we justify complicating the fill selection process by specifying the need to consider deformation resistance? Is there a cost benefit?

The failure in the past to consider fill deformation resistance in the fill selection process was justified. There was no ready means of determining cost benefits. The available numerical models could not accommodate failure and predict realistic levels of convergence. Now, however, new models are becoming available which can accommodate failure and should be able to demonstrate the benefits of deformation resistant fills.

With this in mind, the authors set out to examine fill deformation resistance and whether or not simple plasticity and strain softening models could be used to identify desirable fill convergence resistance properties.

SOIL MECHANICS CONSIDERATIONS

The general perception of hydraulically emplaced fills is that they are in a loosely packed state with porosities in the 45% range. Marbles of uniform size and stacked in a square pattern have a calculated porosity of 47.6%. It is also commonly assumed that an unconsolidated fill will not begin to load up until compaction is complete with convergence values in excess of 10%. Terzaghi & Beck (1967) in a graph showing void ratio versus log pressure results for compression tests on laterally confined laboratory soil aggregates describes a different picture. Their dense sand has a void ratio of 0.8, equivalent to a porosity of 44%. Loose sand has a void ratio of 1.2 or a porosity of 55%. Our so called loose sand is in reality close to being Terzaghi's dense sand, a conflict in terminology.

The grains in the dense sand begin to crush at a load of 9 MPa (1300 psi) when the void ratio has dropped to 0.6 (10% convergence) and continue crushing until the void ratio has been reduced to 0.35 (+20% convergence) at which time the load is 100 MPa. In other words, compaction does not begin until 10% convergence and ends when convergence exceeds 20%. This is contrary to the common perception for mine fills that compaction ends at about the 10% convergence point. The stress at the onset of compaction is also much higher than is commonly perceived for mine fills.

CONFINED COMPACTION TESTS

Fill, when it is resisting convergence, is normally confined. The laterally confined compression test (Oedometer) should be representative of in-situ fill behaviour during periods of opening convergence. On this premise, a simple program was set up to test 30:1 and 10:1 cemented classified mill tailing fill mixes using a standard soil mechanics oedometer. A total of 10 samples of each of the two fills were tested after 28 days of curing.

The capacity of the equipment limited test loads to just under 2 MPa. Maximum strains were in the order of 2% for the 10:1 CSF and 5% for the 30:1 CSF. Table 1 shows the calculated increase in apparent Young's Modulus with increased pressure. There appears to be an exponential increase in the resistance to convergence as the loads and strains increased.

Table 1. Consolidation test results for 10:1 and 30:1 CSF

10:1 Cemented Classified Mill Tailings
(Average of nine tests)

| Stress Range MPa | Incremental Average Young's Modulus MPa |
|---|---|
| 0.048 to 0.335 | 59.9 ( 8,700 psi) |
| 0.335 to 0.766 | 64.1 ( 9,300 psi) |
| 0.776 to 1.676 | 130.0 (18,900 psi) |

30:1 Cemented Classified Mill Tailing
(Average of ten tests)

| Stress Range MPa | Incremental Average Young's Modulus MPa |
|---|---|
| 0.072 to 0.335 | 23.6 (3,400 psi) |
| 0.335 to 0.766 | 35.3 (5,100 psi) |
| 0.766 to 1.676 | 57.6 (8,400 psi) |

A plot of the log of Young's Modulus versus compressive stress for the two fills generated two straight line projections of equal slope but with different intercepts. The instantaneous Young's Modulus for 10:1 CSF could be calculated using the relationship $E = 54.83 \times 10^a$, where a = 0.3437 x stress (MPa). For 30:1 CSF the equation changed to $E = 22.41 \times 10^a$. The results appeared to be significant; however, to be useful, they must be extrapolated to beyond the test limits and extrapolation of exponential values is an extremely questionable practice.

Figure 1 is a plot of the oedometer test result projections for the 10:1 and 30:1 fills. Terzaghi's dense sand curve is also included, but is in terms of percent strain, not void ratio, to be consistent with the consolidation test predictions. The curves show that the oedometer results are consistent within the range of test values, but that the predicted behaviour is questionable when extrapolated. This may have been due to friction between the piston and the confining ring in the oedometer test equipment introducing an artificial stiffening component. The results cannot be trusted.

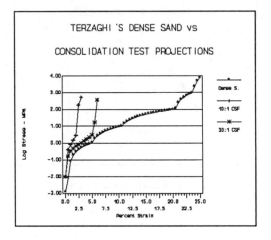

NUMERICAL MODELLING

In elastic models, fill is almost totally ineffective towards influencing the results because of the small displacements in the opening walls. This problem should be rectified with the newer programs for mining which can accommodate failure and generate realistic opening closures. We therefore decided to test this premise using two basic failure models, one a Mohr-Coulomb plasticity model in which cohesion, friction and dilation are assumed to remain constant and the second a Mohr-Coulomb plasticity model (Strain-Softening) where the values for cohesion, friction and dilation can be varied as the plastic strains increase.

The purpose for the models was to test the fill response to convergence. There was no point in including the added complication of the interaction between fill and a surrounding mine opening. We therefore decided to simulate the oedometer test because of the total control over both convergence and confinement.

We had measured values for Young's Modulus, density, strength, and friction, but nothing for Poisson's Ratio for

cemented backfills. All that could be found were assumed values which ranged from 0.10 to 0.45. We could only assume that people did not consider Poisson's Ratio to be significant. We tested this assumption by running three Mohr-Coulomb models and three Strain-Softening models in which only Poisson's Ratio was changed. The values selected were 0.10, 0.25 and 0.40.

Table 2 is a listing of the fill properties used for the models as well as the table in the Strain-Softening models which specified the changes in cohesion, friction and dilation as plastic strains increased. The compressive stresses applied to the top surface of the model were recorded for each 1% strain up to a maximum of 20% strain.

Table 2. Fill properties used in the fill model studies

| Young's Modulus | 34.48 MPa (5000 psi) |
|---|---|
| Poisson's Ratio | 0.10, 0.25, 0.40 |
| UCS | 0.25 MPA (37 Psi) |
| Cohesion | 0.069 MPa (10 Psi) |
| Density (S.G.) | 1.8 (112 lbs/cu.ft.) |
| Friction Angle | 33 Degrees |

Strain Softening Table

| Plastic Strain | Cohesion | Friction | Dilation |
|---|---|---|---|
| 0.000 | 0.069 MPa | 33 Deg. | -15 Deg. |
| 0.001 | 0.000 | 33 | -15 |
| 0.025 | 0.000 | 33 | +15 |

The results were surprising and Poisson's Ratio proved to be significant to a large degree. Failure could only be induced when Poisson's Ratio was 0.10. Poisson's Ratio also had a significant effect on the overall resistance of the fill to convergence.

Figure 2 is a plot of the log Stress versus percent strain for the simple plasticity model with Poisson's ratio set to 0.10. Included is the curve for Terzaghi's dense sand. The dense sand is overall stiffer than the modelled fill over the full strain range. At 10% convergence the modelled stress in the fill is 3.28 MPa (475 psi) vs 9 MPa (1300 psi) for the dense sand.

Figure 3 is a plot showing the effect of the variations in Poisson's Ratio on fill stiffness. At 10% convergence, the stress in the fill ranges from 3.28 MPa (475 psi) for a Poisson's Ratio of 0.10 to 7.79 MPa (1100 psi) for a Poisson's Ratio of 0.40. There is no observable difference in the shapes of the curves even though failure was modelled

435

for the case where Poisson's Ratio was 0.10.

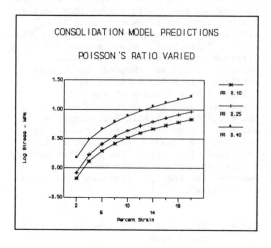

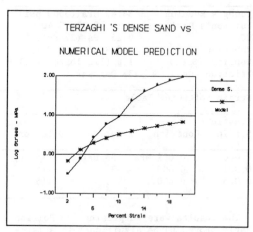

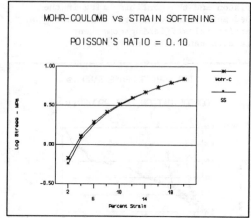

must be known about fill properties and behaviour before fill resistance to convergence can become a part of the fill specification process. Whether or not the simpler failure models used in this study will prove to be adequate for most mining situations will remain an unknown until it has been established that convergence induced fill compaction either does or does not have to be considered for the general case. In the event fill compaction is a common occurrence, more complex "cap" models will have to be employed to properly explore the interaction between fill and the rocks surrounding the filled openings.

Specifically, the measurement of both Young's Modulus and Poisson's Ratio should become a routine part of fill testing procedures. A high stress oedometer, similar to the one proposed by Mitchell (1979), should be developed and used to determine the stress path for high strains and where fill compaction begins and ends. Lastly, field data on the convergence of filled openings and the resultant fill stresses is sadly lacking. More in situ measurements are required.

Figure 4 is a comparison between the Strain Softening and Mohr-Coulomb models where Poisson's ratio was equal to 0.10. There is almost no difference in the results.

There was no point in comparing the differences between the Strain Softening and Mohr_Coulomb models for Poisson's Ratios of 0.25 and 0.40. There was no failure and therefore the behaviours were elastic and identical for common Poisson's Ratios.

## CONCLUSIONS

The results were predictable for a first look at the problem. There were more questions raised than were answered. More

## REFERENCES

Terzaghi, K. & Beck, R.B. 1967. Soil Mechanics in Engineering Practice, Second Edition, p 65-66. New York: John Wiley and Sons Inc.

Mitchell, R.J. & Smith J.D. 1979. Mine Backfill Design and Testing. CIM Bulletin. 72 vol 801:82-89.

*Innovations in Mining Backfill Technology, Hassani et al. (eds), © 1989 Balkema, Rotterdam. ISBN 90 6191 985 1*

# Ground movement prediction as an aid to back-fill mine design

K.V.Shanker & B.B.Dhar
*Department of Mining Engineering, Banaras Hindu University, Varanasi, India*

**ABSTRACT:** The basic need for back-filling in a mining operation is to achieve safe mining conditions and economic recovery of coal/ore deposits. The ground control conditions therefore are the prime requirements for the economic design of a mining system. To avoid high pressure and/or wall closure either at deep depths or at shallow depths located in structurally active geological zones, the need of proper ground control conditions are essential.

Back-fill mining is one such ground control measure that to a great extent, ensures safe recovery of coal/ore deposits from underground mines. Therefore, to establish the compaction characteristics of the fill material in tabular deposits, prediction of ground movement could be useful.

In this paper, an attempt has been made to predict ground movements using numerical modelling technique so that the results could be useful in the design of an efficient and economic back-fill mining system.

## 1 INTRODUCTION

In recent years steady progress has been made in numerical modelling techniques for the prediction of subsidence (Dhar et al.,1982 and Shanker et al.,1983) with reasonable success and the attempt is now being made to extend this model in the design of back-filling, stopes and mining faces. This paper, presents the development of a numerical technique to predict the superincumbent ground movement resulting from the compaction of goaved (broken strata) material. The technique can equally be applicable to stowed or back-filled material, by replacing the goaved material properties with stowing material characteristics. Khoda and Dhar (1972), Dhar et al.,(1983) gives some of the characteristics of back-fill materials and the design procedures. The technique is applied to a number of coal mine panels worked out with caving method and where field surface subsidence observations are available. The predicted profiles are compared with the measured profiles. A reasonable degree of correlation has been achieved between the measured and predicted profiles.

The technique not only predicts surface subsidence, but can also predict sub-surface subsidence profiles and is capable of predicting asymmetrical subsidence profiles by simulating a crack in the main roof in the model.

## 2 THE METHODOLOGY

Herein, the main roof resting over the goaved material is treated as a thin plate resting over an elastic foundation. The main roof is uniformly loaded with cover load all along the transverse section of the main roof. Necessary boundary conditions are then applied and the resulting problem is solved using finite difference technique and vertical deflections of main roof obtained. With these deflections as input data, surface subsidence is calculated separately, treating the superincumbent strata as one of the three different elastic models, namely, (i) isotropic, (ii) transversely isotropic, and (iii) multi-membrane models.

Further, to accommodate the asymmetric nature of the subsidence profiles a crack has been imposed on the surface of the main roof at the point where surface subsidence observed is maximum. Figure 1 shows the bending of main roof over the compacted goaved material and its elastic representation.

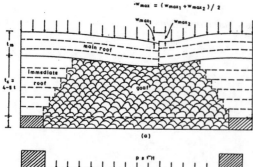

$$w_{max} = (w_{max_1} + w_{max_2})/2$$

main roof

immediate roof

goaf

(a)

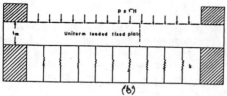

p = rH

Uniform loaded fixed plate

$t_m$

k

(b)

Fig.1 (a) Bending of main roof (plate) showing asymmetrical deflection (b) Elastic representation

## Plate bending-theory and concept

The equation defining small lateral deflections of the middle surface (represented by 'i j k l' in Figure 2) of a thin plate, subjected to lateral loads can be formulated by utilising the basic assumptions made in the theory of beams to obtain directly the equations for thin plates. The governing partial differential equation defining the later deflections of the middle surface of the plate in terms of transverse load is given as:

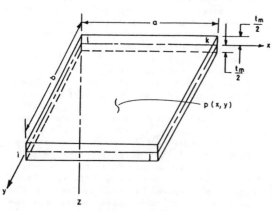

Fig.2 Rectangular plate coordinates and dimensions.

$$\frac{\partial^4 w}{\partial x^4} + \frac{2\partial^4 w}{\partial x^2 \partial y^2} + \frac{\partial^4 w}{\partial y^4} = \frac{p(x,y)}{D} \quad (1)$$

where $D = E \, t_m^3 / 12(1 - \nu^2)$, and

x, y are the co-ordinates in the plane of the plate, w is the deflection along the transverse section of the plate, p(x,y) is the load intensity distribution on upper surface of the plate, $t_m$ is the thickness of the main roof, E is the Young's modulus and $\nu$ is the Poisson's ratio (Macfarland, et al.,1972).

For a main roof resting over a goaved material, as shown in Figure 1, can be treated as a system of thin plates placed on an elastic foundations, as a first approximation. In essence, it is assumed that the elastic foundation is a large continuous spring with the relationship

$$F = \delta k$$

where F is the foundation force per unit foundation area, $\delta$ is the foundation deflection, k is the foundation stiffness constant of goaf or stowed material resistance constant. If the reactions of the goaved material is incorporated in Equation (1), then the partial differential equation for the equilibrium of plates resting on an elastic foundation is obtained as

$$\frac{p'(x,y)}{D} = \frac{p(x,y) - kw(x,y)}{D}$$

Once the governing equation for plate bending is developed, the boundary conditions are then applied in terms of lateral deflections of the plate.

For a Fixed Edge Boundary Condition, the deflections and slope of the middle surface must vanish at the boundary. For example, the boundary conditions on a fixed edge parallel to the Y-axis is

$$w = 0 \text{ at } x = a$$
$$\partial w / \partial x = 0 \text{ at } x = a$$

Similarly parallel to the X axis, is

$$w = 0 \text{ at } y = b$$
$$\partial w / \partial y = 0 \text{ at } y = b$$

For a Free Edge Boundary Condition, which is present at the points along the imposed crack, the bending moment and transverse shear force attains zero value. Then the boundary conditions on a free edge parallel to the x-axis at y = b are:

$$M_y = - D [ \nu \frac{\partial^2 w}{\partial x^2} + \frac{\partial^2 w}{\partial y^2} ] = 0 \text{ at } y = b$$

$$V_{yz} = - D[ \frac{\partial^3 w}{\partial y^3} + (2-\nu) \frac{\partial^3 w}{\partial x^2 \partial y}] = 0 \text{ at } y = b$$

Similarly the boundary conditions on a free edge parallel to y-axis at x = a are:

$$M_x = - D [ \frac{\partial^2 w}{\partial x^2} + \nu \frac{\partial^2 w}{\partial y^2} ] = 0 \text{ at } x = a$$

$$V_{xz} = -D[\frac{\partial^3 w}{\partial y^3} + (2-\nu) \frac{\partial^3 w}{\partial x^2 \partial y}] = 0 \text{ at } x = a$$

where $M_x$ and $M_y$ are bending moments, $V_{yz}$ and $V_{xz}$ are shear forces.

**Application of finite difference method:**
The solution of plate problems via closed form classified methods, such as Navier solution, Levy solution etc. is limited to simple geometries like square, recta-ngular, circular or elliptical shapes. However, for complicated problems, such as in the case of present model studies having irregular shape of extraction, the application of numerical methods is essential. The method of finite differ-ence is an approximate numerical technique used to solve complicated geometry, load distribution and boundary conditions, and yields a simplified form in which differential equation of equilibrium and the boundary conditions are replaced by a set of algebraic equations.

**Prediction of subsidence:** After computing the deflection or closure of the main roof, the next step is the calculation of subsidence resulting from the former by adopting suitable elastic model.

Salamon (1977) made extensive investi-gation on the theory of elasticity for tabular excavations. He has suggested mathematical expression to calculate the ground movement. These equations with suitable modifications have been used here to predict the surface subsi-dence. For example, using multi-membrane model, to calculate vertical displacement, the equation is given as

$$S(x_o, y_o) = a^2/4\pi \iint_A [ \frac{z_1}{(z_1^2 + a^2 R^2)^{3/2}} - $$

$$\frac{z_2}{(z_2^2 + a^2 R^2)^{3/2}} ] w(x,y) \, dx.dy$$

where $R^2 = (x_o-x)^2 + (y_o-y)^2$ ,

$z_1 = z + H$ , $z_2 = z - H$ ,

$S(x_o, y_o)$ is the surface subsidence at a point $(x_o, y_o)$, H is the depth of the seam, $w(x,y)$ is the closure or deflection of the main roof at a point $(x,y)$ in the seam level, A is the area of integra-tion to be performed over a given mine excavation. Here $a^2$ is multi-membrane model constant, and has been assigned a constant value for a given coal mine/ coalfield, after trial and error attempts.

A computer program (Fortran language) was developed to calculate subsidence using all the three elastic models for a given type of irregular mining geometry. The details of the technique and program has been given elsewhere (Shanker 1987).

### 3 CASE STUDY

The mine simulated and discussed here is located in a major coalfields of India. In this mine seam-3 was under extraction when field investigations were carried out (Mozumdar, 1985). The salient mine and model parameters for the case study are as follows.

**Case study 1: mining and simulation para-meters:**

| | | |
|---|---|---|
| 1 | Depth of the mine | 47 m |
| 2 | Length & width of the panel | 230x170 m |
| 3 | Inclination of the seam | 9.5 degrees |
| 4 | Thickness of the seam | 3.0-9.0 m |
| 5 | Thickness of extraction | 2.5 m |
| 6 | Percentage of extraction | 84% |
| 7 | Thickness of main roof (assumed) | 8.7 m |
| 8 | Thickness of caving(assumed) | 10.0 m |
| 9 | Simulation scaling factor | 1:1760 |
| 10 | Stiffness of the goaved material(assumed) | 0.085 MPa |
| 11 | Young's modulus of main roof(assumed) | 8.16 GPa |
| 12 | Poisson's ratio(assumed) | 0.25 |
| 13 | Nature of overlying strata | 80%sandstone 20%coal & shale |

The seam was developed upto 2.55 m height by bord and pillar method, forming 24 x 24 m (Centre to centre) pillars. The developed pillars were extracted upto 2.55 m by conventional splitting and stock method while maintaining a diagonal line of extraction. Figure 3 shows the layout of the Mine-1 workings

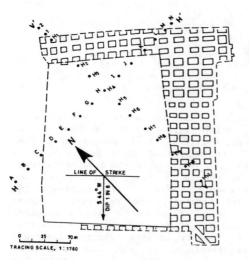

Fig.3 Plan showing the excavations of Mine-I.

while Figure 4 shows the simulated workings and grid patterns selected.

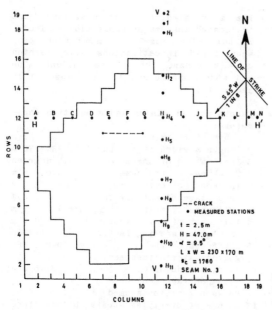

Fig.4 Plan showing the simulated excavations of Mine-I

## 4 RESULTS AND DISCUSSION

Figure 5 and 6 show the comparison of predicted and measured subsidence profiles along the two sectional line H-H'

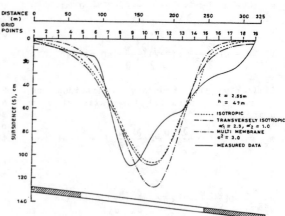

Fig. 5 Comparison of predicted subsidence profiles with measured profile along sectional line H-H' (Mine-I)

and V-V' respectively for the case study Mine-I. The profiles have been drawn for all the three models viz., (i) isotropic, (ii) transversely isotropic, and (iii) multi-membrane.

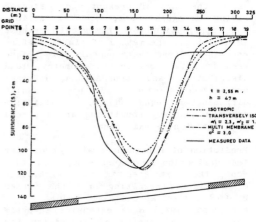

Fig.6 Comparison of predicted subsidence profiles with measured profile along sectional line V-V'(Mine-I)

It can be observed that the two models transversely isotropic and multi-membrane models predicted closer to the measured profiles. Whereas isotropic profile showed a lower bound values of subsidence.

**Prediction of subsurface subsidence:** The technique predicts not only surface subsidence but also sub-surface subsidence at any given horizon above seam level. To demonstrate this, a hypothetical mine panel of 150 x 100 m was chosen. The seam is assumed to be horizonal, the depth being 200 m. Then the subsidence has been predicted for different horizons (i.e. Z=0, 50,000 and 175 m) above the seam level using multi-membrane model. The subsidence values have been calculated and plotted along the two sectional lines, H-H' and V-V', as shown in Figures 7 and 8. It can be inferred from these two figures that as the depth

increases, initially at a lower rate, later accelerating. This is true above the excavated region. However, above the panel boundaries, (beyond abutment regions) the trend is reversed, i.e. surface subsidence being maximum, while sub-surface subsidence keeps on decreasing with increasing depth of horizon.

## 5 CONCLUSION

The technique developed to predict surface and subsurface subsidence over a mining region can help to visualise the nature of the strata behaviour as a result of mining. Therefore when a mining process is planned for a given coal seam, the numerical modelling makes it possible to determine the displacement of strata resulting from mining. Once this is predicted it is possible to resist it by offering a suitable support mechanism. In large mine excavations back filling becomes essential to safe guard the surface features and make mining safe. The characteristics of fill material can thus be simulated in the model at the planning stage and design parameters established. At times to reduce the impact one could even reduce the over all size of the mining panel.

The numerical modelling technique is perhaps a first logical step that could help to plan back-filling for a mining sequence at any depth below ground. Though the technique is developed to predict strata movement for a caving panels, the approach is still valid for back-filled mining faces/stopes by replacing the goaved material characteristics with back-fill material characteristics.

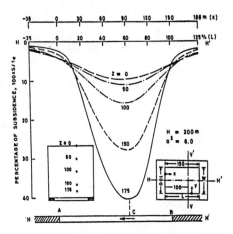

Fig.7 Predicted subsidence profiles at various horizons above the seam level along sectional line H-H' using multi-membrane model.

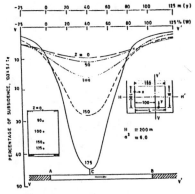

Fig.8 Predicted subsidence profile at various horizons above the seam level along sectional line V-V' using multi-membrane model.

of the horizon, z (depth from the surfaces) increases, the subsidence also

## REFERENCES

Dhar, B.B., Shanker, K.V. and Verma, S.K. 1982. Effect of mine excavations above the workings-a computerised approach to subsidence prediction-paper presented in the symposium on 'Recent trends in mining and drilling techniques', May 13-15, Bangalore.

Dhar, B.B., Shanker, K.V. and Sastry, V.R. 1984. Hydraulic filling-an effective way of ground control-paper presented in the symposium on 'Back fill technology', Luela, Sweden.

Khoda R. and Dhar, B.B. 1972. Cemented hydraulic back fill-Metals and Minerals Review, Vol.11, N 8, 3-9.

MaFarland, D., Smith, B.L. and Bernhart, W.D. 1972. Analysis of plates. Spartan

Books, The Macmillan Press Ltd., London 273

Mozumdar, T.K. 1985. Subsidence in Indian coal mines, M.Tech. Dissertation submitted to Indian School of Mines, Dhanbad, India.

Salamon, M.D.G. 1977. The role of linear models in the estimation of ground movements induced by mining tabular deposits. In: J.D.Geddes (ed.), large ground movements and structures-proc. conf. The Univ. Wales Inst. Sci. Tech., Cardiff, July.

Shanker, K.V., Dhar, B.B. and Sharma, D.K. 1985. Analysis and prediction of ground movement-a laboratory model. Proc.6th Int.cong.: Int. society for mine surveying, Vol.2, Sept. Harrogate, England, 921-927.

Shanker, K.V. 1987. Subsidence prediction using three dimensional numerical modelling technique, PH.D. thesis, Dept. of Mining Engg., Banaras Hindu University, Varanasi, India.

*Innovations in Mining Backfill Technology, Hassani et al. (eds), © 1989 Balkema, Rotterdam. ISBN 90 6191 985 1*

# Double-wall hardened abrasion resistant steel pipeline systems for hydraulic and pneumatic mine backfill conveying

H. Fuss
*SRT Solidresist Rohrtechnik GmbH, Soest, FR Germany*

M. Jost
*Solidresist Pipe Technology Limited, Calgary, Alberta, Canada*

ABSTRACT: A wear resistant pipeline for abrasive solid materials conveying has to have a hardness on the inside of the pipe of approx. 600 Brinell. At the same time, the outside of the pipe should be soft and ductile to withstand dynamic stresses. Moreover, the pipe should be lightweight for easy handling and installation. These physical features are built into the Solidresist 600 pipe by manufacturing the pipe from two different steel grade materials. After the hardening of this double-wall pipe, the result is a lightweight pipe, which combines two extremely diverse physical properties – high hardness and ductility – in a thin pipe wall. This paper presents a description of the manufacturing process and illustrations of the pipeline systems including special fittings and accessories for mine backfill operations.

Wear of steel pipelines used in underground mines for hydraulic or pneumatic conveying of solids – such as backfill or building materials aggregate – can be reduced by increasing the hardness of the steel pipes through a hardening process. The question is, what hardness is required to significantly improve the wear resistance and thereby the operational life of the steel pipe.

Research data indicate the following relationship between hardness of steel pipes and sliding wear – the main factor in pipeline abrasion. 'Loss of material' of a steel pipe as a measure of sliding wear versus the hardness of the pipe measured in Brinell is plotted in Fig. 1. The diagram shows a certain level of wear for mild steel at 100 Brinell, the hardness of mild steel. If the hardness of the steel pipe is increased to 500 Brinell, the 'loss of material' or wear decreases, resulting in a noticeable but relatively minor improvement in the wear resistance of the pipe. However, after a further increase of the hardness to 600 Brinell, a sudden significant decrease in pipe wear can be observed. Fig. 1 shows that the operational life of a steel pipe with a hardness of 600 Brinell is approx. 5 to 6 times longer than that of a mild steel pipe with a hardness of 100 Brinell. Thus, in order to achieve a truly significant increase

in the operational life of a steel pipe exposed to abrasion, it must have a hardness of 600 Brinell on the inside.

For abrasive mine applications, in particular hydraulic or pneumatic backfill conveying, the ideal steel pipe must have the following properties:

1. For genuine wear resistance, as explained before, the pipe must have a hardness of 600 Brinell on the inside.

2. On the other hand, the pipe must be capable of withstanding dynamic stresses in form of external physical impact during normal handling as well as internal pressures. The pipe must therefore be soft and ductile on the outside. At the same time, the pipe must be lightweight for easy handling, i.e., it must have a relatively thin pipe wall.

In other words, the ideal steel pipe for abrasive backfill handling would be a lightweight pipe, which combines two extremely diverse physical properties – high hardness on the inside and ductility on the outside – in a thin pipe wall.

Steel pipes commonly used in mine backfill pipelines range in size and wall thickness from 4" to 6"(Sch 40 or Sch 80) which means, wall thicknesses range from .237" (6 mm) to .432" (10.95 mm). If a single-wall 6" Sch 80 pipe with a .432" (10.95 mm) wall thickness is hardened to 600 Brinell hardness on the inside of the pipe, the hardness on the outside would

Figure 1. Relationship between pipe hardness (in Brinell) and 'loss of material' (sliding wear).

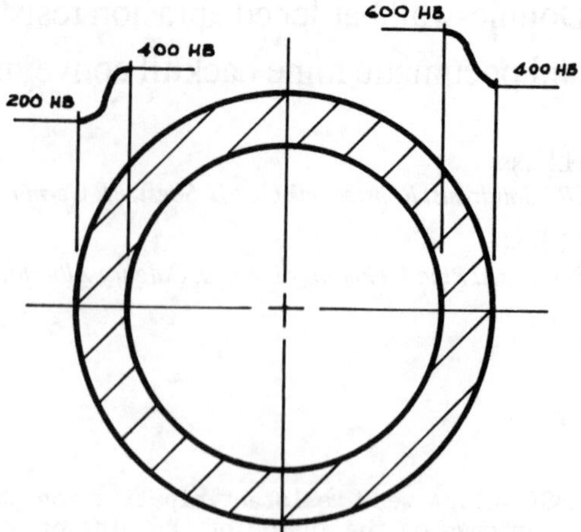

Figure 2. Hardness profiles across the wall of a single-wall hardened steel pipe.

be in the order of 400 Brinell (Fig. 2). Such a pipe would indeed have the necessary hardness on the inside to provide good wear resistance. However, due to the high hardness on the outside, this pipe would be too brittle to withstand dynamic stresses such as external impact, under which it would easily crack.

If, on the other hand, this same pipe would only be hardened to a degree, that the hardness on the outside does not exceed 200 Brinell, then the hardness on the inside would only reach approx. 400 Brinell. Such a hardened pipe would have the necessary ductility on the outside to withstand dynamic stresses, but the hardness on the inside would be too low to offer good wear resistance (Fig. 1).

If one applies the same model to a single-wall 4" Sch 40 pipe with a wall thickness of only .237" (6 mm), the hardness differential between pipe inside and pipe outside would be even more unfavourable.

The conclusion to be drawn from these examples is, that an industrial fabrication of a steel pipe with a single thin pipe wall, which is both 600 Brinell hard on the inside  - as required for satisfactory wear resistance - and ductile on the outside - to absorb dynamic stresses - is not possible.

The Solidresist pipe manufacturing process solves this problem by dividing the pipe wall into two chemically different materials (Fig. 3). This is done by fabricating the Solidresist 600 pipe from two steel materials of different grade, using a hardenable steel for the inner pipe layer and a normal mild steel for the outer pipe mantle. Mild steel, due to its analysis, is not hardenable. Therefore, when this double-layer Solidresist pipe undergoes the hardening process, the inner material, because of its chemical composition, will be completely hardened to 600 Brinell, whereas the outer mild steel mantle will remain soft. The result of this special manufacturing technique is a steel pipe with extreme hardness and ductility built into a thin pipe wall.

The hardening procedure is illustrated in Fig. 4. Both the inner and outer pipe layers are heated from the inside of the pipe with a high-temperature gas/oxygen burner. The energy output is calculated to ensure that the inner steel layer reaches the hardening temperature of approx. 900 degrees C and the outer steel pipe is heated to at least 650 degrees C. The pipe is then quenched. Heating and quenching constitute the hardening process. There are water cooling systems both on the inside and the outside of the pipe. The hardening process, which affects only the inner pipe layer because of its chemistry, consists in a textural conversion from an original ferritic-perlitic texture into a martensitic texture which, at the same time, results in a volumetric expansion of the inner steel layer by approx. .2 per cent by volume. Simultaneous external cooling of the pipe affects the outer soft mild

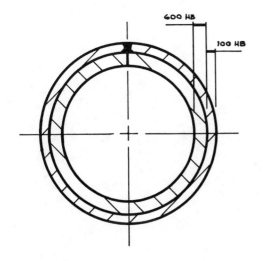

Figure 3. Hardness profiles across the wall of a hardened double-wall Solidresist 600 pipe.

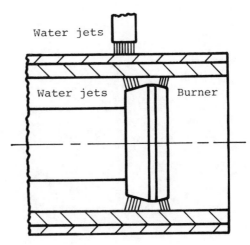

Figure 4. Flame hardening device for Solidresist 600 double-wall pipes.

steel mantle by causing it to shrink onto the expanding inner steel layer, thereby pressing both layers tightly together.

This tight fit over the entire contact surface between the inner, now hardened steel layer and the outer, still ductile mild steel layer, is a key factor which gives the Solidresist 600 double steel wall pipe the ability to withstand operational stresses including physical impact from the inside or the outside of the pipe. In order to ensure a uniform hardening of the inner steel layer it is necessary that both steel layers are firmly pressed together over the entire contact surface prior to hardening. The Solidresist manufacturing process therefore uses two stacked flat steel plates, with the hardenable steel plate for the abrasion resistant inner pipe liner placed on top of a mild steel plate for the outer pipe mantle. This double-layer 'sandwich' is then first formed in a bending press into a double-wall open can which is subsequently closed and seam-welded in a separate welding press.

As shown in Fig. 3, only the outer mild steel layer of the pipe is seam-welded with a specially developed fully automatic submerged arc welding procedure. This welding technique guarantees that the edges of the outer pipe layer are faultlessly welded together, and further, that the welding arc does not penetrate the inner hardenable layer, which would

result in a brittle weld.

Based upon this unique pipe manufacturing technology, Solidresist has developed complete wear resistant steel pipeline systems for mine backfilling, consisting of straight pipes, different types of wear resistant elbows, pipe reducers, telescopic pipes, pipe tensioners, pipe switches, etc., complete with specially designed self-centering flanges and coupling or butt-welding connections to eliminate destructive turbulence caused by pipe misalignment. All Solidresist 600 wear resistant mine backfill pipeline systems are custom engineered according to individual client specifications.

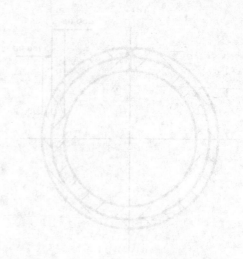

*Innovations in Mining Backfill Technology, Hassani et al. (eds), © 1989 Balkema, Rotterdam. ISBN 90 6191 985 1*

# An evaluation of binder alternatives for hydraulic mill tailings

P.H.Hopkins & M.S.Beaudry
*Falconbridge Limited, Sudbury, Canada*

ABSTRACT: Underground mining often requires that the void left from extracting the ore be filled with a backfill material. This material can be many things but the most common fill in use today is mill tailings. In many mining applications these mill tailing backfills need to be consolidated with some sort of binding material. At the present time, backfills are most commonly consolidated with Portland cement. This consolidation of mine backfill can be an expensive procedure due to the cost of the binder. In many cases the binder cost alone is equal to the cost of preparing, transporting and placing the fill in the mined out stopes underground. This paper examines five of the most attractive binder alternatives used to consolidate mill tailings backfill in place of Portland cement. An optimum binder was selected and an underground trial was completed to confirm the laboratory test results. Using this optimum binder an economic evaluation was made comparing the cost of using the binder alternative versus Portland cement in a typical primary blast hole stope underground. Recommendations are made as to how an operating mine may want to test this optimum binder in their own backfill.

## INTRODUCTION

Over the past decades waste materials have been sent back underground, in one form or another, to fill the void left by mining. This backfill material has varied in composition from crushed or broken aggregate to finely ground mill tailings and has been delivered to underground sites in the form of slurries, solids and, more recently, pastes. In addition, the method of transport has also varied from boreholes and pipes, waste skips and LHD equipment to waste passes and conveyor belts (and many combinations of the above).

Although crushed rock and dense fill systems are becoming popular, the most common mine backfilling method used in Canada is conventional hydraulic mill tailings. This finely ground mill waste material is sent underground as a slurry into an empty stope where the excess water is drained away and the material is left behind as fill. Depending on the requirements for this backfill, a binder or cementing agent may be desired. Once again, various different cementing agents are in use today such as Portland Cement,

Blast Furnace Slag Cement and Fly Ash. The most common binding agent is Portland Cement. This paper addresses less expensive alternatives to Portland Cement and improvements possible to their effectiveness when used as a binder for hydraulic fill.

Various binding agents have been tested in the past to reduce backfilling costs but have required that the mine either accept a loss in early strength development or final fill quality. In these cases scheduling or safety concerns often arose. It became evident that a less expensive binding agent was required that would not sacrifice fill quality.

## EXPERIMENT OUTLINE

The majority of the laboratory and field test work was centered around the following four fundamental backfill binder alternatives:

1 – Type-10 Normal Portland cement.
2 – Iron ore blast furnace slag from Stelco Limited.
3 – Iron ore blast furnace slag from

Algoma Steel Limited.

4 – A mix of Detroit Type-C fly ash and Type-10 cement.

Alternative binders such as Anhydrites, Silica Fumes, Monolithic packing materials and Type-F Fly Ash were reviewed but were not evaluated in detail for reasons such as: poor availability, handling difficulties, poor initial test results or costs.

Strathcona Mill classified tailings were used as the basis for all binder tests. The mineralogy and size distribution of this material is shown in Figures 1 and 2 respectively. Laboratory tests were completed under the following conditions:

* 68% Pulp density.
* Curing humidity = 100%.
* 1-56 day curing period with a 24 hour open air drying time prior to testing.
* Samples tested were 3" x 6" undrained cylinders.
* Binder addition levels of
  - 8:1 (tailings:binder) or 11.1%
  - 16:1      "          "       or 5.9%
  - 32:1      "          "       or 3.0%
* Unconfined uniaxial compressive strength testing.

Lime addition was used as an activating agent with the slag cements to enhance early strength. A full range of lime addition rates were tested with only the most beneficial being presented in this paper.

Various combinations of Type-C Fly Ash and Portland Cement were tested. This paper presents only the most effective mix.

| Mineral | % |
|---|---|
| Quart | 24 |
| Feldspar | 33 |
| Pyroxene | 22 |
| Amphibole | 12 |
| Phyllosilicates | 4 |
| Epidote (other silicates) | 3 |
| Magnetite, Chromite, Ilmenite | 1 |
| Sulphides (mostly pyrrhotite) | 1 |
| Total % | 100 |

Fig.1 Mineralogy of Strathcona classified mill tailings.

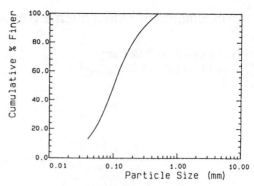

Fig. 2 Size distribution of Strathcona classified mill tailings

INITIAL LABORATORY TESTING

The 28 day unconfined uniaxial compressive strength test results are shown in Figure 3. To evaluate the data in Figure 3, two aspects of consolidated backfill strength should be considered. These are, 1) the early strength obtained after four days and 2) the ultimate fill strength obtained after the fill has fully cured.

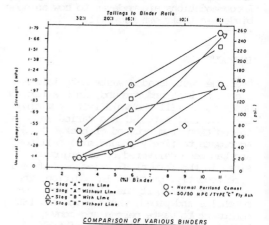

COMPARISON OF VARIOUS BINDERS

Fig.3 Binder alternative 28 day strengths

Short term backfill strength is important to those mines which need to fill a given area and then return quickly to this area with men and/or machines. This early strength is also important to mines that pour backfill plugs in the bottom of blasthole stopes. These plugs then retain the bulk of the stope fill. Early strength in this plug allows the mine to

448

complete the remainder of the stope backfilling on a shorter schedule.

Final strength is often the basis for design of the composition of the consolidated backfill. It is the final fill quality that will allow a mine to efficiently mine next to a previously backfilled stope. High fill quality will minimize fill dilution and can improve local ground conditions in areas of high stress. These two points involving the effectiveness of consolidated mine backfills have been studied by many mining groups and companies around the world. This topic is out of the scope of this paper.

Slag "A" with lime proved to have the highest ultimate strength and was selected for further detailed lab tests. This work is described in the following section and addresses the early strength capabilities of Slag "A" with lime.

Various companies and research groups that have previously tested Portland cement alternatives were contacted and asked for their opinions on the different strength gains indicated in Figure 3. None were able to provide a definitive answer. One difficulty may be the low pulp density (68%) common in pouring backfill in comparison to the higher densities that exist in the concrete industry. It should be noted that there are very few "experts" in the area of backfill technology and only a limited amount of work has been done on replacements for Portland cement in hydraulic mill tailings backfill.

The Appendix lists some of the physical and chemical properties of the four binder alternatives tested. These properties may play an important role in the development of backfill strength.

FOLLOW-UP LABORATORY TESTING

Blast Furnace Slag "A" versus Type-10 Portland Cement

Blast furnace Slag "A", a bi-product of steel production, appeared to be the most attractive binder alternative to Portland cement. In order to verify the initial data obtained and to confirm that activation of this slag cement was possible with minor additions of lime, detailed laboratory study was performed. The results of this work are shown in Figure 4.

Slag "A" had a higher ultimate strength in all three of the binder addition rates tested. In the cases of three and six

percent binder addition this final strength was as high as three times that of Portland cement after one month and more than four times the strength after two months.

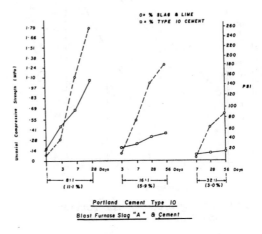

Fig.4  Comparison of Portland cement versus Slag "A"

An early backfill strength that allows men and machines to work on the fill and allows bulk stopes to continue bulk pours, is an important consideration for consolidated fills. A fill strength build up rate equivalent to that obtained with Portland cement is achieved by slag "A" when 6% or greater binder is used. Approximately eight days are required when less than 6% binder is added. The effect of this eight day curing period would have to be considered by any user in conjunction with its effect on mine scheduling.

Additional data obtained from these tests is shown in Figure 5. This data will be used in the section on economic comparisons.

UNDERGROUND FIELD TEST

Laboratory tests are the most common method of comparing the various backfill designs. Because underground conditions vary from those in the lab a full scale field trial is mandatory to confirm that any improvements indicated from a lab study are obtained underground.

449

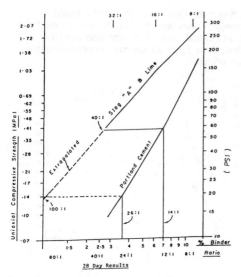

Fig.5  Strength curves for Portland
cement and Slag "A"

To evaluate Slag "A" as a binder
alternative a typical cut and fill stope
on the 2700 level of Falconbridge's
Strathcona Mine was chosen for a field
trial.  This ore zone is steeply dipping
(approximately 70 degrees), roughly 15
feet wide and 120 feet in strike length.
Two layers of hydraulic fill were poured.
Each layer contained a different binder
addition rate.  The first or bottom three
feet was filled with backfill containing
3% (32:1 tailings:binder) Slag "A" and the
top or final three feet was filled with
6% binder addition.  The trial layout is
shown below.

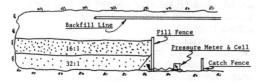

Fig.6  Underground Slag cement trial –
layout

The two layers were poured with one day
between the pours.  Each pour was
observed by the mine operating personnel
for their subjective comments.  A
pressure meter cell was installed at the
bottom of the fill fence to monitor the
hydraulic head resulting from hydraulic
filling.  A low measurement of pressure
would indicate effective excess fill
water drainage.  Diamond drilling of this
fill to obtain core samples was reviewed

but considered to be too costly.

The pouring of both layers of the test
fill was observed by a mine captain,
shift supervisor and the backfill
technology engineer.  No special design
changes were made to the backfill
plant on surface or to the stope to be filled
underground, in order to complete this
trial.  During the first test pour (3%
Slag) the mine captain was able to walk
on the fill within two feet of the
backfill discharge point and commented,
"I have not seen backfill of this quality
in many years".  After one month of
curing technical personnel returned to
the test site to extract a large test
sample.  This sample was extracted using
hand tools.

A sample from the more highly
consolidated top layer was to be taken.
However, hand tools could not break away
a reasonable sample of the fill and
sampling efforts were discontinued.

The results from the pressure cell
located at the base of the fill fence
showed that there was no hydraulic
pressure build up during or after
backfilling.  This fact and the ability
of men to walk directly beside the
backfill pour point during the pour,
something that is not common for our
consolidated fill, indicated good fill
drainage.

No underground design changes appear
necessary in order to use slag "A" as a
binder.

AVAILABILITY OF BINDER ALTERNATIVE

Figure 7 shows where in Canada each of
the four binder alternative can be found.

Each of these binder alternatives are
bi-products from other industries, and
are most easily obtained in the specific
areas where the primary industry is
located.  Type-C Fly Ash can be found
wherever sub-bituminous coal is burned
for electrical energy (Quebec, Ontario,
and most of western Canada) and Slag
cement is found in locations where waste
slag from the iron refining process
exists (Ontario).

ECONOMIC COMPARISON

Between Slag "A" and Portland cement for
a blasthole stope

The advantages of using Slag "A" as a
binder alternative come from two sources:
    1.  The money saved from using Slag "A"

instead of the more expensive Portland cement and,

2. Being able to use less Slag "A" per ton of backfill to achieve strength equivalent to Portland cement.

For economic comparison, the price used for each binder was:

Portland cement.................105 $ / ton
Slag "A"....................... 75 $ / ton

Note ** These prices do not represent contract prices but are an approximate price for the northern Ontario region.

Fig.7 Location of binder alternatives in Canada.

A typical blasthole stope at Falconbridge's Lockerby Mine in Sudbury was used (Figure 8) to make the economic comparison. This stope has dimensions 175' high, 36' wide and 50' deep and requires 2250 tons of consolidated 6.6% (14:1) fill as a plug. The remaining 13,500 tons of fill contain 4.8% (20:1) of binder and constitute the bulk of the pour. These are common consolidated ratios for a stope of this type that will encounter mining on both sides. Backfill consolidation is designed to minimize fill failure during the subsequent adjacent mining operations and thereby reduce the dilution.

Figure 9 presents the results of the economic comparisons. Three options are compared:

a) Only Portland cement.
b) Only Slag "A" with a 1:1 replacement ratio of slag:cement.
c) Only slag "A" but with an

approximate replacement rate of 0.5:1 slag:cement.

Case (c) is considered to be conservative because data indicates that, as a result of the much higher 28 day strength of slag over cement (up to four times), considerably less slag could be used and still equal the 28 day strength of cement.

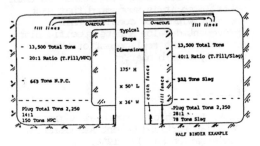

Fig. 8    Stope Data

Binder Ratios, Costs & Savings

| BINDER USED | AREA | RATIO FILL:BINDER | STRENGTH AT 28 DAYS (Psi) | TONS BINDER (Tons) | TOTAL TONS (Ton) | BINDER $/TON ($) | TOTAL COST ($) | SAVINGS ($) |
|---|---|---|---|---|---|---|---|---|
| a) Cement | Plug | 14:1 | 58 | 150 | | | | |
| | Stope | 20:1 | 35 | 643 | 793 | 105 | 83265 | |
| b) Slag "A" | Plug | 14:1 | 170 | 150 | | | | |
| | Stope | 20:1 | 105 | 643 | 793 | 75 | 59475 | 23790 |
| c) Slag "A" | Plug | 28:1 | 83 | 78 | | | | |
| | Stope | 40:1 | 48 | 322 | 400 | 75 | 30000 | 53265 |

Binder Costs and 28 Day Strengths

| STOPE | TONS OF BINDER | PORTLAND CEMENT (a) | | Full Replacement (b) | | Half Replacement (c) | |
|---|---|---|---|---|---|---|---|
| | | Cost ($) | Strength (Psi) | Cost ($) | Strength (Psi) | Cost ($) | Strength (Psi) |
| Plug Pour | 150 | 15750 | 58 | 11250 | 170 | 5625 | 83 |
| Bulk Pour | 643 | 67515 | 35 | 48225 | 105 | 24113 | 48 |
| Total | 795 | 83265 | | 59475 | | 29738 | |

Binder Cost Per Ton of Fill and Savings

| Binder | Binder Cost/Ton Fill ($/Ton) | Savings Over Portland Cement ($/Ton Fill) | ($/Stope) |
|---|---|---|---|
| a) Portland Cement | 5.30 | - | |
| b) Slag "A" | 3.79 | 1.51 (28%) | 23850 |
| c) Slag "A" | 1.89 | 3.41 (64%) | 53662 |

Note-Savings are for a fill with 5% binder added

Fig. 9  Cost comparison using Slag "A" vs. Portland Cement

CONCLUSION

Both laboratory and field tests indicate that there are considerable cost savings and quality improvements possible to the hydraulic tailings backfill from Strathcona Mill when slag cement "A",

451

activated with lime, is used as a binder in place of Portland cement. The binder cost savings can be in excess of 65% and fill strength improvement of up to 350%.

One consideration in using slag cement as a binder is that it takes up to four days for the fill to obtain a strength equivalent to fill consolidated with Portland cement.

At this time there are few mines in the world which are using slag cement in their backfill, so it is suggested that each mine that wishes to consider this improvement develop a site specific plan for testing the replacement of the currently used binder.

REFERENCES

Hajduk, L., Senior Structural Engineer, Canadian Portland Cement Association, Conversation May 17, 1988.

Sallert, M., Cut and Fill Mining with Cemented Mill Tailings at Boliden Metal AB, 12th Canadian Rock Mechanics Symposium, Sudbury, Ontario, May 22, 1978.

Campbell, P., Ames, D.W., Graham, C.B., Backfill Practices and Trends, Proceeding from the 87th Annual General Meeting of the C.I.M., April 1987.

Beaudry, M., Falconbridge Internal Reports: Laboratory Trials Using Type-C Fly Ash, April 1987. Laboratory Tests Using Reiss Lime Slag, November 1988. Laboratory Trials Using Standard Slag Cement, May 1987.

Hopkins, P.H., Falconbridge Internal Report: Evaluation of Backfill Binder Alternatives, March 1987.

APPENDIX

### Chemical and Physical Properties of Four Hydraulic Backfill Binder Alternatives

Chemical Composition:

| Chemical | Portland Cement (%) | Slag "A" (%) | Slag "B" (%) | Type-C Fly Ash (%) |
|---|---|---|---|---|
| CaO | 65 | 33 | 35 | 17 |
| SiO2 | 20 | 38 | 40 | 47 |
| Al2O3 | 4 | 8 | 7 | 19 |
| Fe2O3 | 3 | trace | 1 | 7 |
| MgO | 3 | 16 | 12 | 3 |
| SO3 | 3 | 1 | 4 | 7 |
| S | 0 | 0 | 0 | 0 |
| Na2O + K2O | 1 | 1 | trace | 0 |
| MnO | 0 | 1 | 0 | 0 |
| TiO2 | 0 | 0 | trace | 0 |
| SrO | 0 | 0 | trace | 0 |
| Mn2O3 | 0 | 0 | 1 | 0 |
| Other | 1 | 2 | 0 | 0 |
| Total % | 100 | 100 | 100 | 100 |

Fineness of Grind:

| | | | | |
|---|---|---|---|---|
| Blain (cm^3/g) | 3500 | 4600 | 4200 | 4000-7000 |